Principles of Biomedical Informatics
Second Edition

Principles of Biomedical Informatics
Second Edition

Ira J. Kalet, Ph.D.
University of Washington
Seattle, Washington

AMSTERDAM • BOSTON • HEIDELBERG • LONDON • NEW YORK
OXFORD • PARIS • SAN DIEGO • SAN FRANCISCO • SINGAPORE
SYDNEY • TOKYO

Academic Press is an imprint of Elsevier

Academic Press is an imprint of Elsevier
225 Wyman Street, Waltham, MA 02451, USA
525 B Street, Suite 1800, San Diego, CA 92101-4495, USA
32 Jamestown Road, London NW1 7BY, UK

Notice
No responsibility is assumed by the publisher for any injury and/or damage to persons or property as a matter of products liability, negligence or otherwise, or from any use or operation of any methods, products, instructions or ideas contained in the material herein. Because of rapid advances in the medical sciences, in particular, independent verification of diagnoses and drug dosages should be made.

Library of Congress Cataloging-in-Publication Data
A catalog record for this book is available from the Library of Congress

British Library Cataloguing-in-Publication Data
A catalogue record for this book is available from the British Library

ISBN: 978-0-12-416019-4

For information on all Academic Press publications
visit our website at elsevierdirect.com

Printed and bound in United States of America

14 15 16 17 10 9 8 7 6 5 4 3 2 1

Dedication

To Terry Steele-Kalet, my wife and best friend for over 40 years, who again has kept me on track when things looked impossible, and stood by me through the most difficult times as well as the best of times.

CONTENTS

Preface to the Second Edition

The first edition of this book was published in 2008. Since then, the number of academic programs in biomedical informatics and related fields has increased rapidly. The field itself has seen huge growth, particularly in clinical informatics, where in the last few years the US Government has invested in incentives for practitioners to acquire and deploy electronic medical record systems, and the American Board of Medical Specialties has, through the American Board of Preventive Medicine, paved the way for a Clinical Informatics Board Certification process. In basic biology, large-scale data generation and analysis continues to grow, and informatics plays an ever more central role in this research enterprise. Public health at all levels also is taking on greater challenges in data collection and management. With so many new problems, variations on algorithms and ideas, new applications and new research questions, it seems more important than ever to identify and focus on the fundamental underpinnings of the field. This revision is an effort to update and expand on the fundamentals, while leaving the task of cataloging and describing all this new activity to others more knowledgeable and smarter than me.

The Same Goals and Style

As stated in the Preface to the First Edition, the main goal of this book is to provide an exposition of important fundamental ideas and methods in biomedical informatics, using actual working code that can be read and understood. The object of the code is not only to provide software tools, but for the code to play the same formal role that equations and formulas play in theoretical physics. That is, the programs precisely embody biological, medical, or public health ideas and theories. The application of these programs to specific problem instances and data is analogous to the solution of the equations of physics for specific practical problems, such as the design of airplane wings. Thus an effective way to explain biomedical informatics ideas is by showing how to write programs that implement them. Conversely, a way to achieve a deep and clear understanding of the foundations of biomedical informatics is to learn how such programs work and to be able to *adapt, modify, and extend* them as well as use them.

Changes in the Second Edition

Plus ça change, plus c'est la même chose. (the more things change, the more they are the same)

—*Alphonse Karr*
Les Guêpes, January 1849

A New Structure

In the first edition, Part I presented the foundational ideas, principles, and methods of biomedical informatics, drawn from the fields of databases, object-oriented data modeling, artificial intelligence, probability and statistics, information retrieval, and related areas. The examples showing how these principles and methods are actually used were organized in Part II as a series of biological and medical topic areas and applications. In this second edition, Parts I and II are merged, and each example is presented in the same section where the relevant ideas and methods are developed.

The organization and sequence is mostly the same as the original Part I, with ideas about representation of and operations on data coming first, then symbolic knowledge representation, followed by representation and use of ideas from probability and statistics, and then the various aspects of access to and search of biomedical information (both data and knowledge). Some topics from Part II do not fall neatly into these four categories, but represent computational methods for manipulation of data and/or knowledge in various forms. These are now grouped in a new chapter, "Computational Models and Methods." It includes algorithms for operating on molecular biology data compilations, methods for search (and application to biological problems), and ideas and methods for dynamic models of biological systems. Finally, the material originally in Chapter 10, "Safety and Security," has been somewhat expanded and appears as a chapter now called "Biomedical Data, Knowledge, and Systems in Context" and the Epilogue, originally a short section, is now a little longer, and provides an expanded retrospective of the author's view of how the field of Biomedical Informatics is evolving.

New Material

The following new or expanded topics are included in this edition:

- X-ray physics and CT image reconstruction, recognizing that this area was one of the most central to the origin of biomedical computing,
- Tree structured data, interval trees, and time-oriented medical data and its use,
- On-Line Application Processing (OLAP), an old database idea that is only recently coming of age and finding surprising importance to biomedical informatics,
- A discussion of nursing knowledge and an example of encoding nursing advice in a rule-based system,
- An application of description logic to answer queries about laboratory test codes,
- An introduction to Markov processes, and
- An outline of the elements of a hospital IT security program, focusing on fundamental ideas rather than specifics of system vulnerabilities or specific technologies.

The first edition was intended to be a monograph that marked a level of maturity of the field. Recognizing that this edition needs to be more useable as a textbook, exercises are provided at the end of each chapter. Some of these simply make explicit the places in the body of the text where you find "… it left as an exercise for the reader." Others come from over 10 years of experience teaching this material in various courses in the University of Washington Biomedical and Health Informatics program. Students over the years have ranged from undergraduates to graduate students to postdoctoral fellows and faculty interested in learning about biomedical informatics. The level of these exercises varies greatly, as does the technical, biological, and medical knowledge required.

Programming Languages

Lisp is the main programming language in this edition, as in the first. It is by far the best for achieving the goals of the book, to present biomedical informatics as a systematic set of formal ideas and methods that are naturally expressed as computations on well-defined representations. However, Lisp is not taught or used as much as (in my view) it should be. Although in Chapter 1 the basic ideas of symbolic computing are introduced gradually as needed, I have expanded the Appendix to include more tutorial material as well as references to aid the reader who is not so familiar with Lisp.

Several colleagues who have used the first edition in teaching have suggested translating the code into Python (or some other language more familiar to many students). I've invited those colleagues to do this translation and send it to me, whereupon I will post any and all such contributions on the book web page. This offer remains open.

Notes on Typography

In this book, code and data are distinguished from the rest of the text by using a `different font`. Blocks of code that define functions, classes, etc., are shown without a prompt. When the intent is to show a dialog, that is, to illustrate what happens at the interactive `read-eval-print` loop, a prompt is shown. Different Common Lisp implementations each use their own style of prompt. Here, we follow the same convention as in Graham [137], the "greater-than" angle bracket, >.

How To Use This Book

There is enough material for a year long course that just focuses on the technical aspects of biomedical informatics. In one academic quarter it is reasonable though challenging to cover the material in Chapters 1 and 2. The exercises vary enormously in technical depth, so the instructor should choose judiciously when assigning them, to get a best match to student skills and background. It is a good idea to have students work through or have available a good basic book on Common Lisp, such as Graham [137], or one of the tutorials available on line. In Appendix A I have listed some of these, and other books that have been useful to students.

Depending on the focus of the course, the material on medical images, including the DICOM protocol, can be skipped. If the intent is to prepare for a next course or research in image understanding, content-based image retrieval, or similar topics, it will be more central and useful.

All the code that appears in this book is available at the author's web site at the University of Washington, `http://faculty.washington.edu/ikalet/`. The complete source code for the original DICOM server and client system on which the development in Section 4.3.3 is based, is also available there, as well as the complete source code for the author's Prism radiation therapy planning system.

Acknowledgments

Some figures in Chapter 1 come from Wikimedia Commons. The URL for each is noted, and contributors' names are noted where available. Full terms of licensing of these figures are available at each referenced URL. They are either released into the Public Domain by their author or available under the Creative Commons Attribution-Share Alike 3.0 Unported license, the full text of which is at URL `http://creativecommons.org/licenses/by-sa/3.0/`. Other figures throughout the book have sources noted and acknowledged, or are the author's original creation.

This book contains content from LOINC® (`http://loinc.org`). The LOINC table, LOINC codes, and LOINC panels and forms file are copyright ©1995-2012, Regenstrief Institute, Inc. and the Logical Observation Identifiers Names and Codes (LOINC) Committee and are available at no cost under the license at `http://loinc.org/terms-of-use`.

In addition to all those wonderful colleagues, friends, and family mentioned in the Preface to the First Edition, included following this Preface, it is a great pleasure to thank additional contributors and supporters.

It is a great pleasure to thank Christine Minihane, Senior Acquisitions Editor at Elsevier, for encouragement and support in the decision to do a second edition. It has been a pleasure to work with new editors Catherine Mullane and Graham Nisbet at Elsevier on the development of the structure of the new edition. Catherine has been extremely helpful in keeping me going when it seemed a much too daunting task. I'd like to thank also the (anonymous) reviewers of the proposal for the second edition for their encouragement, and their very helpful suggestions, which I have tried to incorporate as much as possible.

Thanks to Linda Shapiro for inviting me to give a guest lecture to her class on image understanding, which was an opportunity to develop new material explaining the physics and computational aspects of medical images. Linda also helped identify some useful review articles on the topics of image segmentation and image database query by content.

Andrew Simms pioneered the application of OnLine Analytical Processing (OLAP) to enable queries on extremely large-scale data on protein folding. He wrote the section describing OLAP, including both a simple aspect of the protein folding data analysis and a medical example involving eligibility for clinical trials, for which I am very grateful. I look forward to Andrew's work to author a book on OLAP in the biomedical sciences and other areas.

Daniel Capurro generously shared his work on representation and query of interval data for secondary use, from Chapter 6 of his Ph.D. dissertation, including his original implementation in java, on which my Lisp version is based.

I have greatly enjoyed many conversations with Todd Johnson on the nature of information in medicine. We see things somewhat differently, but both points of view are valid and worthy of attention from students. I hope that inclusion of a few key references provided by Todd will make up for the fact that I did not have the space to do justice to the work of Scott Blois and other early pioneers in medical informatics on this topic.

Catherine D'Ambrosio was a pleasure to work with during her nursing Ph.D. studies and the project described here. Thanks to Catherine for being willing to share her work on representing nursing knowledge in rules, and for keeping me posted on her prototyping of a robot nursing assistant.

I am grateful to Tomasz Adamusiak and Olivier Bodenrider for permission to include examples from their publication and presentation at the 2012 AMIA Annual Symposium illustrating how a Description Logic representation can enhance the use of LOINC codes and related terminology.

Thanks to Jamie Pina for providing useful material on public health and to Anne Turner for additional insights into the rapidly changing field of public health informatics. Peter Anderson, while a student and postdoctoral fellow in the University of Washington Biomedical and Health Informatics Training Program, provided me with useful insights and pointers to the use of information theory in molecular biology and about codon usage in particular.

Thanks to Yoshi Kohno and his students for educating me on security and privacy issues, and particularly Adrienne Andrew for assistance in teaching the Life and Death Computing class.

Several students from the most recent Life and Death Computing class wrote outstanding project reports. Two of these contain material I cite in appropriate places or are actually incorporated with permission from the authors, and some editing by me. Nancy Tang did some very illuminating tests of the DNA sequence matching code in Chapter 5 and her results are mentioned there. Edward Samson wrote a nice introduction to Markov Chains and sketched an application to medicine, which I included also in Chapter 5, with his permission.

Although it is not possible to name all my students individually, I owe a great debt to them for their engagement with this material, and for their feedback. I thank all my students for tolerating and struggling with sometimes ill-formed homework assignments and for helping me to add more precision. This edition includes the fruits of that dialog. Any remaining problems, ambiguities, or other issues with the exercises in this book are fully my responsibility. Student project work is always incredibly impressive. I could not incorporate more than a few gems (acknowledged above) from these projects, but all the students' work has taught me a lot.

At the beginning of my work in biomedical informatics in 1983 through the present, Larry Fagan was a huge help to me and my students, through his early work on which mine was based, his personal advice and collaboration, and many years of friendship.

Wolf Kohn has been a dear friend for decades, throughout the writing of both the first edition and this one. It is a pleasure to thank him for his continuing unbridled enthusiasm and encouragement. Many other colleagues, already named in the Acknowledgments to the First Edition, also have provided continued support and encouragement, whose value is immeasurable.

Family of course is key to surviving any big writing project. I've tried to incorporate my son Nathan's wry sense of humor where appropriate, and hope you enjoy it too. My son Alan is working as I did in both medical physics and biomedical informatics. His assistance in proofreading the manuscript is gratefully acknowledged. Discussions with him have kept me grounded and focused. My son Brian, who did the design for the first edition cover, is on his way to becoming an internationally renowned graphic designer. I've learned a lot from him about design and quality. My grandson Brandan, now 4, loves discovering new things, and it is great to see his unrestrained joy of learning. My granddaughter Morrigan, 11, did a drawing of Albert Einstein carrying the first edition of this book. I can't imagine a stronger vote of confidence than that. But the most thanks go to my wife, Terry, who, once again, is acknowledged in the Dedication.

Kirkland, Washington, 2013

Foreword from the First Edition

Biomedical Informatics (BMI) is an extraordinarily broad discipline. In scale, it spans across genes, cells, tissues, organ systems, individual patients, populations, and the world's medical ecology. It ranges in methodology from hard-core mathematical modeling to descriptive observations that use "soft" methods such as those of sociology. It studies, models, and designs data, processes, policies, organizations, and systems.

No wonder it is difficult to find (or write) a text or reference book that encompasses the field. Yet BMI has become a discipline, gradually moving from a variety of *ad hoc* collections of interdisciplinary programs toward a more formally defined specialty, recognized both as a subspecialty option for many fields of medicine and as a concentration in engineering programs. What distinguishes this field is its focus on a fundamentally computational approach to biomedical problems. Of course every field in today's age of information uses computation as a tool, but in other disciplines, other methods such as laboratory experimentation or systems design take pride of place.

Adding to the challenge of presenting the fundamentals of BMI in a book is the varying depth to which concepts need to be understood (and thus explained) depending on what role the reader expects to play. For example, a high level manager such as a chief information officer needs a good layman's understanding of *all* of the issues but generally cannot know the details of all the technologies he or she needs to deploy. Yet the software engineers building those systems need to understand intricate details of databases, communication mechanisms, human-computer interface design, image processing, design for safety, decision support, etc. Similarly, a researcher who directs a clinical trial must focus on a precise understanding of the biomedical question to be answered, the data and methods needed to answer it, the study design, and factors that help assess confidence in the results. Another researcher, studying problems in bioinformatics, needs a deep knowledge of molecular biology and the specific data being analyzed, graph theory, probability and statistics, and data mining and machine learning. Perhaps with time, we will develop a complex ecology of (sub-)subspecialties corresponding to these various clusters of required expertise, but our field is young enough that we have not yet defined these clusters, and indeed many individuals play multiple roles and need multiple types of expertise even at one time, or certainly throughout their careers.

Another source of difficulties in teaching BMI is the large variety of backgrounds from which students are drawn. Many are MDs who choose to specialize in BMI after their formal medical training is complete, perhaps as part of a Fellowship program. Many others are engineering students who find in the challenges posed by health care an appealing outlet for their talents and interests. Some come from backgrounds as librarians, in public health, health economics, management, or law. The intersection of all the subjects studied by people from such diverse backgrounds is close to nil. Therefore, developing classes to serve this population is highly challenging, because it is impossible to rely on almost any specific background.

Yet despite the diversity of applications and backgrounds, common themes have emerged in this discipline. Methods of data collection, encoding, standardization, storage, retrieval, visualization, exchange, and integration pervade the field. Model construction, validation, and analysis underlie most applications.

Pedagogically, many biomedical informatics courses of study have adopted an approach that begins with a broad introductory subject that surveys the field, but cannot at the same time present the details that will be needed in its various specialties. That is then followed by a Chinese menu of far more specialized courses, often drawn from the faculties of more established departments such as Computer Science, Mathematics, Statistics, Economics, Public Health, etc. Yet in most such courses, the topics of special significance to biomedical informatics are scattered among a variety of interesting, but perhaps not highly relevant other material. As a result, many students must still piece together their BMI education from such an *ad hoc* assembly of classwork, readings in journals and conference proceedings, and mentoring by their research supervisors and colleagues. Although this is by no means a bad way to learn, as the field becomes more popular and therefore attracts more students, it is inefficient. It also demands a high level of pedagogic skill and inspiration from all teachers, which makes it difficult to establish new programs and exposes students to the risk of poorly thought out curricula.

Professor Kalet, in this book, tackles the question: Where should one turn after a broad introductory treatment of BMI? In contrast to the introductory books, he takes a frankly technical, computational approach, identifying those underlying methods that cut across the many goals of BMI training and research. Thus, the book begins with a discussion of the nature of biomedical data, and how it can be represented in a computer, described, communicated, and analyzed. It then moves onto the representation and use of symbolic knowledge that can form the basis for building intelligent tools. Next, it covers probabilistic methods that are crucial for both representation and interpretation. Then it introduces methods appropriate to build information retrieval systems, which focus on the vast literature of BMI and related fields, and, at least today, are the only organized systems that contain most of the knowledge gleaned from decades of biomedical investigations. The second half of the book takes up a handful of interesting topics, both because they are important to the advancement of BMI research and because they serve as wonderful illustrations and challenges to the methods presented earlier. These include computational methods critical to bioinformatics, such as computing with gene and protein data, the representation and search for biological networks, representation of structures such as those that define anatomy, and the search for a reliable system to analyze and record drug interactions.

Other chapters touch on standards and medical data communications, cancer radiotherapy planning, and system safety and security.

What unifies these topics is a commitment to the idea that simple familiarity with such concepts is not enough—instead, the serious student and practitioner must develop the skills to program and implement the various methods being described. This book provides a coherent and consistent presentation, and all of the examples and concepts are worked out in an extraordinarily expressive and flexible programming language, Lisp. Although the book is by no means a book about programming, it demonstrates the fundamental reliance of complex BMI ideas on elegant programming implementation. As Benjamin Whorf famously argued,[1] (human) language influences the ways in which we think. Because Lisp uses a very regular, simple syntax and supports powerful methods for defining and using abstractions, it is particularly well suited to the task at hand here. As a result, many examples that could require distractingly tedious and complex implementations in more common programming languages can be done simply in Lisp and retain their logical structure in the implementation.

Programming is essential to BMI. It forces us to specify ideas and methods in precise enough ways that a computer can execute them. It is the manner in which we enlist the help of computers to support the management and exploitation of data, and to interact with us to support the human processes that we undertake. In the past decades, we have moved from an era of unimaginably expensive computers to ones that are cheap enough to embed in every appliance. As a result, the equation defining the costs of projects has shifted completely from the "expensive computer/ cheap programmer" model to the opposite. Thus, development of abstractions and tools that make it easy to conceptualize and implement new techniques provides more leverage than ever, and some of the most important advances are those that improve the capabilities and efficiency of BMI practitioners. With today's tools, it is still *much* easier to conceptualize a new study or a new process than to implement it. However, groups that traditionally separate conceptualization from implementation have not accrued a good track record. Instead, it seems that programmers need to understand the domain for which their programs are being built, and designers need to understand the constraints imposed by programming realities. Indeed, the high costs of programming staff dictate that many junior BMI researchers must act as their own programmers. Therefore, BMI cannot be studied in isolation from computer programming methods and techniques.

Professor Kalet brings to this book extraordinarily broad experience and a deep understanding of the topics he presents. He received his Ph.D in theoretical physics, taught physics, mathematics education, and even liberal arts, has worked professionally in radiation oncology, held professorial appointments in radiation oncology, medical education and biomedical informatics, computer science, biological structure, and bioengineering, has been responsible for security and networking for a hospital IT network, and has created and led a graduate program in biomedical and health

---

[1] John B. Carroll, Ed. Language, Thought, and Reality: Selected Writings of Benjamin Lee Whorf. MIT Press, 1964.

informatics. Thus, he represents in one person the interdisciplinary threads of the field this book is helping to define. In addition, he is a lucid and careful expositor, so the lessons of the book are within reach of the diverse audience it seeks to serve.

Early in our new millennium, despite great advances in health care and the research that supports it, we are surprised by the slow development of methods that improve actual clinical practice, we are frustrated by pathologies in the way health care is delivered and paid for, and we are both daunted and challenged by the vast opportunities that are being opened by new biological insights, improved measurement methods, and our greatly enhanced abilities to collect and analyze vast amounts of empirical data on everything from how molecules interact within a cell to how medical services interact in a population. We are, indeed, in a time of ferment, excitement, and challenge for biomedical informatics. As Alan Perlis said,

"I think that it's extraordinarily important that we in computer science keep fun in computing."[2]

The same is surely true of BMI. This book will contribute to that fun.

Peter Szolovits
Cambridge, Mass.
May 2008

---

[2]Alan J. Perlis. The Keynote Address. In Perspectives in Computer Science, Anita K. Jones, ed. Academic Press, 1977.

Preface from the First Edition

One of the great challenges of biomedical informatics is to discover and/or invent ways to express biomedical data and knowledge in computable form, and to thus be able to automate the analysis and interpretation of a wide range of biomedical data, as well as facilitate access to biomedical data and knowledge. Biomedical informatics textbooks [378,423] typically present a survey of such ideas and applications, but do not discuss how these ideas and applications are actually realized in computer software.[3] Bioinformatics books typically have either narrowly focused on sequence analysis algorithms and other statistical applications, or on how to use a wide variety of existing software tools. Some of the bioinformatics books on closer inspection are really about the underlying biology with little or no reference to computerized data or knowledge modeling except as a (usually mysterious) means to arrive at biological results, reported as text and/or diagrams, rather than in computable form. Others, both biological and medical, focus on sifting through the arcane and problematic syntaxes of various data sources, and the trials and tribulations of extracting useful data from natural language text.

---

[3]Recent exceptions include [28,414,415], which include some perl scripts, and argue for the importance of learning something about programming.

A recurring theme in discussions with biologists, medical researchers, health care providers, and computer scientists is the question, "What is biomedical informatics?" The practical answer, that we are going to provide the software tools to meet biomedical research and clinical practice computing needs, addresses a real need, but seems inadequate to define the intellectual content of the field of biomedical informatics. Thus, the main goal of this book is to provide an exposition of important fundamental ideas and methods in biomedical informatics, using actual working code that can be read and understood. The object of the code is not only to provide software tools, but for the code to play the same formal role that equations and formulas play in theoretical physics. That is, the programs precisely embody biological, medical, or public health ideas and theories. The application of these programs to specific problem instances and data is analogous to the solution of the equations of physics for specific practical problems, for example, the design of airplane wings.

This is not a book about Artificial Intelligence (AI), or at least it does not take the viewpoint typical of AI books. My viewpoint is that of a scientist interested in the creation and study of formal theories in biology and medicine. Whether such theories embody some notion of intelligence is not relevant here. Some of the topics covered strongly resemble those of an AI textbook, because the ideas and languages from that branch of computer science are the right ones for modeling a wide range of biological and medical phenomena, just as differential equations, vector spaces, abstract algebra, and related areas of mathematics are the right ideas and languages for creating theories of physical phenomena.

One implication of this view is that researchers in biology, medicine, and public health will learn computational science (including programming languages and how to write programs) as a normal part of their training. Understanding, writing, and using computer programs will be an essential part of understanding the biology, medicine, or public health phenomena, not just a bunch of machinery to be purchased and used.

While the predominant view in this book is that of computer programs as implementations of formal theories, it is not the only way that computable knowledge in biomedicine is used. In regard to *formal* theories, biology and medicine are relatively young (compared to the physical sciences, particularly physics). Much knowledge is in small chunks like jigsaw puzzle pieces. Independently of whether you believe the pieces can fit together to form a coherent theory, many people find it useful to organize, index, and search collections of such chunks to solve important practical problems. This, too is an important part of biomedical informatics. Thus, Chapter 2 includes some discussion of knowledge as a collection of assertions that may or may not be a complete theory of a biomedical domain. It is developed more fully in Chapter 4, on biomedical information access, as well as in some example chapters in Part II.

There are precedents for using computer programs as renditions of subject matter, not just examples of programming techniques or programming language syntax and semantics. They are an inspiration for this work. Norvig's book, *Paradigms of Artificial Intelligence Programming* [305], explains fundamental ideas of artificial intelligence by showing how to build the programs that implement these ideas. Abelson and Sussman's well-known *Structure and Interpretation of Computer Programs* [1] is really about computation ideas and principles, and not about how to do programming in Scheme. More recently, Sussman and Wisdom have written a remarkable new treatment of classical mechanics [404], not from the point of view of traditional numerical approximation of solutions of differential equations, but from the point of view of expressing the foundational concepts themselves directly in a computable form. My goal similarly is to write about biomedical informatics ideas by showing how to write programs that implement them.

Structure of the Book

Defining what is essential, and for what purpose, is an intractable problem

—Cornelius Rosse and Penelope Gaddum-Rosse
from the Preface to "Hollinshead's Textbook of Anatomy"

The chapters in Part I, Foundations of Biomedical Informatics, develop the principles, ideas, and methods that constitute the academic discipline of Biomedical Informatics. These can be organized loosely into biomedical data representation, biomedical knowledge representation, organization of biomedical information (both data and knowledge), and information access. Biomedical knowledge representation, as inherited or adapted from the field of Artificial Intelligence, can be loosely divided into categorical or symbolic methods and ideas, and probabilistic methods and ideas. Thus, the book begins with representation and transformation of biomedical data. Next, ideas and methods for symbolic and probabilistic biomedical knowledge are presented. This part concludes with an introduction to ideas and methods for biomedical information access, addressing both data and knowledge. Thus, Part I roughly corresponds to the "Biomedical informatics methods, techniques, and theories" at the top of Ted Shortliffe's well-known diagram, found on page 32 (Figure 1.19) of [378].

Part II, Biomedical Ideas and Computational Realizations, provides a series of applications of the principles, illustrating how to express biomedical ideas, data, and knowledge easily in computable form. The chapters include examples from various biomedical application domains, but are not in any way a survey or comprehensive in coverage. The areas touched on include molecular and cellular biology, human anatomy, pharmacology, data communication in medicine and radiology, and radiation oncology. Thus, Part II traverses the horizontal axis of the Shortliffe diagram mentioned above.

Some additional topics are included in Part II. Chapter 5 develops ideas for how to simulate systems that change with time, using state machines and other formalisms. Chapter 10 treats issues around the design of medical device software, and the deployment of electronic medical record systems and medical devices in a network environment. Standards are briefly mentioned in Part I, and two particular examples, HL7 and DICOM, are developed in more detail in Chapter 8.

The Appendix provides reference material on the programming languages and environments used in the book. It also includes pointers to resources for pursuing some of the engineering aspects of building a practical system in more depth.

As Rosse and Rosse suggest (quote above), not everything can be included in a single book. I've left out topics in bioinformatics and computational biology which appear to be well covered elsewhere and are somewhat far from the overall themes of this book. These include the more complex sequence analysis algorithms, searching sequences for motifs, generating phylogenetic trees, to mention a few obvious omissions. Although I have included in Chapter 5 an introduction to simulation using state machines, I don't go into much depth about the details of Systems Biology Markup Language or the many projects that are active. The references in that chapter should be useful as a link to get more deeply into these topics. Another notable omission is the vast subject of electronic medical record systems, and the representation and use of clinical guidelines. Nursing and Public Health were in the original plan for the book, but time did not allow development and inclusion of examples from these areas other than the use of a public health problem in Chapter 2. Finally, the topic of data mining, or machine learning, while growing rapidly in importance, is not covered here, in the interest of keeping the book a manageable size and finishing it in a finite time.

Software Design and Engineering

There is a crisis in biomedical software today. The NIH has recognized the importance of computing in biomedical research through the Biomedical Information Science and Technology Initiative (BISTI). The cost of building custom (one of a kind) programs of any scale, for each research team, is usually prohibitive. Yet, commercial software packages are often not adaptable to the specific needs of research labs, let alone the complexities of clinical and public health practice. It is certainly not simply a matter of taking advantage of existing technology that has been produced by computer scientists. The problems of data integration, design of extensible and flexible software systems and programming environments, matching different data models and modeling systems, are all important research questions in computer and information science. The emerging partnership that BISTI promises to promote is critical for the future of biomedicine.

Beyond BISTI, the NIH is also calling for researchers to include plans for dissemination, education, and support in proposals that include substantial software development. This is an encouraging development, as few places can do such work without significant investment of staff time.

Significant issues in biomedical informatics that are sometimes overlooked are code reuse and software engineering. While the enthusiasm for "open source" software development is growing rapidly, there is little discussion of best practices in software design, and the result is a large body of freely available code that is notoriously difficult to extend or modify. Many advocates of "open source" seem to believe that by itself, this approach will produce (by some evolutionary process) high quality software. This is insufficient for the crisis we face today, especially in safety critical applications such as clinical medicine. The examples I develop in Part II are intended to be exemplars of good software design as well as clear illustrations of the underlying principles.

The challenge is even greater in health care than in biomedical research. Software inadequacies, errors, and failures have been shown to be directly or indirectly responsible for serious injury as well as death to patients [250]. Regulatory agencies such as the US Food and Drug Administration (FDA) face a daunting job to deal with this on the scale of very complex electronic medical record systems with huge numbers of hardware and software components. Vendors of medical software are not experienced in designing multiuser or real-time systems (computerized medical devices), and they are especially inexperienced at designing systems that are robust and safe to operate with connections over the open Internet. Chapter 10 explores these questions from the perspective that something can indeed be done to prevent or minimize the hazards. In this chapter, we explain the necessity of Internet connectivity for medical devices and systems, and address the dangers as well as some preventive measures in connecting computerized medical equipment and data systems to the Internet.

Programming Languages

A language that doesn't affect the way you think about programming is not worth knowing

—Alan J. Perlis [324]

The main programming language that is used in this book is Common Lisp. Previous exposure to Lisp is *not* assumed. To help the reader who may be unfamiliar with Lisp, Chapter 1 provides an introduction to the essential ideas, along with discussion of biomedical data and their representation. However, the reader who is already familiar with the general culture and tools of programming will have a much easier time than one who is not familiar with text editors (the best of which is, in my opinion, emacs), operating system command interpreters, file systems, basic network operations and facilities. Previous experience with a specific programming language can be a help or a hindrance, depending on the reader's willingness to see the programming idea in radically different ways (or not). In teaching this subject, I occasionally find that I have to help a student unlearn ideas picked up from an exposure to C or java, in order to learn very different styles of abstraction.

A goal of this book is to refocus thinking about biomedical informatics programming, not as a technical skill for manipulating bits and bytes, but as a natural, logical, and accessible language for expressing biomedical knowledge, much as mathematics is used as the universal language for expressing knowledge in the physical sciences. Lisp is used in this work, because it was designed for building these abstractions, higher level languages, and extensions. As bioinformatics in particular moves from the data-intensive stage of searching and correlating massive amounts of raw data to the knowledge-intensive stage of organizing, representing, and correlating massive numbers of biological concepts, the expressive power and efficacy of Lisp will become even more important.

From the very beginning of informatics research in medicine, Lisp has been a prominent choice of programming language. In medicine and biology, much of the knowledge (as well as data) is non-numeric. Lisp is particularly well suited to this. However, Lisp is also fine for numeric computation, user interface programming, web applications, and many other things as well. So, the importance of Lisp in biomedical informatics should not be underestimated by thinking of it as only the language people (used to) use for Artificial Intelligence research. Its power comes from the ability to easily create abstraction layers and programs that transform expressions before evaluating them. This makes it easy to define higher level languages in Lisp, that provide just the right tools for efficient problem solving. Also in Lisp, functions can be manipulated as data: new functions can be written by other functions, and then used. While this may seem mysterious and useless to the programmer new to Lisp, once you see how it works, you will wonder how people could write programs without it (cf. Graham [137], Chapter 1).

As biomedical data and knowledge become increasingly complex, and the software to manage it grows in scale, it is almost a truism to emphasize how important it is to effectively use abstraction and layered design. Lisp is also important because it has proven to be the best suited to these goals. I chose Common Lisp (one of the two major contemporary dialects of Lisp) because it has all the technical support to write "industrial strength" programs that perform as well or better than programs written in more popular programming languages such as C++, java, and perl. Common Lisp has a very solid and powerful object-oriented programming facility, the Common Lisp Object System (CLOS). This facility provides a flexible means to write extensible code; indeed, CLOS is built on its own ideas, and includes a meta-object protocol (MOP), which provides ways to extend and augment CLOS itself without needing or modifying any CLOS source code. Common Lisp also provides a macro facility that makes it easy to write code that transforms expressions, that is, it supports writing interpreters for higher level (but embedded) languages for special purpose tasks.

Lisp is a fine general-purpose programming language. It has been shown to be easier to use than C or java, and compilers are available to generate code that runs as fast or faster than equivalent code written in C. It certainly supports higher performance than java. A recent study by Ron Garrett (formerly Erann Gat) [125] demonstrated these and other important points. Some characteristics that distinguish the Lisp language include:

1. Lisp programs generally take fewer lines of code than other languages to solve similar problems, which also means less development time, fewer programmers, lower cost, less bugs, and less management overhead,
2. Lisp has lots of built-in tools and facilities for tracing execution of programs, testing small and intermediate components, and thus programs written in Lisp tend to be easier to debug than programs written in other languages,
3. Lisp has lots of libraries and a large freely available code base, including support for high performance 3-D graphics, user interfaces, and network socket programming, plenty of support from Lisp software vendors, and a significant number of Lisp programmers,

4. many myths about Lisp have circulated over the years but are untrue (for commentary on this, see [137,305]),
5. and most important, Lisp provides a superb interactive programming and execution environment.

Use of Lisp for solving problems in biomedical informatics has experienced slow but steady growth in recent times. There are many web sites and numerous code libraries to perform specific tasks, but it remains daunting to the new informatics researcher and programmer to become familiar with the advanced programming ideas and the skills needed to utilize existing facilities and contribute to this field. Too often the well trodden but less powerful path of programming in the conventional languages and styles is adopted and taught. This is done despite the availability of several good introductions and working environments. Fortunately several good books are available. They are described in Appendix A.1.4.

Some examples also use the Prolog [68] programming language. Prolog is a programming language designed to represent logic systems, principally first order logic restricted to Horn clauses. Simple Prolog-like interpreters have been written in Lisp, and indeed Lisp has also been used for logic programming and constraint programming, but the availability of excellent (and free) Prolog systems also makes it attractive to use Prolog directly. Prolog is important for the same reasons as Lisp, because it provides a more abstract symbolic language in which to directly express biologic concepts, rather than encoding them in more machine-like data structures.

Many people have asked, "Why use Lisp and Prolog? Why not use java or perl (or Python or Ruby, etc.)?" The answer above should be sufficient, that the best language to represent abstract ideas is one that supports abstract symbols and expressions (lists) directly. It is possible to mimic some of this in java, perl, etc., but it is makeshift, at best. I would ask (and have asked) such a questioner the opposite question, "Why would anyone use java or perl, when Lisp seems more appropriate?" The responses always include exactly two reasons: first, everyone else is doing it, and second, there is a lot of support for those languages. This is not sufficient justification to make a technically inferior choice, especially in view of the fact that there are plenty of choices of robust Lisp and Prolog implementations (both free and commercial), and a critical mass of users of both.

Significant bioinformatics work is done with Lisp and Prolog. Appendix A, Section A.3 provides a brief summary of some of the more noteworthy projects with pointers to publications and web sites.

Biology, Medicine, and Public Health

What should the reader know about biology, medicine, and population health? A basic background in modern biology will help considerably. Although many concepts are introduced as needed, the treatment is brief and not really a substitute for basic textbooks in molecular biology, genetics, anatomy, physiology, clinical medicine, or public health. The reader should be familiar with basic concepts of modern biology (genes, proteins, DNA, basic functions of living organisms), and have some experience with disease, treatment, and the health care system.

Mathematics

Traditional mathematics does not play a prominent role, except in Chapter 3, which includes a modest introduction to statistics. For that, and as an excellent background for logical thinking and problem solving, a basic background in Calculus is highly desirable. An understanding of the mathematical idea of a function in the abstract sense is a considerable help in understanding computer programs. The idea of procedural abstraction, or defining layers of functions, is essential to building large and powerful programs.

Notes on Typography

In this book, code and data are distinguished from the rest of the text by using a `different` `font`. Blocks of code that define functions, classes, etc., are shown without a prompt. When the intent is to show a dialog, to illustrate what happens at the interactive `read-eval-print` loop, a prompt is shown. Different Common Lisp implementations each use their own style of prompt. Here, we follow the same convention as in Graham [137], the "greater-than" angle bracket, >.

How To Use This Book

My main objective in writing the book is to provide a principled exposition of the foundational ideas of biomedical informatics. It provides a solid technical introduction to the field, and a way to gain deeper insight into familiar biomedical computing ideas and problems. I hope that experienced researchers and practitioners in biomedical and health informatics will find it a useful synthesis of the technical aspects of our field.

The intended audience also includes:

- biologists, clinicians, and others in health care, who have little previous exposure to programming or a technical approach to biomedical informatics, for whom an introduction to symbolic computing and programming is provided as needed in the first few chapters,
- computer scientists, mathematicians, engineers, and others with technical backgrounds, who have an interest in biomedical informatics but not much exposure to biology or medicine, for whom I have introduced some basic ideas in molecular and cellular biology, human anatomy, pharmacology, clinical medicine, and other biomedical areas where needed, and
- students interested in biomedical informatics as a field of study and potential career, including undergraduates, medical and health professional students, graduate students in biomedical

and health informatics, and in related fields (computer and information sciences, basic biological sciences).

The book can thus serve as a text for a course in several contexts. It has been prototyped in a course I have taught for several years, "Knowledge Representation and Reasoning in Biology, Medicine, and Health." There is much more material in the book than can be reasonably covered in one quarter. Two quarters would be comfortable. The KR course covers the material in Chapters 1, 2, 3, and parts of Chapters 5, 6, 7, and 9. Topics from the other chapters have been included in another course I taught, called "Life and Death Computing." Other combinations are possible too.

Acknowledgments

The idea for this book originated in a discussion with Pete Szolovits at the 1999 American Medical Informatics Association (AMIA) Fall Symposium. I had just started to teach a course for medical and health professionals and students who might be interested in biomedical informatics but who had no prior computing background. The regular introduction to programming taught by the Computer Science Department was not a very satisfactory option. I learned from Pete that at MIT the biomedical informatics students took a version of the famous "Scheme" course, 6.001, which also was very challenging. We talked about the idea of having a book that introduced biomedical informatics along with programming in some suitable language (obviously Lisp), along the lines of Norvig's AI text [305]. A long time passed and many more discussions took place before that idea became concrete enough to attempt to write such a book. Pete not only encouraged me to do it, but provided significant example code, lots of references, and lots of reassurance when the project seemed intractable. His editorial advice has been invaluable.

My interest in computing started during my freshman year in college, 1961, when I found that the Cornell University computer center provided free student access to its Burroughs 220 computer (the picture on page xxiii shows me in 1962 at the operator's console). I taught myself Burroughs 220 assembly language and began my first programming project. The goal of this project was to replicate some experiments in automatically composing music according to rules of harmony, voice leading, and also some more modern composition techniques. The experiments I wanted to replicate and build on were done by Lejaren Hiller and Leonard Isaacson [157], whose work resulted in a composition, "Illiac Suite for String Quartet," named for the computer at the University of Illinois that they used. I soon realized that neither assembly language nor languages like `Algol` and `FORTRAN` were suitable for this task. I did not know at the time that a very powerful programming language called Lisp had just been invented to express exactly these kinds of symbolic, abstract ideas. I also did not know that this kind of programming was called "artificial intelligence." Nevertheless, it was my start. It is a pleasure to thank my high school classmate, Eric Marks, who taught me music theory and got me interested in that project. It set the stage for my later interest in software, which led me to get involved in the development of radiation therapy planning systems, artificial intelligence in radiation therapy planning, and from there to biomedical informatics more generally.

Funding for this "Scholarly Works" project was made possible by Grant 1 G13 LM008788 from the National Library of Medicine, NIH, DHHS. The views expressed in this and any related written publication, or other media, do not necessarily reflect the official policies of the Department of Health and Human Services; nor does mention of trade names, commercial practices, or organizations imply endorsement by the US Government.

Illustrations noted as "courtesy of National Human Genome Research Institute" (NHGRI) are in the public domain and are from the NHGRI Talking Glossary web site.[4] The illustration from the National Institute of General Medical Sciences (NIGMS) is from the Protein Structure Initiative image collection web site.[5]

It is a pleasure to acknowledge the encouragement and advice of Valerie Florance, Deputy Director, Extramural Programs, National Library of Medicine, from whom I first learned about the NLM Publication Grant program, as well as many stimulating conversations with Milton Corn, Director, Extramural Programs, National Library of Medicine.

Many other people have contributed ideas and code to the book. They are gratefully acknowledged here. However, any errors or inaccuracies are solely the responsibility of the author. The book is typeset using $\LaTeX 2_\varepsilon$ [247,287]. The figures showing the formulas for the molecules appearing in various chapters were produced with Shinsaku Fujita's $X^Y\!M\!T\!E\!X$ package set [120], a system for typesetting chemical structural formulas.

- Larry Hunter, Peter Karp, and Jeff Shrager all provided much editorial advice, as well as encouragement. Larry is the source of the amino acid representations in Chapter 5. The biochemical pathway search idea and the cell cycle simulation in that chapter are based on Jeff Shrager's tutorials on his BioBike web site [380].
- Ted Shortliffe, Michael Kahn, and John Holmes also provided strong encouragement, especially in support of my application for the NLM Publication Grant that has supported much of the writing.
- Jesse Wiley suggested the example application of DNA sequence processing introduced in Chapter 1 and developed further in Chapter 5. Thanks also to Alberto Riva for permission to use code from his CL-DNA project [342], which provided some ideas for how to implement the examples.
- Mark Minie provided pointers to the NIGMS and MHGRI image collections, as well as much interesting and provocative discussion about the foundations and future directions of molecular biology.
- The chapter on the Foundational Model of Anatomy comes out of a long-standing collaboration with the UW Structural Informatics Group. Thanks go specifically to the

---

[4]http://www.genome.gov/10002096.
[5]http://www.nigms.nih.gov/Initiatives/PSI/Images/Gallery/.

Digital Anatomist project for providing the brain figure in Chapter 1 and the heart figure in Chapter 6. It is a pleasure to thank especially Jim Brinkley, Cornelius Rosse, and Jose L. V. (Onard) Mejino for help with the FMA and the CTV project described in Chapter 9. Onard also contributed the figure in Chapter 6 showing the lymphatic tree parts.

- Thanks to Richard Boyce, John Horn, Carol Collins, and Tom Hazlet for permission to incorporate material from our drug interaction modeling project in Chapter 7. This work began as a simple exercise for a class I taught on biomedical programming (the Life and Death Computing course mentioned earlier). Richard Boyce then a Biomedical and Health Informatics Ph.D. student, took it up, refined it, and added the idea of managing evidence by categorizing it and using a Truth Maintenance System. John Horn, Carol Collins, and Tom Hazlet contributed much pharmacokinetic knowledge as well as modeling advice.

- Thanks also to Richard Boyce for providing the example and introduction to decision modeling in Section 3.3 of Chapter 3.

- It is a pleasure to thank Dave Chou for the sample HL7 data in Chapter 8, as well as many helpful discussions about HL7, LOINC, and EMR systems.

- The design ideas as well as almost all of the implementation of the Prism DICOM System at the University of Washington were the work of Robert Giansiracusa in collaboration with the author [200], as part of the Prism Radiation Therapy Planning System [203]. What I present in Chapter 8 is a simplified version with some variations, to explain the implementation. Appendix A includes a pointer to the original DICOM source code. The development of the DICOM code took place during a period of many visits to Israel, to the Soroka Medical Center Oncology Department, affiliated with Ben Gurion University. I owe a great deal of thanks to my host and friend, Yoram Cohen, for his kindness, encouragement, and support. Many others at Soroka and BGU helped in this project, notably Drora Avitan and Yanai Krutman.

- The Prism project described in Chapter 9 had many contributors, but especially I would like to acknowledge the contributions of Jon Unger and Jon Jacky, who were co-authors of both the code and the documentation. Jon Jacky and I collaborated for approximately 20 years on three major radiotherapy treatment planning projects, of which Prism was the third. Jon Jacky was also a contributor to the DICOM system, which is part of Prism. The ideas for modeling simple safety issues for radiation therapy machines in Chapter 10 also originated with Jon Jacky, as a guest lecturer in one of the classes I taught.

- It is a pleasure to thank Mark Phillips for help with the ideas in Chapter 3, and for being a wonderful colleague for many years, including working on Prism and related projects described in Chapter 9.

- The Clinical Target Volume project described in Chapter 9 is an ongoing project with many collaborators. In addition to Jim, Cornelius, and Onard, mentioned above, Mary Austin-Seymour, George Laramore, Mark Whipple, Linda Shapiro, Chia-Chi Teng, Noah Benson, and Casey Overby deserve special mention.

- The Planning Target Volume project described in Chapter 9 was the work of Sharon Kromhout-Schiro, for her Ph.D. dissertation in Bioengineering at the University of Washington.
- The UW Medicine IT Services Security Team has taught me a lot about hospital IT security, what works and what does not. It has been a pleasure to work with you all.
- Thanks to Brian Kalet for the cover design.

The support of the faculty of the Biomedical and Health Informatics program at the University of Washington is much appreciated, particularly Fred Wolf, Chair of the Department of Medical Education and Biomedical Informatics, Peter Tarczy-Hornoch, Head of the Division of Biomedical and Health Informatics, and George Demiris, Graduate Program Coordinator. Thanks also go to Wanda Pratt for her encouragement of this project and her willingness to make her DynaCat program available. I was not able to include the DynaCat program because of limits of space and time, but hopefully there will be a second edition.

Thanks to George Laramore, Chair of my home department, Radiation Oncology, and to colleagues in Radiation Oncology for their encouragement and interest.

Many students over the last few years have read and commented on drafts of the manuscript and have been subjected to assignments that eventually became material in the book. Although there are too many to name you all individually, I greatly appreciate all your feedback, and involvement with learning this material.

Steve Tanimoto and Alan Borning welcomed me as an adjunct faculty member to the University of Washington Computer Science and Engineering department in 1983. Their hospitality and that of their colleagues has been nothing short of wonderful.

I would like to acknowledge the kind advice, support, and guidance of the late Peter Wootton, Professor of Radiation Oncology (Medical Physics), who was my mentor from 1978 when I joined the University of Washington as a postdoctoral fellow in Medical Physics, and a dear friend until his passing in 2004.

My advisor from graduate school, Marvin L. ("Murph") Goldberger, has kept in touch over the years and offered valuable reassurance even though I did not stay with theoretical physics. His kindness is gratefully acknowledged.

Finally, I would like to acknowledge the editors at Academic Press, Luna Han, Kirsten Funk, and especially Christine Minihane and Rogue Shindler, who helped enormously in pushing the manuscript to completion.

List of Figures

List of Tables

Biomedical Data

Chapter Outline

Principles of Biomedical Informatics. http://dx.doi.org/10.1016/B978-0-12-416019-4.00001-9

> *Just the facts, ma'am...*
> *– Sergeant Joe Friday*
> ***Dragnet TV series, ca. 1956***

It's a truism to say that biology, medicine, and public health are flooded with facts. The complete genetic sequences of many organisms from bacteria to human beings are now available. We also have many petabytes of data on the structures of the proteins in these organisms. The biomedical literature is rich with laboratory results, case studies, and clinical trials reports. Repositories exist that contain large amounts of public health surveillance and epidemiologic data. Electronic medical record systems are being deployed in many hospitals. It is a wonderful time for biomedical research and for the potential improvement of health care practice, except for one big problem: there is no coherent system for relating all or even part of the data. Worse yet, every repository and source of information seems to have its own arcane syntax and semantics for organizing and representing the data. It makes the Tower of Babel, with only 70 different languages,[1] look like a simple problem.

Biomedical data come in many forms. They include data about the fundamental molecules of life, the small molecules that are used by living organisms for energy and as building blocks for reproduction and growth, and the large molecules that encode all the information about molecular processes such as metabolism and regulation of cellular activity. At a higher level in complex organisms like humans, we have a large amount of data about physiological processes and the behavior of organs such as the heart, data about individual patients in hospital electronic medical record systems, and public health surveillance data, to mention just a few. In the area of public

---

[1]The story, in Genesis, Chapter 11, names the 70 nations descendant from Noah, initially all speaking the same language, but one could infer that the "confounding of language" that happened gave each their own language.

health especially, the potential for enriched data useable for surveillance, for epidemiologic studies, and for understanding the impact of public health planning and intervention is huge.

The Physiome Project (`http://www.physiome.org/`) is a systematic effort to create comprehensive collections of data and models of physiological processes. The project web site defines the Physiome as follows: "The physiome is the quantitative and integrated description of the functional behavior of the physiological state of an individual or species." Indeed, we now have the ability to accurately model components and in some cases whole organs. The sophistication of these models is sufficient in some cases to be useable for interactive simulation systems for teaching clinical cognitive skills [368,369]. The ideas behind simulation modeling will be introduced in Sections 5.4 and 5.5.

As our understanding increases of the entities and processes these data describe, new possibilities will emerge. Discussions are already taking place about the eventual widespread use of DNA sequencing and other high throughput data acquisition in the practice of clinical medicine. Much research is directed to identifying genetic markers that will help determine the prognosis of disease and guide selection of treatments. Medical imaging techniques are developing not only toward higher spatial resolution, but also toward diversifying the kinds of information that an image represents. We already now have a variety of functional imaging techniques, such as Positron Emission Tomography (PET), where the images indicate levels of certain kinds of metabolic activity rather than anatomical structure. To combine these data sources in a way that makes sense for medical decision making is already a huge computational and conceptual challenge.

The growth of capacity, speed, and availability of the Internet has also opened new challenges. For researchers, it has become imperative but very difficult to make large data sets available to other researchers, and similarly to make it possible to compute correlations and combine data from diverse sources. For medical practitioners, integration of access to patient data in electronic medical record systems has been called for at the highest levels of medical community leadership and the US federal government. It remains an elusive challenge. Similarly, for patients to get access to their own information is widely acknowledged as very important, but systematic solutions are not forthcoming.

The varieties of biomedical data have some biological and technical aspects in common. Some of the technical aspects include: how to encode something that is not a numeric measurement (gene and protein sequences, organ names and relationships, descriptions of clinical findings, names of diseases, etc.), how to organize both numeric and non-numeric data for easy retrieval and interpretation, and how to represent hierarchical relationships or groupings of data that have an *a priori* relationship (for example, a collection of genes whose expression levels are being measured, or a set of related blood tests). Some of the biological aspects include: how to represent higher level phenomena (such as phenotypes, diseases, and therapies) in relation to lower level ones (genotypes, cellular and organ level processes, and body structure), and how to systematically and consistently assign names and classifications to all the entities, processes, and relationships [350]. All this depends on the assignment of a particular meaning for each datum. The numbers, symbols, and text by themselves have no meaning.

The representation and organization of biomedical data require consideration of three dimensions of ideas.

Context   Data will need to be represented in the context of running programs, where the data are volatile. When the program terminates, the data disappear. In order to be able to preserve the results of data input and computations, it is necessary to have a second context, a representation of data in files or other forms of non-volatile storage, so that they have long-term persistence, and can be searched, retrieved, and updated. Finally a third context is that of network communication, where the data need to be represented inside network messages. Many standards have been (and are being) developed for storing biomedical information in files, and transmitting it over a network. Standards are needed for application programming interfaces (APIs) for access to biomedical information from within running programs.

Encoding   The choice, for each data element, of how it will be represented in context in machine readable form is called an "encoding." Each encoding of information requires that some description exists, specifying the encoding. This can be implicit, described in technical documents and implemented in the source code of computer programs. It can also be done with metadata, where the encoding and structure are specified in a data description language (which itself is implicitly specified). Another option is to create so-called *self-describing* data, where the encoding and structuring are at a sufficiently low level that the metadata can be incorporated with the data. The encoding problem has generated considerable effort at standardization as well as innovation. All of these strategies have a place in biomedical informatics. Looking ahead, Section 1.1.6 presents some examples of the meaning and use of metadata.

Structure   For each kind of data in the various contexts, the structuring or grouping of data elements into larger logical units, and the organization of these logical units in a system needs to be decided. Appropriate structuring of biomedical information depends (among other things) on its intended meaning and use. Many ideas are applied here, including tables consisting of records, each with identical structure and encoding, object-oriented styles of structure, with many complex data types, often hierarchically related, logical networks, logical expressions with symbols, and others. The organization of such structures ranges from relational databases to object-oriented databases to semi-structured systems of (so-called) self-describing data, as well as biomedical ontologies with their own external forms.

In this chapter we present a series of biomedical examples illustrating these dimensions. Depending on the design, the dimensions can be orthogonal (independent) or very dependent on each other. Work in all of these areas is guided by some considerations of ease of use. For example, is it easy to retrieve the data with a computer program on behalf of a user (through a suitable user interface)? However, another important consideration is to facilitate sharing of data between laboratories. For clinical practice, too, sharing of medical record data between different health care providers for an individual patient is important. The goal of a data interchange standard is to decouple the details of representing data in transmissible form (such as network messages) from the details of the storage format of a data source. An interchange standard enables data to be shared between different laboratories or health care organizations, even if they choose to use different local storage schemes. These are sufficiently difficult problems to occupy a significant amount of research effort in biomedical informatics. Section 4.3.1 provides an explanation of how network communication between computers and programs works, and develops some aspects of the two currently most important medical data exchange protocols for clinical data, HL7 and DICOM.

In the following sections, examples will show how to structure and use biomedical data, including symbols for atoms and molecules, sequences representing DNA, RNA and proteins, and structures representing medical laboratory data, and clinical data from electronic medical record systems. An example problem, representing patient data, serializing and storing it in files and reconstructing the data from a file, shows how to represent metadata in various ways, starting with tags and values, then using declarative structures in program code (classes, in the usual object-oriented style), and using program code that itself is accessible from within a program. A general framework for describing the structure of data is called a "data model." The most widely used data model is the relational database model, but other kinds of data models have been useful as well. This chapter presents the relational model, the Entity-Attribute-Value (EAV) model and the use of arrays, also called On-Line Analytical Processing (OLAP). In all of these approaches, maintaining data quality is important. The chapter concludes with some observations about the nature of data, information, and knowledge.

1.1 The Nature and Representation of Biomedical Data

The first things to consider are the various forms of biomedical data, and how such data can be represented in a computer and manipulated by a computer program. Looking at a typical biology textbook, you will find a rich collection of photographs (including many taken using a microscope), diagrams, drawings, chemical formulas, and lots of description. Many medical textbooks have the same characteristics. The textual descriptions are about the attributes of biological entities such as cells, organs, tissues, fluids, chemical compounds found in all those, and about the relations between these entities and properties. Some of the properties of biological objects are numerical, such as the concentration of certain chemicals in blood, the size of a tumor, the pH (degree of acidity) in a cell, tissue, organ, or body fluid. Other properties are qualities, that can only be named, but not quantified. Examples of such qualities are: the type of a cell or tissue, the protein(s) produced by gene transcription, the presence or absence of an organ in an organism, the parts of an organ. These all have names, not numerical values.

Living organisms have structure at many levels. The basic *processes* of life include ingesting food from the surroundings, processing it for building blocks and for energy (metabolism), growth, reproduction, waste disposal, etc. The basic *unit* of life is the cell. A huge variety of living organisms are single celled organisms. The organisms we are most familiar with, however, including ourselves, are multicellular, made up of billions of cells of many different types. In this section, we will consider structure at the cellular level and even smaller, at the level of the organic molecules that are found in every cell, and in intercellular spaces as well. Figure 1.1 shows a schematic drawing of the inside of an animal cell.

All cells are characterized by having a *membrane*, a boundary layer that defines the inside of the cell, and gives it identity as a unit. Plant cells have in addition a more rigid structure, a cell wall. All cells contain DNA as their genetic material. Cells are classifiable into two major categories according to whether or not the DNA is contained in a subunit called a *nucleus*. Organisms that consist of cells that have no nucleus are called *procaryotes* and the cells are called procaryotic cells. These include mainly the bacteria. Cells that have nuclei containing the DNA of the cell are called eucaryotic cells. The eucaryotes include all the animals, plants, and some additional single celled organisms.

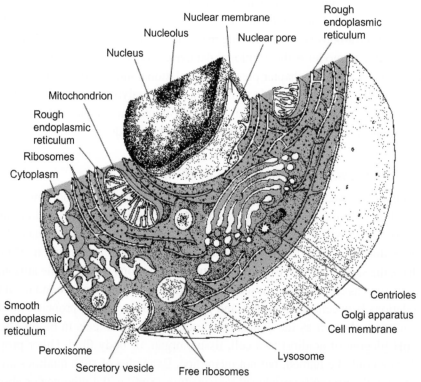

Figure 1.1 A drawing illustrating the interior components of a typical animal cell.
(Courtesy of National Human Genome Research Institute.)

One of the biggest differences between procaryotic and eucaryotic cells is that in procaryotes, the DNA subsequence that codes for a protein is *contiguous*, that is, all the bases from the start of the gene to the end are included in determining the amino acid sequence of the corresponding protein. In eucaryotes, the DNA of a gene that codes for a protein consists of subsequences of non-coding DNA that are interspersed with the subsequences that combine to be translated into the protein. The parts of a gene that code for amino acids are called *exons* and the parts that do not code for amino acids, but are between the exons, are called *introns*.

Examples of cell types include: squamous cells, epithelial cells, muscle cells, blood cells. Those are further divided, as for blood cells, there are red cells and white cells, with several varieties of white cells. These categorical attributes are also related back to numerical values. Since we can count cells, it is possible to report for an individual the concentration of different types of blood cells in the blood. This is done in a clinical laboratory, and the results are reported back to the physician as an aid to decision making for his/her patient. The type of a cell is represented by a name, not a number, since it does not mean anything to do arithmetic with cell types. We would not add one cell type to another to get some third cell type.

This section describes some ways in which large biomolecules can be represented in computable form. These sequences encode the primary structure of DNA, RNA, and proteins in terms of their components, nucleotides, and amino acids. Such large polymer-like molecules naturally are represented by lists of symbols. From the sequences we can predict some properties of these entities, and we can also compare sequences to see if they are similar. It is now routine to conduct a similarity search comparing a newly sequenced protein or gene with known ones. If a very close match is found to a known protein or gene, it is a reasonable guess or hypothesis that the function and other properties of the unknown or new sequence are the same as or closely related to those of the closely matching known sequence.

Such sequences are built up by assembly processes in cells, from their basic building blocks. The supply of building blocks comes from ingesting food (proteins and other material from other organisms) or from a nutrient environment. So, in most cases, there must also be a process for breaking up proteins into amino acids, and similarly to process smaller molecules to generate the energy needed for the synthesis of new proteins, etc. These processes are called *metabolic* processes. The role of proteins in the biochemical transformations of sugars and fat molecules is that of enzymes. An enzyme is a catalyst for a reaction. It provides a mechanism for the reaction to proceed more rapidly than in the absence of the enzyme. In some cases, a reaction will simply not happen at all without the enzyme that mediates it. Many such enzymatic reactions occur in cells. An example is the typical first step in the breakdown of glucose, which involves attaching a phosphate group to the glucose molecule. Here is a schematic description of this reaction.

```
glucose + ATP ---[Hexokinase (HXK)]--> glucose-6-phosphate + ADP
```

The enzyme that mediates this is called "Hexokinase." In attaching the phosphate group to the glucose, a molecule of ATP is converted into ADP (that is the source of the phosphate group). ATP is adenosine triphosphate, a variant of the nucleotide built from the base, Adenine, with three phosphate groups in a chain instead of just one. ADP, adenosine diphosphate, is similar, except that it has just two phosphate groups connected together instead of three.

These enzymes, reactions, and the menagerie of biological molecules they involve are thoroughly described in molecular biology books such as [10]. The Enzyme Commission database is a compilation of basic data on many thousands of enzymes that mediate such reactions (see Figure 1.5 on page 11). Many other data and knowledge resources exist, and provide the option of downloading copies of files containing this information.

Many hundreds of thousands of proteins found in living organisms have been named, catalogued and their properties recorded. These properties include the function(s) of the protein, the gene that codes for it, diseases that may relate to its absence or mutation. Proteins have many different roles in cells and tissues. They can serve as enzymes to facilitate biochemical reactions, such as metabolism of glucose. They can regulate other processes by serving as signals. Proteins also are components of cellular and tissue structure.

Proteins are large molecules made up of long sequences of small units. In proteins, the small units are from a set of 20 different molecules called *amino acids*. Figure 1.2 shows the general structure of an amino acid, and a specific example, Alanine. These molecules are called amino acids because of the presence of the NH_2 group, an *amine* group, and the presence of a COOH group, which typically gives an organic molecule properties associated with acids. The COOH group is easily ionized, losing a proton (Hydrogen ion). Such proton "donors" are acidic in behavior. For more background or a review, see any general chemistry textbook, such as [449].

Each amino acid can be symbolized by a name, which could be its full name, an abbreviated name, or a single letter, so the amino acid sequence of the protein can be represented by a sequence of names or letters. This sequence is called its *primary structure*. It is not a sequence of numbers. One possible way to encode this information would be to assign a number to each of the 20 different amino acids, and then the protein would be a number sequence. However, it is meaningless to do arithmetic operations with these numbers. They are just names. What *is* meaningful is to compare two sequences to see if they are similar, meaning that they have the same or almost the same amino acids in the same order. This is a comparison operation between lists of named things or objects, not a comparison between numbers.

Naturally occurring proteins are found in cells usually in a complex folded configuration. Sections of the sequence of amino acids can form helical structures, where successive amino acids

Figure 1.2 The general structure of an amino acid is shown on the left, and the structure for alanine on the right, with R replaced by a methyl group, CH_3.

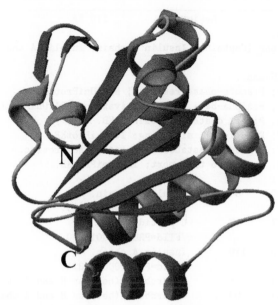

Figure 1.3   A model of an enzyme, PanB, from *M. Tuberculosis.*
(Courtesy of National Institute of General Medical Sciences, Protein Structure Initiative).

form hydrogen bonds with each other. Such a structure is called an α helix. Another common structure is a β sheet, in which the polypeptide forms a series of linear segments, folded back and forth in a weave-like pattern, again, held together by hydrogen bonds. Connections between such structures are called *turns*. These are all examples of *secondary* structure. Sets of such structures can form more complex aggregations called *domains*. These higher level structures, including structures of entire single protein units, are called *tertiary* structures. Finally, protein molecules can often be associated in pairs (dimers) or larger units, *quaternary* structures, that can then aggregate to form crystal structures. Figure 1.3 shows a diagrammatic illustration of protein secondary and tertiary structure.

Large digital data repositories are available containing information about proteins, such as the Uniprot/Swiss-Prot Knowledge Base, a project of the Swiss Institute for Bioinformatics. The Uniprot data are downloadable in a variety of formats from the ExPASy web site (`http://www.expasy.ch`), maintained by the Swiss Institute for Bioinformatics. Figure 1.4 shows an abbreviated excerpt from a Swiss-Prot entry for the precursor protein from which human insulin is formed. In Chapter 5, Section 5.1 we show how to extract sequence information from one such formatted file type, and how to compute some useful things from the sequence data. Here we are only considering what *kinds* of data may be found in the files.

For a complete explanation of the individual field items, the reader should consult the documentation available at the ExPASy web site. Here we describe just a few of these. The DR records are cross references, in this case to the Gene Ontology (GO). It is useful to be able to have a computer program look up these cross references, so that information can then be used

```
ID    INS_HUMAN                    Reviewed;           110 AA.
AC    P01308; Q5EEX2;
DE    Insulin precursor [Contains: Insulin B chain; Insulin A chain].
GN    Name=INS;
OS    Homo sapiens (Human).
DR    GO; GO:0006953; P:acute-phase response; NAS:UniProtKB.
DR    GO; GO:0046631; P:alpha-beta T cell activation; IDA:UniProtKB.
DR    GO; GO:0008219; P:cell death; NAS:UniProtKB.
DR    GO; GO:0007267; P:cell-cell signaling; IC:UniProtKB.
DR    GO; GO:0006006; P:glucose metabolic process; TAS:ProtInc.
DR    GO; GO:0015758; P:glucose transport; IDA:UniProtKB.
FT    SIGNAL          1      24
FT    PEPTIDE        25      54         Insulin B chain.
FT                                      /FTId=PRO_0000015819.
FT    PROPEP         57      87         C peptide.
FT                                      /FTId=PRO_0000015820.
FT    PEPTIDE        90     110         Insulin A chain.
FT                                      /FTId=PRO_0000015821.
FT    DISULFID       31      96         Interchain (between B and A chains).
FT    DISULFID       43     109         Interchain (between B and A chains).
FT    DISULFID       95     100
FT    VARIANT        34      34         H -> D (in familial hyperproinsulinemia;
FT                                      Providence).
FT                                      /FTId=VAR_003971.
FT    HELIX          35      43
FT    HELIX          44      46
FT    STRAND         48      50
FT    STRAND         56      58
FT    STRAND         74      76
FT    HELIX          79      81
FT    TURN           84      86
FT    HELIX          91      95
FT    HELIX         105     108
SQ    SEQUENCE      110 AA;   11981 MW;   C2C3B23B85E520E5 CRC64;
      MALWMRLLPL LALLALWGPD PAAAFVNQHL CGSHLVEALY LVCGERGFFY TPKTRREAED
      LQVGQVELGG GPGAGSLQPL ALEGSLQKRG IVEQCCTSIC SLYQLENYCN
```

Figure 1.4 An excerpt of an entry from the Swiss-Prot database of proteins.

in combination with the data shown here. The FT records are *feature* descriptions. Some of the features shown are the places in the amino acid sequence where different types of structures occur, such as α-helix structures, β strands, and turns. In this record, the locations of disulfide linkages are also reported. The names following the FT tags are the feature types. The numbers are the sequence start and end points for each feature. This particular entry describes a polypeptide that is a *precursor* for the insulin protein. The molecule folds up, forming the disulfide bonds indicated, and the section marked PROPEP in the FT records is spliced out, leaving the two PEPTIDE sections, linked by the disulfide bridges. Finally the last part shown, the SQ section, contains the actual sequence of amino acids, one letter for each.

```
ID    1.1.1.39
DE    Malate dehydrogenase (decarboxylating).
AN    Malic enzyme.
AN    Pyruvic-malic carboxylase.
CA    (S)-malate + NAD(+) = pyruvate + CO(2) + NADH.
CC    -!- Does not decarboxylates added oxaloacetate.
PR    PROSITE; PDOC00294;
DR    P37224, MAOM_AMAHP;  P37221, MAOM_SOLTU;  P37225, MAON_SOLTU;
```

Figure 1.5   An excerpt of an entry from the Enzyme database.

Many proteins function as *enzymes*, chemical compounds that facilitate chemical reactions while not actually being consumed. They are not reactants or substrates (the inputs to the reaction), and they are not products (the results of the reaction). However, many biologically important reactions do not proceed without the presence of the corresponding enzymes. So, an important piece of information about a protein is its function. Is it an enzyme, and what type of enzyme is it? If it is an enzyme, what reaction(s) does it facilitate? Figure 1.5 shows an example, a small excerpt from the Enzyme database, another resource provided by the Swiss Institute for Bioinformatics at the ExPASy web site.

The line beginning with ID is the Enzyme Commission number, unique to each entry. The DE line is the official name, and the AN lines are alternate names, or synonyms. This enzyme catalyzes the reaction that removes a carboxyl group from the malate molecule, leaving a pyruvate molecule, and in the process also converting an NAD+ molecule to NADH. The NAD+ and NADH molecules are *coenzymes*, molecules that participate in the reaction, change state somewhat, but are later reused in other reactions, and generally not consumed overall in the process.[2] The reaction catalyzed by malate dehydrogenase is described in a kind of stylized symbolic form on the line beginning with CA. The CC line is a comment, not meant for use by a computer program. The PR line is a cross reference to the Prosite database, which has other information about proteins, and the DR line is a set of cross references to entries in the Swiss-Prot database where the sequences corresponding to various versions of this protein may be found.

So, it would be useful to be able to represent molecules and reactions, for example to determine if a cell is able to process some starting substrates and produce from them the necessary products. Similarly, having the enzymes associated with the reactions, one can look up what processes would be blocked if an enzyme was missing, for example because of a genetic defect or mutation. We will show how the data above can be put in a form so that we can do just such computations.

It is evident from visual inspection that all living organisms can be described in terms of structural components. These structures may differ in detail from one type of organism to the next, but also share properties across large families of organisms. Indeed the name "organisms" suggests things that are made up of things called "organs." Roughly speaking, the human body

---

[2]These molecules are all variants in a family. The abbreviation NAD stands for *Nicotinamide adenine dinucleotide*. This molecule and its derivatives serve as hydrogen transporters in many different metabolic reactions (oxidation and reduction). More about NAD and its biochemical functions can be found in standard biochemistry [300] and molecular biology [10] texts.

(as well as many other types of living bodies) is made up of organs, organized into distinct spaces, and subsystems, with definite spatial relationships and connectivity. Organs in turn are made up of well-defined parts, with their own spatial relationships to each other as well as the organ as a whole. This hierarchy of structure continues down in size to the level of individual cells that make up tissues, organ parts, and organs. Cells also have identifiable parts with relationships to each other and the cell as a whole. This is the domain of the science of anatomy. Anatomy textbooks describe all of this in exquisite detail.

Anatomy is one of the foundational subjects that all medical, dental, and nursing students (and many others) must master early in their training. It is essential to clinical practice, and equally, to biomedical research. Structure and function (physiology) are the dual framework for understanding health and disease, diagnosis and treatment. They are also the essential framework for describing biological knowledge from molecular biology to ecology and evolution.

However, the descriptions are informal. The student learns to reason about the structures, but the possibility of automating such deductions or inferences requires formalizing the knowledge about the structures. Formal representations of the body, organs, organ parts, tissues, etc., including the properties and relationships, are known as *anatomy ontologies*. In Chapter 2, Section 2.3.4 we will examine one such anatomy ontology in detail, the University of Washington Foundational Model of Anatomy (FMA). In addition to describing its principles and content, we will look at how to write programs that retrieve entities and compute results from the information stored about each entity. In order to serve as a basis for sound reasoning, the FMA must satisfy some consistency criteria. We will see how these constraints can be checked, which should also shed some light on the general problem of automated auditing of large ontologies (see Figure 1.6).

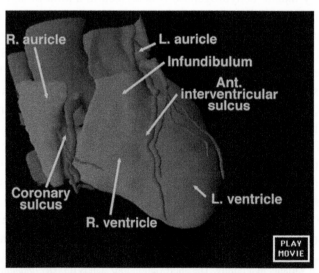

Figure 1.6 An image from the Digital Anatomist Thorax Atlas, showing an anterior view of the heart, with various parts labeled.
(Courtesy of the Digital Anatomist Project at the University of Washington.)

```
NAME: Heart
PARTS:
  Right atrium
  Right ventricle
  Left ventricle
  Left atrium
  Wall of heart
  Interatrial septum
  Interventricular septum
  Atrioventricular septum
  Fibrous skeleton of heart
  Tricuspid valve
  Mitral valve
  Pulmonary valve
  Aortic valve
  ...
PART OF:
  Cardiovascular system
  ...
ARTERIAL SUPPLY:
  Right coronary artery
  Left coronary artery
VENOUS DRAINAGE:
  Systemic venous tree organ
NERVE SUPPLY:
  Deep cardiac nerve plexus
  Right coronary nerve plexus
  Left coronary nerve plexus
  Atrial nerve plexus
  Superficial cardiac nerve plexus
LYMPHATIC DRAINAGE:
  Right cardiac tributary of brachiocephalic lymphatic chain
  Brachiocephalic lymphatic chain
  Left cardiac tributary of tracheobronchial lymphatic chain
ATTACHES TO:
  Pericardial sac
...
```

Figure 1.7   An excerpt from the University of Washington Foundational Model of Anatomy (FMA).

Organs and tissues also have a lot of data and knowledge associated with them. Organs, for example, have names, lists of their constituent parts, location within the body of the organism, and possibly other information as well. This information is also symbolic and consists of items that need to be grouped together in lists or more complex structures. Figure 1.7 shows some information about the human heart, taken from the FMA.

Some of this information can be used for spatial reasoning about anatomy, as applicable to surgical procedures, consequences of bodily injury, and so on. Information about the connectivity of the lymphatic system and lymphatic drainage of various organs and anatomic locations makes it possible to construct computational models for spread of tumor cells. This in turn helps to determine the target for radiation therapy for cancer, resulting in improved accuracy of treatment. This is described in detail in Section 5.5.3.

Electronic medical records are rapidly becoming the standard of practice for managing clinical data, in medical offices and hospitals, small and large. Some of the information stored in these systems is numeric, such as the results of many laboratory tests. Examples include: counts of the number of various types of blood cells (red cells, white cells, and their subtypes), concentration of drugs, sugars, important proteins, etc. Some non-numerical information is also important in these laboratory tests, such as the units used for the tests, as well as tests for the presence of bacteria and the identities of the bacteria (blood cultures and bacterial cultures from other parts of the body). Electronic medical record systems also include enormous amounts of textual information, such as physician's dictation of findings about the patient (history and physical examination).

Gene and protein sequence data use a small "alphabet" of symbols, four for elements of gene sequences and 20 for the amino acids that make up proteins. This encoding problem is simple by comparison with the problem of representing clinical laboratory test data. Laboratory data are the results of tests that the clinical laboratory performs when a sample of a patient's blood is drawn from a vein, and put in a tube or tubes to be sent to the laboratory. Some examples of blood tests are: red and white blood cell counts, hemoglobin, creatinine level, glucose level, etc. These tests are usually grouped into *panels* that can be ordered as a unit. Table 1.1 shows some values from a complete blood count for a patient, along with the standard (normal) range for each test. Note that the units in which the values are reported are different for the different tests. White blood cells (WBC) and red blood cells (RBC) are both counted, but there are so many more RBC that the number is reported in millions of cells per microliter rather than thousands as for WBC.

Table 1.2 is another example of a panel, and lists the names, values, and standard ranges for some of the tests in a comprehensive metabolic panel ordered on a patient. You can see by comparing the values with the ranges that all the values are within the standard ranges, except for the creatinine level. This is not cause for alarm in this case, since the patient has only one kidney and a slightly elevated creatinine level is not unusual for this patient. The creatinine level is an indicator of how well the kidneys are filtering the blood and getting protein breakdown products out of the body.

Many other kinds of samples of body tissues also can be used to do tests, including urine and bits of material obtained by swabbing your throat, ear, nose, eye, etc. There are thousands of types of laboratory tests, each requiring its own identifier or code, as well as its own conventions

Table 1.1   Complete Blood Count Panel

Component	Result	Standard Range
WBC	6.14	4.3–10.0 Thou/μL
RBC	4.80	4.40–5.60 Mil/μL
Hemoglobin	13.8	13.0–18.0 g/dL
Hematocrit	40	38–50 %
Platelet count	229	150–400 Thou/μL
...	...	...

Table 1.2 Comprehensive Metabolic Panel

Component	Result	Standard Range
Sodium	139	136–145 mEq/L
Potassium	4.1	3.7–5.2 mEq/L
Chloride	107	98–108 mEq/L
Glucose	94	62–125 mg/dL
Urea Nitrogen	21	8–21 mg/dL
Creatinine	1.5	0.3–1.2 mg/dL
Protein (Total)	6.8	6.0–8.2 g/dL
Bilirubin (Total)	0.7	0.2–1.3 mg/dL
...	...	...

for representing the test result itself. Tests can include determining what chemical substances are present in the sample, and inoculating agar gel containers (called "cultures") to determine if bacteria are present in the sample.

Electronic medical record (EMR) systems are complex database systems that attempt to integrate and provide ease of access to all the data needed for care providers and patients in the medical setting. They include facilities for acquiring data from instruments, for data entry by practitioners in the clinic and on the wards, and generation of reports for use by managers and others. We've already touched on the complexity of representing such a huge and diverse data set in discussing laboratory data. Information such as pathology and radiology reports often take the form of dictated text. Although highly stylized, such reports are difficult to represent in as structured a form as laboratory tests.

A highly specialized area of medicine where computers were utilized very early on for clinical data is radiology. The era of digital imaging began with the introduction of computed tomography head scanners in the 1970s, followed by digital radiography and many other digital imaging modalities. The storage, retrieval, and analysis of digital images is sufficiently complex to warrant specialized electronic medical record systems originally called "Picture Archiving and Communication Systems" (PACS). Entire journals are devoted to digital imaging researchfootnoteJournals that focus on medical images in particular include "BMC Medical Imaging," published by BioMed Central, "Computerized Medical Imaging and Graphics," published by Elsevier, and "Journal of Digital Imaging," published by Springer. A picture, or image, is represented as an array of picture elements, called *pixels*. Each pixel is represented by a number (for gray scale images) or a set of numbers (for color images) that represent intensity (of white light, or separately red, green, and blue), or in some systems, hue, saturation, and value (roughly the color, the relative mix of the color and white, and the brightness). Other kinds of representations are used as well. A full treatment of digital representation of images can be found in standard textbooks on computer graphics such as [111]. A brief introduction to medical imaging and digital images is provided here in Section 1.1.5.

The core challenges in designing EMR systems are to come up with logical and consistent ways to organize the data on all the patients, procedures, facilities, providers, etc., ways to easily retrieve the particular subsets of data needed at different places and times in the health care setting, and ways to organize and present the information so providers and patients can easily use it. However, there is now growing interest in being able to compute some useful results from the information, not just retrieve it for human examination. For example, it may be possible to predict the onset of adverse events by applying logical reasoning procedures to data on a patient, or alternately, to identify the onset of such events early on. If such things can be predicted or at least flagged early, then possibly the adverse events can be prevented or successfully treated by timely intervention before they happen. Some of the data and procedures will be numeric in nature, but others will be reasoning steps applied to symbolic information.

One of the most difficult problems is to match up the various activities around health care practice with the proper codes for billing for services provided. The complex rules around billing procedures depend on accurately categorizing the diagnoses as well as the procedures. As a result, considerable research has been conducted to devise methods of automatically extracting structured data from free text reports, using natural language processing techniques. Many other places exist in the EMR for such "free text" besides pathology and radiology.

The encoding of patient data seems to be a good match to an "object-oriented" style of data modeling. A simple example of this is developed in Section 1.2.2. However, the more common approach is to create a tabular design for storing and accessing medical data, based on the *relational database* model. An introduction to databases in general and the relational model in particular is given in Section 1.3. In some cases, the flexibility needed to store data with highly variable complex structure is gained from the Entity-Attribute-Value (EAV) model, described briefly in Section 1.3.4.

Electronic medical record system design is evolving rapidly, and huge investments are being made in deploying commercial systems, as well as in studying the effectiveness of different design approaches. Expenditures for EMR installations at medium-sized institutions can run in the range of $100–$200 million dollars (2012).

Along with the rapid deployment of institutional systems, there is a growing interest in "personal health records." Although it is an unsettled legal question as to who actually owns the medical information about a patient, there are many initiatives to provide patients with ways to store, access, and carry with them their own complete health record. More background may be found in a review article [407] and a web page `http://www.nlm.nih.gov/medlineplus/personalhealthrecords.html` at *Medline Plus*, a service of the National Library of Medicine.

At the National Library of Medicine's annual research training program directors meeting in July, 2005, Donald Lindberg (Director of the National Library of Medicine) commented on the future directions for institutional and personal health records, and what they should contain. He observed that the electronic health record should be a lifelong record, including some things that are not now part of existing systems. These are:

- genome analysis,
- family relationships (pedigree),
- ethical conventions, and
- the patient's own observations.

Of these, the first, incorporation of genomic data, is getting the most attention. Genomic variant information can play a significant role in medication selection and dosing [312].

It is also important to have EMR systems that can easily support anonymization of data for research use. The opportunities that widespread adoption of EMR systems present for discovery are enormous, but there are many complications. Data collected and recorded in a non-research clinical setting may not have the same quality and predictive power as data collected in a more focused setting such as a clinical trial. A study of warfarin dosing algorithms [361] reports in detail on one such exploration.

Public health activities provide communities with essential health assurance measures which allow the members of the population to live healthy lives. Without recognizing it, most people living in the U.S. benefit from the services of public health departments. For example, the management of communicable diseases is managed by local public health departments. The focus is on population, prevention, and control rather than treatment. The responsible organizations are mostly government rather than private.

Public health practitioners constantly work to assure that the communities they serve are able to live healthy, safe lives. While there is no possibility of removing all health risks from the lives of individuals, public health organizations are responsible for leveraging resources to target the most pressing health concerns within their communities. The majority of decisions that public health practitioners must make require the collection and analysis of health-related data. Such data is acquired from multiple sources and can take many forms. To support public health practitioners in the management and analysis of data, informational tools are becoming ever more necessary. Many of these informational tools are delivered in the form of electronic information systems. As increasing computational power has become decreasingly expensive, public health groups have turned to computer-based solutions to support their data management and data analysis efforts. The development of such informational tools and the subsequent integration of them into the public health environment require an understanding of the needs of public health organizations, the methods they use to complete their work, the political context in which public health organizations are situated, and the underlying technologies which make these informational tools possible.[3]

Patrick O'Carroll [309, page 5] has defined *public health informatics* as "the systematic application of information and computer science and technology to public health practice, research, and learning." In the domain of public health, as with biology and medicine, we often have to deal with categorical or non-numeric information. We collect information about disease prevalence, and assign numbers to cities, states, and other geographic entities. These entities have names, and the data come from governmental or health care provider agencies, which also have names, but are not themselves numerical values. For example, one might ask for a list of all the cities in Washington state that have seen an abnormal incidence of kidney cancer, in order

---

[3]This brief introduction to public health practice was kindly provided by Jamie Pina.

to test a hypothesis about carcinogens in the environment. The incidence rates are numbers, but the list of cities is a list of names of objects or entities.

The basic science underlying much of public health practice is epidemiology, the study of the spread of diseases and environmental toxic substances throughout populations and geographic areas. In addition, public health studies the organizational aspects of health care practice and how it affects the health of populations. Many kinds of data are used in public health research and practice. Examples include:

- time series of various sorts, used for things like syndromic surveillance, outbreak detection, long-term trends in morbidity and mortality, etc.,
- vital statistics, such as birth and death records and other related data,
- immunization records,
- reportable disease records such as tumor registries, STD registries, etc., and
- risk factor data.

An additional example of data that are significant for public health is the FDA's Adverse Event Reporting System (AERS). This system accumulates data on adverse events related to drug interactions. The subject of drug interactions from a strictly pharmacologic point of view is developed in Chapter 2, Section 1.2.5.

Like medical and biological data, these are diverse types, coming from many disparate sources. There are many unsolved problems around defining and using appropriate vocabulary, syntax, and semantics. The social and political issues around the organization of public health practice also strongly affect the success rate of informatics projects. Challenges also abound because of the difficulty of coordinating among jurisdictions and levels, since public health agencies exist at local, county, state, regional, national, and international levels. The background and context of these challenges is treated in depth in a book edited by Patrick O'Carroll [309]. Chapter 11 of that book briefly discusses data standards relevant to public health data. A more detailed treatment will be presented here in Section 1.1.4 and later in Section 4.3.2.

1.1.1 What Can Be Represented in a Computer?

In order for computer programs to deal with biomedical data, the data must be *encoded* according to some plan, so that the binary numbers in the computer's memory represent numeric data, or represent text, character by character, or possibly more abstract kinds of data. Many encoding schemes have been used for biomedical data. We describe a few here, to give a flavor for how this might work. We also show how small programs can express or embody the encodings, and in some cases convert from one encoding scheme to another. Examples include biomolecular sequence data (DNA and proteins), laboratory data from tests done on blood samples and other substances obtained from patients in a clinic or hospital, and medical image data.

In addition, everyone needs methods for searching and sorting through the vast collections of bibliographic reference data now available such as the MEDLINE database of journal articles and other informational items. The entries in such bibliographic databases are definitely not numeric, but are indexed by keywords, and the search methods used depend on being able to manipulate lists and complex structures. Chapter 4 illustrates how bibliographic information can be structured, retrieved, and organized.

The digital computer has only one way to represent information, and that is in terms of collections of *bits*. A bit is a single unit of information, represented by the state of an electrical circuit component in the computer. This circuit can be either on or off, but no other value is possible. We talk about this in digital computer terms by assigning the "off" value the number 0 and the "on" value the number 1. With combinations of multiple bits, we can represent as many different things as we have circuits for. The collection of circuits is called the *memory* of the computer. It is usually organized into *bytes*, each consisting of 8-bits. With 8-bits, we can represent 256 different things. Each bit combination would stand for something. We could decide for example that the combination 01101110 would stand for the category "squamous cell" and 01101111 would be "epithelial cell." Clearly, 256 categories is not nearly enough to deal with what we know about biology, medicine, and public health. However, modern computer memories have many billions of bits, so we really can use this idea to encode lots of things. However, writing instructions for creating lists of such codes and sorting and comparing them is not very efficient, so we instead have many layers of electronics and computer programs to create and manipulate these encodings in a more reasonable and more abstract way.

Since we are already familiar with natural language text, we use a method adopted centuries ago by mathematicians. We use letters and words as *symbols* that name the various biological entities. These symbols (whether their names are single letters or long words) can be manipulated by computer programs, in just the abstract ways we will need. So, we assign names (symbols) to each of the entities and their relationships, and we will see how to organize these so that we can compute interesting and useful things.

We'll start with proteins, and then extend the ideas to the other biological molecules, especially DNA. Later subsections will then briefly describe some other kinds of biomedical data.

A protein is a long polymer or sequence of simpler units, amino acids. Thus, representing proteins requires some way to represent sequences. This can be done by writing the names of the amino acids in order, starting with a left parenthesis, separating each name by some space (one space character is sufficient, but any number is permitted, including continuing on the next text line), and finally marking the end of the sequence with a right parenthesis. This structure is called a *list*. So a sequence of amino acids that make up a (very short) protein might look like this[4]:

```
(tryptophan lysine aspartate glutamine aspartate serine)
```

The parentheses conveniently mark the beginning and end of the protein sequence and spaces suffice to cleanly separate the names. Provided that we don't use spaces or other non-printing characters in the names, no other complex punctuation or syntax is needed. A biologist familiar with amino acids (only 20 different kinds appear in proteins) might prefer shorter names to be

---

[4]In this book, to identify text that constitutes a computer program or part of a computer program, or data, we use a different font than that of the main text.

able to encode much bigger proteins in a more compact way. So, each amino acid also has a short name, only three letters long, and a single letter code, standardized by the International Union of Pure and Applied Chemistry (IUPAC).[5] Table 1.5 lists the full names, short names, and single letter abbreviations of the standard amino acids. Using the short names, the above sequence could then also be written as (trp lys asp glu asp ser), or using the single letter codes as symbols for each (W K D E D S).

A computer by itself cannot impute meaning to any of these symbols. We use these particular ones so that people can read the data based on agreed on meanings. The meaning that they have for a computer program is operational, as data to be manipulated by algorithms. Here is a simple example. A computer program can compute the length of a list, because each symbol is a single entity and the list can be traversed by built-in computer operations. This idea is independent of what the symbols may mean to us. In this particular context, however, we can interpret the length of the list as a measure of the size of the protein. The computer program does not "know" what a protein or amino acid is. It only deals with symbols and lists. But to the extent that the symbols and lists faithfully map to the corresponding properties of proteins, the computer program computes something that does have meaning for us.

Going further, the sequence of amino acids is not the only thing we know about a protein. Many proteins have known functions, such as serving as components of cellular structures, serving as enzymes for biochemical reactions, and as regulators of activity of genes in cells. These functions can also be given symbolic names, and we can then create a data structure, a collection of information about a protein, again just using the symbols and parentheses to group the symbols into lists. A list could certainly contain other lists in addition to symbols, so we can make up nested structures, which is a very useful way to keep things organized. Similarly, we could combine such symbolic information with numeric information by using the normal representation of numbers. Here is an example of how to encode the Enzyme database excerpt shown in Figure 1.5 in terms of symbols, strings, and lists:

```
(enzyme
    (id "1.1.1.39")
    (name "Malate dehydrogenase (decarboxylating)")
    (sym-name malic-enzyme)
    (reaction (reactants s-malate nad+)
              (products pyruvate co2 nadh))
    (prosite-ref PDOC00294)
    (swissprot (P37224 MAOM_AMAHP)
               (P37221 MAOM_SOLTU)
               (P37225 MAON_SOLTU)))
```

---

[5]IUPAC is an international scientific organization for advancement of the chemical sciences and their application to benefit humankind. For more information on IUPAC, see their web site, http://www.iupac.org/.

Here we have used standard Latin text characters and chosen symbol names that would be somewhat familiar to practitioners, and used digits and decimal points to represent numbers. The fragments of text enclosed in double-quote (") characters are *strings*. A string is just text, to be read by people, printed, etc., while symbols, the words that are not enclosed in quotes, are like mathematical variables, names that a computer program can use and manipulate directly.

Can a computer program do anything with this? Yes, certainly. We need to be able to represent the Latin letters and digits as binary codes. Such a representation is called a character encoding and the collection of characters that are assigned codes is called the character set.

In the most commonly used character set, known as ASCII, each code is represented in the computer memory as a single byte. The first 32 codes are non-printable, and are considered "control" characters. This dates from the use of these codes with teletype devices, where the teletype message needed special codes indicating start of message, end of message, and "carriage control" to indicate a carriage return, a line feed (advance the printer paper by one line), a form feed (advance the paper by a whole page length), and the like. These control characters are still in use and recognized by computer programs to have special interpretations, not exactly the same as with the teletype, and (unfortunately) not even the same for every program. For example, "control-C" is often interpreted to mean "interrupt the currently running program," but it also can be used to mean "end of transmission." The printing characters begin with the value (decimal) 32 for the space character. Both upper case and lower case Latin letters are included. Table 1.3 shows a sampling of some of the codes.

Table 1.3   Table of ASCII Codes (Only a Sample Shown)

Decimal	Hexadecimal	Character
0	0	NUL (null)
1	1	SOH (control-A)
2	2	STX (control-B)
3	3	ETX (control-C)
...	...	...
32	20	(space)
33	21	!
34	22	"
...	...	...
48	30	0
49	31	1
...	...	...
65	41	A
66	42	B
...	...	...
120	78	x
121	79	y
122	7A	z
etc.		

With this encoding, the characters of the text can be read by a program, usually from a disk file, but sometimes from a keyboard or a network connection to another computer. Now an important step needs to take place. The symbols and lists represented by the characters need to have some representation in the computer program memory. Thus, the character sequences (text) need to be translated into a representation in the computer where each symbol has its own encoding, not as a sequence of the characters that make up its name. Similarly it is necessary to have a representation of a list, so that the list contents can be manipulated directly. This process of translating text into encoding that a computer program can manipulate is sometimes called *parsing*. For now, we will assume that we have an interactive computing environment, a computer program that can read such text, do the translation so that we can apply operations to the data and get results.

Computer programs in the form we call "source code" are plain text, using character sets such as the printable characters from the ASCII character set. Character attributes such as font style, type size, or display attributes such as boldface are irrelevant. The meaning of the program is entirely in the plain text. Computer programmers typically use text creation and editing tools that directly display and manipulate plain text. These programs are called *text editors*. Two popular ones available for a wide range of operating systems (including Linux) are called vi and emacs. The emacs text editor also runs on just about any kind of computer or operating system available today, including Microsoft Windows PC and Apple Macintosh systems.

Program source files are read by computer programs called *interpreters* or *compilers*, which translate the text into computer machine instructions that can be executed by the machine itself. Word processing document files contain a lot of other proprietary data that carry text formatting information and other information that interferes with the processing of the program text by the interpreter or compiler. So, the text of a computer program is really a set of symbols whose meaning can be understood by a suitably configured computer system. Humans can of course read such text also, since we write it. Good text editors such as emacs provide some aids for specific programming languages, like Lisp, to make the text easy to understand. Examples are: automatically coloring (in the text editor display window) some of the words that the editor program recognizes as special to the programming language, and doing some display formatting specific to the programming language (automatically indenting some lines).

Now that we have a way to represent each symbol and the various lists in a way that a computer program could operate on these things, what operations would we want to do? For the protein amino acid sequences, one thing we might do is to have a computer program count the number of amino acids (for larger proteins you would not want to do this by hand). Another important and more complex procedure is to count how many amino acids of each kind are in a particular protein sequence. Yet another is to extract particular items from the complex structures above. For simplicity we will start with just the amino acid sequence.

In order to describe and actually perform such computations, we need a computing environment in which we can create files containing the text representation of the above. We also need an interpreter or compiler program that can convert the text representation to an internal

```
> (+ 276 635 1138)
2049
> (/ (+ 182 178 179 183 182 181 181) 7)
1266/7
>
```

Figure 1.8   A command window running an interactive Lisp session.

representation that the computer can use. We'll start by describing an interpretive environment that can process symbols and lists like the ones we describe here. In this environment, you can use many built-in operations on such data, and create your own more complex operations that you can then use as if they were built in.

The programming language best suited for the data and computations we will be doing is called *Lisp*, an abbreviation for "List Processing," since its focus is exactly on representing and processing things represented as symbols and lists. Lisp is the second oldest computer programming language in use today.[6] It was invented by John McCarthy in 1958 while he was on the faculty at MIT. McCarthy's landmark publication [272] is well worth reading, not only for historical reasons but for the insight it provides into the meaning of functions and computation.

A session with a computer display, keyboard, and a Lisp interpreter will consist of typing expressions at the keyboard, in response to a *prompt* that the interpreter displays on the screen, in a window, the interpreter window, or command window. The interpreter translates the typed text into an internal representation of the expression, evaluates the expression, and prints out the result. Figure 1.8 shows a command window with a prompt, a typed expression, and the resulting response, followed by a reprompt, another typed expression, the response, and a third prompt.

An expression can be a symbol, a list, or some other type of data represented in text form. In addition to symbols and lists, numbers are included in the language, and operations can be performed on numbers. We'll start with arithmetic, since that is simple and familiar. Suppose we want to add together a bunch of numbers. After the interpreter prompt, we type an expression (a list) beginning with the name of an operation. The rest of the list contains the operands, or inputs to the operation. The operators are usually *functions* in the mathematical sense, that they are defined for some domain of valid inputs, and produce results in some set called the *range*. Following is a more detailed explanation of the examples in Figure 1.8 (the first set of numbers are the number of calories in each of three meals eaten recently, perhaps by the

---

[6]The oldest is FORTRAN, invented a few years earlier by John Backus at IBM, and first implemented in 1957. The focus of FORTRAN, by contrast, was on expressing numerical formulas and algorithms in order to do high-speed arithmetic with large data sets.

author). The > is the interpreter's prompt; the user types the expression.[7] Then the interpreter looks up the name + to see if it is defined as an operator. It is, so the values of the rest of the list elements are provided to the + operator, it returns the result, 2049, and prints it on the display, then redisplays the prompt, and waits for the next expression. This cycle of interaction with the interpreter program is called the `read-eval-print` loop.

```
> (+ 276 635 1138)
2049
```

Lisp uses *prefix* notation with mathematical functions, just as with any other operators. For any expression, the operator is first and the operands follow. This contrasts with most other popular programming languages (FORTRAN, C, C++ Java, BASIC, perl, etc.), which use *infix* notation, more like mathematical expressions. In infix notation, operators appear between operands, and thus need to be repeated to have expressions in which there are multiple operands even though the operator is the same. For example, to add up a bunch of numbers, the mathematical expression might be $a + b + c + \ldots$ while in Lisp it is (+ a b c ...), with the operator first, then the inputs.[8]

Before the operation is performed, each of the inputs is also *evaluated*. A number evaluates to itself, that is, numbers are constants. We could, however, put yet another expression as one of the inputs. This evaluation of inputs makes *nested* expressions possible. Again, we use numbers to keep the examples simple. Suppose we want to compute the average weight of a person over some number of days. In this case, we want to divide the sum of the daily weights by the number of days. The expression we type has three items in the list. The first is the / symbol, the second is a list (an expression beginning with the + symbol and then some numbers), and the third is the number 7.

```
> (/ (+ 182 178 179 183 182 181 181) 7)
1266/7
```

The inner expression was evaluated to give 1266 and then the result of the division was displayed as a ratio. Numbers in Lisp can be integers, ratios, so-called *floating-point* numbers (corresponding to decimal numbers in some range), and even complex numbers. The arithmetic operators do some amount of type interconversion automatically, so that if we provide any of the weights the above expression in decimal form, the result will be a decimal, or floating point number, instead of a ratio.

```
> (/ (+ 182.0 178 179 183 182 181 181) 7)
180.85715
```

---

[7]Many Lisp interpreter systems are available, both free and commercial, and each has their own style of prompt. For simplicity we just use >, following the same convention as Graham [137] and Norvig [305]. See Appendix A for information on how to obtain a working Lisp system.

[8]This is sometimes also called Polish notation. The opposite, where the inputs come first and the operator comes last, is called *reverse Polish* notation, or RPN. This style or arrangement is used by Hewlett-Packard handheld calculators. Infix is used by Texas Instruments calculators. Each has their proponents.

Note that the type conversion happened for all the numbers even though only one was in decimal form. Now you might wonder, what about evaluating the numbers themselves? The interpreter does this. A number evaluates to itself, as noted earlier. The value is exactly the number you typed, represented in some appropriate way inside the computer system. So, the rule of evaluating inputs is uniform, with only a few exceptions. Exceptions to the input evaluation rule are called *special forms* or *macros*, and we will take this complexity up a little later in this section. Operators that are special forms or macros often also have the property that they cause *side-effects*, meaning that they don't simply return a result like a function does, but also cause some change in the computational environment.

Symbols can also be used as inputs to expressions, in place of the numbers above, if the symbols have values assigned to them. In addition to being names of things, symbols can be used as variables, in the same way as variables in mathematics. Before we show how to do this, we'll first consider just using the symbols by themselves, as names of things, since some of our biomedical data are not numbers, but symbols and lists. In this case, it won't make sense to do arithmetic on a list of names. What operations do we perform on them? For a sequence of amino acids making up a protein, we could ask, "what is the first amino acid in the sequence?" and indeed there is a built-in function, called `first`, that takes a single input, a list, and returns the first thing in the list. However, writing

```
> (first (trp lys asp glu asp ser))
```

will cause an error. Following the evaluation rule, the interpreter will try to evaluate the expression (`trp lys asp glu asp ser`). It will look up the name `trp` to see if there is an operation by that name. Of course there is not. That is the name of one of the amino acids, but it is not the name of an operation. What we want the interpreter to do here is to take the list literally, not treat it as an expression. To do this, we need a *special form*, named `quote`. Any expression beginning with the symbol `quote` will not be evaluated but used literally as data. So we need to write

```
> (first (quote (trp lys asp glu asp ser)))
TRP
```

This works as we intended.[9] The result is the symbol `trp`. This same special form, `quote`, is needed when we operate with symbols as well as with lists. Suppose we wish to construct a list containing the symbols for the amino acids. The `list` operator will take any number of inputs and make them elements of a list. But

```
> (list trp lys asp glu asp ser)
```

won't work. Each of the symbols will be evaluated by the interpreter. As noted earlier, a symbol can be used like a variable in mathematics and be assigned a value. To evaluate the symbol, the

---

[9]You might wonder about whether there is a distinction between upper case and lower case letters. There is, but in most Lisp environments, as the Lisp interpreter reads expressions, it converts any lower case letters appearing in symbol names to upper case. So, while the convention is that we write our programs using mostly lower case, it all gets converted as it is read in.

interpreter just looks up its value, which is recorded along with the symbol's name and other information about it. If no value has been previously assigned, an error occurs. Here we don't want values of symbols, we just want the symbols themselves, so we use `quote`.

```
> (list (quote trp) (quote lys) (quote asp) (quote glu)
        (quote asp) (quote ser))
(TRP LYS ASP GLU ASP SER)
```

The `quote` operator is the first example of several such operators that do not follow the evaluation rule. There are only a few other important ones. As they are needed they will be introduced. Some of these, like `quote`, are *built in* primitives, and are called special forms. Others are implemented as *macros*. A macro first performs some operations on its inputs *without* evaluating them, and constructs an expression to be evaluated. The resulting expression is then evaluated. Ordinary expressions are just small programs that are run by the interpreter. A macro, on the other hand, is in effect an operator that first *writes* a small program and *then* runs it. We will also introduce macros as they are needed and even show how writing your own macros will enable you to describe and process biomedical data and knowledge at a very high level.

This use of `quote` can get verbose as we see here. The `quote` special form is so often and widely needed, it is part of the Lisp standard to allow an abbreviated version of such expressions. Instead of an expression beginning with the `quote` operator, it is allowed to simply put a single quote ' character in front of a symbol or list. So the above expressions would be written as follows:

```
> (first '(trp lys asp glu asp ser))
TRP
> (list 'trp 'lys 'asp 'glu 'asp 'ser)
(TRP LYS ASP GLU ASP SER)
```

It is almost exactly parallel to the natural language idea of quoting. When we refer to a person by name, we use the name, and it "evaluates" to the person himself. For example, we would write that the *author* of this book is Ira Kalet, not "Ira Kalet." However, when we want to refer to the name itself, we quote it, so we would write that the author's *name* is "Ira Kalet," not Ira Kalet.[10] The quoted form is the name, while the unquoted form refers to the actual person.

In English typography, most often the double quote convention is followed, as here, but in Lisp, the abbreviation for the `quote` operator is the *single* quote character. In Lisp, the " or double quote character is used for a different purpose, to delimit strings, or literal text data. So, `"enzyme inhibitor"` is a sequence of characters making up text data, not a symbol or list of symbols.

We have been writing symbols by name using ordinary text characters. Aren't they just *strings*? A string is a sequence of *characters*, and the characters are in turn available from some standard character set, usually the common Latin characters, but some Lisp systems also support

---

[10] A much more entertaining version of this contrast can be found in the dialog between Alice and the White Knight, in Lewis Carroll's famous work, *Through the Looking Glass*. A formal treatment can be found in [292].

more extended character sets, even Unicode. A single character is a different kind of datum than a one-character string, just as a symbol is different from a list containing one symbol.

The difference between symbols and strings is that a symbol is a single indivisible unit, that may have properties associated with it, such as a *value*, while a string cannot be assigned a value. Strings are constant values that evaluate to themselves, just as numbers do. It is also the difference between being easily able to do computation directly with meaningful concepts on the one hand, and having to do complicated analysis of letter sequences to do even simple things with the meanings of the text. As an example, if the amino acid sequence were treated as text instead of a list of symbols, the operation of finding the first amino acid would be a matter of analyzing character after character until you found a blank space character, then assembling the characters up to that point into a separate string. The symbol and list representation eliminates that by providing a higher level language.

It is important to note that the operators *, +, and so on, are also symbols, each of whose name is just one character long. These particular symbols have predefined function definitions, just like the `first` function and others we have mentioned. There is no special syntax for arithmetic operators. In general, there is nothing special about the choice of the names for operators. The multiplication operator could just as well be named by the symbol `mult` and we would write expressions like (`mult 3 4`). When we create a programming language, we choose which symbols will refer to which operations. Their meaning in the context of running the computer program is operational. They do what they are defined to do, nothing more or less.

In addition to the `first` function, there are also `second`, `third`, and so on, up to `tenth`. However, if you need to look up a list element at an arbitrary position in the list, a more general function, `elt`, is available.[11] This function takes *two* inputs, because you need to specify which position you are looking for, as well as the list in which to look. Note that in Lisp as in many computer languages, indexing of sequential data like lists starts at 0, so, (`first seq`) is equivalent to (`elt seq 0`), and the following expression could also be written (`fourth` '(`trp lys ...`)).

```
> (elt '(trp lys asp glu asp ser) 3)
GLU
```

A very useful function complements the `first` function, by returning the remainder, or *rest*, of the list, which is the list containing everything *but* the first element. This function is called `rest`, as you might expect.

```
> (rest '(trp lys asp glu asp ser))
(LYS ASP GLU ASP SER)
```

---

[11]For lists, the `nth` function also does this, but `elt` is more general in that it will work with any sequence, including strings and vectors.

Lists are actually linked chains of two part structures called `cons` cells. The two parts are called the *first* part and the *rest* part.[12] The first part in the cons cell is a pointer to the first element of the list. The rest part in the cons cell is a pointer to the rest of the list, the next cons cell, or the empty list (at the end). So, `(rest '(glu trp lys asp glu asp ser))` is the sequence consisting of everything after the first element. It begins with the symbol `trp` in this example.

Rounding out basic manipulation of lists, the `cons` function *constructs* lists.[13] It takes two inputs. The first input becomes the first element of the list, and the second input becomes the rest of the list.[14] You can think of `cons` as creating a new list by putting its first input on the front of the list that is the second input.

```
> (cons 'glu '(trp lys asp glu asp ser))
(GLU TRP LYS ASP GLU ASP SER)
```

You might think that using `cons` to put back together the parts returned by `first` and `rest` would give back the original list. It's not quite the same. It gives a new list, that looks like the original, and shares a part (the *rest* part), but is not identical to the original. Each use of the `cons` function creates a new cons cell.

If we wanted to do something with every element of the list, we could use the `first` function to get the first element, process it, then use the `rest` function to get the remainder (or rest) of the list, and just repeat (recursively) the same process on what is left. The only thing to be careful of here is to know when to stop. This requires a check to see if the list is empty. The last cons cell in a list must have the empty list as its rest part. The empty list is represented as `()` and is also represented by the symbol `nil`. The function `null` returns the value "true" (represented by the symbol `t`) if its input is the empty list. It returns "false," if the input is anything else. The value "false" is also represented by the symbol `nil`.[15]

In addition to using symbols to name and refer to biomedical entities, we can also use them as *variables*, in the mathematical sense. We can assign values to them and then put the variables into expressions where the value may vary, but the computation will always be the same. Assignment of a value to a symbol is done with another special form, using the `setf` operator. The `setf` operator takes two inputs. The first input is the name of a place to put a value. Typically this would be a symbol, which will *not* be evaluated. Other kinds of place names are possible too and will be introduced later. The second input is the value to put there (which *will* be evaluated). An expression using `setf` is still an expression and just returns its second input as theresult. This

---

[12]Originally these were called `car` and `cdr`, respectively. The names come from the way in which the first Lisp implementation worked on an old IBM mainframe computer. In Common Lisp, these terms are no longer preferred, but they are still in use, notably in the other popular Lisp dialect, Scheme [1,100].

[13]Strictly speaking, `cons` constructs *conses*. The *rest* part of a cons could be a symbol or number or some other structure other than `nil` or another cons. For a more thorough discussion, see [137, Chapter 3].

[14]If the second input to `cons` is a list, the result is a list.

[15]What I describe here is actually a subset of the possible cons cells, those that form *proper lists*. Refer to [394, page 29] or [137, Chapter 3] for more details.

causes the *side effect* of assigning a value to the symbol or place named in the first input. For a symbol, we can find out its value by just typing it as input (*without* quoting) to the interpreter.

```
> (setf polypeptide-1 '(trp lys asp glu asp ser))
(TRP LYS ASP GLU ASP SER)
> polypeptide-1
(TRP LYS ASP GLU ASP SER)
```

Once a symbol is assigned a value it can be used anywhere that the value itself would be used. The symbol is *not* preceded by a quote because we want the value, not the symbol itself. Symbols can also be reassigned values as a program proceeds in its execution.

Another useful function for creating lists from other lists is the append function, which creates a new list from input lists. The new list contains all the elements of the original lists, in the same order as the originals.

```
> (first polypeptide-1)
TRP
> (setf polypeptide-1 (cons 'lys polypeptide-1))
(LYS TRP LYS ASP GLU ASP SER)
> polypeptide-1
(LYS TRP LYS ASP GLU ASP SER)
> (setf polypeptide-2
     (append polypeptide-1 '(cys tyr tyr ala cys ser ala)))
(LYS TRP LYS ASP GLU ASP SER CYS TYR TYR ALA CYS SER ALA)
```

It is important to keep in mind that functions like list, cons, and append do *not* change their inputs. The lists they produce are new. To preserve a reference to the results for later use, it is necessary to use setf to make the result be the value of some variable or be stored in some other place where data can be referenced.

Although assignment is familiar to programmers exposed to most languages, it is really unnecessary for many computations. Very large programs that do important calculations have no assignment expressions in them. They may have hundreds or thousands of functions defined, and when they are used, the user invokes one top level function, which uses the results of the other functions by nested calls and finally returns a result.

What other kinds of operations would be useful to be able to perform on data organized in lists? A useful operation is to search a list to determine if something is present. The built-in function, find, does this. For example, one might ask if a particular protein contains any instances of the amino acid, Cysteine, because it has a -SH group, which can form a link with other Cysteines in the sequence, and thus affect how the protein folds up into a three-dimensional structure. These links are called *disulfide bonds*.

```
> (find 'cys polypeptide-1)
NIL
```

```
> (find 'cys polypeptide-2)
CYS
```

The `find` function returns the first successful match, if found, or `nil` if no match is found. In addition to finding things, it is useful to know where in the list they are, and it would be especially useful to be able to locate *all* the occurrences of an item. The `position` function returns the index or position of the first match with an item if there is a match, or `nil` if unsuccessful.

```
> (position 'cys polypeptide-2)
7
```

How will we locate all the instances of Cysteine? We will need to be able to locate the first one, then search the remainder of the sequence. The `position` function would be much more useful if we could provide a starting index for it, so that it would not necessarily start at the beginning. This is so useful, it is indeed part of the standard definition of the `position` function. This function has *optional* inputs, specified by *keywords*. The idea of keyword arguments or inputs is that you provide a label for each input, rather than depending on the position of the input in the expression. These keyword arguments must come after all the required and optional arguments, but they can be in any order after that, since each one is a pair consisting of the keyword and the value. Keywords are symbols beginning with the : character. They have the special property that, like numbers, they evaluate to themselves, and their values cannot be changed. So, for example, the value of the keyword symbol `:test` is the keyword symbol `:test`, just as the value of the number 7 is just 7.

Many functions, such as `find` and `position`, that take sequences (such as lists) as inputs have optional keyword arguments. Keyword arguments appear in the parameter list after required and optional arguments, and consist of keyword-value pairs. For example, if we want to specify a different comparision for the `find` function instead of `eql` which simply tests if two things are the exact same entity, we would provide an input labeled with the : `test` keyword. Here is an application of `find` to look up a topic in a list of name-topic pairs.

```
> (find "gene clinics"
        '((ira "radiotherapy planning")
          (peter "gene clinics")
          (brent "ngi mania"))
        :test #'equal
        :key #'second)
(PETER "gene clinics")
```

It uses two keyword arguments to `find`, an alternate test, `equal`, and an alternate lookup, or key, function to apply to each item before comparing. In this case instead of comparing the string "gene clinics" with each list, which would fail, it compares the string with the *second element* of each list.

These arguments or inputs are *functions*. There needs to be some way to refer to the function associated with a symbol (as distinct from its *value*). This is done by the interpreter when it finds a symbol at the beginning of a list expression. However, here we are passing parameters.

Just as `quote` gives the symbol or list itself, the `function` operator gives the function object associated with the symbol or function expression. Just as `quote` is a special form that does not evaluate its argument, `function` is a special form as well, and does not evaluate its argument. In the expression (`function plugh`), the symbol `plugh` will not be evaluated, so it should not be quoted. Instead its function definition will be looked up and returned as a function object. As with `quote`, this idiom is used so often that it also has a special abbreviation, just as `quote` does: (`function plugh`) can be abbreviated as `#'plugh`.

The use of functions as data is so important and useful that in Lisp functions are implemented as first-class data, just like numbers and other data types. So, you can construct a list of functions, and even make a function object be the value of a symbol, or store the function in a cons cell.

The result of application of `find` is the entire item, not just the part used for the comparison. This extension makes `find` very powerful. Table 1.4 lists the standard keyword arguments that can be used with `find`, `position`, and several other functions.

Using the result from the first call to the `position` function, above, to search for the next Cysteine, we would specify starting at position 8, using the `:start` keyword argument.

```
> (position 'cys polypeptide-2 :start 8)
11
```

There is no built-in function to accumulate all the positions where a given item appears in a sequence, but we could easily define one. The power to define new operators beyond those provided as the basic elements of a programming language is essential for doing anything really complex. This is a fundamental mathematical idea, not just a programming convenience. We can choose any name we like for our new operator, as long as it is one that is not already in use for something else. Then the new operators can be used as basic elements in more complex expressions. This is the idea of creation of *abstraction layers*.

This brings us to the next special form we will need, a means for defining our own functions. To do this we need to be able to specify what the inputs should be to our function, by giving them temporary or local names, and we will need to specify what expression or expressions to evaluate to get the results for those inputs.

Table 1.4 Standard Keyword Arguments in Sequence Functions

Parameter	Purpose	Default
:test	function for comparision	eql
:key	"lookup" function	identity
:start	starting position	0
:end	ending position	nil

A simple example is a function to compute the length of the hypotenuse of a right triangle, given the two other sides.[16] We'll call the sides x and y, and the function name will be hypotenuse. The operator that creates new function definitions and attaches them to names (symbols) is the defun macro (the name comes from "define function").

```
> (defun hypotenuse (x y)
    (sqrt (+ (* x x) (* y y))))
HYPOTENUSE
```

The result returned by defun is the symbol naming the new function definition. Usually a Lisp program consists of a collection of such expressions using defun. Function definitions can use the names of other functions that are defined in this way, just as if they were built in. So, one can build up a very large and complex program, by defining functions that are useable as building blocks in more complex functions and so on, until you have a few very powerful functions that you use directly to achieve significant results. This general approach of writing many smaller functional units and combining them to make bigger functional units, and so on is called *layered modular design*.

A function defining expression starts with the symbol defun. The first argument (not quoted) is the symbol to be associated with the new function. In the example above, it is the symbol hypotenuse.

The next argument or input to defun is a list containing symbols naming formal parameters that are inputs to the function. These symbols can be used in the expressions in the body of the function. They will have values assigned to them when the function is called. In the example above, there are two inputs, denoted by the symbols x and y. The list (x y) in our hypotenuse function example is called a *lambda list*. It not only specifies local names to use to refer to the inputs, it can also specify that the inputs are keyword inputs and other kinds of inputs as well. The function definition expresses an abstract idea, a computational procedure that can be performed on some kind of inputs. It encapsulates the details of the computation, just as a power tool incorporates complex mechanisms to achieve effects. The user of the power tool has to be careful, but does not have to build the tool over and over each time it is needed for different materials, as long as the materials are of the type for which the tool was designed.

The remaining arguments are expressions to be evaluated as the *body* of the function. The value of the last expression is the value returned by the function. So, as you can infer, if there is more than one expression, all except the last are there to produce temporary or permanent *side effects*. In the example above, there is only one expression, (sqrt (+ (* x x) (* y y))), and its value is the value returned by the function when it is used.

If the first of these expressions in the body is a string (and there are one or more expressions following), that string becomes the *documentation* associated with the function. This is *not* the same as a "comment" as used in other programming languages (Common Lisp provides for comments,

---

[16]A right triangle has one angle that is a *right* angle, or 90°. The hypotenuse is the diagonal, the side opposite the right angle.

too). The documentation string, if provided, can be retrieved by code in a running program. Thus it is possible to provide interactive local documentation. The describe function, a built-in Common Lisp function, returns the documentation associated with its argument. Here is an example.

```
> (defun hypotenuse (x y)
      "returns the hypotenuse of the triangle
       whose sides are x and y"
      (sqrt (+ (* x x) (* y y))))
HYPOTENUSE
> (describe 'hypotenuse)
HYPOTENUSE is a SYMBOL.
   It is unbound.
   Its function binding is #<Interpreted Function HYPOTENUSE>
      which function takes arguments (X Y)
   Its property list has these indicator/value pairs:
DOCUMENTATION "returns the hypotenuse of the triangle
          whose sides are x and y"
```

Now we can implement a function to accumulate all the items. The idea we will use is *recursion*. In a recursive algorithm, we split the problem into a small piece and a remainder. We assume that the function we are defining can be applied to the remainder just as well as the original. The function will have to first determine if it is done, meaning that there is nothing left to compute, in which case it returns a result. If there is work to be done, it does the small piece, and then combines that result with the result of using the same (original) function on the remainder. We'll start with a very simple example, the factorial function, again using numbers.

The factorial function takes as an input a non-negative integer, and its value is the product of all the integers from 1 up to that integer, except for 0; the factorial of 0 is defined to be 1. In mathematical formulas, an exclamation point is sometimes used to denote the factorial of a number. For example, 5! is 120, and n! represents the product of all the positive integers up to and including n. This function is defined mathematically more precisely as follows: if n is 0, the result is simply 1 by definition, otherwise it is the product of n and the factorial of one less than n.

$$f(n) = \begin{cases} 1 & \text{if } n = 0, \\ nf(n-1) & \text{otherwise.} \end{cases}$$

This is doing the product in reverse, recursively, from n down finally to 0. Although this may be counter to a natural inclination to start from 0 and go up, such recursive definitions can solve very powerful computing problems, especially where the size or structure of the problem is unknown at the start and only becomes apparent when the computation proceeds.

To implement the factorial function as a procedure in Lisp, we need to be able to test whether two numbers are equal, and we need to be able to evaluate different expressions depending on the result of such a test. The = function returns t or nil depending on whether its two inputs are numerically equal or not, and the `if` special form provides conditional evaluation. An `if` expression has three inputs, a test expression, an expression to evaulate if the test returns t (or anything else non-nil), and an expression to evaluate if the test returns nil. The implementation of the `factorial` function in Lisp follows. It is a literal translation of the mathematical definition above.[17]

```
(defun factorial (n)
  (if (= n 0) 1
    (* n (factorial (- n 1))))))
```

Each time a function is *called*, or used in evaluating an expression, some information is needed about that function and its inputs. The link to the function definition and the links to the inputs make up a *stack frame*. These stack frames are accumulated in a list called a *stack*. As more and more recursive calls are made, the stack becomes larger. When a function evaluation completes, the particular stack frame associated with that use of the function is erased, and the result is returned to the previous function invocation, or *caller*. We can use the `trace` macro to see how this works. The `trace` macro takes a function name or list of function names and arranges for extra things to happen. When a traced function is called, the Lisp system will print on the screen the traced function name and its inputs. When the function finishes and returns a result, the result is shown on the screen as well. Using `trace`, we can see a visual record of the part of the computation that involves the traced function.

```
> (factorial 5)
120
> (trace factorial)
(FACTORIAL)
> (factorial 5)
 0[2]: (FACTORIAL 5)
   1[2]: (FACTORIAL 4)
     2[2]: (FACTORIAL 3)
       3[2]: (FACTORIAL 2)
         4[2]: (FACTORIAL 1)
           5[2]: (FACTORIAL 0)
           5[2]: returned 1
         4[2]: returned 1
       3[2]: returned 2
     2[2]: returned 6
   1[2]: returned 24
 0[2]: returned 120
120
```

---

[17]From here on, we will not show the prompt of the Lisp interpreter with `defun` expressions, but will assume that you can accumulate these in a file to become your collection of program code.

We'll apply this same idea, recursion, to solve the problem of accumulating all the places where an item can be found in a sequence. The function, which we will call `all-pos`, will have to find the first position, from some starting point, and then use the same method to find the remaining instances by specifying the next position as the new start and recursively calling itself. Since the function will need to be able to start at different places as the search progresses, we include a third input, the place in the sequence to start.

When we obtain the position of the next instance of the item, we'll need to reuse it, so we make it the value of another local variable within the body of our function. This requires a way to create local variables within a function definition, or generally within an expression. The `let` special form does this. In the following code, the symbol `pos` serves as a temporary storage or variable that has as its value the result of the `position` function expression. Finally, how will we accumulate the results? The recursive call should give us a list of the remaining positions, so the first one found should just be put on the front. We already have a function to do this, the `cons` function.

```
(defun all-pos (item seq start)
  (let ((pos (position item seq :start start))
    (if pos
        (cons pos
              (all-pos item seq (+ 1 pos)))
      nil)))
```

Note that every recursive function must have some way to determine if it is done, otherwise it may run forever (or until it runs out of resources in the computing environment). In this case, the `if` expression checks if the `position` function returned `nil`, meaning that no more references to the item were found in the sequence. The function will return an empty list, `nil`, as the result for this case, and if it is a recursive call to the function, the empty list will then be the input to the `cons` function. So, the result is built up using `cons` as the recursions complete, just as the `factorial` function used the `*` (multiply) function to build up its result.

We use `all-pos` by providing an amino acid symbol, `cys` for Cysteine, to look for, a protein to search (`polypeptide-2` on page 29), and we specify 0 as the start point. As expected, both Cysteine locations are returned in a list.

```
> (all-pos 'cys polypeptide-2 0)
(7 11)
```

A good example of the importance of disulfide bonds is the insulin molecule. Insulin is an important protein that regulates the metabolism of sugar. The disease known as diabetes is directly related to a deficiency in the body's ability to produce insulin as needed. The stability of the insulin molecule depends on disulfide bonds that form in producing insulin from a precursor protein called proinsulin. The proinsulin folds up, forming the disulfide linkages, then a portion of the proinsulin is cut out by cleavage enzymes. If the disulfide bonds break, the insulin molecule splits into two pieces and can't refold because the section in the proinsulin that enabled the folding is gone. Polypeptides similar to proinsulin that have the Cysteines located differently will not fold up to make insulin.

Before moving onto DNA and other data, we will look at two additional ways to implement a function like the all-pos function. These methods will be useful in other applications as well.

The fact that the all-pos function requires as an input the starting position from which to search can be a useful feature, but in many cases, such parameters are not useful, and almost always should start out with some fixed well-known value (0 in this case). A simple way to handle this is to just write a wrapper function, that provides the fixed input.

```
(defun all-positions (item seq)
  (all-pos item seq 0))
```

Sometimes the inner function is really only needed locally for the wrapper to use, and it would be better not to globally define a separate function. In large programs with many functions, it can happen that the same name is used for different definitions in different parts of the program. This will cause serious errors and will be hard to track down. We need a way to define local functions, which have names that are only valid within a limited scope. The labels special form provides this capability. The labels form creates local function definitions with no effect on the outside environment. Thus we can rewrite all-pos as a local function which we will call all-pos-aux, and then call it, all within the new wrapper definition of all-positions.

```
(defun all-positions (item seq)
  (labels
      ((all-pos-aux (item seq start)
         (let ((pos (position item seq :start start)))
           (if pos (cons pos
                         (all-pos-aux item seq (+ 1 pos))))
               nil))))
    (all-pos-aux item seq 0)))
```

Now we can make one more refinement that is often useful. Because we combine the result of the recursive call with the local value of pos, the recursive process will need to keep track of all the stack frames as described for the factorial function on page 33. We can make a slight modification that allows the Lisp compiler to implement our function as an iterative process, where there is no need for the stack frames. We do this by rearranging so that the result accumulates on the way in instead of on the way out. Then when the last call returns, the result is available without threading back through the previous calls. Since no computation is needed on the way out, all those stack frames are unnecessary. Such a function is called a *tail-recursive* function. To do this, the auxiliary function (the one defined locally as all-pos-aux) needs one more parameter, accum, a variable to hold the list of found positions, which starts out empty (nil). Thus we have

```
(defun all-positions (item seq)
  (labels
```

```
((all-pos-aux (item seq start accum)
    (let ((pos (position item seq :start start)))
        (if pos
            (all-pos-aux item seq (+ 1 pos)
                            (cons pos accum))
            (reverse accum)))))
(all-pos-aux item seq 0 nil)))
```

The only tricky part here is that accum accumulates the values in reverse order, putting each new one on the front of accum, so the result has to be the reverse of the contents of accum. The built-in function reverse does just what we need, taking a list as input and returning a new list, with the items in the reverse of the original order.

1.1.2 Cells, DNA, Proteins, and the Genetic Code

At the beginning of this section, we introduced proteins as sequences of amino acids and stressed the important roles of proteins as units of control, and significant parts of the structure of cells. One of the greatest discoveries in biology in the 20th century was that much of the key information that controls the operation of cells, the basic building blocks of living organisms, is in a huge molecule, generically called "DNA," short for "Deoxyribo-Nucleic Acid." One of the functions of DNA is to encode (and transmit to the next generation) information from which the cell can produce proteins. The information sections in the DNA that correspond to and encode for proteins are called "genes." Some of these correspond to the genes of classical genetics, though we now know that the operation of inheritance and expression of genetic characteristics is very much more complicated than the initial ideas discovered by Mendel much earlier. A review article [128] explains and discusses in more detail this question of what is a gene, and the challenges of how to describe the workings of DNA.

Like the proteins, DNA is itself a sequence of small molecules connected together in a kind of polymer. DNA is a long double-stranded helix, where each strand is a sequence of units called nucleotides. Only four different kinds of nucleotides are found in DNA. These nucleotides are composites of a sugar molecule (β-D-2-Deoxyribose, see Figure 1.9), a phosphate (PO_3) group, and one of four compounds called "bases," from the purine or pyrimidine family. The bases are

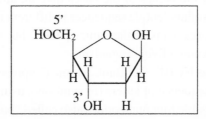

Figure 1.9 β-D-2-Deoxyribose, the sugar component of the nucleotide units of DNA, with labels to show the locations of the 3' and 5' carbon atoms, where the phosphate groups attach.

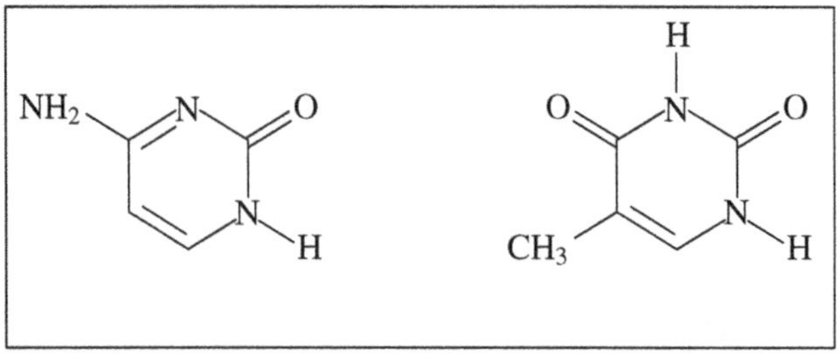

Figure 1.10 The pyrimidine bases, Cytosine and Thymine, two of the four bases that are constituents of the nucleotide units of DNA.

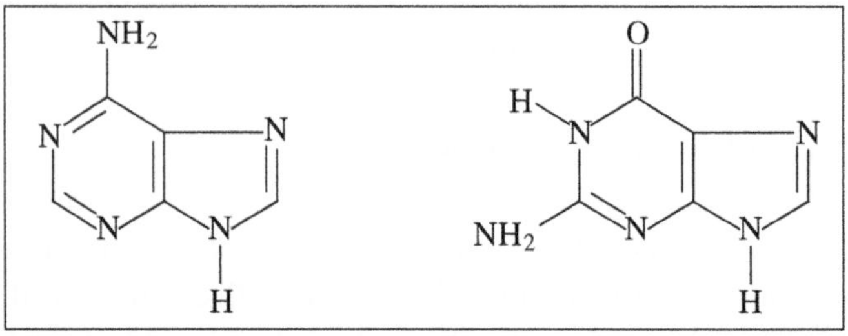

Figure 1.11 The purine bases, Adenine and Guanine, the other two bases that are constituents of the nucleotide units of DNA.

Guanine, Adenine, Cytosine, and Thymine. Their chemical formulas are shown in Figures 1.10 and 1.11. The DNA sequences are typically described by letter sequences naming the bases at each position in one of the two strands making up the DNA molecule. The nucleotides themselves are sometimes also called *bases* because they are uniquely identified by which base each includes.

The "backbone" of the DNA molecule is a sequence consisting of alternating phosphate groups and sugar (Deoxyribose) molecules. The phosphate group connects the sugar molecules together by bonding at the carbon atoms labeled 3′ and 5′ in Figure 1.9. The OH at the 3′ carbon connects to the phosphate group (PO_3), and the OH at the 5′ carbon connects to the other end of the phosphate group, splitting out a water (H_2O) molecule in the process. The bases connect to the sugar molecules at the 1′ position, at the right in Figure 1.9.

Nucleotides are labeled by the names of the bases that are attached to the sugar-phosphate "backbone" units. The pattern of nucleotide sequences is different in different kinds of organisms and even between different organisms in the same species, but these four units are the same ones used in the DNA of all known living organisms.

In most data sources, such as NCBI's GenBank, a typical representation of a DNA molecule or sequence would consist of a sequence of letters (or symbols), using the four letters G, C, A, and T, to represent each of the possible four nucleotide (also called "base") pairs that could appear in a double helix DNA strand. Although the DNA molecule is double stranded, the bases are paired uniquely, A with T and G with C, so that only the bases on one strand need to be represented.

```
CACTGGCATGATCAGGACTCACTGCAGCCTTGACTCCCAGGCTCAGTAGATCCTCCTACCTCAGCCTCTC
GAGTAACTGGGACCACAGGCGAGCATCACCATGCTCAGCTAGTTTTTGTATTTGTAGAGATGAGGTTTCA
CCATATTGCCCAGGCTGGTCTTGAACTCCTGGGCTCAAGCAAGCCACCCACCTTGGCCACCCAAAGTGCT
```

Figure 1.12 An excerpt of a DNA sequence from the BRCA1 gene.

Figure 1.12 shows a small fragment of the region around a well-known gene associated with breast cancer, called BRCA1.[18] The figure shows the nucleotides of only one of the two strands, because (as noted) the bases in corresponding positions in the two strands pair uniquely with one another, so specifying one strand determines the other.

1.1.2.1 The Fundamental Dogma of Molecular Biology

The relation between DNA and proteins is called the "Genetic Code." Each amino acid corresponds to one or more patterns of three nucleotides. For example, the nucleotide sequence GGA corresponds to the amino acid glycine, and TTC corresponds to the amino acid phenylalanine. Each combination of three nucleotides is called a codon. With four possible nucleotides in three places, there are 64 (4x4x4) codons altogether. Not all codons correspond to amino acids; there are three special codons that signal the end of a sequence, TAA, TAG, and TGA. For most of the amino acids, there are several codons that represent the same amino acid. For example, the amino acid Lysine is represented by two codons, AAA and AAG, and Leucine is represented by any one of six. Table 1.5 shows the standard correspondence between codons and amino acids.

Table 1.5 The Genetic Code by Amino Acid

Letter	Abbrev	Full Name	Codons
A	Ala	Alanine	GCA GCC GCG GCT
C	Cys	Cysteine	TGC TGT
D	Asp	Aspartate	GAC GAT
E	Glu	Glutamate	GAA GAG
F	Phe	Phenylalanine	TTC TTT
G	Gly	Glycine	GGA GGC GGG GGT
H	His	Histidine	CAC CAT
I	Ile	Isoleucine	ATA ATC ATT
K	Lys	Lysine	AAA AAG
L	Leu	Leucine	TTA TTG CTA CTC CTG CTT
M	Met	Methionine	ATG
N	Asn	Asparagine	AAC AAT
P	Pro	Proline	CCA CCC CCG CCT
Q	Gln	Glutamine	CAA CAG
R	Arg	Arginine	AGA AGG CGA CGC CGG CGT
S	Ser	Serine	AGC AGT TCA TCC TCG TCT
T	Thr	Threonine	ACA ACC ACG ACT
V	Val	Valine	GTA GTC GTG GTT
W	Trp	Tryptophan	TGG
Y	Tyr	Tyrosine	TAC TAT

---

[18]Obtained from NCBI GenBank web site, contig NT_010755.15 starting at position 4859744.

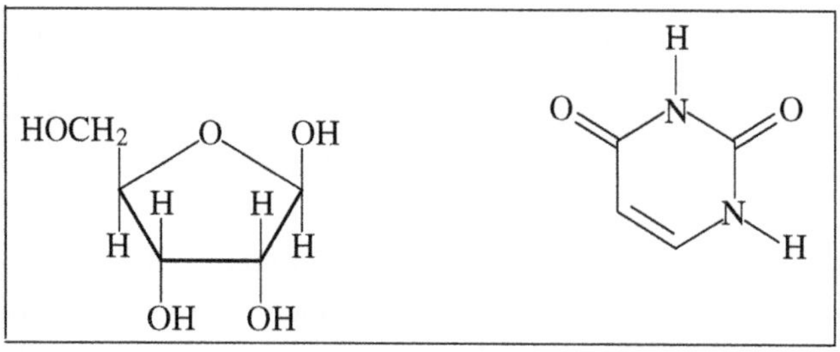

Figure 1.13   The sugar, β-D-Ribose, and the base, Uracil, which are constituents of the nucleotide units of RNA.

This correspondence is approximately the same in all known organisms. At the NCBI web site, in the taxonomy section[19] there is a tabulation of the Standard Code (shown here) and other known codes, many of which are used in mitochondrial DNA.

Sections of DNA that code for proteins consist of *genes*, which, when activated, are transcribed (copied) into a similar kind of molecule, RNA, which differs only in that the base Thymine is replaced by another base, Uracil, denoted by the letter U, and the sugar component, β-D-Ribose, differs from β-D-2-Deoxyribose in that the ribose form has an OH group instead of just an H in one of the ring positions. Figure 1.13 shows the chemical structure of β-D-Ribose and Uracil.

The codon for Methionine, ATG, is also a "start" codon, indicating where the transcription of DNA into RNA will begin, so most proteins have a Methionine at the beginning of their chains. The ATG by itself is not sufficient to initiate transcription (copying of DNA into RNA) or translation (synthesis of a protein using the encoding in a strand of RNA). It is also necessary to have a promoter region where a protein called RNA polymerase can bind to the DNA strand. Identifying promoter regions and other elements of the transcription process is a challenging problem.

In prokaryotes, the process of mapping from DNA to RNA to proteins is simpler than in eukaryotes. In prokaryotes, the RNA strand containing coding information, that is transcribed from the genetic DNA, is translated directly into an amino acid sequence. In eukaryotes, much of the DNA in the region of so-called *genes* actually does not code for proteins. The RNA that is transcribed consists of a sequence of coding regions and non-coding regions. The coding regions are called "exons" (because they are *external* or *expressed*), while the non-coding regions are called "introns," since they are *internal* and not expressed as protein sequences. The identification of genes, or where along a DNA strand the transcription starts, is a very hard problem, about which much knowledge exists, but for which there is no single algorithmic method. Even the notion that there is a static assignment of subsequences to genes seems to be an oversimplification of what is now known about molecular biology [128]. Similarly, distinguishing between exons and introns is also a very difficult problem. The initial RNA that is produced in the transcription step (in eukaryotes) is spliced to remove the introns. The resulting "mature" RNA, which then leaves the nucleus, is called messenger RNA or mRNA.

---

[19]URL http://www.ncbi.nlm.nih.gov/Taxonomy/Utils/wprintgc.cgi?mode=c.

However, some of the RNA strands generated in the transcription process are important molecules by themselves. In particular, there are special RNA molecules called *transfer RNA*, or tRNA, whose function is to provide links in the construction of proteins from the mRNA. Each amino acid corresponds to at least one tRNA. For some amino acids there are several tRNA types. The tRNA molecules are highly folded into a more or less double-ended shape. One end has a codon that is complementary to the codon in the mRNA that will be translated to an amino acid, and the other end binds the specific amino acid corresponding to the codon.

The mRNA molecule is picked up in the cytoplasm of the cell by a *ribosome*, a complex structure which mediates the translation of the mRNA into a protein. As the ribosome moves along the mRNA, the right tRNA binds to the next codon in the sequence. The tRNA molecules are rather like jigs, or tools, that hold together parts on an assembly line, so that the parts can be joined together. Once the join of the new amino acid is made to the existing chain, the tRNA disconnects, and the ribosome moves onto the next codon to be translated. This process is shown diagrammatically in Figure 1.14.

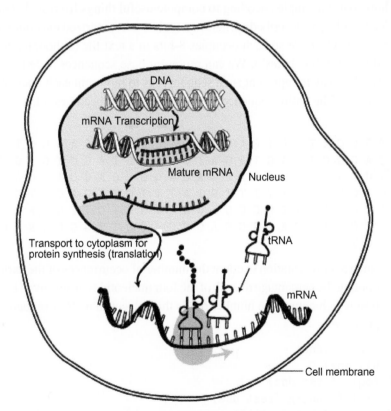

Figure 1.14 Illustration of the process of transcription and translation from DNA to mRNA to proteins
(Courtesy of National Human Genome Research Institute, NHGRI).

The proteins in turn regulate many of the activities of the cell, including the expression or activity of transcription and translation. Some aspects of the transcription and translation process can be modeled by simple computations. This is described in Chapter 5, Section 5.1.

The complete DNA sequence for a given organism is called its *genome*. The genomes of many organisms contain over a billion such nucleotides. Biologists have determined the genomes (partially or completely) for hundreds of organisms. Moreover, much information is now available for hundreds of thousands of individual genes and proteins, including how they interact with each other, and how activity levels of genes and proteins vary with different experimental conditions.

More background on molecular and cellular biology can be found in an article by Larry Hunter in the AI Magazine [168] and the opening chapter in an earlier book [167] on which the article is based. A broader treatment of this subject is available in a more recent text by the same author [169]. For the ambitious reader, standard textbooks such as [10] go into even more comprehensive detail.

1.1.2.2 *Representing DNA in Computer Programs*

To illustrate how to use a simple encoding to compute useful things from such data, consider a sequence of nucleotides, as described earlier. In a typical computerized encoding, each letter is represented by its ASCII code so each occupies 8-bits in a text file (although ASCII is a 7-bit code it is usual to use 8-bit "bytes"). We can represent base sequences as letter sequences, or long strings. Another way to represent such sequences is to represent each nucleotide (or base) as a symbol in a list. The entire sequence in Figure 1.12 then becomes a list of symbols and looks like this:

```
(C A C T G G C A T G A T C A G G A C T C A C T G C A G C C T T G A C
 T C C C A G G C T C A G T A G A T C C T C C T A C C T C A G C C T C
 T C G A G T A A C T G G G A C C A C A G G C G A G C A T C A C C A T
 G C T C A G C T A G T T T T T G T A T T T G T A G A G A T G A G G T
 T T C A C C A T A T T G C C C A G G C T G G T C T T G A A C T C C T
 G G G C T C A A G C A A G C C A C C C A C C T T G G C C A C C C A A
 A G T G C T)
```

We will use such a representation to count the number of occurrences of the various symbols, as a way of computing the percentage of each of the four nucleotides in any given DNA sequence. Here is a function which counts the number of Gs that appear in a DNA sequence such as the BRCA1 excerpt above.

```
(defun count-g (dna)
  (if (null dna) 0
    (if (eql (first dna) 'g)
        (+ 1 (count-g (rest dna)))
      (count-g (rest dna)))))
```

This function checks if the sequence, named dna here, is an empty list. It does this by calling the null function. The null function returns t if its input is nil and returns nil if its input is not nil. If the sequence is empty, there are certainly no Gs, so the function returns 0 as its result. If the list is not empty, a nested if expression checks the first base in the list, and either adds 1 or does not add one to the result of computing the Gs in the *rest* of the list. We don't need to know how long the list is. When the program reaches the end of the list, it does not go any further; there are no more recursive calls.

Now, it turns out that counting items in a list is so useful that there is a built-in function in ANSI Common Lisp, called count, so another way to do the above is simply to use this built-in function. Instead of (count-g seq) we would write (count 'g seq), where seq has a value which is a list of the bases in a DNA sequence. The code for count-g is valuable anyway, to introduce you to the idea of being able to write your own counting functions. This will be very important for more complex situations where there is no pre-existing facility.

The DNA in a cell normally consists of two complementary strands, in which each of the bases in one strand is bound to a corresponding base in the other strand. Adenine binds to Thymine and Guanine binds to Cytosine. This pairing occurs because Adenine and Thymine each have two locations where a hydrogen bond is readily formed, and they match each other, while Guanine and Cytosine each have three such locations that match up. Thus one would expect that G-C bonds will be somewhat stronger than A-T bonds. This is important in biological experiments involving the determination of the sequence of a DNA strand. In many cases, the biologist has a DNA sample that does not have enough copies of a DNA strand to be able to determine its sequence. In this case, a very clever method called the Polymerase Chain Reaction, or PCR, is used to amplify the DNA, so that enough is available for the next steps. In each step of the PCR process, each of the two complementary strands is replicated by synthesizing a new complementary strand for each, thus doubling the number of strands. Many doublings are possible. After only 30 doublings, the number of strands is increased a billion times.[20]

The way the complementary strands are generated is to provide short sequences of nucleotides, called *primers*, that bind to the right starting places on each of the original strands. Since we want high efficiency, choosing binding sites and primers that are high in GC content usually is preferred. So it is useful to be able to compute the percentage of the sequence that consists of G and C. It is again a simple counting problem.

We used recursion to count the Gs in the earlier example. Here we need to accumulate multiple results. While this can be done by calling the count function four times, it will involve traversing the sequence four times. By customizing, we can reduce this to a single traversal, accumulating counts of all four bases simultaneously. There is no built-in function to do this, but we can easily write this function, using other built-in facilities of Common Lisp. This time, rather than recursion, we will write the process in terms of iteration. Common Lisp includes iteration operators, since this is very useful. The dolist function will iterate over a list, setting the value of a local variable (in this case, the symbol base) to each successive element in turn. Four local variables created by a let expression will hold running counts of the four bases,

---

[20]Such is the simple power of exponential growth: 2^{30} is about a billion.

and we will use the `incf` operator to count up for each base we find. The `incf` operator takes as input a symbol or other name of a place where a numeric value is stored, and increases it by 1. Like `setf`, `incf` modifies the value associated with the place name it gets as input. One approach would be to selectively count the G and C symbols while keeping a running total of all the bases. However, it is just as easy to count each of the four bases and then compute the ratio from that. To count all the bases in a list such as the one above, just go through one by one, using four variables that accumulate the numbers.

```lisp
(defun dna-count-simple (seq)
  "does an explicit count of the G A C and T in seq"
  (let ((ng 0)
        (na 0)
        (nc 0)
        (nt 0))
    (dolist (base seq)
      (case base
        (g (incf ng))
        (a (incf na))
        (c (incf nc))
        (t (incf nt))))
    (list ng na nc nt)))
```

The symbol `base` is a local variable within the `dolist` expression. For each element of the list, `seq`, the variable `base` is assigned that value, and the body of the dolist is executed. In this particular function, the `dolist` body uses the `case` operator to determine which one of the four counters, `ng`, `na`, `nc`, or `nt`, to increment, depending on the value of `base`. The result is a list of four numbers in a predetermined order, number of G, number of A, number of C, and number of T.

To compute the GC ratio from a list of four numbers in the order returned by `dna-count-simple`, divide the sum of the first and third numbers by the sum of all four. This function in mathematical notation is as follows, where R is the GC ratio, T_0, T_1, etc., refer to the elements of the list, zero indexed, and the $\sum$ means add up all four T numbers.

$$R(T) = \frac{T_0 + T_2}{\sum_{i=0}^{3} T_i}.$$

You might think we can do a straightforward translation into Lisp, as follows:

```lisp
(defun gc-ratio (freq-table)
  (/ (+ (first freq-table) (third freq-table))
     (+ freq-table)))
```

However, we can't just write (+ `freq-table`) because the value of `freq-table` is a list, and we need the numbers as individual inputs, not a list. We need (+ 23 65 13 45), not (+ '(23 65 13 45)). The `apply` function arranges for this. The inputs to `apply` are a function and a list of arguments.[21] The `apply` function will rearrange its inputs so that the list, the second argument to `apply`, becomes the rest of an expression in which the function appears in the right place. In order to do this we need to be able to treat functions as pieces of data, as inputs to other functions. Thus, we need the function whose name is the + symbol. As explained earlier, there is a special operator for this, called `function`. The expression (`function` +) evaluates to the function whose name is +. We can then pass that function as input to another, or use it with `apply`. Then,

```
(apply (function +) '(23 65 13 45))
```

will be the same as if we wrote just (+ 23 65 13 45). Now we can create lists of arguments, assign such lists to variables, and use the lists where otherwise we would have no way to write out the individual elements in the right form. So, if the variable `freq-table` had the list of base counts and we want to add them up, we write

```
(apply (function +) freq-table)
```

and it gives the desired result. As noted earlier, passing functions as arguments is so often needed that the `function` special form has an abbreviation that is widely used, just as there is an abbreviation for the `quote` special form. We write (`function` x) as #'x, so the expression above would become simply (`apply` #'+ freq-table). In the rest of the book, we will use the #' notation, rather than writing out the `function` form. So, here we use the #' notation to specify that we intend the function named + as the first input to the `apply` function.[22]

```
(defun gc-ratio (freq-table)
  (/ (+ (first freq-table) (third freq-table))
     (apply #'+ freq-table)))
```

To do the entire operation, we *compose* the two functions. (Composition means that the result of one function will be the input to the other.) Of course we also need a way to read the data from a file and convert it into a list of symbols. That file reading problem will be solved later. For now, just imagine that we have already written a function called `get-data` that takes a filename as input and returns the kind of list we have described above. Here is a function that combines the operations. Given the filename, it returns the GC ratio, a single number. This illustrates the idea of *functional composition*, in which we build up powerful complex functions from layers of simpler functions.

```
(defun gc-ratio-from-file (filename)
  (gc-ratio (dna-count-simple (get-data filename))))
```

---

[21] Actually, `apply`, like many other built-in operators, has a much richer semantics than described here. To learn more about this and other functions, standard books on Common Lisp such as [137,394] are highly recommended.

[22] The Common Lisp standard allows just the quoted symbol as well, but in my opinion, if you intend to provide a function, using this notation makes the intention so much clearer.

A more general solution to the problem of counting occurrences of items like bases would add table entries as new items are found in the list. That way, the same code can be used to count frequencies for bases, for amino acids in polypeptides, or any other sequences. Each item's entry in the resulting table would be a pair, with the item first, then the number of occurrences. If a table entry already exists for the item, just increment it, otherwise add a new entry with the initial count set to 1. Here we just use a list to hold the pairs. Each pair is a list containing the item (usually a symbol) and the running count. This code illustrates that the incf operator can increment any number stored in a named place. Here, the expression (second tmp) is used as the name of the place where the count is stored, and it is directly altered.

```
(defun item-count (seq)
  "returns a frequency table of all the items in seq, a list
  of items, tagging the count by the item itself"
  (let ((results nil))
    (dolist (item seq results)
      (let ((tmp (find item results :key #'first)))
        (if tmp (incf (second tmp))
            (push (list item 1) results))))))
```

Now we show how to implement get-data. It reads a file containing only the letters of the sequence, possibly with some whitespace characters and numbers labeling the lines in the files. Actually, the files typically available from bioinformatics databases such as GENBANK and Swiss-Prot are more complex. These files contain additional data about the sequences, such as who submitted them, what functions or identification might be assigned to them, cross reference identifiers to other databases, and other properties of the biomolecule that the sequence represents. There are many formats in use and the databases have overlapping but not identical structure and content. Extracting information from such files and converting data from one form to another takes up a large percentage of the programming effort in developing bioinformatics software. Initially, we will ignore all these issues. Then we will add a little more code to be able to read one of the simpler file formats, the so-called FASTA format, named for a sequence alignment package first developed in 1985 by David Lipman and William Pearson [259,320].

A straightforward recursive implementation to read a sequence from a file would be to read a single character, make it into a string, turn it into the corresponding symbol, and put it on the front of a list that is returned by recursively reading the rest of the file. The resulting list is built up from the end back to the beginning as each recursive call returns a successively larger list. The final resulting list is in the same order as in the file (you should walk through the code by hand with a small example, to convince yourself that the order is right). The string function returns a one-character long string from its input, and the intern function creates or just returns the symbol whose name is the string that is the input. You could just create a list of characters instead of converting them to symbols, but it's a good idea to get used to using symbols to represent abstract entities. For more complex operations that we will want to perform later, characters or strings will not be suitable.

```
(defun naive-read-from-file (strm)
  (let ((char (read-char strm nil :eof)))
    (if (eql char :eof) nil
      (cons (intern (string char))
            (naive-read-from-file strm)))))
```

This function takes a *stream* as input. A stream is a source of (or sink for) sequential data, with indefinite length. In using a stream, you can keep asking for more items with each successive access, or keep writing items out to it. Files are typically opened for reading and writing as streams. Data can be written to a stream (using the `print` function and others) as well as read from it. The `read-char` function reads a single character from a stream that is an open connection to a text file. In the above, the variable `strm` is the open stream, the `nil` argument specifies that reaching the end of the file should not be treated as an error, and the `:eof` symbol specifies what the `read-char` function should return in case the end of the file is reached.[23]

As explained earlier, for each recursive function call, a stack frame must be allocated to hold the information for that invocation of `naive-read-from-file`. Space for the stack is large but limited in most computing environments. Even in very large programs this is not a problem, since most functions will do their work and return a result without too much calling of themselves or other functions. However, for big files this particular function will fail, because the number of stack frames required is so large that the program will likely run out of stack space. As an example, the full BRCA1 file downloaded from NCBI has a sequence of about 200,000 bases. A list of 200,000 cons cells is not very big and easily accomodated in most any Lisp working environment. But a stack that has to grow to 200,000 stack frames is a very large demand, because each stack frame is much larger than a cons cell, and typically much less space is available for the stack than for working data in a program.

This is another example like our first attempt at finding all the positions of an item in a sequence (see page 36) where recursion is logically sound and an elegant solution, but impractical. However, as before, by rearranging the code so that the list is built on the way in, rather than on the way out, a good compiler can turn the recursive code into an iterative process. As before, the result list is in reverse order from what is in the file, so we have to return the reverse of the accumulated result.

```
(defun read-from-file (strm accum)
  (let ((char (read-char strm nil :eof)))
    (if (eql char :eof) (reverse accum)
```

---

[23] Symbols that begin with the colon (`:`) character are constant symbols. Like numbers and strings they evaluate to themselves, so the value of the symbol `:eof` is the symbol `:eof`. Unlike ordinary symbols, their values cannot be changed by assignment, just as you cannot change the value of the number 7.

```
            (read-from-file strm (cons (intern (string char))
                                        accum)))))
```

One other wrinkle with this tail-recursive function, `read-from-file`, is that it takes *two* parameters, the input stream and an accumulator. In most uses the accumulator variable, `accum`, should start with an empty list and accumulate items onto it as the recursive calls proceed. Here is how the `read-from-file` function would typically be used:

```
(defun get-data (filename)
  (with-open-file (strm filename)
    (read-from-file strm nil)))
```

The `with-open-file` operator opens the file named by the variable `filename`, whose value should be a string or pathname, and assigns the resulting stream as the value of the variable `strm`. That stream is then the input to the `read-from-file` function defined previously. When `read-from-file` finishes, the `with-open-file` operator closes the stream and the file.

Finally, the file may contain space characters, end-of-line or newline characters, or other punctuation, beyond the basic G, C, A, and T. As well, we might want to read files that have either upper case or lower case base letters. So, we just add a line to check what character was read in, and if it is one of the ones we are looking for, we retain it, otherwise we skip it. A single character by itself will be interpreted as a symbol with a one character name, so we need a way to designate a character as a character data type. The way this is done is to precede it with a prefix, #\, so the character *c* is denoted as #\c and so on. Upper case characters are distinct from lower case so we list both.

Thus, the final version of `read-from-file` has to handle three conditions, and we replace the `if` expression with one using `cond`. The `cond` operator is similar to `if` but instead of alternatives, it has a series of clause forms as its inputs. The first expression in each clause is evaluated, until one evaluates non-nil, or the end of the `cond` expression is reached. The first clause with a non-nil first expression result is then examined. If there are any further expressions they are evaluated in sequence, and the value of the last one is the result of the `cond`. The remaining clauses are ignored. If no first expressions evaluate non-nil, the result of the `cond` is `nil`.

The three conditions are:

1. end of file reached, so just return the result,
2. a valid character was read, so add it to `accum` and continue via a recursive call,
3. a non-valid character was read, so just continue with the recursive call

and the code then follows the description.

```
(defun read-from-file (strm accum)
  (let ((char (read-char strm nil :eof)))
    (cond ((eql char :eof) (reverse accum))
          ((member char (list #\g #\c #\a #\t #\G #\C #\A #\T))
           (read-from-file strm
```

```
                           (cons (intern (string-upcase
                                           (string char)))
                      accum)))
          (t (read-from-file strm accum)))))
```

This works fine if the file contains a single sequence, and we wish to get all of it into a list (or other form) for computing from it. However, typical sequence data consists of multiple records, and of course the organization of the data is not always as simple as proposed here. In the next section, we show how to read an entire file of several hundred thousand records, the Uniprot Knowledge Base FASTA file, downloadable from the ExPASy web site (`http://www.expasy.ch`) of the Swiss Institute for Bioinformatics. From such a file one can then process the sequences to get some theoretical characteristics of the proteins, such as molecular mass. We also show how to compute the percent amino acid content of each sequence, and compare them for closeness of match. More sophisticated variants are left as exercises, for example reading through until you find a record with a specified accession number, or reading records until a partial sequence match is found with some unknown sequence. Section 5.2 of Chapter 5 introduces the idea of searching through networks or complex structures. The methods of search can be used to solve the sequence matching problem, and that application is also described in Chapter 5.

1.1.2.3 FASTA Files and Reading Multiple Records

The FASTA file format is described at `http://www.ncbi.nlm.nih.gov/blast/fasta.shtml`. Each sequence in a FASTA file begins with one line of descriptive data beginning with the right angle bracket character, (>) and terminating with a newline. Starting with the next line of text, the actual sequence follows. Aside from newlines, the characters of text must be those representing the elements of the sequence, the single-letter codes defined by the IUPAC (International Union of Pure and Applied Chemistry, a worldwide standards body that maintains naming conventions for chemical compounds, among other things). The table on page 39 conforms to the IUPAC standard codes for amino acids in protein sequences. The following is a sample entry from the UniProt/Swiss-Prot protein sequence database in FASTA format.[24]

```
>P38398|BRCA1_HUMAN Breast cancer type 1 susceptibility protein
- Homo sapiens (Human)
MDLSALRVEEVQNVINAMQKILECPICLELIKEPVSTKCDHIFCKFCMLKLLNQKKGPSQ
CPLCKNDITKRSLQESTRFSQLVEELLKIICAFQLDTGLEYANSYNFAKKENNSPEHLKD
EVSIIQSMGYRNRAKRLLQSEPENPSLQETSLSVQLSNLGTVRTLRTKQRIQPQKTSVYI
ELGSDSSEDTVNKATYCSVGDQELLQITPQGTRDEISLDSAKKAACEFSETDVTNTEHHQ
PSNNDLNTTEKRAAERHPEKYQGSSVSNLHVEPCGTNTHASSLQHENSSLLLTKDRMNVE
KAEFCNKSKQPGLARSQHNRWAGSKETCNDRRTPSTEKKVDLNADPLCERKEWNKQKLPC
```

---

[24]I have added an extra line break after the word "protein" to fit the page width. The first two lines here are actually a single line in the original file.

```
SENPRDTEDVPWITLNSSIQKVNEWFSRSDELLGSDDSHDGESESNAKVADVLDVLNEVD
EYSGSSEKIDLLASDPHEALICKSERVHSKSVESNIEDKIFGKTYRKKASLPNLSHVTEN
LIIGAFVTEPQIIQERPLTNKLKRKRRPTSGLHPEDFIKKADLAVQKTPEMINQGTNQTE
QNGQVMNITNSGHENKTKGDSIQNEKNPNPIESLEKESAFKTKAEPISSSISNMELELNI
HNSKAPKKNRLRRKSSTRHIHALELVVSRNLSPPNCTELQIDSCSSSEEIKKKKYNQMPV
RHSRNLQLMEGKEPATGAKKSNKPNEQTSKRHDSDTFPELKLTNAPGSFTKCSNTSELKE
FVNPSLPREEKEEKLETVKVSNNAEDPKDLMLSGERVLQTERSVESSSISLVPGTDYGTQ
ESISLLEVSTLGKAKTEPNKCVSQCAAFENPKGLIHGCSKDNRNDTEGFKYPLGHEVNHS
RETSIEMEESELDAQYLQNTFKVSKRQSFAPFSNPGNAEEECATFSAHSGSLKKQSPKVT
FECEQKEENQGKNESNIKPVQTVNITAGFPVVGQKDKPVDNAKCSIKGGSRFCLSSQFRG
NETGLITPNKHGLLQNPYRIPPLFPIKSFVKTKCKKNLLEENFEEHSMSPEREMGNENIP
STVSTISRNNIRENVFKEASSSNINEVGSSTNEVGSSINEIGSSDENIQAELGRNRGPKL
NAMLRLGVLQPEVYKQSLPGSNCKHPEIKKQEYEEVVQTVNTDFSPYLISDNLEQPMGSS
HASQVCSETPDDLLDDGEIKEDTSFAENDIKESSAVFSKSVQKGELSRSPSPFTHTHLAQ
GYRRGAKKLESSEENLSSEDEELPCFQHLLFGKVNNIPSQSTRHSTVATECLSKNTEENL
LSLKNSLNDCSNQVILAKASQEHHLSEETKCSASLFSSQCSELEDLTANTNTQDPFLIGS
SKQMRHQSESQGVGLSDKELVSDDEERGTGLEENNQEEQSMDSNLGEAASGCESETSVSE
DCSGLSSQSDILTTQQRDTMQHNLIKLQQEMAELEAVLEQHGSQPSNSYPSIISDSSALE
DLRNPEQSTSEKAVLTSQKSSEYPISQNPEGLSADKFEVSADSSTSKNKEPGVERSSPSK
CPSLDDRWYMHSCSGSLQNRNYPSQEELIKVVDVEEQQLEESGPHDLTETSYLPRQDLEG
TPYLESGISLFSDDPESDPSEDRAPESARVGNIPSSTSALKVPQLKVAESAQSPAAAHTT
DTAGYNAMEESVSREKPELTASTERVNKRMSMVVSGLTPEEFMLVYKFARKHHITLTNLI
TEETTHVVMKTDAEFVCERTLKYFLGIAGGKWVVSYFWVTQSIKERKMLNEHDFEVRGDV
VNGRNHQGPKRARESQDRKIFRGLEICCYGPFTNMPTDQLEWMVQLCGASVVKELSSFTL
GTGVHPIVVVQPDAWTEDNGFHAIGQMCEAPVVTREWVLDSVALYQCQELDTYLIPQIPH
SHY
>
```

The first datum after the > is usually the accession number, a unique identifier for this sequence, and the rest of the line is a text description. Both items are optional. No space is allowed between the > and the identifier. It appears there is no standard for these items, but the various data sources follow a convention that vertical bar characters separate fields in the description, and each data source has a standard set of items that are included in the description. The Swiss-Prot description has an accession number followed by a vertical bar, then an identifier (not necessarily stable), a space, and then some descriptive text.

The FASTA format requires two additions to the file reading code described on page 48. One is the detection of the > and processing (or ignoring) the line on which it appears. The second is to handle the possibility that a file may contain multiple sequence records, so rather than reading to the end of the file, the code must detect the next > and stop there if one is found before the end of the file. If possible, this new version of the read-from-file function should be written in such a way that it can be called iteratively to process any number of records in a single file. It would also be a good idea to be able to use the same code for reading both DNA and protein sequences, which calls for a way to handle the letter checking in a more general (parametrized) way.

Two straightforward modifications to `read-from-file` can handle the > and the letter checking. For generalized letter checking, we provide a list of allowed characters as an input to the function. To handle the description line, add a clause to the `cond` form after the end-of-file check, to see if the character read is >. If so, it is either at the beginning of the record to be read, in which case the `accum` variable will still be `nil`, or at the beginning of the next record (`accum` is not `nil`). If at the beginning of the record to be read, just read past the description and proceed. Otherwise, return the reverse of `accum`, which contains the completed sequence in reverse order. A call to `reverse` is needed because the `read-fasta` function builds up the data backwards, by putting new amino acids at the front of the result list.

```
(defun read-fasta (strm allowed-chars accum)
  (let ((char (read-char strm nil :eof)))
    (cond ((eql char :eof) accum)
          ((eql char #\>)
           (if (null accum) ;; at beginning of sequence
               (progn
                 (read-line strm) ;; skip description and cont.
                 (read-fasta strm allowed-chars accum))
             (reverse accum)))
          ((member char allowed-chars)
           (read-fasta strm allowed-chars
                       (cons (intern (string-upcase (string char)))
                             accum)))
          (t (read-fasta strm allowed-chars accum)))))
```

Now for convenience we define two constants. One is the list of allowed letters for protein sequences and the other is for DNA sequences. We don't include here *all* the letter codes (there are some special ones for unknowns, gaps, or ambiguous assignments). Those could be added if appropriate. We allow for both upper case and lower case letters, and generate these lists using the map function. This function takes a sequence result type and a function to map, and generates the result sequence from application of the function to each element of the input sequence in turn. The `identity` function just returns each letter unchanged, and the `char-downcase` function returns the lower case version of each letter. Then append just puts the two lists together.

```
(defconstant +aa-codes+
    (let ((valid-letters "ACDEFGHIKLMNPQRSTVWY"))
      (append (map 'list #'identity valid-letters)
      (map 'list #'char-downcase valid-letters))))

(defconstant +base-codes+
    (let ((valid-letters "GCAT"))
      (append (map 'list #'char-downcase valid-letters)
      (map 'list #'identity valid-letters))))
```

Then a wrapper function, `read-first-protein`, shows how `read-fasta` would be used. It takes a filename and one of the letter lists and reads the first record from the specified file.

```
(defun read-first-protein (filename codes)
  "returns the first protein sequence in filename"
  (with-open-file (strm filename)
    (read-fasta strm +aa-codes+ nil)))
```

One would use `read-first-protein` by passing a filename, perhaps of a file obtained from Swiss-Prot. The list of characters used as allowed characters will default to the value of constant `+aa-codes+`.

```
> (setq a-protein
    (read-first-protein "swiss-prot-dump.fasta")
```

This does not allow for reading more records, or for searching for a record, but it could be used in principle to process a downloaded single record from an NCBI or other database. Further, we have ignored all the information on the first line. To process multiple records it is important to obtain at least the accession number from each record.

To process multiple records, we need something in addition to or alternative to `read-char`. When reading the > that indicates the beginning of the next record, it is necessary to somehow move the file position back before it, so that a repeat call to `read-fasta` can start with the >. Common Lisp provides two ways to do this. The `unread-char` function puts the last character read from a stream back onto the stream so that the stream is in the same state as if the character had not yet been read. Another function, `peek-char`, returns the next character to be read but does not advance the stream pointer, so it can be used to do a little lookahead. The simpler way is to call `unread-char` in the right place in the code, namely, when finished reading the current sequence record. Using `peek-char`, which returns the next character in the stream without moving the stream pointer forward, would be more tangled.

So, here is an implementation of `read-fasta` that leaves the file pointer in the right place when reaching the end of a sequence. It can thus be called repeatedly to read and return successive sequences from a single file.

```
(defun read-fasta (strm &optional (allowed-chars +aa-codes+)
                                   (ac-parser #'sp-parser))
  (labels
      ((scan (accum)
         (let ((char (read-char strm nil :eof)))
           (cond ((eql char :eof) (reverse accum))
                 ((eql char #\>) (unread-char char strm)
                                 (reverse accum))
                 ((member char allowed-chars)
                  (scan (cons (intern (string-upcase (string char)))
                              accum)))
                 (t (scan accum)))))))
```

```
      ;; the ac-parser call will return nil if end of file is reached
      (let ((accession-number (funcall ac-parser strm)))
        (if (null accession-number)
            nil
          (list accession-number (scan nil)))))))
```

This version has additional features that make it more convenient and more efficient. One is to return the sequence in the right order instead of reversed. There is no reason to expect the caller to do that. There are two places in the above code where the function returns a result. One is where it finds the next > record marker, and the other is when it reaches the end of the file. In each of these two places, the `accum` variable has the sequence from the file in reverse order, so we return the reverse of it, which restores the order that matches what was in the file.

The second is to return the accession number for the sequence entry along with the sequence. It is usually the first thing after the >. At the beginning of the process of reading a sequence entry, a function is called to obtain the accession number (parametrized here as `ac-parser` and defaulting to a simple version for Swiss-Prot). The `read-fasta` function then returns a two element list containing the accession number and the sequence obtained by the local function, `scan`.

Finally, since the recursive calls to process the sequence are done in the `scan` function, we can make the character codes and accession number parser be optional and have default values that work for proteins. The `scan` function is tail recursive, and we want it to compile to an efficient process, so we can't have any optional parameters.

Examining a download of the Swiss-Prot FASTA format data, it appears that the accession number is always just six characters and is the very next thing after the >. It is not actually a number, since it includes alphabetic letters as well as digits. No matter, our function for this, `sp-parser`, will just turn it into a string and return the string. The `sp-parser` function returns the Swiss-Prot accession number from the open stream `strm`, assumed to be connected to a FASTA file. The function returns `nil` if the end of the file is reached before getting the six characters, which will happen if it has already read the last record in the file.

```
(defun sp-parser (strm)
  (let ((ac (make-string 6))
        (ch (read-char strm nil :eof))) ;; get rid of the >
    (if (eql ch :eof) nil ;; at end of file!
      (progn
        (dotimes (i 6)
          (setq ch (read-char strm nil :eof))
            (if (eql ch :eof) ;; shouldn't happen!
                (return-from sp-parser nil)
              (setf (aref ac i) ch)))
        (read-line strm nil :eof)
        ac))))
```

After calling this version of `read-fasta`, one can make another call with the file still open, so that it will read the next sequence in the file. Here is how to use it to read a specified number of records. The result is an association list in which the sequence accession number (a string) is the key. This is an example where repeated calls to `read-fasta` can accumulate a list of sequence records. This function reads the first n records in a file.

```
(defun read-proteins (filename n)
  "returns the first n protein sequences in filename"
  (let ((sequences nil))
    (with-open-file (strm filename)
      (dotimes (i n (reverse sequences))
        (push (read-fasta strm)
              sequences)))))
```

With the amount of working memory available in typical desktop and notebook computers (1 GB or more) it is possible to read the entire Swiss-Prot sequence data (from the FASTA file version) into a list in memory. Then further computations can be performed on the list of sequences. However, other data sets may be too large to fit even several GB. In such cases, you would have to write programs that process a record at a time, keeping only the most relevant information. In Section 5.1.2 an example of this kind of processing is presented, namely, the generation of amino acid frequency profiles and comparision of an "unknown" protein sequence with the known tabulated profiles.

1.1.3 Anatomy

The process of mapping out the structure of the human body has been going on for several thousand years. As early as 1600 B.C.E., the ancient Egyptians apparently knew how to deal with a considerable range of injuries by surgical procedures, as well as other basic medical knowledge, such as the role of the heart in circulating body fluids. This was documented in an ancient artifact called the "Edwin Smith Surgical Papyrus," discovered in Egypt, translated by James Henry Breasted [49], and currently held by the New York Academy of Medicine. It is one of several such documents that record much knowledge of anatomy. The ancient Greeks, in the 5th and 4th centuries B.C.E., continued to study anatomy as well as physiology. Prominent contributors were Aristotle, whose work on logic was a starting point for the symbolic logic to be developed in Chapter 2, and Hippocrates, to whom we owe one of the widely used oaths taken by graduates of medical schools at the time of conferring of their medical doctor degrees. In the second century C.E., Galen practiced and wrote about surgery and anatomy. A well-known system of nomenclature for biomedical concepts and entities is named after him, and will be described also in Chapter 2. The study of anatomy continued in Europe and the Middle East through the Middle Ages, and now worldwide, through modern times.[25]

A huge boost technologically was the development of X-ray imaging at the beginning of the 20th century, and even more so, the development of cross sectional imaging using X-rays (the

---

[25] A brief history of the development of the science of anatomy may be found in Wikipedia under "history of anatomy."

Computed Tomography, or CT, scanner, and later the Magnetic Resonance Imaging, or MRI, scanner). The ability to produce images of living animals (including humans), without invasive surgical procedures, has revolutionized diagnostic medicine, as well as opened new possibilities for visualization of anatomy for teaching purposes. The Visible Human Project of the National Library of Medicine [293, 296] has provided three-dimensional image data for two entire human bodies, male and female, for use in both research and teaching.

Two separate but related kinds of information about anatomy are useful to represent in computerized form. First, the logical relationships discovered in ancient times, and revised many times, express what kinds of things are contained in a human body, and how these things are related. This is the kind of information found in the FMA, introduced earlier (see page 12) and excerpted in Figure 1.7. The representation of these anatomic entities and their relationships in the FMA is based on the idea of *frames*, which will be developed in Section 2.3 of Chapter 2. An application of the FMA to prediction of the spread of cancer cells is described in Section 5.5.3. Network access to the FMA contents, including searching and traversing its many relationships, will be covered in Chapter 4. The essential idea here is to be able to represent anatomic entities with symbols and lists, so they can be grouped into entities, attributes, and relations. Thus, the heart information could become a list structure with the attributes labeled by symbolic tags and the values implemented as lists of symbols, as follows:

```
(heart
   (part-of cardiovascular-system)
   (contained-in middle-mediastinal-space)
   (has-parts (right-atrium left-atrium
                right-ventricle left-ventricle
                mitral-valve tricuspid-valve ...))
   (arterial-supply (right-coronary-artery
                      left-coronary-artery))
   (nerve-supply ...)
   ...)
```

In addition to the logic of anatomy, however, two additional very important kinds of data need representation as well. One of these is the image data mentioned earlier, both X-ray projected images and cross sectional images, and photographic images, for example of skin lesions or open views into a body during surgery. Microscope images provide views of anatomy at the tissue and cellular level. So we need a representation and methods for computing with (and display of) image data. Images are usually represented as arrays of numbers, with each number or set of numbers corresponding to a color or gray level for a particular spot, or *pixel*, in the image. The image is then considered to consist of a rectangular array of such spots. If the resolution is high enough, a considerable amount of detail can be discerned. In Section 1.1.5 we show how to represent cross sectional images, do gray-scale mapping, and compute sagittal and coronal images from a set of transverse cross sectional images. Section 1.2.4 illlustrates how to store and retrieve image data in a relatively simple system using a combination of text and binary files.

Finally, the anatomic objects that appear in image data sets often need to be identified and delineated with their exact shapes represented, for purposes of illustration, and more importantly to perform measurements on these organs and other components (such as tumors). The exact location of such objects in a particular cancer patient's body is crucial information for designing and delivering accurate and effective radiation treatment, for example. Here it will suffice to say that some kind of three-dimensional coordinate system, with X, Y, and Z axes, and a well-defined origin point is needed. Typically, organs and tumors are represented as collections of planar *contours*, or polygons. A simple but effective representation of a contour is a list of points, where each point in the polygon is a two element list consisting of the X coordinate and the Y coordinate, assuming that the points all are in the same Z plane. To make the data more self-contained, we can include the Z value along with the list of points, in a larger list, using symbols to label these two components. The points are non-redundant; it is assumed that in rendering the contour or computing area or volume from it, the last point is connected back to the first.

```
(contour (z 2.0)
         (vertices ((−3.41 6.15) (−3.87 6.47) (−4.01 7.01)
                    (−3.87 7.51) (−3.87 7.61) (−3.46 8.11)
                    (−2.91 8.33) (−2.41 8.20) (−2.05 8.15)
                    (−1.68 7.79) (−1.64 7.10) (−2.00 6.47)
                    (−2.55 6.10) (−3.00 6.06))))
```

These polygons are usually created from the images manually, using a computer-aided drawing tool. This is an extremely time-consuming process, but is so important that many decades of research have been devoted to the development of algorithms for automating this delineation of objects from images, usually called *image segmentation* [99,326]. Many such algorithms are in use today in clinical systems, though there is still much room for improvement in accuracy, success rate (sometimes the algorithm simply fails to find a contour), and speed. Making useable tools based on successful algorithms is also challenging, and a subject of research in human computer interaction.

1.1.4 Medical Laboratory Data

Here is a way to represent laboratory results such as those in Tables 1.1 and 1.2, for example, the red and white cell counts in the blood of a patient.

```
(cbc
  (wbc 6.14 thou-per-microltr)
  (rbc 4.80 mill-per-microltr)
  (hemoglobin 13.8 grams-per-dl)
  (hematocrit 40 percent)
  (platelets 229 thou-per-microltr))
```

This expression encodes part of the results of a complete blood count (labeled by the symbol cbc) and consists of the white cell count, with its symbolic label, wbc, the numeric value, and the units, thousands per microliter, represented by the symbol, thou-per-microltr, along with a red cell count, similarly encoded. Here, again, we have used standard Latin text characters and chosen symbol names that would be somewhat familiar to practitioners, and used digits and decimal points to represent numbers. To make this encoding easy for people to read, I have started the inner lists on separate lines, and indented them slightly, just as for the enzyme data on page 20. For the laboratory tests, one would want to look up a particular value by name, such as the white blood cell count, which will give some indication of the presence of infection. Although it would seem that these laboratory tests are all numerical values and have a certain regularity, in fact there are thousands of such tests, many with non-numeric result types, and with highly variable structure. Moreover, although the names above seem straightforward they leave out a lot of important information about the test, such as how it was performed, under what conditions, and what constitutes a normal range of results for the particular laboratory.

A simple example of a computation on laboratory data would be to generate (as needed, for display or alerting) a special flag identifying a test result that is outside a prespecified "normal" range. In order to do this, a table of normal ranges or *reference values* is needed in addition to the test result itself. This could be done with numeric data in a straightforward way. The predicate out-of-range returns t if the value is out of range, and nil if within range. The EMR system would then use this result to set some indicator in a display of the data.

```
(defun out-of-range (value low high)
   (or (< value low) (> value high)))
```

If the value is non-numeric, perhaps a string naming one of many possible results, we would need to have a list of the names of normal results. In this case, the predicate just checks if the name is in the normals list. If so, it returns nil, meaning the result is not out of range. If not in the normals list, it returns t, meaning the result *is* out of range.

```
(defun named-result-out-of-range (value normals)
   (not (member value normals)))
```

Reference ranges are not universal, and may vary from one institution to another or from one electronic medical record system to another. What is acceptable may depend on the particular patient's condition. For example, a blood creatinine level of 1.3 mg/dL might be considered abnormal in general, but for someone with only one functioning kidney, it is a reasonable and acceptable value.

It might seem that these functions don't do much. They are building blocks, from which you would build larger functions, and so on, just as large buildings are made from combining many small components.

In a typical clinical laboratory, the instruments produce results in digital form, and these results are stored in computers running clinical laboratory system software. These systems have in the past been made accessible through local area network access but that required clinicians and other staff on the wards to learn how to use the specific laboratory system software, and of course required the clinical laboratory to manage access by issuing user IDs and passwords to everyone. The next step in evolution was to provide some way to transmit the laboratory data to another system, the electronic medical record, which (it was hoped) would provide an integrated view of all the sources of medical data, as well as provide a central system for registration of patients, scheduling clinic visits, tracking inpatients, accepting orders for medications, and generating the bills to the patients or their insurance plans. At present the only generally accepted standard for constructing, sending, and interpreting messages with this data in them is the Health Level 7 (HL7) standard, developed and managed by a consortium known as (oddly enough) HL7. In Section 4.3.1 we show how network communication of medical data works, and in Section 4.3.2 we provide some details about the HL7 standard.

The HL7 standard does not specify the names, encoding of values, grouping of tests into panels, or other low level details. Many systems have been put into use, and the definitions of these panels may vary from one institution to the next, or even from one device manufacturer to the next. So, to integrate many such systems even in a single institution, it is necessary to translate one set of definitions into the others. Electronic medical record systems cannot therefore be standardized and simply installed. Each hospital needs to define its own tables of tests, conventions for displaying results, and display screens that organize the data the way it is needed for that institution. To exchange data between institutions is of course very much more difficult.

To achieve meaningful data interchange, one of the first steps needed is to create a standard system of nomenclature for the tests and for the encoding of the test result values. A well-known scheme for labeling and encoding laboratory test results is the Logical Observation Identifier Names and Codes (LOINC)[26] system [113,273]. The LOINC system proposes a set of names and codes for labeling laboratory results and clinical observations. This involves achieving agreement about the type of a datum, its external representation, the units in which it is measured, and other factors that affect the interpretation of the result being reported. The goal of LOINC is to assign unique codes and names to each of the thousands of observations and laboratory tests in use in clinical practice, so that the associated data for a particular patient can be labeled in a computerized message transmitting the data between various electronic medical record (EMR) systems and other application programs using medical data. Similarly, these same codes can be used by an electronic medical test order entry system to identify what a care provider has ordered for a patient. Although LOINC started out with a focus only on the clinical laboratory, today it encompasses a broad range of tests and observations throughout medical practice. The current set of identifiers has over 50,000 entries.

The LOINC table is available from the Regenstrief Institute as a text file. The entries are organized into one record per line of text where lines are delimited by a <CR><LF> two character sequence. Within a record or single line, individual fields are delimited by <tab> characters, used as separators, so that two successive <tab> characters indicate a blank or omitted field.

---

[26]"LOINC" is a registered trademark of Regenstreif Institute, Inc.

Strings are represented as text surrounded by double quote characters, as in Lisp. The few fields that are not strings are integers. This makes it easy to read the LOINC codes into a list for processing in a program, except for handling the <tab> characters, which is a little tricky.

This problem of reading data files in various formats and transforming the data into data structures that can be used by computer programs is widespread throughout biomedical informatics. Tab-delimited text is used frequently enough to warrant writing some code that handles it in general, and the LOINC file in particular. The strategy will be to write a function, parse-line, that takes as input a line of text containing tab-delimited data and returns a list of the individual items. We will assume that the items are readable by the Lisp read function. We won't deal with parsing or interpreting the items themselves. Then the parse-line function will be applied a line at a time to each line in the LOINC file, by another function, parse-loinc.

The strategy for parse-line will be to write an inner function (in a labels expression, of course), called read-items, which just processes a single item at a time, puts it on an accumulator list, and calls itself recursively to process the rest of the items. The tricky part is in handling the tab characters.

Our recursive inner function starts out the same way all such functions do, which is to check if it is at the end of the line, in this case by examining the input variables, str and pos. The variable pos is the position of the next character to be read, so if it is greater than or equal to the length of the string, we have reached the end. In this case, if the previous character read was a tab character (when tab is t), that means there is a blank or empty field at the end of the line, so we add a blank string to the accumulator, accum, otherwise we return the accumulator value as is (actually it returns the reverse of the accumulator list since it has been built backward).

If we are not at the end, there are several possibilities. Either the first character to be read is a tab, or it is not. If it is, we are going to recursively process what is left. If the previous character read was a tab, here too, we have to add in a blank string, since two successive tab characters indicate an empty field.

If the first character is not a tab, we have some data to process. The built-in Common Lisp function read-from-string can read this data and convert it into an internal Lisp item, perhaps a string or a number. We need to specify as inputs not only the string to read (the variable str) but also that read-from-string should not signal an error when it reaches the end of the string, and it should start at a position not necessarily the beginning (this is the value of the pos variable). We need back from read-from-string two things, the result of reading the data item, and the position at which the next character should be read.

Indeed the read-from-string function provides exactly those two values. Until now, we have been using and writing functions that each return a single value. The read-from-string function is an example of a function that returns *multiple values*. To use them we use the multiple-value-bind operator to make them the values of local variables. It is like let, in that these variables can be used within the scope of the expression, but not outside.

The item variable contains the item that was read, or the symbol :eof if the end of the string was reached. If it is :eof, then the function read-items will return the accumulated list of items to that point (actually the reverse of the list as we noted earlier). If the end was not reached, the item is put on the accumulator and a recursive call is made to process what remains. In the recursive call, the value for the position variable, pos, is the value of next, since that is the next character to process.

Once we have the inner function, read-items, the main function just calls it, with an empty accumulator, starting at position 0. The tab variable starts out with a value of t because we are going to start by processing a field. If the line *starts* with a tab, it means the first field is blank.

```lisp
(defun parse-line (line)
  "parses tab delimited text, assumes no other whitespace between
objects"
  (labels ((read-items (str accum pos tab)
              (if (>= pos (length str))
                  (reverse (if tab (cons "" accum) accum))
                  (let ((first (char str pos)))
                    (if (eql first #\Tab)
                        (read-item str
                                   (if tab (cons "" accum) accum)
                                   (1+ pos) t)
                        (multiple-value-bind (item next)
                            (read-from-string str nil :eof :start pos
                                              :preserve-whitespace t)
                          (if (eql item :eof)
                              (reverse (if tab (cons "" accum) accum))
                              (read-items str (cons item accum)
                                          next nil)))))))))
    (read-items line nil 0 t)))
```

Now, to deal with a multi-line file, we just need to open the file, and process records until we reach the end of the file. The following code will handle the LOINC data file available from Regenstrief, and process it into a list of lists, one list per record, with each field a separate string in the list. Once again, we define an inner tail-recursive function, this time called read-loinc-records, to process a record, and check for end of file. If at the end it returns the reverse of the accumulated records, and if not at the end, it adds the new record result to the accumulator list and recursively processes the rest of the file. The main function opens the file and calls the read-loinc-records starting with an empty list as the accumulator. It is not quite that simple, because the first record in a LOINC file is a list of column labels, so a separate call to parse-line gets that data, and *then* the parse-loinc function calls read-loinc-records. The result is a two element list, the column headers and a list of the records.[27]

---

[27]The parse-loinc function here uses format, a Common Lisp facility for producing character (text) output to text streams, such as a terminal window or a file. The first argument to format is the output stream, where t means *standard-output*, and nil means just return a string. The rest is a format control string and data for it. Rather

```
(defun parse-loinc (filename)
  (let ((n 0))
  (labels ((read-loinc-records (strm accum)
              (let ((line (read-line strm nil :eof)))
                (format t "Reading record ~A~%" (incf n))
                (if (eql line :eof) (reverse accum)
                  (read-loinc-records strm
                                      (cons (parse-line line)
                                            accum))))))
      (with-open-file (strm filename)
        (list (parse-line (read-line strm))
              (read-loinc-records strm nil))))))
```

These two functions use tail recursion, so it should be possible for the compiler to generate very efficient code. Indeed it turns out that compiling is necessary in order to avoid stack overflow.

The following is a convenience function for printing the LOINC data after it is processed into a big list. For each item in a LOINC record, this function prints the label for each field together with its value. In this representation, each record is a list of strings, and each field is a string delimited by double quote characters, with the field name, an = and the field value. Where the field value is a string, it is *not* further surrounded by double quote characters, but just printed literally. Here what we want to do is take a function that will print a header label and an associated item, and iterate it over a list of header labels and a list of corresponding items. There are many ways to do this. One commonly used idiom is the idea of *mapping*. The `mapcar` function takes a function and some lists, and applies the function to items from each of the lists in turn, until it reaches the end. As it goes through, it accumulates the results of the function in a new list. Here instead of defining a named function and then using it in `mapcar`, we use a *lambda expression*, which is an anonymous function.[28]

```
(defun print-loinc-record (headers record)
  (mapcar #'(lambda (hdr item)
              (format nil "~A = ~A" hdr item))
          headers record))
```

Figure 1.15 shows an excerpt of the contents of one record in the LOINC table of test names and descriptions, reformatted from the original file. This is exactly what is produced by `print-loinc-record` when applied to a single LOINC table entry.

---

than explain it here, the reader is referred to Graham [137, Chapter 2] and for a complete reference on formatted output, Steele [394, Chapter 22].

[28]The special form, `lambda`, is used to tag a description of an anonymous function. The expression, `#'(lambda (x) (* x x))`, for example is an unnamed function that takes one input and returns the result of multiplying it by itself. Anonymous functions can be used anywhere that named functions can be used.

```
("LOINC_NUM = 14743-9"
 "COMPONENT = Glucose"
 "PROPERTY = SCnc"
 "TIME_ASPCT = Pt"
 "SYSTEM = BldC"
 "SCALE_TYP = Qn"
 "METHOD_TYP = Glucometer"
 ...
 "CLASS = CHEM"
 "SOURCE = OMH"
 "DT_LAST_CH = 19980116"
 "CHNG_TYPE = ADD"
 ...
 "CLASSTYPE = 1"
 "FORMULA = "
 "SPECIES = "
 "EXMPL_ANSWERS = "
 "ACSSYM = CORN SUGAR, D GLUCOPYRANOSE, D GLUCOSE, D GLUCOSE,
 DEXTROSE, GLU, GRAPE SUGAR,"
 ...
 "RelatedNames2 = Glu; Gluc; Glucoseur; Substance concentration;
 Level; Point in time; Random; Blood; Blood - capillary; Blood cap;
 Blood capillary; Bld cap; Bld capillary; cap blood; cap bld;
 Capillary bld; Capillary blood; Quantitative; QNT; Quant; Quan;
 Glucomtr; Chemistry"
 "SHORTNAME = Glucose BldC Glucomtr-sCnc"
 "ORDER_OBS = BOTH"
 ...)
```

Figure 1.15 An excerpt from a record in LOINC.

Some fields are blank, because the LOINC system has not yet assigned a value, or for a particular test, there is no relevant value for that field. Every LOINC test name is assigned a unique identifier, the LOINC_NUM field. It consists of a number, a hyphen, and a check digit. The COMPONENT field is part of the LOINC name for the test. This name consists of subfields separated by carat (ˆ) characters. This and the next five fields together comprise the complete LOINC name. The standard specifies that these fields are combined into a composite name by adjoining them with the colon (:) character, so none of them should contain the colon (:) character. The SHORTNAME field is provided for use in HL7 messages. The test described in Figure 1.15 is a common test for glucose blood level using a glucometer.

A computer program that processes a message containing a test result labeled with the LOINC system can then look up information about that test from its label, and have branching code also to decide how to display it. However, this is rarely useful or practical, not only because of the need to interface systems that do not use LOINC, but also because of the lack of context. The test label by itself does not have sufficient information for the practitioner to interpret the result. Other information is needed, including the normal range limits, which may be institution or population specific, as well as specific information about the patient, that may not be on the same display screen as the test result.

1.1.5 Medical Images

The use of X-rays to obtain images of the internal structure of the human body has a venerable history, dating from the discovery of X-rays by Wilhelm Conrad Roentgen [346]. In recent years, the availability of small detectors for X-rays, and mathematical methods for computing two-dimensional data from one-dimensional projections, has made possible the direct production of digital images by the Computed Tomography (CT) scanner [163]. As detector technology advanced, it became possible to directly produce digital X-ray images that were previously done with planar sheets of film (Computed Radiography). Other kinds of digital imaging techniques as well were developed during this period, the last three decades of the 20th century. The repertoire of the modern radiology department now also includes magnetic resonance imaging (MRI), ultrasound imaging, positron emission tomography (PET), and various kinds of nuclear isotope scans. A good introduction to the physical aspects of medical image production and processing is the textbook of Hendee and Ritenour [154]. The following is a brief sketch.

X-ray images are *projection* images, in which a beam of X-rays passes through a person's body, with some amount of the initial beam being absorbed and some transmitted to a sheet of film or a digital detector (which we refer to as the "imager"). Along each ray line from the X-ray source to the imager X-ray photons are absorbed, scattered, or transmitted according to well-known processes. The amount of scatter or absorption will depend on the energy of the X-rays as well as the density and composition of the material through which the beam passes. The process is schematically illustrated in Figure 1.16.

The image that is produced will be light where many X-rays were absorbed, and dark where many were transmitted. This is how film works—the exposed areas turn dark because the silver

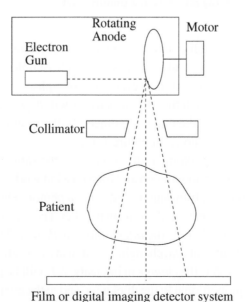

Film or digital imaging detector system

Figure 1.16 The process of X-ray image production.

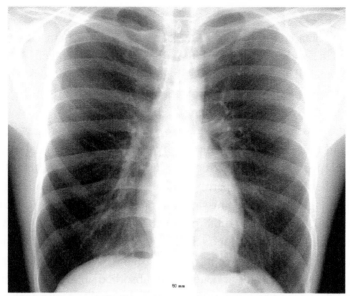

Figure 1.17   An example X-ray image of a person's chest *(from Wikimedia Commons, URL* `http://` `commons.wikimedia.org/wiki/File:Chest.jpg`*).*

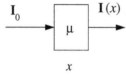

Figure 1.18   Attenuation of X-ray photons in a uniform slab.

ions in the film are reduced to metallic silver. Digital imaging systems are built to produce images with similar gray-scale properties as film images, to enable radiologists to easily transfer their experience with film to the interpretation of digital radiographs.

The intensity of X-rays that reach the film or detector will depend only on the total material along each ray from the source to the image, so objects will appear as *overlapped* or obscured. An example of such an image is shown in Figure 1.17.

Assuming that all the X-ray photons in the beam have the same energy, it is fairly easy to describe mathematically how the attenuation and transmission works. At any location in a material, there is some probability of absorbing or scattering a photon. Photons get absorbed in proportion to the number and cross section of atoms in the slab. These properties can be represented for any material by a single number, μ, the *attenuation coefficient*. The diagram in Figure 1.18 shows a ray starting out with intensity I_0, changing to intensity $I(x)$ after traversing a thickness x.

For a very small thickness, Δx, the *change* in intensity, ΔI, will be proportional to the current intensity, I. It will also be proportional to the thickness, and to the attenuation coefficient (which represents the density and type of material). So, the intensity change for a small thickness can be written as

$$\Delta I = -\mu I \Delta x. \tag{1.1}$$

For continuous media, the intensity, I, is a continuous function of the depth, x, and the ratio $\frac{\Delta I}{\Delta x}$ becomes the derivative, $\frac{dI(x)}{dx}$, sometimes also written as $I'(x)$, so the equation becomes

$$\frac{dI(x)}{dx} = I'(x) = -\mu I(x). \tag{1.2}$$

This is a differential equation, to be solved for the (unknown) function $I(x)$. The solution can be found in the introductory sections of books on differential equations, such as [160, page 26]. It is

$$I(x) = I_0 e^{-\mu x}. \tag{1.3}$$

When the density μ varies as the X-rays traverse a thick body, the total intensity will be the product of successive infinitesimal attenuations, and in this case, in the exponent there will be an integral over the varying attenuation along the ray. The product of exponentials becomes an exponential of a sum that becomes an integral in the limit of an infinite collection of infinitesimal intervals. If the total thickness is w, the limits of the integral will be 0 and w.

$$I(y, z) = I_0 e^{-\int_0^w \mu(x,y,z)dx}. \tag{1.4}$$

This result can be obtained by modifying Eq. (1.2) to have μ as a function of x. The equation then becomes:

$$\frac{dI(x)}{dx} = I'(x) = -\mu(x)I(x). \tag{1.5}$$

The equation can be separated, rewriting in the form

$$\frac{dI}{I} = -\mu(x)dx, \tag{1.6}$$

and by integrating both sides, you should get

$$\ln I - \ln I_0 = -\int_0^w \mu(x, y, z)dx. \tag{1.7}$$

In Eq. (1.2) the integral on the right simply gave μx since μ was a constant. Here the integral remains in explicit form. The term $\ln I_0$ is the constant of integration, determined by the condition that with no attenuation, the intensity is the original intensity, I_0.

Scattering and absorption of photons (X-rays) happens by three well-understood processes, the Photoelectric Effect, Compton Scattering, and Pair Production. The attenuation coefficients are different for each process, and they vary with the energy of the incoming photons as well as ρ, the density of the tissue.

- In the Photoelectric Effect a photon is absorbed by an atom and an electron is emitted. The attenuation depends on both the atomic number, Z (the number of protons in the nucleus), characterizing which element it is, and the energy, E, of the photon.

$$\mu_p \propto \rho Z^3 E^{-3}. \tag{1.8}$$

- In Compton Scattering a photon collides with an orbital electron and is deflected. The attenuation coefficient depends on the atomic mass, A (the total number of nucleons in the nucleus, protons plus neutrons), as well as Z and E.

$$\mu_c \propto \rho \left(\frac{Z}{A} \right) E^{-1}. \tag{1.9}$$

- In Pair Production a photon is absorbed in a collision with a nucleus or other heavy charged particle and an electron-positron pair is produced.

Here are some specific numbers indicating which processes are most significant for different energies of X-rays used in diagnostic radiology:

- X-ray tube voltages are in the range of 70,000–160,000 volts, producing photons of 70 KeV to 160 KeV.
- The cross section (and thus attenuation coefficient) of each photon interaction depends on the photon energy.
- Pair production only occurs above 1.022 MeV so is irrelevant to diagnostic X-ray imaging.
- At 70 KeV the probability of Photoelectric absorption is much greater than that of Compton scattering, and is highly dependent on the kind of atom (actually atomic number)
- At 160 KeV Compton scattering predominates and is not dependent on composition, only tissue density (actually electron density, but that is more or less proportional to tissue density, and not as dependent on the chemical composition as the Photoelectric effect).

Film images are produced by the photons causing a reduction action in the silver compound (s) in the film, leaving metallic silver where the photon struck the film. The silver density at a spot increases with the number of photons that hit the film at that spot. Digital images are produced by the photons being absorbed in a scintillation crystal that gives off light flashes when photons hit it. The light flashes are detected by photomultiplier tubes, so in essence, the photons are counted as they hit the scintillators.

Digital images consist of two kinds of information. The most apparent is the image itself, consisting of an array of numbers, which can be integer or decimal. Each number represents a picture element, or *pixel*. At that spot in the image, the display may show a monochrome brightness corresponding to the value of the number, or a color with brightness, hue, and saturation qualities (or equivalently, a mixture of red, green, and blue intensities). The size of the array may vary for a given modality. For cross sectional images, such as Computed Tomography (CT) or Magnetic Resonance (MR) images an array of 512 rows by 512 columns is common, while chest X-ray images can be 2000 by 2000, or more.[29] The array may not be square; computed radiographs that correspond to chest films are typically portrait style, with more pixels in the vertical dimension than in the width, which corresponds to the typical viewport shape of an X-ray film.

The second kind of information associated with a digital image is the descriptive data that specifies what kind of image it is, how it was obtained, who the subject (patient) is, where in some patient or machine-centric coordinate system the image is located, its orientation, and many other possibly useful items. These are encoded in many different ways. Manufacturers of digital imaging systems used magnetic tape or removable disk media to archive images. The format and representation of the data in the files on these media was (and still is) proprietary and manufacturer (and model) dependent. Typically the descriptive data come first in sequence (the image "header") followed by some binary representation of the image pixels. People who wished to use the images in systems other than the original would need to write custom programs to decipher these formats. By the early 1990s the radiology community recognized the need for some more standardized ways to transmit and receive data, resulting eventually in a network protocol, DICOM, which is discussed in detail in Chapter 4, Section 4.3.3.

Further, images may be aggregated into sets, and the set of images may have its own attributes, which apply in common to all the images. One very important concept is that of a position related set, where all the images share the same coordinate system, and are oriented and located in that coordinate system, relative to each other. This allows the possibility that the set of two-dimensional images can be reformatted to provide two-dimensional images in other planes through the volume that the set represents. The images can be coalesced or interpolated to form a regular three-dimensional array of image data. Solid models or surfaces that represent the three-dimensional structures seen in cross section in the images can then be computed from it, as well as projections to simulate conventional X-ray images from arbitrary directions without further radiation exposure to the patient.

The advent of X-ray transaxial computed tomography (CT) scanners in the 1970s marked the beginning of the digital era of medical radiology. The first scanners produced transverse cross sectional images of heads, an example of which is given in Figure 1.19.

These images are produced by projecting a very thin "fan" X-ray beam through a patient's head, from each of many angles. The transmitted X-ray intensities are measured by photon (X-ray) detectors on the opposite side of the patient from the X-ray tube. Each intensity is related approximately to the integral of the X-ray linear attenuation coefficients of the tissues along the line from the X-ray source to the detector. The attenuation coefficients are in turn related to the tissue densities, since the attenuation of the X-rays is mostly due to Compton scattering, which depends only on the electron density, not on the composition of the tissue.

---

[29]A common aphorism is "A picture is worth a thousand words," but considering an average word length of 10 characters, a thousand words is a mere 10 K bytes, while typical CT or MR images are 500 K bytes or more. It is less clear that the simple data size is a good indicator of how much information is there.

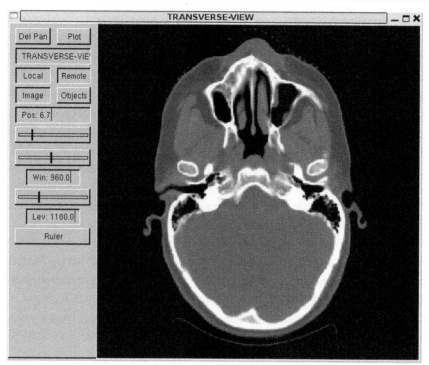

Figure 1.19   A CT image of a person's head, showing the skull, cerebellum, sinus cavities and nose.

The image data, or equivalently, the tissue densities, can be reconstructed from the projected data, using a method called *filtered back-projection*. This method relies on a property of the Fourier Transform called the projection-slice theorem, which relates the one-dimensional Fourier Transform of the line integral data mentioned above to the two-dimensional Fourier Transform of the original function. So, if you can measure the projections in all directions, you can recover the original function. For a detailed discussion of digital medical imaging, see [154, 343]. The following is a brief introduction to the mathematics of cross sectional image generation.

The essential problem in visualizing anatomic or other spatial information using projection X-rays or similar systems is that the images (or digital data collected) represent line integrals through the volume of the patient's anatomy. Thus, objects obscure other objects and it is difficult, if not impossible to determine exactly where in space things are. If we could somehow untangle these projections to get full three-dimensional data, it would be much more useful than just the projection images, so much more so that when it was finally accomplished it merited a Nobel Prize [77, 163].

Fast, small computers and digital detectors made it practical to consider two powerful mathematical ideas to accomplish the generation of cross sectional images from projections: the Radon transform and the Projection-slice theorem. The mathematical problem is: how to recover a two-dimensional function from its one-dimensional line integrals? In formula 1.4 the intensity is the exponential of a line integral, but from this point on it will be sufficient and simpler to just consider log I, not I.

The intuitive solution to the problem is the same one that might be used for many perceptually similar problems, such as seeing what is behind a screen, or in a crowd. You look in from many different angles, and correlate these different views to determine what produced them.

These ideas were first used for radio astronomy by Bracewell and others to map the Sun [42–45,319]. Later, Cormack, Hounsfield, and others showed how to use them for medical X-ray images [76–78,163].

Every sufficiently smooth function can be considered as made up of sine (or cosine) waves with varying weights. The weights as a function of the wave frequency define a function called the Fourier transform.

- If f is a function of x, the Fourier transform of f is

$$\hat{f}(\omega) = \int_{-\infty}^{\infty} f(x)e^{-ix\omega}dx. \tag{1.10}$$

- The two-dimensional version is just a double integral with x and y, giving a function of ω_x and ω_y

$$\hat{f}(\omega_x, \omega_y) = \int_{-\infty}^{\infty}\int_{-\infty}^{\infty} f(x, y)e^{-ix\omega_x}e^{-iy\omega_y}dxdy. \tag{1.11}$$

- The inverse transform looks similar, except the sign in the exponent is changed.

The Radon Transform of a function of two variables is the integral of that function along a line in the two-dimensional plane. It is a projection, something like the X-ray formula (see Figures 1.20 and 1.21).

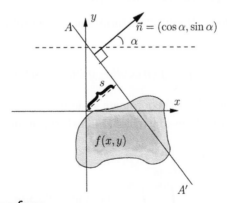

Figure 1.20 The Radon Transform
(from Wikimedia Commons, URL `http://commons.wikimedia.org/wiki/File:Radon_transform.png`*).*

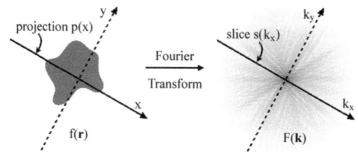

Figure 1.21 The Projection Slice Theorem
(from Wikimedia Commons, URL `http://commons.wikimedia.org/wiki/File:ProjectionSlice.` `png`*).*

For simplicity just project onto the x axis, giving

$$p(x) = \int_{-\infty}^{\infty} f(x, y)dy. \tag{1.12}$$

The slice through the Fourier transform of f corresponding to this is the function of ω_x obtained by setting $\omega_y = 0$.

$$s(\omega_x) = \hat{f}(\omega_x, 0) = \int_{-\infty}^{\infty} \int_{-\infty}^{\infty} f(x, y)e^{-ix\omega_x}dxdy. \tag{1.13}$$

This is just the one-dimensional Fourier transform of the projection.

$$\hat{p} = \int_{-\infty}^{\infty} p(x)e^{-ix\omega_x}dx. \tag{1.14}$$

To get the entire $\hat{f}$, just rotate the projection, which also rotates the Fourier space coordinates, and then transform back.

Here is a more complete description of the steps:

1. Collect all the projections (Radon transforms) at all the different angles (only 180° are needed),

$$R[f(x, y)](p, \theta) = \int_{-\infty}^{\infty} f(p\cos\theta - q\sin\theta, p\sin\theta + q\cos\theta)dq, \tag{1.15}$$

    where p, q are the rotated (linear) coordinate system,

2. Fourier transform each one (one-dimensional transform) in the corresponding p direction,

$$F[R](\omega, \theta) = \int_{-\infty}^{\infty} R(p, \theta)e^{-i\omega p}dp, \tag{1.16}$$

$$= \int_{-\infty}^{\infty} \int_{-\infty}^{\infty} f(x, y)e^{-i\omega p}dpdq, \tag{1.17}$$

$$= \int_{-\infty}^{\infty} \int_{-\infty}^{\infty} f(x, y)e^{-i\omega(x\cos\theta + y\sin\theta)}dxdy. \tag{1.18}$$

3. The collection of 1-D Fourier transforms is (by inspection) the two-dimensional transform with respect to new variables, $u = \omega \cos \theta$ and $v = \omega \sin \theta$,

$$F(u, v) = \int_{-\infty}^{\infty} \int_{-\infty}^{\infty} f(x, y)e^{-i(xu+yv)}dxdy, \tag{1.19}$$

so, Fourier transform back the entire collection (two-dimensional transform) in terms of u, v,

$$f_{recon}(x, y) = \int_{-\infty}^{\infty} \int_{-\infty}^{\infty} F(u, v)e^{i(xu+yv)}dudv. \tag{1.20}$$

An alternative but equivalent formulation is possible, which made the image computation possible to do in parallel, which in the early CT scanners was very important. Instead of first collecting all the data and computing all the Fourier transforms, each projection can be "back-projected" as it is obtained. To do this, rewrite in terms of integration over ω and θ. This change of variables gives a factor of $|\omega|$,

$$f_{recon}(r, \theta) = \int_{0}^{\pi} \int_{-\infty}^{\infty} |\omega| F(\omega, \theta)e^{-i\omega p}d\omega d\theta. \tag{1.21}$$

The integrand is the (inverse) Fourier transform of $|\omega| F(\omega, \theta)$ which can be rewritten as the convolution of the individual (inverse) Fourier transforms, as follows,

$$f_{recon} = \int_{0}^{\pi} F^{-1}[|\omega|] \otimes R(p, \theta)d\theta. \tag{1.22}$$

The convolution of the projection at each angle can be computed while the machine collects the projection data for the next angle. The function $F^{-1}[|\omega|]$ is called the *filter* function, and this version is called *filtered back-projection*. It is also referred to as *convolution back-projection*.

The above formulas are mathematically exactly correct for the ideal simplified case with an infinite number of infinitesimal angle steps, the continuous case. However, many different filter functions have been tried in addition to the basic one, to correct for the reality of finite, limited data sets, and other problems.

Yet another method that was exploited is to treat the problem as an algebraic inverse problem. The projection set is produced by applying an unknown transformation (which depends on the densities of matter in the body being scanned) to a known X-ray beam, to produce the measured projections. So the problem is to find this transformation function by starting with a guess and then making corrections, iterating to get a "good enough" image. Here are the steps:

1. Start with a uniform image.
2. Compute its projections, and the difference between them and the measured projections.
3. Adjust the image points in proportion to the difference.
4. Iterate until the estimated error (the difference between the actual and computed projections) is low enough to accept.

The equations for attenuation, projection, and reconstruction assume that the X-ray beam consists of photons of a single energy. If the beam is not monoenergetic, the transmitted intensity is obtained by integrating over energy.

$$I(y) = \int_0^{E_{max}} I_0(E) e^{-\int_0^w \mu(x,y,E)dx} dE. \tag{1.23}$$

Then $\log I$ is no longer the simple line integral or Radon transform of a two-dimensional image. Applying the basic reconstruction methods produces artifacts such as streaking. Correcting for this effect is a complex engineering problem, but has largely been solved in modern CT scanners. However, the presence of material that is very different from bone or soft tissue, such as metal implants or artificial joints, or surgical clips, can still produce such artifacts. Radiologists are trained to recognize these (streaks) as distinct from actual tissue or other material.

Figure 1.22 shows a CT scanner with cover removed to show the X-ray tube, the detectors, and the ring on which both rotate synchronously.

As noted earlier, the first person to demonstrate the effectiveness of this idea as a practical medical device was Godfrey Hounsfield [163]. This in turn was based on theoretical work and experiments done by Alan Cormack much earlier [76,78]. Hounsfield and Cormack were awarded the Nobel Prize in Medicine in 1979 for this work. Hounsfield's research and development was supported by EMI, the same company that produced and marketed the first commercial

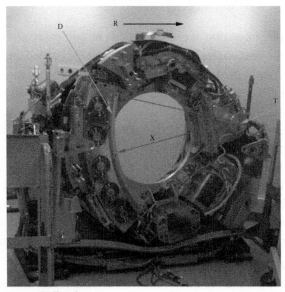

Figure 1.22 CT Image Data Collection
(from Wikimedia Commons, URL `http://commons.wikimedia.org/wiki/File:Ct-internals.jpg`*).*

CT scanner. This company was also the publisher of the music of The Beatles. It has been noted[30] that the CT scanner is the Beatles' greatest legacy, since Hounsfield's work was funded by EMI's profits from the Beatles' phenomenal record sales.

The images are two-dimensional arrays of numbers representing the image intensity; the spots in the image that the numbers represent are called *pixels* (for "picture elements"). In a typical CT image, each pixel may represent a 1–2 mm square cross sectional region. The image pixel numbers represent average tissue densities over some thickness (in the direction perpendicular to the image), roughly the width of the fan-shaped X-ray beam used to produce the data from which the image is computed. This width is called the *slice thickness*. When the slice thickness is large, it will be difficult to detect small spots, as they will be averaged with the surrounding tissue and not stand out so much in contrast. As the ability to produce images with thinner and thinner X-ray beams developed, it became possible to get high resolution in the direction perpendicular to the image plane, and this made it possible to construct images in orthogonal planes from the transverse images by selecting pixels from a series of transverse images as if they were stacked into a three-dimensional array, and reorganizing them for display. Figure 1.23

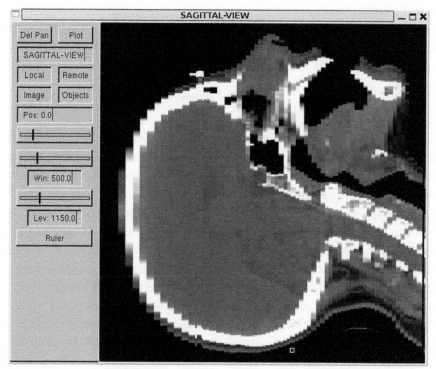

Figure 1.23   A sagittal cross section through the midline of the same person, constructed from a series of 80 transverse images.

---

[30]In a presentation by Dr. Ben Timmis, Consultant Radiologist at the Whittington Hospital, in the UK; see URL http://www.whittington.nhs.uk/default.asp?c=2804.

shows a sagittal cross sectional image constructed in this way.[31] Later in this section we will develop the explicit code for this.

In order to properly display the image data, a computer program needs substantial information about the image pixel array and the image it represents. This includes such things as the dimensions of the array (how many pixels are in a row, how many rows are in the image), and what the image pixel numbers represent (what numbers should correspond to maximum brightness, where should the midrange be, etc.). To reorganize the transverse image data into sagittal images like the one shown, it is also necessary to know the table index of each image, so that the pixels from that image can be placed properly along the perpendicular axis. In the image shown, the successive transverse images are about 5 mm apart, but the pixels represent spots about 1.5 mm in size, so the pixels are replicated to form strips, representing the thickness of the X-ray beam (which usually but does not always correspond to the inter-image spacing). Other information that is important of course is the name of the patient and other identifying information, as well as the reason for getting the images. Different techniques will be used (such as different X-ray tube voltages, and use or omission of intravenous contrast) depending on whether the problem is to look for signs of pneumonia, or small lesions that might be tumors, or evidence of a broken bone.

The numbers computed for each pixel are normalized to a scale known as "Hounsfield Units," after Godfrey Hounsfield. The Hounsfield scale is defined so that the Hounsfield number of water at standard temperature and pressure is 0, and the Hounsfield number of air is -1000. The formula is:

$$H(x) = \frac{\mu(x) - \mu_{\text{water}}}{\mu_{\text{water}} - \mu_{\text{air}}} \times 1000. \tag{1.24}$$

Water and soft tissue will then have values near 0, typically within 100 units or so. Typical bone values range from 400 to 2000 HU. Although most articles and texts refer to μ as a "linear attenuation coefficient," the CT reconstruction does not exactly produce such numbers, despite the theory in the preceding equations.

In a typical image, the array of numbers has a wide dynamic range, with much more granularity than can be discerned by the human eye as gray intensities. Image display programs include facilities for the user to control how the image data are converted, or *mapped* into gray levels displayed on the screen. The two most common controls are called *window* and *level*. The level setting is the pixel number that will be displayed as a half intensity gray spot. The window specifies the range of pixel values that are mapped from black to full white.

The range of Hounsfield Units from air to hard bone is over 2,000 units (bone is around 1,000 H). This is too many gray levels to be useful. Typical gray-scale displays offer only 256 gray levels. Thus, a subset of the full range is usually selected to map between 0 and 255 in display intensities. The *window* refers to the width of this subset and the *level* refers to its center.

$$H_l = L - W/2, \tag{1.25}$$

---

[31]This image set is rather old and does not represent the kind of resolution achievable today.

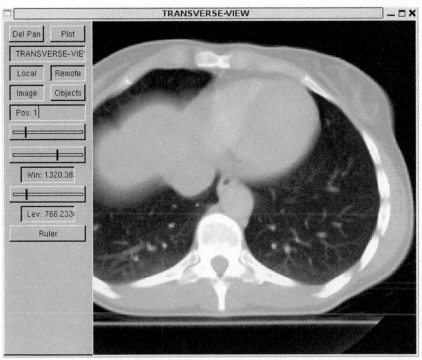

Figure 1.24 A lung cross section with window and level set to show some details of lung tissue. With the right settings, and careful comparison between adjacent images, it is possible to detect very small lesions as distinct from bronchial trees and blood vessels.

$$H_u = L + W/2, \tag{1.26}$$

$$G(H) = \begin{cases} 0 & \text{if } H \leqslant H_l, \\ (H - H_l)/W \times 255 & \text{if } H_l < H < H_u, \\ 255 & \text{if } H \geqslant H_u. \end{cases} \tag{1.27}$$

Although the Hounsfield scale ranges from -1000 upwards it is more usual for machines to shift this scale by 1000 so that only non-negative integers are used. For example, a window of 400 and a level of 1100 will show all pixels below 900 as black, all above 1300 as white, and everything between will be displayed proportionally at intermediate gray levels. To see details in the lungs, the lower range, near 0, should be in the gray range, as in Figure 1.24, while to visualize bone details, the level must be set somewhat higher, since bone pixel values are more in the range from 1500 on up.

One use of cross sectional images is to construct organ models and identify tumor locations for radiation therapy planning (see Chapter 2, Section 2.2.8 for more on this). Much work has been done on image processing algorithms to automate the process of finding organ boundaries, so that three-dimensional models can be constructed. Figure 1.25 shows a cross section of an abdomen, where the patient's left kidney can easily be identified (the view is from feet to head

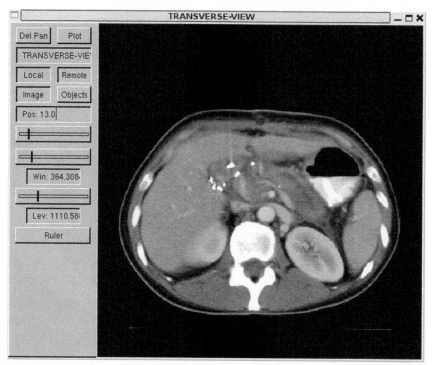

Figure 1.25 A cross section of a patient's abdomen, showing the left kidney pretty clearly (on your right) while the liver and right kidney appear fused as if they were a continuous entity. The vertebral column is plainly visible at the bottom center, as well as the stomach on the patient's left side, shown half full of contrast material. The small bright spots in the upper middle of the image are metal surgical clips that were put in during surgical removal of a large tumor.

and the patient is on his/her back, so the patient's left is to your right). However, the boundary between the right kidney and the liver is very blurred, as the two organs are very close in density at the points where they touch. The current state of practice in radiation therapy today is a combination of automatic and manual techniques, with much research remaining to be done.

Typically, the image pixel values are represented in files and in computer programs directly as binary numbers, not as text. This saves space, and is also important because image processing algorithms do arithmetic on these data. Later in this chapter, we will show how image data can be represented in files and programs and read in (and written out) efficiently.

Radiology departments today are almost completely filmless. The development of high-resolution X-ray detector arrays has made digital radiography a reality. In addition, many other digital imaging systems are part of the standard of practice, including ultrasound, magnetic resonance imaging (MRI), and various systems using radioisotopes, such as positron emission tomography (PET). Although it is not possible to give even a short survey here, we will delve a little into the question of representation and manipulation of image data in computer programs, using CT images as an example.

The image pixel values for a CT image are typically small integers, in the range from 0 to 4095, or sometimes larger, up to 65,535 (thus they can be represented in 12-bit or 16-bit binary words). A typical image consists of a 512 row by 512 column two-dimensional array of such numbers. A simple way to encode these is as nested lists of lists, or a list of numbers for each row of pixels and a list of such lists for all the rows of an image. Columns correspond to the individual elements of the row lists.

```
> (setq image-5 '((0 0 5 6 0 3 678 737 929 ...)
         (0 0 0 4 3 2 734 456 ...) ...))
((0 0 5 6 0 3 678 737 929 ...)
 (0 0 0 4 3 2 734 456 ...) ...)
```

One could use the `elt` function described earlier to access any pixel. To get the pixel value at column 233, row 256, we would use `elt` twice, in nested expressions.

```
> (elt (elt image-5 256) 233)
1037
```

We could even create abstraction layers, by writing named functions and hiding the fact that these arrays are implemented with nested lists. However, access to arbitrary elements of lists is very slow. In effect, in order to get to the 233rd element of a list, you have to look up the `rest` part of each cons cell, following the chain until the right place. Since computer memories are actually arrays, and organized so that each location is directly addressable, it would make sense to have a more efficient way to represent arrays. Of course there is, and Common Lisp has a built-in data type, the `array`, which can be constructed to take advantage of the computer's direct addressing, and give very high speed of access. Most programming languages have this kind of facility, since many other kinds of data besides images also take this form.

An array is constructed by the function `make-array`, which just requires that you specify the size (dimensions) of the array. To create an array which we can fill with image pixel values, we would specify two dimensions, each of size 512.

```
> (setq image-5 (make-array '(512 512)))
#2A((NIL NIL NIL ...)
    (NIL NIL NIL ...)
    ...)
```

Since we did not specify the initial values for the array elements, they are all set to `nil`. To look up a value in an array, the function `aref` is used. It needs the array and the index values, as many as there are dimensions of the array. So, to do the same lookup as before, we have

```
> (aref image-5 233 256)
NIL
```

Of course this time we get `nil` because we have not set any values. To set a value in the array, you combine `aref` with the `setf` macro. Although we have mainly used `setf` to attach values to symbols, it is really a general-purpose assignment operator. Its inputs are: an expression naming a place that can have a value, and a value to put there. So, to set the pixel (233, 256) to the value 1037, we evaluate the following expression:

```
> (setf (aref image-5 233 256) 1037)
1037
```

Now that value is stored in that place in the array. Now, we consider the construction of the sagittal image in Figure 1.23. Suppose we have a list of transverse images, that is the value of a variable, image-list. We want to pick out the 256th column of each transverse image and make it a column in a new image array. We'll need to iterate over the list, with dolist, and we'll need to iterate over the rows in each image, picking out the 256th pixel from every row and making it a column value in the sagittal image. The dotimes function is similar to dolist. It takes a local variable, the index variable, starts it at 0 and evaluates the body of the dotimes expression, then increments the index by 1 each time through, until it reaches the end or stop value. We also need to keep a count of how many columns in we are, which is the same in this simple example as how many images we have processed.

```
(defun make-sagittal-image (image-list)
  (let ((new-img (make-array '(512 512)))
        (image-num 0))
    (dolist (image image-list)
      (dotimes (i 512)
        (setf (aref new-img i image-num)
          (aref image i 256)))
      (incf image-num))
    new-img))
```

The order of indexing in aref for two-dimensional arrays can be thought of as row index first, then column index. It really does not matter which you choose, as long as you are consistent throughout your program.[32] The code above assumes that each image in image-list is only one pixel thick, otherwise the proportions of the sagittal image will not be square. It is left as an exercise to solve the problem of thick transverse images. In this case, one needs an additional parameter, the thickness, or spacing (in pixels) of the transverse images. You would use this parameter to determine how many identical columns of pixels must correspond to a single column from the transverse image set. For example, if the pixels were 1 mm square, and the slice thickness 5 mm, each column computed must have four adjacent copies, for a total of five, to make the image proportions right. Real image data sets include the spatial sizes in actual patient space, in mm or cm, for all the data, so you must also be able to keep track of conversion between pixel numbers and positions in real space.

---

[32]In terms of efficiency, when optimizing the code for high performance, one would actually consider which index will be varying most, if you are iterating over both indices. Optimization is a complex subject and is discussed in more detail in Section A.1.2 of Appendix A.

Arrays are the right way to represent the mapping of the pixel numbers to gray levels, too, as mentioned earlier. In a typical computer display system, there are no more than 256 gray levels, though we indicated earlier that the image pixels can range over much larger values. The window and level mapping described earlier is also a relatively straightforward operation on an image array. We take a single image array as input, along with a window and a level, and create a new array with values in the reduced range of 0 to 255. The way we will do this is to create a short mapping array of a specified number of elements (called range-top here, either 4095 or 65535, or perhaps some other upper limit), and each entry in this array will be the gray value for that pixel value. Then to get the gray value for a given pixel value we just look it up in the map array.[33]

```
(defun make-graymap (window level range-top)
  (let* ((map (make-array (1+ range-top)))
         (low-end (- level (truncate (/ window 2))))
         (high-end (+ low-end window)))
    (do ((i 0 (1+ i)))
        ((= i low-end))
      (setf (aref map i) 0)) ;; black
    (do ((i low-end (1+ i)))
        ((= i high-end))
      (setf (aref map i)
        (round (/ (* 255 (- i low-end)) window))))
    (do ((i high-end (1+ i)))
        ((> i range-top))
      (setf (aref map i) 255))
    map))
```

The low-end variable is the value below which all pixels will be displayed as black, and the high-end variable is the value above which all pixels will be white. So, up to the low-end location in the map array, the values will all be 0, from low-end to high-end they will be proportional, up to 255, and from high-end to the end of the map array, the values will all be 255.

This code uses a more general iteration, the do macro, because we are not always starting from 0, so dotimes is too simple. The do macro creates any number of local variables (here we are just using one, i), gives them initial values, and then assigns them new values each time through the iteration. The execution terminates when the test expression is satisfied, which is in a list following the local variable forms.

Now that we can make a map array for any value of window and level, we can use the map to convert an image to a mapped image. This is simple. Just iterate through every pixel, making the new image pixel value be the entry in the map array whose *index* is the value of the original

---

[33]Note that let* is like let except that it initializes the variables *sequentially* instead of in parallel, so that you can refer to a variable in the initialization expression of a variable that comes after it. In this case, we are initializing high-end by using the initial value of low-end.

pixel. So, if the value in the image pixel is 1246, the new image will have the value in the 1246th location of map, which will be a number between 0 and 255.

```
(defun map-image (raw-image window level range)
  (let* ((x-dim (array-dimension raw-image 1))
         (y-dim (array-dimension raw-image 0))
         (new-image (make-array (list y-dim x-dim)))
         (map (make-graymap window level range)))
    (dotimes (i y-dim)
      (dotimes (j x-dim)
        (setf (aref new-image i j)
          (aref map (aref raw-image i j)))))
    new-image))
```

We are using the number that aref looks up in raw-image as an *index* for the aref that looks up a value in map. It's simple and can be made to run very fast.[34] There is one new generalization in the above code. In make-sagittal-image we explicitly did arrays of dimension 512 by 512 (and computed the sagittal image through the midline of the transverse images), but here we arranged to handle arrays of any dimension. The array-dimension function will look up the dimensions of an array, so we can write code that handles arbitrary sizes. The dimensions are indexed starting at 0, up to a maximum limit depending on your Lisp system, but in no case less than seven, so you can be sure to have support for up to seven-dimensional arrays, and each dimension can have a large number of elements, also implementation dependent.

Later in this chapter we will show how to store image data in files and retrieve it as part of a very general method for serializing medical and other kinds of data. In Chapter 2, Section 2.2.8 there is a little more discussion of the uses of image data in radiation therapy planning. Section 4.3.3 in Chapter 4 describes how to send and receive image data over a network connection using the DICOM protocol. Implementation details for a DICOM system using a state machine are provided in Section 5.4.4 of Chapter 5.

Many other kinds of digital images are commonly produced and used in medicine, for diagnosis and treatment planning. One such modality, Positron Emission Tomography (PET), uses essentially the same mathematical ideas as CT, to reconstruct cross sectional images from line integrals. In this process, the patient receives an injection of a radioactive isotope that decays by emitting low energy positrons. The positrons interact with electrons in the immediate vicinity, destroying both the electron and the positron, producing two photons of equal energy, emitted in opposite directions. The patient's head or body is surrounded by photon detectors. Detecting and counting the photons emitted along a given line through the scanned volume gives a measure

---

[34] Actually, the code here will not be fast unless some optimizations are included. However, once you have written code you are sure is correct, it is easy to make it run fast. It is much harder to do this in reverse, to write very fast but broken code, and expect to fix it later to run correctly. In Section A.1.2 of Appendix A we show how to optimize this code as well as use optimization techniques more generally.

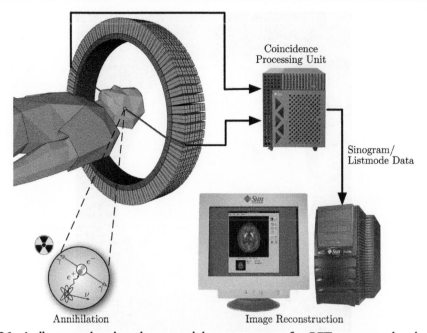

Figure 1.26 A diagram showing the essential components of a PET scanner, showing also the positron annihilation process that produces two photons in opposite directions.
(Courtesy of Jens Langner, from Wikimedia Commons, URL `http://commons.wikimedia.org/wiki/File:PET-schema.png`*).*

of the integral of the positron emitting isotope along that line. Once a complete set of such line integrals is obtained, the same procedure (filtered back-projection) can be used to compute the cross sectional density of the isotope as a function of position. Figure 1.26 shows a schematic of a typical PET imaging system. In order to reduce the effect of noise, photons are only counted if both photons register simultaneously in the corresponding opposite detectors along the same line (coincidence counting).

Usually the computed pixel values are displayed as different colors rather than gray intensities. The red and yellow areas represent more uptake of the radioisotope, while the green and blue areas represent less. There are a few differences in PET imaging vs. CT. The amount of radioisotope that can be injected safely is small, so the data acquisition is much slower, to get enough counts for a useful image. Typically a PET scan can take 30–60 min, while X-ray CT scans are done in seconds. The resolution is much lower for a PET image, and the numeric values have less precision. However, an additional difference makes these images extremely useful. The image pixel values are related to the amount of uptake of the radioisotope labeled compound, not the tissue density. Typically this is a compound related to glucose (Fluoro-deoxyglucose, FDG, is commonly used). Thus, a PET image gives a measure of *metabolic activity*. Abnormal tissue such as a malignant tumor or a focus for epilectic seizures may have a density almost the

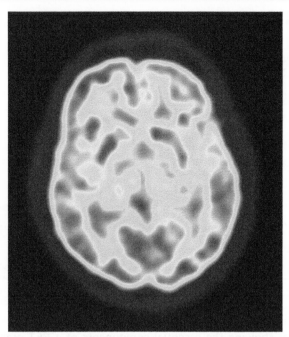

Figure 1.27   A cross sectional image of a person's head, generated using Fluorine-18 labeled FDG. The data for this image took 20 min to obtain. *(Courtesy of Jens Langner, from Wikimedia Commons, URL* `http://commons.wikimedia.org/wiki/File:PET-image.jpg`*)*.

same as normal tissue but very different activity, so may show up better on a PET image than on CT. Figure 1.27 is a typical PET image of a person's head.

Magnetic resonance imaging (MRI) uses a large magnet and a radio frequency signal generator to produce cross sectional images in a somewhat similar fashion. This process uses the fact that some atomic nuclei will absorb and re-emit radio signals at frequencies that are characteristic for those nuclei. The resonant frequency also depends on the strength of the magnetic field. So, by having a magnetic field that varies in one direction, a given frequency of radio signal will represent the sum of all the responses from atoms along a line in the orthogonal direction. By varying the frequency and rotating the field gradient direction, one can obtain a set of line integrals and again reconstruct a cross sectional image from them. Magnetic resonance images largely show anatomic structure, as do CT images, but the image will show some dependence on the chemical composition of the tissue as well, since different nuclei have different resonant frequencies. Thus MRI gives yet another kind of visual information about the patient's anatomy. For some purposes (particularly planning for radiation treatment for cancer) it is important to have two different kinds of images, usually PET and CT, or MRI and CT, to get a more precise definition of the spatial extent of disease than could be done with only one kind of image set.

Finally, it should be noted that in many areas of medicine besides radiology, images are being digitized, stored, transmitted, and displayed. This facilitates remote diagnosis (telemedicine) and makes expertise available in places where it would otherwise not be available. Some specialties in which this is becoming common are pathology, dermatology, and ophthalmology.

An example of how this can be surprisingly cost-effective is an experience of a patient that was taking a chemotherapy agent for cancer, who needed to consult his oncologist while he (the patient) was traveling several thousand miles from home. The medication can cause blistering of fingertips and toes. The patient took pictures of his fingers with the camera on his mobile phone, and sent them to the oncologist by electronic mail. The oncologist responded that the blisters looked quite typical and not to be concerned unless they became very uncomfortable. With this information the patient was comfortable with continuing the medication to the end of the treatment cycle. The whole process took only a few minutes of the patient's time and only a few minutes of the oncologist's time. The patient got the information needed very quickly.

1.1.6 Metadata

In the previous sections we showed several ways to organize biomedical data, including the use of *tags* to label items and the use of structure to organize data. Although the tag representation seems appealing, and is easy for a person to read and understand, the tags have no particular meaning to a computer program. Data are meaningful only when accompanied by some interpretation, that defines what the symbols mean. While this may seem obvious, it is easy to overlook in the rush to create repositories of "self-describing data." For example, what does the following list of numbers mean?

```
5984107
8278719
2214646
8220013
5433362
4274566
. . .
```

Are they telephone numbers? Perhaps they are patient identification numbers assigned by a hospital. They might also be identification numbers assigned to terms in a terminology system such as the Gene Ontology (see Chapter 2, Section 2.3.7) or the Foundational Model of Anatomy (see Chapter 2, Section 2.3.4). They could equally well be a series of pixel values in a medical image, a microscope image, a packed, encoded representation of expression array data, or a very compact encoding of laboratory test data. Some conventions are needed as well as labeling to clearly identify the meaning.

1.1.6.1 Tags As Metadata

The description, interpretation, or identification of the meaning of a particular set of data can take many forms. The data set above may be accompanied by a written document specifying the meaning and format of the data. Another approach is to include with the data some identifying tags that label the elements, so-called "self-describing" data. Here is how the numbers above might appear using such a scheme:

```
<phone numbers>
   <work>5984107</work>
   <home>8278719</home>
</phone numbers>
<phone numbers>
   <work>2214646</work>
   <home>8220013</home>
</phone numbers>
<phone numbers>
   <work>5433362</work>
</phone numbers>
<phone numbers>
   <cell>4274566</cell>
</phone numbers>
   . . .
```

Now we (humans, not computers) can see that indeed they are telephone numbers, and they are organized so that numbers belonging to a single person or entity are grouped together. Of course there must be some agreement as to the meaning of the tags, which is still outside the context of the data and tags. The words "phone numbers" and the other tags have no intrinsic meaning to a computer program.

This idea of grouping items and providing tags is such a powerful way to organize information that an internationally popular (and rather more complicated) syntax called XML was invented to support the idea. We will deal with it later in the chapter, in Section 1.2.6. The XML standard provides two methods for specifying the structure of the data labeled by a tag, XML Data Type Definitions and XML Schemas. One might then say, "We just include schemas or data type definitions that define the meaning of the tags." Such schemas do *not* define the meaning of the tags. These only define the allowed or expected *syntax* or structure of the data. A computer program can use a schema or other such "metadata" to check if a document or data structure is correctly constructed but it cannot discern what to do with the data.

In any case, there must be some external document describing the agreement on the meaning of the elements of the data type description language. And so it goes. There is always some boundary beyond which a separate documented agreement about meaning must exist. The ongoing debate about self-describing data is not about whether data should be self-describing (which is impossible), but about where this boundary will be, and what description languages will be created for use inside the boundary (and documented outside the boundary).[35]

---

[35]This is reminiscent of the problem of infinite regress in describing what is holding up the (flat) Earth. One version is in Hawking [150] as follows: 'A well-known scientist (some say it was Bertrand Russell) once gave a public lecture

In order for a computer program to process the data, the program itself must recognize the tags, so the meaning of each tag (at some level) is encoded in the program. Using human readable tags can mislead the person unfamiliar with how computer programs work. A computer program does not understand the meaning of the phrase "phone numbers." It can be written to take some particular action when it encounters that text string as a piece of data, but the way it recognizes the tag is by comparing with some other information containing the tag. So, a computer program is also in a sense a data description. The designer(s) of such systems and data repositories will have to choose how this matter is addressed.

Computer programs are notoriously intolerant of misspellings or other variations. A slight change in the tag creates an entirely different distinct tag. Humans have the ability to handle variant words and phrases, but in general computer programs do not. If the tag reads "phone number" (with the "s" left out), a program written to process the tags used above would likely break. A program can be written to handle a range of such variations, but that too must be completely and precisely specified, and the resulting more sophisticated program will also have brittle boundaries.[36]

1.1.6.2 Source Code As Metadata

The data that describe the tags, that is, how their associated data elements should be used, what are the allowable value types, etc., are called "metadata," or data about data. Similarly, if a record structure is defined without tags, in which the data follow some fixed conventions for format and encoding, as in DNA and protein sequence data, there may still be metadata. In a relational database, the kind of data stored in each field of a record is specified by a statement in the standard relational database language, SQL.[37] This SQL statement also is a piece of metadata. The distinction is important in designing systems. One can design a system that has a high level of generality by making the system use the metadata as a guide to how to handle the data. This idea makes a database system possible - you write a description of your data in SQL, and the database system interprets the SQL to facilitate entry, storage, and retrieval of your data, using the tags you specified.

---

on astronomy. He described how the earth orbits around the sun and how the sun, in turn, orbits around the center of a vast collection of stars called our galaxy. At the end of the lecture, a little old lady at the back of the room got up and said: "What you have told us is rubbish. The world is really a flat plate supported on the back of a giant tortoise." The scientist gave a superior smile before replying, "What is the tortoise standing on?" "You're very clever, young man, very clever," said the old lady. "But it's turtles all the way down!" In our case of the infinite regress of description of data, we may say, "It's turtles (metadata) all the way up!"

[36] An impressive early step in the direction of expanding the boundary and making programs able to be self-correcting is the DWIM facility for Interlisp, an earlier version of Lisp designed by Warren Teitelman when he was at the XEROX Palo Alto Research Center. The DWIM facility, when it encountered a name for which a function definition was needed but absent, would look through the symbols to see if there was one that might be the right one, so it could guess how to fix spelling errors, and some other problems. Some of DWIM's fixes were a bit arcane and the facility was sometimes described as "Do what Warren, not I, would Mean."

[37] Later in this chapter we introduce the Relational Model, relational databases, SQL, and related ideas.

The use of structured data types in ordinary programming languages is an even more generalized version of the same thing. However, in this case, we are referring to parts of the program itself as metadata, although normally we think of structure definitions as code. The distinction between code and data in a program is not always as one conventionally learns in introductory C or java programming. The code that defines structures in a program becomes metadata when the program can actually use it *explicitly* during execution. The possibility that a piece of code in a computer program can itself be metadata is very important, as we will see later in this chapter.

1.1.6.3 Data, not Variables, Have Types

In a conventional programming language, data items are stored in memory locations associated with variables. The variables are declared to have values of certain types, and the storage allocated for each variable's value is associated with that type, perhaps something like four bytes for integers, one byte for each character in a string, etc. When your program runs, it must have the right instructions to process these bytes, because the bytes themselves have no information about what type of data they represent. Sometimes, a program has errors in it or data get brought into a running program that are ill formed, and a mysterious malfunction occurs. For example, a floating point arithmetic instruction attempts to operate on a 4-byte sequence that is not a valid floating point number, but is a valid integer. The error message generated by the system will be quite obscure. The 4-bytes don't have information saying they are an integer instead, and a C program would not use such information anyway. In this example the program can signal an error because some binary numbers are not valid floating point numbers. In a worse scenario, the data are simply mistaken for the wrong type, just processed without notice, and a very inappropriate result is obtained, without any idea why the strange results happened. Manifestations could be things like garbled text appearing in a display of patient data, or a graph having strange anomalies, or colors coming out wrong.

Alternately, the internal binary representation of data in a running program could carry with each piece of data a (binary) tag, identifying its type. The running program could then use these tags to determine what to do with the data. It makes possible the idea of "run-time dispatch." This means that an operator can have many different implementations, one for each type of input it might receive. When it executes, it checks the type of the data and chooses the right method for that type of data. For functions with multiple inputs, there could be methods for all the combinations of different types for each of the inputs. Such functions are called *generic* functions, because their code is organized to perform a generic operation, with different details depending on what kind of data is input.

In a system like Lisp, where *data* have types associated with them, we can still have variables. Symbols can serve as variables because they can have values or data items associated with them. However, the value that a symbol is assigned can be *any* type. All the symbol does in this context is hold a pointer to the actual data item. Symbols can even have values that are other symbols, as we have seen, and of course in that case, the type of such a datum is symbol.

A system where the data themselves carry type is the basis for key ideas of object-oriented programming. The association of types with data is consistent with the idea of a type hierarchy, with general types, subtypes, sub-subtypes, and so on. The specialized types inherit the properties associated with their parent types. A generic function method applicable to a type will also be applicable to all its subtypes. You will note that we have not said anything about classes here. The ideas of object-oriented programming depend on having type hierarchies, and user-definable types that extend the type hierarchy. The role of the idea of classes is that classes are a way to implement user-defined *complex types* with component parts.

Finally, you might be wondering if this handling of type information in a running program is very inefficient. It is, so implementations have ways to cache information and generally optimize the operations to make the efficiency higher. In general, this is adequate, and the additional computation time is not going to be noticeable, but the savings in ease of writing the programs are often enormous. In those rare cases where performance is extremely important, a good object-oriented programming language provides type declaration capabilities to help the compiler do an even better job of generating efficient code. This is discussed in more detail in Section A.1.2 of Appendix A.

1.2 Objects, Metadata, and Serialization

The code we develop in this section will show how to store (in files) and represent (in a program) patient data that might be found in a typical electronic medical record system. The examples use text files for external storage, and tags (also called "keywords") to label the data items. The patient example developed here is simple and unconventional, but it illustrates some ideas that can help to manage complexity. Many electronic medical record (EMR) systems are built on relational database technology. The conventional wisdom is that a relational database is currently the most efficient, scalable architecture. However, the complexity of medical data suggests that the relational model may not be the best fit. The challenge is to be able to store and retrieve heterogeneous collections of data that have highly variable structure. Although the implementations we show here might be hard to scale up, they illustrate important ideas which can guide a more scalable design. Later in this chapter, we will also examine the relational model to see how it compares with these methods.

Three methods will be developed, to show how structured data may be encoded and manipulated. The first is a simple arrangement of tagged data, where the structure of a complex record is represented by hierarchical grouping of items in nested lists. The second method, using standard object-oriented programming ideas, shows how the *class definition* facilities of typical object-oriented programming languages provide the abstraction layers that had to be done explicitly in the first method. The third method, using *metaobjects*, shows how you can use the implementation of the object-oriented system itself to write even more abstract (and therefore more powerful and general) procedures and programs.

In this chapter we are representing and manipulating *data* but the structures that describe the data could well be argued to be a representation of some *knowledge* about the data. Later in this chapter (Section 1.2.6) we will introduce the syntax and semantics of XML, which provides methods and vocabulary for tagging data and representing structure in a standard way. In Chapter 2 we introduce RDF (Resource Description Framework) and OWL (Web Ontology Language), which use XML syntax along with some standard tags that represent some forms of knowledge, such as the idea of *classes* and logical relations between classes. These are important because, like the examples we develop in the next sections, they use plain text (usually encoded as ASCII) and thus are a means for encoding data and knowledge for export and import from one system to another. Further, the syntax of XML has become very popular, so there is now a large code base of programs that implement XML generators and parsers. We will illustrate the use of such code to store and retrieve information from such external formats.

1.2.1 A Simple Solution Using Tags (Keywords)

Patient data are a good match to hierarchical representation, where structures include elements which themselves have structure. Using symbols, strings, numbers, and lists, we can organize information about a patient in terms of keyword-value pairs. (Here, we are using "keyword" to mean the tag that labels a datum, which could itself be structured as a list of tagged elements, and so on.) Here is an example of an information record on a single patient.

```
(patient (name "John Q. Smith")
         (hospital-id "123--45-6789")
         (age 72)
         (address (number 211)
                  (street "Willow Street")
                  (city "Browntown")
                  (zip-code 90210))
         (diagnosis (name "prostate cancer"))
         (lab-tests (psa (("23-Aug-2003" 4.6)
                          ("12-Mar-2004" 7.8))))
         ...)
```

This keyword-value representation is easy for a person to read, especially with the indentations. The indentations and line breaks have no impact on the logical structure. Nor does the amount of space between items (though at least one space character is needed, it does not matter how many there are). The extra "whitespace" is simply to make the data more easily human readable.

Although earlier we mentioned XML as a syntax for defining tags and structure, it is complex. Parentheses and symbols are much simpler to read and much simpler to process in a computer program, and will suffice for now. For this simple example, we represent all the data as text in a file consisting of nothing but printable characters.

If the data above appeared in a text file, a single call to the Common Lisp `read` function would return a list structure exactly as shown, which could be examined by a computer program, by standard list operations such as `first` and `rest`. Rather than having to define data structures and access methods, and then decode and parse the input to assign the values to the right places in the data structures, the above text representation is already a valid representation of a Lisp data structure.

This format takes advantage of the fact that the Lisp reader and printer can do some of the work for us. All elementary Lisp data types have well-defined external or printed representations, even including user-defined structured types (defined with `defstruct`) and arrays. For reading and writing these simple data types from/to files, we can just use `read` and `print`. We don't need to do any lexical analysis of the text on input or pay much attention to formatting on output (though some formatting would enhance human readability of the data).

However, writing a program to manipulate the above structure could get tedious. Some work would be involved in creating abstraction layers, so that the code that works with data elements goes through accessor functions. For example, to get the diagnosis, we would define an accessor, named `diagnosis`:

```
(defun diagnosis (patient-data)
   (find 'diagnosis (rest patient-data) :key #'first))
```

and similarly for each of the other fields and subfields. Like the Swiss-Prot files and other keyword labeled records, this scheme has the advantage that records do not need to have uniform structure, and so the structure does not need to be predefined. This is workable if we impose some restrictions on the structure of the data, such as for example that certain items are always found at the top level of the list of data for the patient. Additional assumptions or restrictions must be made about items that themselves have structure. A disadvantage of this approach is that changing the structure of the data, perhaps by adding another field or subfield, could require changing many of the accessors. However, it confines the problem to rewriting the accessors, not rewriting explicit code scattered throughout a large program.

The design of such abstraction layers comes up in so many software systems that it seems reasonable to build an abstraction layer for support of abstraction layers. What this means is to have some functions or operators that can create the abstraction implementations without you, the programmer, being concerned about how the implementation actually works. This is just an extension of the idea of a high level programming language like Lisp. A high level programming language is a way to describe your computations without being concerned about how the abstractions actually translate into binary numbers that represent basic computing machine operations. To create these structure abstractions, many programming languages provide a way for you to define your own *structured types*. So-called *object-oriented* programming languages (including Lisp) go further and provide yet higher level entities called *classes*. In the next section we show how this works in the Common Lisp Object System (CLOS), and then how you can effectively use the metadata of CLOS to write structured data to files and read it back to reconstruct equivalent instances in programs.

1.2.2 An Object-Oriented Design

In using a simple scheme with tags, we had to write lots of accessor functions and keep in mind some definition of the structure of the patient record. This work will be required for each kind of structure we introduce into our medical record system. An object-oriented programming language provides a standardized collection of syntax and semantics that make this work much easier. We can write definitional expressions that specify the structure and the names of the accessors, and the object-oriented programming environment automatically generates the support code.

In effect we are creating an abstract model of medical information about patients, and letting the modeling system translate the formal model into lower level code. When the model changes, the program need not change except to accommodate the specific additions or deletions. Aggregated data units become, in a typical object-oriented programming language, instances of classes, whose definitions represent the model implemented in an actual program. Figure 1.28 is an example of a part of an object-oriented model for patient information. The figure uses a notation called Entity-Relationship (ER) Modeling. Each box in the figure represents a *class* specification, including the class name and the names of the attributes of that class. Each instance

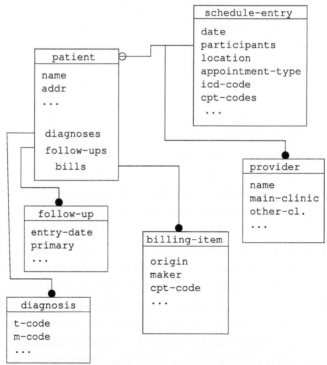

Figure 1.28 An Entity-Relationship model for a subset of patient data.

of a class can have different values assigned to the attributes. The lines connecting the boxes represent relationships between classes, indicating that one class is a subclass of another, or that an attribute of one class can have values which are instances of another class.

The connections between the boxes represent the typical kinds of relationships found in object-oriented models. A more thorough discussion of object-oriented design may be found in [358].

The model can be implemented with a set of class definitions. For each box we have a class, and we implement families of classes using inheritance. In such families some classes are defined directly as subclasses, or specializations, of a parent class. Thus, in the patient data model, the `patient` entity and the `provider` entity could each be defined as a subclass of a more general entity, `person`.

As a first try, we use *structures* as a simple version of the idea of user defined types. For each entity we define a structure type. The `defstruct` macro creates structure definitions. Its inputs are the name of the structure type, a list of possible "supertypes" of which this structure will be a subtype, and then the names of the slots that will appear in any instance of the structure type. The idea is that each instance will have these named slots, or fields, to which values can be attached. Such user-defined structures are a way to create complex objects with attributes (the slot values), and treat each such object as a single entity.

```
(defstruct person ()
   name birthdate telephone email ...)
```

The `defstruct` macro does more than just record a definition for the structure. It also creates several new functions and adds them to the working environment of the program. These functions are:

- a constructor function that can be called to create a new instance of the structure. By default it is called `make-<structname>` where `<structname>` is the name you specified. In the above example, the constructor function will be called `make-person`.
- accessor functions for the slots, each named with a prefix using the structure name, so in the above example, the accessors will be `person-name`, `person-birthdate`, and so on. There will also be `setf` methods so that expressions like `(setf (person-email x) <some data>)`, where x is a `person` instance, will also work.
- and others (see Graham [137], Chapter 4, Section 4.6 for details).

Instead of just specifying the name of the new structure type, you can specify that this structure is a subtype of another (already defined) structure. This is done by making the structure name into a list, with an `:include` option, naming the parent structure type. Here are examples.

```
(defstruct (patient (:include person))
   diagnosis appointments ...)

(defstruct (provider (:include person))
   specialty title office patients ...)
```

Specifying that `patient` and `provider` are subtypes of `person` will automatically make any slot defined in the `person` type also be present in both the `patient` type and the `provider` type. This is one aspect of inheritance. These specializations also add their own slots to the inherited ones. Similarly any accessors for `person` are inherited and applicable to both subtypes.

Here are some example functions that might be made available through some user interface in an EMR. Although in a real system, one would store records in some kind of database, here we just keep a list of patient structures in a global variable, `*patients*` and similarly a list of providers in a global variable, `*providers*`.

```
(defvar *patients*)

(defvar *providers*)
```

The `add-provider` function adds a new physician or nurse practitioner or other patient care provider, given this person's name, specialty, and title. The other information would be updated later.

```
(defun add-provider (name specialty title)
  (push (make-provider :name name
                       :specialty specialty
                       :title title)
        *providers*))
```

Here is how one might look up the specialty of a provider, given the name. The `find` function does the lookup, using the `person-name` function as a lookup key. This function is applicable since a provider is also a person. Note that the accessor for `specialty` is `provider-specialty`, not `person-specialty`, since that slot is particular to the `provider` structure type.

```
(defun lookup-specialty (name)
  (provider-specialty (find name *providers*
                            :key #'person-name
                            :test #'string-equal)))
```

Here is how updates can be done to any field in a structure. Once again, we use a lookup key function, and this time we can also use the more general `person-telephone` accessor, since the `telephone` field is defined in the `person` structure. However, we have a specific function for patients (and would need a separate function for providers) since we need to know in which list to look. The name by itself does not specify which kind of record. One cannot just look in both lists, since a provider can also be a patient, at the same institution, so it is best to keep separate records for these roles.

```
(defun update-patient-phone (name phone)
  (let ((record (find name *patients*
```

```
                        :key #'person-name
                        :test #'string-equal)))
    (setf (person-telephone record) phone)))
```

Inheritance in relation to the type/subtype relationship provides a way to build a program system incrementally and reliably. Adding the subtypes takes advantage of all the support code that is already written for the parent types, and does not in any way impact or require revision of code that supports the parent types. This code reuse has been one of the most attractive aspects of object-oriented programming.

The idea is not without its problems and limitations. In an electronic medical record system that supports appointment scheduling and billing as well as recording clinical data, the `provider` class might have a slot for a list of `patient` instances, the provider's patients. The model is complex because a patient typically has a relation to multiple providers, and vice versa. In principle, one only need represent one relation and the other can be computed, but this would be very inefficient for a large hospital or clinic. For efficiency, redundant references should be maintained in the system. However, this makes updating entries more complex, because the redundant information has to be kept consistent. For example, when a patient sees a new provider, multiple modifications are made, to the patient record, and to the provider record. Similarly, when a relationship ends, the relevant information should be removed everywhere, so the system has to have code to do the removal in all the redundant locations. What is really needed is a transaction system layered on this structure model, a so-called "object-oriented database," so that the entire history of relationships can be recorded somehow. This does not really solve the problem of what the currently valid relationships are, but it is necessary in order to meet requirements for record keeping in a medical setting.

Managing this kind of complexity is a big challenge for system designers, as it requires the implementation of consistency constraints, and results in a system that is no longer made of relatively independent parts. Designing such systems involves considering the trade-offs among simplicity, reliability, efficiency, and ease of use.

Structured types have been a part of many programming languages, including Pascal [193], C [224], and their successors. They do not implement the full capabilities of *classes*. The idea of classes is to extend data type ideas even further in abstraction. A significant difference between structures and classes is that structures provide an efficient but very restricted system for defining slots, accessors for slots (functions that return and set the slot values), and other properties of each structure, but classes provide a much wider range of possibilities. In particular, the accessor for a slot is named uniquely for each structured type, even if the slot has the same or a related meaning for different structures. We may have several kinds of things in our program, each part of a human body, and each is a structure such as heart, kidney, tumor, etc. Each will have a location (of its center, perhaps), specified by coordinates x, y, and z.

```
(defstruct heart ()
  size beat-rate x y z ...)
```

```
(defstruct kidney ()
   side x y z ...)

(defstruct tumor ()
   size grade tissue-type x y z)
```

The accessors, however, will be named differently, `heart-x`, `kidney-x`, `tumor-x`, because the storage arrangement for each may differ and there has to be some way to distinguish what to do in each case. So, you have to write different code for each type, in order to get and store the data.

One might think that defining a parent or base type with these shared slots would solve this problem. First, it becomes a design nightmare, because any time you use a name, you would need to check to see if it is defined in some base type, before defining a duplicate base type. Moreover, structures are limited to *single inheritance*, that is, a structure type can have no more than one parent type. So, it is not possible to use such base types to solve the problem when the shared slots are not the same set in every case. In Common Lisp, *classes* provide a nicer way to handle this problem.

1.2.2.1 Object-oriented Programming in Lisp

Implementation of an object-oriented design can be done with almost any programming language [179,181], but the work is facilitated by using a programming language with the right abstractions already built in. The essential abstractions are: the generalization of types to include abstract (or structured) data types (including user-defined types) that have attributes, the idea of inheritance or specialization of types, and the idea of type-based dispatch in implementing functions or procedures. Usually the idea of having different blocks of code to execute for different types of inputs is described by saying that functions are implemented as collections of *methods*, where a particular method is associated with a particular type of input (or inputs).

Here we provide a brief introduction to these ideas as implemented in the Common Lisp Object System (CLOS). A concise introduction to the basic ideas of CLOS may also be found in Graham [137, Chapter 11]. A more complete and detailed discussion is given in Steele [394, Chapter 28].

Types, Structures, and Classes

Every programming language provides some way to represent *types*. A type definition specifies what kinds of values an entity of that type may have. The types that are supported by almost all programming languages include: numbers of various sorts, characters, and strings (sequences of characters). More complex types, such as arrays, enumerated types, and user-defined structures, are supported by many programming languages. Lisp is distinguished by supporting symbols, lists, and functions as built-in types, in addition. Numbers are usually subdivided into *subtypes*, such as integer, rational, float, complex, and others.

There is no need in Lisp for what are called "enumerated types" in other languages, where one defines a sequence of names to be allowed values of a type. An example of this would be to define the names of colors as forming an enumerated type called "color." In Lisp, we would just use symbols naming the colors, and there would be no need for a statement defining them.

In Lisp, the types are arranged in a type hierarchy, so that any datum belongs to the overall type, *Lisp object*, sometimes referred to as type t, and may belong to a series of successively more specialized types. For example, the number whose text representation is 3.14159 is of type `single-float`, which is a subclass of `float`, which is a subclass of `real`, which is a subclass of `number`, which (finally) is a subclass of t. Similarly, in Lisp, strings are implemented as character vectors, thus the `string` type is a subclass of `vector`, which is a subclass of `array`, which is a subclass of t. The hierarchy of types is important for defining *inheritance*, and for dealing with *function dispatch*.

User-defined types are created using either `defstruct` or `defclass`. We have already seen examples of how `defstruct` creates *structure* types, and they become a part of the type hierarchy just as if they were built in. Similarly, `defclass` creates user-defined *classes*, which also become part of the type hierarchy. Each of these facilities provides a way to create a "record-like" structure, in which a datum contains named components, called *slots*, each of which can be referred to by name, rather than relying on the position of the component in an array. A structured type may have different kinds of values in different slots. The programmer need not be concerned with how the data are actually stored in memory.

In most discussions of object-oriented programming, the focus is on user-defined types, so one might get the impression "Object-oriented programming = classes," but this is most certainly *not* true. In the Smalltalk programming language, everything is an instance of a class of some sort, and the built-in classes are treated the same as the additional user-defined classes. This is also true in Lisp. On the other hand, although in java it is claimed that everything is a class, in reality, much of java's syntax and semantics are about managing namespaces.

User-defined types (defined either by `defstruct` or `defclass`) share the possiblity of utilizing *inheritance*. Built-in types also exhibit inheritance, where there is a type-subtype relation. Examples include numbers and their various subtypes, and sequences, whose subtypes include strings, vectors, arrays, and lists.

The type system and the idea of a class are closely integrated in Common Lisp. Every type corresponds to a class, and every class definition results in a type definition as well. There are three kinds of classes in Common Lisp, *built-in* classes, *structure* classes, and *standard* classes. The built-in classes have restricted capabilities and cannot have user-defined subclasses. Many of the standard Common Lisp types correspond to built-in classes, such as `number` and its subtypes, `array`, `sequence`, `cons`, and others. Each structure type defined by `defstruct` corresponds to a structure class. Similarly, classes defined by `defclass` define types that are subtypes of `standard-object`, which is a subtype of the general type t, or Lisp object. When using "class" to refer to a built-in class or structure class, the context should make the meaning clear.

Generic Functions and Methods

The idea of a generic function is that the function's body (the code that executes when the function is called) is made up of methods (defined by `defmethod`). Each method is identified with a particular combination of the types of the arguments to the function. So, the same function call with different types of inputs will result in different methods being used.

A program can add methods to any generic function, though one should not do this unless the generic function's author has specified how this should be done. If a function has been defined using `defun`, it will be an ordinary function, and one cannot subsequently define methods for it.

An example from the author's radiation treatment planning system, Prism, will illustrate the idea. The Prism system provides interactive control panels and graphic visualizations of patient specific anatomy, radiation treatment machines, and the radiation dose distributions that the machines produce with various settings of the controls. The program calculates the dose using well-known formulas. A snapshot of a Prism screen is showin in Figure 2.5 in Chapter 2, Section 2.2.8. The program can produce several kinds of cross sectional and projected views of the anatomy and other objects. The renderings are computed by a generic function called `draw`. The `draw` function takes two inputs, an object to draw and a particular view in which to draw it. For each combination of a type of object, and a type of view (coronal, transverse, sagittal, beam's eye view, or other), there will be a method. As new kinds of objects were added to Prism, new `draw` methods were written. Similarly as new kinds of views were added, new methods needed to be written. To insure against errors that could occur if the program attempted to draw objects in views where the particular combination had no method, a fall-back or default method was included. Here it is.

```
(defmethod draw ((obj t) (v t))
  "DRAW (obj t) (v t)
This is a default or stub method so we can build and use the various
functions without crashing on not yet implemented draw calls."
  (format t "No DRAW method for class ~A in ~A~%"
          (class-name (class-of obj))
          (class-name (class-of v))))
```

The `defmethod` macro has a form similar to `defun`, except that any of the inputs can have a type specifier, as illustrated. In the case above, the type specified is type `t`, which is also the default. Any type can be a type specifier, not just the name of a class defined by the `defclass` macro. The `defmethod` macro associates methods, or code blocks, with an existing generic function, or if no generic function of that name exists yet, one is created, and the body becomes a method for it. If an ordinary function already exists with the same name, it is an error. Any other method for the `draw` function must have the same inputs, an object to draw and a view to draw it in. The details of what happens will vary, but any more specific method that the programmer provides will supercede this method.

This method uses two functions that are part of the Common Lisp Object System. One, `class-of`, returns the class of its input. The class is an actual piece of data in the system, not just a name that appears in your code. Classes are objects created by the `defclass` macro, or in some cases, are built in classes that are predefined. The class-name function takes a *class* (NOT an instance) as its input, and returns the class name, usually a symbol. We will use this idea that classes are actual objects in the system, not just definitions in source code that are used by a compiler. This is part of the CLOS Meta-Object Protocol (MOP).

In CLOS, methods are associated with functions, not with classes. This is different from some other object-oriented programming languages in which methods are associated with classes. Those object-oriented languages implement the capabilities of operations with instances of the corresponding classes. However, that makes it very awkward to have functions that dispatch based on the type of more than one argument. In CLOS, dispatch can be based on any number of function arguments. You just write a method for each combination of argument types.

Since classes are just an extension of the Common Lisp type system, one can write generic functions with methods for *any* types, whether those types are built in, defined by defstruct or defined by defclass. Thus one could define a generic function that performed an abstract operation on a variety of data types, implementing that abstract idea with different methods according to each data type. It is not necessary for the classes that specialize methods to have a common "base class" in CLOS, since everything already is a subclass of the class t.

Some of the built-in functions of Common Lisp are actually generic functions, and you can augment them for new data types or classes that you defined, to provide any special behavior needed. Often used examples include the CLOS initialize-instance function, which your code should never call (it is called by make-instance), but for which you can provide a method to do extra work of initializing new instances of a class you have defined. See Graham [137, Chapter 11] for more details.

Generic functions support inheritance, just as classes and slot definitions do. So, if a generic function has a method for some combination of input types, that method will *also* be applicable for any combination of input types in which each argument is a *subtype* of the type for which the method was defined.

Here is an example. The Prism Radiation Treatment Planning system (described more fully in Chapter 2) provides a wide variety of control panels for drawing anatomy, positioning radiation beams, viewing the simulation geometry, etc. There is a generic-panel class, with a method for the generic function destroy. Each panel is a subtype of generic-panel so this method is applicable to each of them. This is a simple consequence of the fact that if an entity belongs to a certain type, it also belongs to all the parent types, so an instance of a beam-panel is also an instance of a generic-panel.

This can be seen by the results returned by the two built-in functions, type-of, which returns the most specialized type an entity belongs to, and typep, which returns t if an entity is of a specified type. The following example also illustrates that the ideas of type and class apply to the entire type hierarchy, not just the complex types defined by defstruct and defclass.

```
> (type-of 3.14159)
SINGLE-FLOAT
> (typep 3.14159 'number)
T
```

```
> (class-of 3.14159)
#<BUILT-IN-CLASS SINGLE-FLOAT>
```

As you can see, the `type-of` function returns a symbol that names the type, while the `class-of` function returns the actual class object itself, not just the name of the class. This is also true of structure classes and standard classes as well as built-in classes.

Finally, an important element of Lisp that is not found in most other programming languages is that functions are also data, and the type `function` is also part of the type hierarchy. Thus, one can have variables whose values are function objects, and your program can store function objects anywhere you might store anything else. Furthermore one can write functions that create new function objects, which can then be used just as if they had been there all along. This will be particularly important as we move from data to knowledge in Chapter 2.

1.2.2.2 A Redesign Using Classes

Here is a rewrite of the organ examples, with classes instead of structures. Note that the slots need some additional specification in order to have the accessors created. The accessors of slots in a class defined by `defclass` are not provided automatically as for structure definitions, but these accessors can be *generic* functions, meaning that they can share a common name, but have different methods depending on the type of their input(s).

```
(defclass heart ()
   ((size :accessor size)
    (beat-rate :accessor beat-rate)
    (x :accessor x)
    (y :accessor y)
    (z :accessor z)
    ...))

(defclass kidney ()
   ((side :accessor side)
    (x :accessor x)
    (y :accessor y)
    (z :accessor z)
    ...))

(defclass tumor ()
   ((size :accessor size)
    (grade :accessor grade)
    (tissue-type :accessor tissue-type)
    (x :accessor x)
    (y :accessor y)
    (z :accessor z)
    ...))
```

In this case, a single *generic* function, x, can retrieve the *x* coordinate of any instance of these objects. Just as defstruct automatically created accessor functions for each structured type, defclass will create *methods* for the generic functions named as the accessors. However, the names can be shared among classes.[38]

Now we can define a patient class, analogous to the patient structure, with slots that will hold the individual attributes for each patient that is an instance of this class. Here, we follow the ER diagram, Figure 1.28, though it would be a better design to include the person class and implement patient and provider as subclasses of person.

```
(defclass patient ()
   ((name :accessor name :initarg :name)
    (hospital-id :accessor hospital-id :initarg :hospital-id)
    (age :accessor age :initarg :age)
    (address :accessor address :initarg :address)
    (diagnosis :accessor diagnosis :initarg :diagnosis)
    (lab-tests :accessor lab-tests :initarg :lab-tests)
    ...))
```

Note that the address and diagnosis slots will contain instances of structured objects. We define classes for them too. These kinds of objects will appear in other parts of the medical record.

```
(defclass address ()
   ((number :accessor number :initarg :number)
    (street :accessor street :initarg :street)
    (city :accessor city :initarg :city)
    (zip-code :accessor zip-code :initarg :zip-code)))

(defclass diagnosis ()
   ((name :accessor name :initarg :name)
    (evidence :accessor evidence :initarg :evidence)
    ...))
```

This definition of the diagnosis class does not correspond exactly to the ER diagram. The fields t-code and m-code are specific to cancer, and are part of a disease classification system. The classification systems for tumors are not uniform, so it is a challenging modeling problem to determine how in general to represent this information.

We would like as before to store the information on a patient or series of patients (instances of the patient class), in a file, in a "serialized" form. By reading the file appropriately we should be able to reconstruct an object that is equivalent to the original instance in our program.

---

[38]It would seem reasonable to have defclass provide such accessors by default, but there are potential complications when the same slot is declared in a subclass. Should a new method be created by default? I don't have a proposal worked out for this.

The read and print functions will not work for data that are instances of classes defined by defclass. The Common Lisp print function does not print an instance of a class in a way that it can be reconstructed.

We could make an instance of this class and try to print it.

```
> (setq temp
     (make-instance 'patient :name "Ira K"
                        :hospital-id "99-2345-6" :age 63)
#<PATIENT @ #x7161b75a>
> (print temp)
#<PATIENT @ #x7161b75a>
```

This instance of the patient class is printed as an unreadable object (the #< reader macro denotes an unreadable object, followed by the class name and some implementation-dependent information). We will need to provide some extension to support serialization of class instances, including classes where slot values are themselves class instances.

A seemingly elegant solution is to use the CLOS facilities themselves to extend the print function's capabilities. According to Steele [394, page 850], the printing system (including format) uses the standard generic function, print-object. We can therefore write specialized methods for it, corresponding to each class we define in our program, so that when instances of those classes are printed using any print function, they will be handled by the specialized method, and not the system generic method. To print something readably, we start by defining a method for print-object when the object is a patient.

```
(defmethod print-object ((obj patient) strm)
   (format strm "(Patient ~%")
   (format strm "  :name ~S~%" (name obj))
   (format strm "  :hospital-id ~S~%" (hosptal-id obj))
   (format strm "  ...etc.) ~%" ...))
```

Now when we try to print our instance, it will use our specialized method instead of the system default. This will happen because the built-in print function calls the generic function, print-object, to do the actual work.

```
>(print temp)
(Patient
  :NAME "Ira K"
  :HOSPITAL-ID "99-2345-6"
  ...etc. )
```

This can also get tedious, verbose, and error prone when a program has many different classes. It does have one nice property. The calls to format will be implemented through the print-object generic function, so if a slot value is another class instance and there is a corresponding `print-object` method, the information will be printed properly, nested *logically*. The resulting text file will not be formatted nicely for the human reader, however.

This output can be read as input to recreate an instance equivalent (`equal`) to the original. We get the entire list with one call to `read`, and then hand it to `make-instance` using `apply`. Unlike `print-object`, there is no built-in function `read-object` for which we can write methods, so we define one here.

```
(defun read-object (stream)
   (apply #'make-instance (read stream)))
```

Unfortunately, unlike `print-object`, if there is a slot whose value is an instance of a class, the `read` function will simply fill that slot with the list that was written out, instead of creating an object from it. So, this `read-object` function only works for classes whose slots do not themselves have values that are instances of classes.

It would seem that we could solve this problem by writing out the call to `make-instance` in the file. Then the nested objects would be created along with the top level one. This is essentially a program that writes a program out to a file, to be executed later. However, the `read` function will just return a list representing an expression to evaluate. So, the `read-object` function becomes

```
(defun read-object (stream)
   (eval (read stream)))
```

or, another way is to just call the `load` function. But this must be considered extremely bad practice. In essence, instead of writing data to a file to be later processed, we are writing an entire program to a file, later to be executed. We have not even begun to look into searching files, which would be very difficult with this arrangement, and one can imagine how inefficient and inflexible the whole scheme might become. Changing the class definition even slightly would involve editing a file or files, in perhaps thousands of places, because all the repeated calls to `make-instance` would change.

Another tempting way to design `read-object` would be to make it a generic function, like the built-in `print-object`. Then you might think you would be able to write methods for `read-object` for all the relevant classes. For a class that has a slot with a value that is itself an object, the method for that class would need to make an explicit recursive call to `read-object`. This simply won't work at all. The input to `read-object` is a stream, not an object instance, so there is no way to determine which methods are applicable. Even preceding it by a call to `read` first won't work because the whole point of this design is to generalize the `read` function to handle objects that are otherwise unprintable and unreadable, so using `read` first is circular.

Aside from the problem of slots whose values are themselves objects, the above scheme has the additional burden that it requires a lot of writing of explicit methods for writing the object data out to files. This is because (in the usual style) the code for writing must explicitly name each slot to be referenced and written, as well as explicitly writing out the slot name itself. In most programming environments, the slot names are part of the source code but are not available to a running program. If the program code could examine its own definitions, this burden might be relieved. In most programming languages it is impossible, but you will see in the next section that in Lisp this is easy to do.

1.2.3 A Better Solution Using Metaobjects

While conventional object-oriented programming is nice, it still requires a lot of work for each new class that you add to the system. What if the information about each class were available to the running program? Then instead of writing explicit code for printing and reading instances of each class, we could write a function that gets the information about the class of the object it is printing or reading, and uses that information. If this worked, we could write one general function that would work for all defined classes, and we would not need to write extra code when adding a new class to the program.

The idea that a running program can gain access to information about its own data structure definitions has been present in Lisp for a long time, and is just now attracting attention in newer programming languages and environments. It is sometimes called "reflection". In Lisp your code can check on the type of an object, retrieve its class, and retrieve properties of classes and objects. Thus the programmer can indeed write code that can access class data, such as the class name, the superclasses (parent classes), lists of slot names, and other information about a class.

The way it works in Lisp is that the Common Lisp Object System (CLOS) itself is implemented using the same machinery that the application programmer uses. The way classes are implemented is that they are themselves objects, instances of a "higher level" class (a *metaclass*) called `standard-class`. The class `standard-class` is called a metaclass since its instances are "class" objects. This and the classes created by our code are examples of "metaobjects." CLOS has other metaobjects as well, for example generic functions, which are instances of the `standard-generic-function` class.

This design is known as the CLOS Meta-Object Protocol (MOP). Kiczales et. al. [229] explain the idea of a meta-object protocol by showing how to build a somewhat simplified version of the MOP. The actual CLOS MOP specification is reprinted in Chapters 5 and 6 of their book. It is not yet adopted as an ANSI standard, but is implemented in most Common Lisp systems.

Here we show how the MOP can help solve the problem of storing complex medical data in the style of keyword-value pairs, without writing a lot of extra code. Our plan is that the code will use the class definition itself as metadata, to decide what to read and write. We will define a function, `slot-names`, that returns a list of symbols naming the slots defined in a class. Applying this to the `patient` class instance on page 100, we should get

```
> (slot-names temp)
(NAME HOSPITAL-ID AGE ADDRESS DIAGNOSIS ...)
```

The MOP idea we will use here is that a *class* is an actual data object in a program, not just a source code construct. Classes have properties that a program can look up when it is running. In particular, it should be possible for a running program to find out what the names are of the slots for a class. The MOP function `class-slots` gives us a list of the slots (actually the `slot-definition` metaobjects which describe the slots), including inherited ones from superclasses, that are available in any particular class. Their properties can be examined using accessors, just as for other objects. From the `slot-definition` objects we can get their names (as symbols), which we can then use with `slot-value` to read the value stored in each slot, and with (`setf slot-value`) to write a value in a slot (as when restoring an object from the serialized data in a file). The MOP function `class-slots` requires as input a class, which is easy to obtain. For any object, the MOP function `class-of` returns its class (the actual class metaobject, not the name of the class). Since slots are implemented with metaobjects, we can use the MOP function `slot-definition-name` to obtain the name of a slot, given the slot itself. To get a list of all the slot names in a class, we iterate over the list of slot definitions using `mapcar`.

```
(defun slot-names (obj)
   (mapcar #'slot-definition-name
       (class-slots (class-of obj))))
```

The functions `slot-definition-name` and `class-slots` may be in a different package than the `common-lisp` package, but these are the names defined in the MOP documentation.[39]

Applying this to the `patient` class instance on page 100, we get what we expected.

```
> (slot-names temp)
(NAME HOSPITAL-ID AGE ADDRESS DIAGNOSIS ...)
```

Thus, the MOP gives a way to find out about a class's slots, iterate over them, writing the contents of each, and recursively doing the same for anything that is itself a class instance rather than a Lisp type that already has a standard printed representation. This can be done in a way that is human readable and allows a symmetric function for reading the data back in and constructing an instance equivalent (`equal`) to the original. These slot name symbols can be used as input to the standard CLOS functions `slot-value` and (`setf slot-value`) to read and set data in the slots of an object. The standard constructor function in CLOS for creating instances of classes, `make-instance`, also can take a symbol as input that is the value of a variable, so one can write code that creates and manipulates class instances without having any class names or details about the class explicitly written in the code.

Now, we will build the code that can read and write a representation of the patient data described above. The serial representation we will use is still based on the idea of keyword-value pairs, something like that of the previous section. Each keyword will be either a class name or a slot name, but we will not use parentheses to group data, except when they are a simple Lisp expression that should go into a particular slot. We will therefore need some way

---

[39]In Allegro CL^{TM}, a commercial Common Lisp implementation from Franz, Inc., these functions are external in the CLOS package. In CMUCL, an open source Common Lisp system, a more complex accessor is needed in place of `slot-definition-name` and `class-slots` is in the PCL package. In CLISP, another open source Common Lisp implementation, the name corresponding to `slot-definition-name` is `slotdef-name`; this and `class-slots` are internal in the CLOS package. Other implementations may have other versions, since the MOP is not yet standardized.

for the reader function to determine that the end of the slot name/value pairs is reached for a given instance. We will use the convention that the keyword[40] symbol :END is never the name of a class or a slot, so it can be used to indicate that there are no more slots for the current instance being read. This simple scheme should work fine for simple objects, whose slot values are printable Lisp data. Just as with our previous attempts, it will be problematic for slots that themselves have objects in them, or lists of objects. However, in this case, we will be able to solve that problem with only a small addition to the supporting code and framework. One might think that XML solves this problem by requiring beginning tags and ending tags. However, the problem here is not to decide where an entity begins and ends, but to distinguish between tags that name classes and tags that name slots of classes.

The get-object function will read from a data stream that looks like this (the ellipses just mean "lots more keyword-value pairs"):

```
PATIENT
  NAME "Ira K"
  HOSPITAL-ID "99-2345-6"
  AGE 63
  ...
 :END
```

The indentations and appearance of each slot on a separate line are ignored by the Lisp read function, but this formatting makes it easier for human inspection. The get-object function starts by reading the first symbol in the input. It expects that this will be the name of a class that has a class definition already in the system. Then it reads slot-name/value pairs, repeating until the :END tag is encountered instead of a slot name. Indefinite iteration (loop) is used here because the code cannot "look ahead" to see how many slots are present in the file, or in what order they appear. This implementation of get-object will handle any ordering of slot data. Not all slots need to be specified in the data file (some might have suitable default values), but any slot appearing in the data file must be included in the class definition. Since the class definition is separate from the data in the file, we would say that the data are *not* "self-describing," but we are using the class definition as metadata.

```
(defun get-object (in-stream)
  (let* ((current-key (read in-stream))
         (object (make-instance current-key)))
    (loop
       (setq current-key (read in-stream))
       (if (eq current-key :end)      ;; no more slots?
           (return object)
```

---

[40]Note that here :END is a keyword in the technical sense that it is in the Common Lisp keyword package. Elsewhere in this section we are using the term "keyword" in the generic sense of "label."

```
        (setf (slot-value object current-key)
              (read in-stream))))))))
```

Note that the `get-object` function will work for *any* class defined in our program, because it uses the class definition rather than having explicit names here. We do not have to write any methods for anything. It does not even have to be a generic function itself, since the same exact code executes for all classes. Thus, the class definition serves as the metadata.

In an industrial-strength program, we might add some error checking to detect malformed or unrecognizable input, to make the code more robust. One problem certainly is that the code will signal an error if it encounters a symbol naming a non-existent slot for the class being read. Similarly, an error will be signaled if at the beginning a class name is encountered for which there is no class definition.

However, in well-controlled circumstances this may not be necessary. In the author's radiation therapy planning system, the code described here has been used for nearly 20 years, to plan radiation treatments for over 5000 patients, and that system includes no error checking code. Error checking is not needed because these data files are only created by the same program, and do not come from any other source. The program code can guarantee that the syntax and semantics are correct, as defined here.

In order to create these files, we need a complementary function to write the data to a file, in the format shown. This function, which we call `put-object`, works with the same machinery we described above. It uses the `slot-names` function defined previously to get a list of the slots, and iterates over them writing out the contents of each.

```
(defun put-object (object out-stream &optional (tab 0)
    (tab-print (class-name (class-of object)) out-stream tab t)
    (dolist (slotname (slot-names object))
      (when (slot-boundp object slotname)
         (tab-print slotname out-stream (+ 2 tab))
         (tab-print (slot-value object slotname)
                    out-stream 0 t)))
    (tab-print :end out-stream tab t))
```

The `tab-print` function provides the indentation and new lines when needed to keep the file reasonably human readable. It puts in as many leading spaces as specified by the `tab` parameter, and includes a trailing space, so that keywords and values will be separated.

```
(defun tab-print (item stream tab &optional (new-line nil))
   (format stream "~A~S "
           (make-string tab :initial-element #\space)
           item)
   (when new-line (format stream "~%")))
```

This set of functions can now write and read back any collection of CLOS class instances, with arbitrary class definitions, even if mixed together in a single list of instances. Adding new classes does not require adding any new code. The class definition itself, through the MOP, provides the details for reading and writing.

This works fine if all the slots have values that are simple Lisp objects, that is, they can be printed readably by the standard printing system (including lists, arrays, and structures). However, it will fail if the slot value is itself an object (instance of a class) or contains objects, for example, in a list. In the example on page 88, the diagnosis slot actually contains a structured object, an instance of the diagnosis class. It would fail because format, like print, will not have any way to deal with an instance of a user-defined class. This is exactly the same problem we encountered in Section 1.2.2. This time, however, we can extend the metadata to enable the code to handle slots whose values are class instances and other ordinarily unprintable data.

A straightforward way to extend get-object and put-object to handle this case is to make them recursive. If the slot contents is an object, then make a recursive call to read or print it. Similarly, if the content of the slot is a *list* of objects, an iteration over the list will be needed. (Iteration is not needed for a list of simple Lisp data, since the Lisp system can handle it directly.)

In order to decide which kind of slot contents we are dealing with, we will need a new function, slot-type which will return one of three keywords: :simple, :object, or :object-list. The slot type :simple means that the contents of the slot are of a printable and readable Lisp data type, one that has a well-defined printed form. The slot type :object indicates that the content of the slot is an instance of some kind of object, and therefore cannot be read by the standard Lisp function read. In this case, as noted, we just make a recursive call to get-object. The slot type :object-list means that the slot contents are a list of objects, which will require an iterative call to get-object.

There is no such function in Common Lisp, CLOS, or in the MOP. We will write this additional code a little later in this section. For now, we will just assume it exists and show how to use it.

A case form takes care of the three cases, depending on the value returned by slot-type. The case operator has as its first input an expression to evaluate to obtain a *key*. The remainder of the case expression contains clauses like cond but instead of evaluating test expressions, each clause has a key or list of key values. The key resulting from the first input is used to find the first applicable clause. If it is found then the rest of the clause is evaluated in sequence. If the key is not found, case returns nil.

Here is get-object with these recursive and iterative calls added. The expression after case, the call to slot-type, is expected to return one of the three keywords. Whichever it is, the following expression will be the one that is evaluated and the others are ignored.

```
(defun get-object (in-stream)
  (let* ((current-key (read in-stream))
         (object (if (eq current-key :end)
                     nil ;; end of object list
```

```
                       (make-instance current-key))))
        (loop
          (setq current-key (read in-stream))
          (if (eq current-key :end)      ;; no more slots?
            (return object)
          (setf (slot-value object current-key)
            (case (slot-type object current-key)
              (:simple (read in-stream))
              (:object (get-object in-stream))
              (:object-list
                  (let ((slotlist '())
                        (next-object nil))
                      (loop
                        (setq next-object
                          (get-object in-stream :parent object))
                        (if next-object
                            (push next-object slotlist)
                          (return (nreverse slotlist)))))))))))))))
```

Indefinite iteration is used also in the inner loop because the code has no way of determining in advance how many objects there might be, if the slot value is a list of objects (in the file). We need a way to indicate the end of the list, and here again, the keyword :END serves the purpose. If :END is detected where a class name should be, it signals that there are no more objects, so get-object should return nil, which terminates the loop when next-object is tested.

Similarly for put-object, we add a case form, to handle each of the three possible slot types. Here is the expanded implementation.

```
(defun put-object (object out-stream &optional (tab 0))
    (tab-print (class-name (class-of object)) out-stream tab t)
    (dolist (slotname (slot-names object))
      (when (slot-boundp object slotname)
        (tab-print slotname out-stream (+ 2 tab))
        (case (slot-type object slotname)
          (:simple
              (tab-print (slot-value object slotname)
                         out-stream 0 t))
          (:object
              (fresh-line out-stream)
              (put-object (slot-value object slotname)
                         out-stream (+ 4 tab)))
          (:object-list
```

```
(fresh-line out-stream)
(dolist (obj (slot-value object slotname))
    (put-object obj out-stream (+ 4 tab)))
(tab-print :end out-stream (+ 2 tab) t)))))
(tab-print :end out-stream tab t))
```

In put-object, the iteration over the list in a slot containing an object list can be definite iteration using dolist or mapc, since the list is available at the outset (the object already exists and the slots have values). These two ways to do iteration over a list are equivalent, and it is a matter of style which you might prefer.

Note that the recursive calls to put-object and tab-print include increments to the amount of whitespace to indent, so that objects that are values in slots get nicely pretty-printed in the data file. The choice of 4 and 2 for the incremental amounts of whitespace is arbitrary but seemed to work well in the author's radiation therapy planning system.

One detail remains. Since there is no built-in function slot-type so we have to write it. This function should return a keyword identifying the contents of the slot as :SIMPLE for readable and printable Lisp objects, :OBJECT for an instance of a user-defined class, and :OBJECT-LIST for a list of such instances. It is the one element that needs to be defined for each class, if the slots are expected to contain other objects or lists of objects. In a typical application of this code, most classes will have simple slots and we only need a default method. For more complex classes, it is not hard to write a method that just returns the non-simple slot types, as compared with writing a method for print-object, which would be many times larger. Here is the default method:

```
(defmethod slot-type ((object t) slotname)
  :simple)
```

Note that in this method, the input variable slotname is not used. Most Lisp compilers will flag this by issuing a warning. It is not an error, but the intended behavior. The definition of the function includes this input, but this particular method does not use it.[41]

For the patient class, the address slot and the diagnosis slot both contain instances of objects defined by CLOS class definitions. For the patient class, we might then have the following method:

```
(defmethod slot-type ((object patient) slotname)
  (case slotname
    ((address diagnosis) :object)
    (otherwise :simple)))
```

In the otherwise clause, if this class is a subclass of a big class or is several layers down in a class hierarchy, it might be onerous for the programmer to look up all its parent classes,

---

[41]The compiler warning can be eliminated by using a declaration, (declare (ignore slotname)), but this is not required. The code is still correct without it.

so all the slots can be accounted for. Since there is always a default method, one only needs to put in the case form any *new* non-simple slots. A simple way to handle the otherwise clause is to use the standard CLOS function call-next-method, which will then use the previously written methods to look up any inherited slot. The call-next-method function can be used in any method. It finds the next most specific applicable method after the one from which it is called, runs that code with the same inputs as the original function call, and returns whatever that next most specific method returns. Here is a version of slot-type for the patient class, using call-next-method.

```
(defmethod slot-type ((object patient) slotname)
  (case slotname
    ((address diagnosis) :object)
    (otherwise (call-next-method))))
```

An alternative way to write these methods would be to write a method for each value of slotname, using eql specializers. This is much less desirable because it requires that any slot with a given name (possibly used in many different classes) must have the same meaning in terms of content type. This restriction creates global constraints on names, making it more work to add new classes.

Finally, one might consider that this information should be encoded in type declarations. This is inappropriate for several reasons. The slot-type categories we are using here are not the same kind of thing as the type hierarchy in the Common Lisp system itself. In general, type declarations are optional for variables and locations, since type is a property of what is stored there. One could indeed expand on the Common Lisp Object System and define new slot-definition meta-objects, with support for declaring and even enforcing restrictions on what is stored in those slots. Such extensions are found in the large and powerful frame systems developed for knowledge representation as opposed to data. In Chapter 2 we will develop the idea and a simple example of a frame system, with pointers to some available implementations.

The get-object and put-object functions only handle one object at a time. However, it is simple to write a next layer of input and output functions that handle a file of many such records on input, or an entire list of objects on output. The input function will be called get-all-objects and the output function will be called put-all-objects.

The get-all-objects function opens a file whose name is the value of the input variable filename, and iteratively calls get-object to accumulate a list of all the objects found in the file, until the end of the file is reached. It returns the list of object instances. If the file does not exist, it returns nil.

```
(defun get-all-objects (filename)
  (with-open-file (stream filename
                          :direction :input
                          :if-does-not-exist nil)
    (when (streamp stream)
```

```
(let ((object-list '()))
  (loop
    (cond ((eq (peek-char t stream nil :eof) :eof)
           (return object-list))
          (t (push (get-object stream) object-list)))))))))
```

The put-all-objects function takes as input a list of objects whose class definitions exist in the environment, and a filename, opens a file whose name is the value of the variable filename, and iteratively calls put-object on successive elements of the list object-list. If a file named filename already exists, a new version is created.

```
(defun put-all-objects (object-list filename)
  (with-open-file (stream filename
                          :direction :output
                          :if-exists :new-version)
    (dolist (obj object-list)
      (put-object obj stream))))
```

Other enhancements are possible, such as adding a :BIN-ARRAY type, for large binary data arrays stored in separate files. To handle such files we could create a function read-bin-array for reading such files and a companion write-bin-array for producing them. Since the binary data are in another file, the text file should contain information for how to read that file, such as the filename and the array dimensions. Medical image data, discussed in Section 1.1.5, are an example of the need for this. In Section 1.2.4 we show how to implement these two functions.

Although it would seem that this is an inefficient solution to storing patient data, it has been used successfully for over 10 years in the author's radiation therapy planning (RTP) system, *Prism* [203], which has been in clinical use at the University of Washington Medical Center, as well as at several other clinics.[42] On the order of 100,000 complete records of patient data and radiation treatment plans are kept online, from the first day in 1994 when the system was put into clinical use. Index files help to find individual patients' data, and retrieval is fast enough for clinical use. Each patient case instance is stored in a separate file in the file system, with a unique filename generated by the system. Thus, each file is small, and access is quick, requiring only a lookup of the filename in the index, and the opening and reading of a relatively small file. It is possible that a relational database back end could be used in this application, but it is not at all clear that this would perform better.[43] A brief introduction to relational database ideas is given in Section 1.3. An object-oriented database would be a good match here, but no reasonable implementations were available in the early 1990s when Prism was designed and implemented. In Section 1.3.1.5 we give a brief introduction to the ideas of object-oriented databases.

---

[42] A brief introduction to cancer, radiation therapy, and radiation treatment planning may be found in Chapter 2, Section 2.2.8.

[43] A commercial RTP product in the mid-1990s tried this, but their system had very unsatisfactory performance.

1.2.4 Medical Images: Incorporating Binary Data

So far, in the patient record example, we have considered only data that are easily and logically stored as text. Medical image data, on the other hand, are not efficiently represented as character code sequences. As noted earlier, the image pixels are represented in a computer program as a two-dimensional array of numbers, where the number at a pixel location represents the brightness (or color) of the image at that spot. The pixel array is understood to represent samples of image data on a regular rectangular grid. The numbers can be integers in the range 0–4095 (typically) for Computed Tomography (CT) images, and small decimal numbers for Positron Emission Tomography images. Such numbers, when represented in the form of text, will take up four or more bytes (one for each digit) plus space characters or other delimiters. In the case of CT and Magnetic Resonance (MR) images, only two bytes are required to represent each number in binary form (because the range of values is limited by the machine design to 16-bit signed or unsigned integers). When reading binary numbers into a computer program data structure, no conversion is necessary. By comparison, reading a character (digit) representation will require some parsing and conversion. So, for image data, the bytes (or multi-byte sequences) should instead be written and read directly as binary numbers representing numerical data, image pixels.

A two-dimensional array of pixel numbers would be useless unless we knew something about the image and the pixels. As described earlier, it is important to know the slice thickness in order to construct sagittal and coronal images and other renderings from the transverse image data. In order to interpret the images, the radiologist (and radiation oncologist) needs to know the position inside the patient corresponding to the image, as well as other data about the machine settings (the X-ray tube voltage, whether or not the patient had received injection of contrast dye, etc.). Medical image data sets therefore include "annotation" data, or information about the images. Typically they include: the patient's name (a character sequence, or text), birth date (also text), the inter-pixel spacing of the grid, in cm (may be binary, floating point format rather than simple integer, or fixed point), and many other attributes. So the parsing of an image data file can be very complex, since the interpretation of the byte stream is mixed and arbitrary according to some software designer's choices. These choices vary widely among the different manufacturers of medical imaging systems. The difficulty of dealing with such arcane file formats led eventually to the creation of a network protocol for exchanging image data, called DICOM (described in Chapter 4, Section 4.3.3). The DICOM protocol still presents the problem of parsing an arcane sequence of heterogeneously encoded data items. The advantage of DICOM is that you only need to learn and implement this one protocol, which is vendor and system independent.

The DICOM protocol only addresses the exchange of data between client and server programs, but not how the individual applications should store it. This allows the application software designer to choose a storage scheme that is best for her application. After presenting a little background about medical images, we show how to extend the keyword-value pair representation to support medical images in binary form.

1.2.4.1 Representing Image Data

A typical encoding of an image as a data structure in a computer program is as an instance of an image class. The following code shows how such a class might be defined:[44]

```lisp
(defclass image ()
  ((uid :type string :accessor uid :initarg :uid)
   (patient-id :accessor patient-id :initarg :patient-id
               :documentation "The patient id of the
               patient this image belongs to.")
   (image-set-id :accessor image-set-id :initarg :image-set-id
               :documentation "The image set id of the
               primary image set the image belongs to.")
   (position :type string
               :accessor position :initarg :position
               :documentation "String, one of HFP, HFS,
               FFP, FFS, etc. describing patient position as
               scanned (Head/Feet-First Prone/Supine, etc).")
   (description :type string
               :accessor description :initarg :description)
   (origin :type (vector single-float 3)
               :accessor origin :initarg :origin
               :documentation "Origin refers to the location in
               patient space of the corner of the image as defined
               by the point at pixel array reference 0 0 or voxel
               array reference 0 0 0.")
   (size :type list ;; of two or three elements, x y z
               :accessor size :initarg :size
               :documentation "The size slot refers to the physical
               size of the image in each dimension, measured in
               centimeters in patient space.")
   ;; ...other slots
   )
  (:default-initargs :id 0 :uid "" :patient-id 0
                     :image-set-id 0 :position "HFS"
                     :description "")
  (:documentation "The basic information common to all types of
                   images, including 2D images, 3D images."))
```

---

[44]In this class definition, we have included a new element, documentation for each slot, and for the class as a whole. In Common Lisp, documentation strings are part of the program, and are accessible to a running program, unlike comments in source code, which are ignored by compilers and interpreters and not available to the program itself.

This general class is then specialized to `image-2d` and `image-3d` which have further special attributes. An `image-2d` depicts some two-dimensional cross sectional or projected view of a patient's anatomy and is typically a single CT image, an interpolated cross section of a volume, or the result of ray tracing through a volume from an eyepoint to a viewing plane. Here is the `image-2d` subclass:

```
(defclass image-2d (image)
  ((thickness :type single-float
              :accessor thickness :initarg :thickness)
   (x-orient :type (vector single-float 3)
             :accessor x-orient :initarg :x-orient
             :documentation "A vector in patient space defining
             the orientation of the X axis of the image in the
             patient coordinate system.")
   (y-orient :type (vector single-float 3)
             :accessor y-orient :initarg :y-orient
             :documentation "See x-orient.")
   (pix-per-cm :type single-float
               :accessor pix-per-cm :initarg :pix-per-cm)
   (pixels :type (simple-array (unsigned-byte 16) 2)
           :accessor pixels :initarg :pixels
           :documentation "The array of image data itself.")))
```

The `pixels` slot for an `image-2d` contains a binary two-dimensional array, in which the value at each location in the array refers to a sample taken from the center of the region indexed, and values for images with non-zero thickness refer to points midway through the image's thickness. The origin of the pixels array is in the upper left-hand corner, and the array is stored in row-major order so values are indexed as row, column pairs, and the dimensions are ordered y, x.

Typically, cross sectional medical images such as X-ray Computed Tomography (CT), Magnetic Resonance Images (MRI), and Positron Emission Tomography (PET) are generated in sets, in which the orientation of the images, their scale factors, and their positions in the patient are all related. Such a collection of images is called a "position related set," or "series." The simplest way to represent such a collection in a program is as a list of `image-2d` objects. The list can be searched and the data extracted for display and other computations, including the creation of sagittal, coronal, and other reformatted images as described earlier. In addition to being useful as two-dimensional images, the set can be reassembled into a three-dimensional image. This kind of object is represented by the class `image-3d`.

```
(defclass image-3d (image)
  ((voxels :type (simple-array (unsigned-byte 16) 3)
    :accessor voxels
    :initarg :voxels
```

```
          :documentation "a 3D array of image data values"))
     (:documentation "An image-3D depicts some 3D rectangular
   solid region of a patient's anatomy."))
```

In the 3D image, `voxels` is the three-dimensional array of intensity values or colors coded in some standard way, such as RGB (Red, Green, and Blue intensity) values. The value at each index of the array refers to a sample taken from the center of the region indexed. The origin of the voxels array is in the upper left back corner and the array is stored in row, then plane major order, so values are indexed as plane, row, column triples, and the dimensions are ordered z, y, x.

Rearranging the set of 2D pixel arrays into a single 3D voxel array makes some image processing calculations simpler. These include volume rendering and the computation of projected images (simulated X-ray films, also called Digital Reconstructed Radiographs, or DRR).

1.2.4.2 Reading and Writing Image Data

Software packages to display and manipulate medical images use many different forms of storage of the image data in files. One way to do this is to store the pixel values as 16-bit binary numbers, and put the other information in a separate text file, formatted as a tagged data stream, just as we did for the patient data example in Section 1.2.3.

Figure 1.29 is an example of a keyword-value arrangement for a text file containing the "annotations" or descriptive information about the images. Each instance begins with a keyword naming the class, `IMAGE-2D`, and then keyword-value pairs for each slot in the class, that is, slot names followed by values in a format acceptable directly to the Lisp `read` function. These slots correspond to the `image-2d` class (including slots inherited from the `image` class). This file can be processed and ones like it produced by the `get-object` and `put-object` functions described in Section 1.2.3. Note that the `pixels` slot does not contain actual image pixel values, but instead carries a list of information describing a separate file containing the pixel values. This information would be used by a new procedure `read-bin-array` to read in the actual array data and store it in the slot of the newly constructed object, as we suggested earlier.

We now turn to the implementation of `read-bin-array` and `write-bin-array` mentioned previously. By separating the pixel array from the annotations, we can read in the pixel array data very efficiently and simply, using the standard Common Lisp function `read-sequence`. This function reads bytes from a binary file until the sequence passed to it is filled. The sequence will contain binary elements as specified by the element type of the file. The `read-sequence` function can determine the size of the sequence passed to it, since in Common Lisp sequences are objects that carry with them information about their size. Here is a version that creates a new array, given the total size in bytes. The array is a one-dimensional array, since that is the only type that `read-sequence` can process. The elements are 16-bit binary quantities, not 8-bit bytes.

```
(defun read-bin-array (filename size)
   (let ((bin-array (make-array (list size)
                     :element-type '(unsigned-byte 16))))
```

```
IMAGE-2D
  DESCRIPTION  "Pancreatic cancer: abdominal scan"
  ACQUISITION-DATE  "24-May-1994"
  ACQUISITION-TIME  "16:40:23"
  IMAGE-TYPE  "X-ray CT"
  ORIGIN  #(-17.25 17.25 0.0)
  SCANNER-TYPE  "AJAX 660"
  HOSPITAL-NAME  "University Hospital, Hackland, WA"
  RANGE  4095
  UNITS  "H - 1024"
  SIZE  (34.504 34.504)
  PIX-PER-CM  14.8388
  THICKNESS 0.500
  X-ORIENT #(1.000 0.000 0.000 1.000)
  Y-ORIENT #(0.000 -1.000 0.000 1.000)
  PIXELS  ("pat-1.image-1-1" 512 512)
 :END
IMAGE-2D
  DESCRIPTION  "Pancreatic cancer: abdominal scan"
  ACQUISITION-DATE  "24-May-1994"
  ACQUISITION-TIME  "16:40:23"
  IMAGE-TYPE  "X-ray CT"
  ORIGIN  #(-17.25 17.25 1.0)
  SCANNER-TYPE  "AJAX 660"
  HOSPITAL-NAME  "University Hospital: Hackland, WA"
  RANGE  4095
  UNITS  "H - 1024"
  SIZE  (34.504 34.504)
  PIX-PER-CM  14.8388
  THICKNESS 0.500
  X-ORIENT #(1.000 0.000 0.000 1.000)
  Y-ORIENT #(0.000 -1.000 0.000 1.000)
  PIXELS  ("pat-1.image-1-2" 512 512)
 :END
IMAGE-2D
  DESCRIPTION  "Pancreatic cancer: abdominal scan"
  ...
```

Figure 1.29 A portion of a sample image set data file.

```
(with-open-file (infile filename :direction :input
                                 :element-type
                                 '(unsigned-byte 16))
   (read-sequence bin-array infile))
 bin-array))
```

Alternately, if the receiving array were already in existence it could be passed in, instead of passing a size parameter. In this case, the function need not return the array as a value, since it directly modifies the array that is passed in.

```
(defun read-bin-array (filename bin-array)
  (with-open-file (infile filename :direction :input
                                   :element-type
                                   '(unsigned-byte 16))
    (read-sequence bin-array infile)))
```

As nice as this is, it is better from the point of view of the program that uses the array if it is a two-dimensional array instead of a one-dimensional array. If it is a 2D array, the program that uses it can explicitly use row and column indices instead of explicitly computing offsets into the linear array. However, as noted, the `read-sequence` function handles only one-dimensional arrays. One solution to this problem is to put a loop in `read-bin-array`, and read a row at a time, but this turns out to be tricky and inefficient.

This problem of needing to index an array in different ways comes up sufficiently often that the Common Lisp standard includes a facility to do this in a standard way. We can create a two-dimensional array, and then create another array which references exactly the same elements but is indexed as a one-dimensional array. This is called a *displaced array*. It is used for other purposes as well, such as referring to a part of an array offset by a fixed amount, as a "subarray." So, here is the 2D version of `read-bin-array`:

```
(defun read-bin-array (filename dimensions)
  (let* ((bin-array (make-array dimensions
                                :element-type '(unsigned-byte 16)))
         (disp-array (make-array (array-total-size bin-array)
                                 :element-type '(unsigned-byte 16)
                                 :displaced-to bin-array)))
    (with-open-file (infile filename :direction :input
                            :element-type '(unsigned-byte 16))
      (read-sequence disp-array infile))
    bin-array))
```

Alternatively, as in the 1D case, if we want to supply the array to be filled as an input,

```
(defun read-bin-array (filename bin-array)
  (let ((disp-array (make-array (array-total-size bin-array)
                                :element-type '(unsigned-byte 16)
                                :displaced-to bin-array)))
    (with-open-file (infile filename :direction :input
                            :element-type '(unsigned-byte 16))
      (read-sequence disp-array infile))))
```

Since the array passed to `read-sequence` is a 1D array, the code works efficiently. In particular, there is no need to read in a one-dimensional array and then copy it to a separate two-dimensional array. The 3D array case is left as an exercise for the reader.

To get the ultimate efficiency achievable in these functions, it would be important to include declarations so that the compiler can generate in-line code and optimize the use of storage for the array values. For this function, the use of declarations is illustrated in Appendix A, Section A.1.

Now we can add this support to the `get-object` function. It only requires adding one more branch to the `case` expression, to handle the `bin-array` slot type. We use here the first version, where the second argument is the array dimensions list. In the file, it is assumed that the slot data are the filename of the binary data and the array dimensions, as shown in Figure 1.29.

```
(defun get-object (in-stream)
  (let* ((current-key (read in-stream))
         (object (if (eq current-key :end)
                     nil ;; end of object list
                   (make-instance current-key))))
    (loop
      (setq current-key (read in-stream))
      (if (eq current-key :end)   ;; no more slots?
          (return object)
        (setf (slot-value object current-key)
              (case (slot-type object current-key)
                (:simple (read in-stream))
                (:object (get-object in-stream))
                (:object-list
                 (let ((slotlist '())
                       (next-object nil))
                   (loop
                     (let ((obj (get-object in-stream
                                            :parent object)))
                       (if obj (push obj slotlist)
                         (return (nreverse slotlist)))))))
                (:bin-array
                 (let ((bin-info (read in-stream)))
                   (read-bin-array (first bin-info)
                                   (rest bin-info))))
                ))))))
```

Now we need the counterpart of `read-bin-array`, which takes an image pixel array and writes it to a specified file. This function, `write-bin-array`, will take as input a filename for the file to write, and the binary array. It is not necessary to also provide the dimensions, since the array has that information available. The idea is to get the array dimensions and in an iteration (using `dotimes` is probably the simplest) just call `write-byte` to write each pixel value to the file. The only thing to be careful of is to open the file with the specification that the element type is `(unsigned-byte 16)`. This is an optional input to `with-open-file` or the open function.

Then, the `write-bin-array` function can be used in an extended version of `put-object`, in the same way as we extended `get-object`. Both of these functions are left as exercises for the reader.

1.2.5 Drug Interactions: Organizing Drugs into Classes

A drug is a substance that can chemically affect living organisms, human, or other. Drugs are most familiar as therapeutic agents, used to treat infections, heart disease, cancer, diabetes, and many other illnesses. Drugs are also used as preventive agents and sometimes as diagnostic tools as well. Some drugs, such as caffeine (as found in coffee and tea) and alcohol (in beverages), have become commonplace in social settings. Some of these (and other more dangerous drugs) have been implicated in antisocial contexts as well. Because of the potency and danger inherent in many useful drugs, some are controlled in that they are available only by prescription from a doctor. Narcotics are even more strictly controlled. Drugs have served in some societies as powerful weapons (for example, so-called "poison darts"). They can also be part of a public health measure, such as fluoridation of drinking water.

The actions of biological systems (organisms) on the drugs, such as drug absorption, distribution, and elimination, are called "pharmacokinetics." These processes all directly relate to how much of a drug is present in the bloodstream or in other body fluids, or available to the target(s) of the drug. Thus, one can think of pharmacokinetics as focusing on the rise and fall of drug levels in the body. The form in which the drug is present is also relevant, as there are some drugs that are not active in their initial form, but are substrates from which the active form is made continually in the body.

The action of the drugs on the organisms, on the other hand, intended or unintended, is called "pharmacodynamics." This includes the effects of the drug on organ function, cells, or other chemical substances. The effects that are useful, therapeutic, or performing some key body function are of course dependent on concentration, or level. However, the effects will depend on other factors too. It may be that the binding site of a drug is occupied because of the presence of other substances, and the effect of the drug will be diminished or absent. The pH of body fluids also can affect how the drug will interact with its intended target. Of course many drugs have undesirable effects as well as the intended ones. When a drug has excessive concentration, to where the undesirable effects reach critical levels, that is an overdose condition. Even the intended function can be harmful in excess. A common example is the use of aspirin as an anti-coagulant to prevent clots from forming in blood vessels (particularly coronary arteries). While this effect can be life saving in preventing strokes, an excess of aspirin or a combination of aspirin with other "blood thinners," such as warfarin, can cause internal bleeding and lead to death.

In addition to the effects just mentioned, the packaging of drugs, the dosages, and routes of administration all are significant in evaluating the potential for problems. The study of these matters along with pharmacokinetics is called "pharmaceutics".

Overall, the study of drugs, their properties and interactions, is called "pharmacology." Several excellent textbooks cover this subject, most notably the encyclopedic work of Goodman and Gilman [148].

This section concerns only one aspect of pharmacology, that of *drug interactions*. The context is that of use by patients of one or more drugs for preventive, diagnostic, or therapeutic purposes. When taking more than one drug at a time, a patient may experience effects that can be ascribed to the actions of one drug on the pharmacokinetics or pharmacodynamics of the other. This phenomenon is called a drug-drug interaction. Multiway interactions are also possible. When talking about binary drug-drug interactions, we refer to the drug whose serum concentration or effectiveness is being changed as the "object drug." The drug whose action *causes* the concentration of the object drug to change is called the "precipitant drug." This relation between object drug and precipitant drug is *not inherently* symmetric, although for some pairs of drugs, each one does affect the other.

There are many kinds of drug interactions. Some come from pharmacodynamic processes. A drug may compete for the same receptors as another drug, for its desired effect to take place. As mentioned earlier, two drugs may augment each other so that their combined effect exceeds a safe limit. Other interactions come from pharmacokinetic mechanisms. One drug causes a decrease in production of an enzyme that is the principal way in which another drug is metabolized, so that the second drug is not eliminated as rapidly as it should be. This results in an excessive and possibly unsafe level of concentration of the second drug. The opposite effect can also happen. Although a lot of data are available about these mechanisms, much remains to be done in terms of building predictive models.

Drugs that are metabolized by enzymes such as those in the Cytochrome P450 (CYP) family may be strongly affected by other agents that interfere with the production of those enzymes. Many drugs are known to inhibit (reduce below normal) or induce (increase above normal) the production of various CYP enzymes, which in turn may increase or decrease the metabolism of the drugs those enzymes metabolize. For many drugs, data exists on their metabolic mechanisms. FDA guidelines encourage detailed investigations into the metabolic mechanisms of a drug and its potential for drug interactions during that drug's early development [257]. These investigations are often followed by clinical trials to determine the significance of potential drug interactions [257,175].

The use of electronic data resources and decision support to screen for drug interactions is now the standard of practice for many pharmacies, from small neighborhood operations to those operated by major health care providers. Software that can predict drug interactions is very important in clinical practice. There are many commercial products that purport to address this need, as well as experimental systems developed in biomedical informatics research groups. Nevertheless, a recent study found that some of the systems commonly in use today have an alarming failure rate, so much so that they are routinely ignored or bypassed in practice [151]. The article cited notes, "The software systems failed to detect clinically relevant drug-drug interactions one-third of the time." Another study, which looked at handheld prescribing guides, found that all the systems studied failed to inform users of at least one life-threatening drug-drug interaction [285]. There is clearly much work to be done.

A large effort has gone into building drug knowledge resources, both online Web accessible systems, and products to be used on personal computers and handheld devices. The question we address here is how to organize this information in a way that supports computing interesting inferences. To start with, one must consider what types of information are available, what is actually needed, and finally what form it should take. In the following sections we present the beginning of a framework for sound theoretical predictions for drug interactions, with potential for improving this situation.

1.2.5.1 Drug Information Catalogs

The amount of information available about large numbers of drugs seems vast. Encyclopedic textbooks have been published and regularly updated. There are many online databases, both proprietary and governmental. An example of the latter is the Drugs@FDA web site produced and maintained by the U.S. Food and Drug Administration (FDA), a branch of the Department of Health and Human Services. At this site, one can look up records of all drugs that have been approved for use in the U.S. The records include: history of the approval process, manufacturer information, dosages available, and in some cases actual drug label documents. Proprietary drug information products such as MicroMedex[45] include similar information. For example, the MicroMedex entry for Simvastatin includes basic physical properties, proprietary names, adverse effects, interactions, pharmacokinetics, uses, and administration. Drug labels include a description, clinical pharmacology, indications (what the drug is used for), contraindications (under what conditions the drug should not be used), warnings and precautions, adverse events or side effects, potential for abuse or dependence, dosage and administration, and other information. These records are most readily accessible by web browsers or proprietary programs that run on handheld devices or personal computers.

Such catalogs of information can easily and obviously be represented in terms of instances of classes in an object-oriented programming language like Common Lisp (using CLOS), or as relations in a relational database, or as frames in a frame system. For example, a drug class could be defined using CLOS as follows:

```
(defclass drug-label ()
  ((name :documentation "The generic name of the drug")
   description
   (clinical-pharm :documentation "The pharmacologic action")
   (indications :documentation "List of uses of the drug")
   (contraindications :documentation "Conditions that indicate the
                                      drug should not be used")
   (precautions :documentation "Interactions usually included here")
   (dosage :documentation
           "Recommended dosage for various situations")
   ...))
```

Instances of this class would be the drug labeling information for each specific drug. One problem with this already is that the contents of the slots in every case of a real drug label are copious text rather than structured data. Such records would work well as a back end data repository for an information retrieval system, as described in Chapter 4. The text in the slots, together with slot names, would be used to construct an index according to one or another

---

[45]MicroMedex is a registered trademark of Thomsom Healthcare, Inc.

IR model, and then queries could be made to retrieve relevant documents. However, it is not possible to use this kind of organization with an inference system such as the logic systems described in Chapter 2. So, a second try might be a class definition like the following:

```
(defclass drug()
  ((generic-name :accessor generic-name
                 :documentation "Commonly used name")
   (drug-class :accessor drug-class
               :documentation "the therepeutic class of the drug")
   (dosage :accessor dosage
           :documentation "a list of approved dosages")
   (form :accessor form
         :documentation "oral, transdermal, or iv")
   (indications :accessor indications
                :documentation "what is the drug supposed to treat,
                                for example, a list of ICD-9 codes.")
   (side-effects :accessor side-effects
                 :documentation "a list of symbols naming side
                                 effects")
   (substrate-of :accessor substrate-of
                 :documentation "list of enzymes this drug is a
                                 substrate of")
   (inhibits :accessor inhibits
             :documentation "list of enzymes drug inhibits, with
                             mechanism, strength")
   (induces :accessor induces
            :documentation "list of enzymes this drug induces")
   (narrow-ther-index :accessor narrow-ther-index
                      :documentation "yes or no")
   (level-of-first-pass :accessor level-of-first-pass
                        :documentation "unknown or [enzyme level]")
   (prim-first-pass-enzyme :accessor prim-first-pass-enzyme)
   (prim-clearance-mechanism :accessor prim-clearance-mechanism
                             :documentation "metabolic or renal")
   (contraindications :accessor contraindications)
   (prodrug :accessor prodrug
            :documentation "is it a prodrug, yes or no?")
   ...))
```

This kind of record is very terse, and would not support rich textual web pages, because the *contents* of the slots in this case are just symbols or lists of symbols, or very short text fragments. However, it could be used to generate assertions in a logic system. This will be taken up in Chapter 2. The symbols that are values in these slots would be inputs to the <- or -> macros described in Section 2.2.1 or Section 2.2.4 of Chapter 2. The challenge then will be to make up the rules that relate these items of information. Most drug information sources do not provide any reasoning services. Their value is in the extent of the textual information they contain, and its effective indexing.

Instances of the drug class defined above could be stored in a big text file, using the serialization code of Section 1.2.3. Here is an excerpt of an example record. Not all slots and values are shown.

```
DRUG
  GENERIC-NAME  CYCLOSPORINE
  PRODRUG  NO
  PRIM-CLEARANCE-MECHANISM  METABOLIC
  PRIM-FIRST-PASS-ENZYME  NIL
  LEVEL-OF-FIRST-PASS  (CYP3A4 INTERMEDIATE)
  NARROW-THER-INDEX  YES
  INDUCES  NIL
  INHIBITS  ((CYP3A4 COMPETITIVE STRONG)
             (P-GLYCOPROTEIN UNKNOWN STRONG))
  SUBSTRATE-OF  (CYP3A4 P-GLYCOPROTEIN)
  ...
 :END
```

An entire list of drug records can be read with a single call to `get-all-objects`, defined in Section 1.2.3. Similarly, a list of such drug objects can be written with `put-all-objects`, defined in the same place.

1.2.5.2 *Procedural Knowledge About Enzymatic Metabolism*

The basic idea of drug interaction based on metabolic inhibition or induction is that a drug will affect another drug if it (the precipitant) significantly modifies the production of enzymes that metabolize the target, or object, drug. Using the class-based catalog scheme, one can write procedural programs that look up the relevant metabolic information and identify possible interactions based on that information. The code will use standard slot accessor functions to do the lookup.

Some drugs, for example warfarin and fluvoxamine, have multiple enzymatic pathways. Warfarin is a substrate of CYP1A2, CYP2C19, and CYP3A4. Fluvoxamine is metabolized by each of those, and CYP2C9. One of these may be a primary pathway, but the existence of others makes the problem complicated. If drug A inhibits enzyme 1, which is only one of the metabolic pathways of drug B, will this be significant? It depends on the importance of enzyme 1 compared to others that metabolize drug B. From looking at these entries it is apparent that taking the antidepressant fluvoxamine while also taking warfarin can cause a significant increase in the blood concentration of warfarin since *all* the metabolic routes of warfarin are inhibited by fluvoxamine. So, in that case, we can draw a stronger conclusion than for the case where only one enzyme is inhibited or induced.

There is yet another complication: some people possess one or more drug metabolizing enzymes whose catalytic function is significantly less than that found in a typical individual. These people are known as "poor metabolizers." In those patients, the effect on the remaining enzyme(s) would be much more significant than for normal subjects, since they do not have the alternative path(s) available. Dealing with this depends on having the information accessible from the patient's electronic medical record (or entered at the time of using an interaction consultation system).

A drug interaction system that relies on pairwise entries cannot represent this information, except to include volumes of text in the abstract. But a sufficiently sophisticated algorithm can compute these cases. If every enzyme in the `substrate-of` list of drug A is in the `inhibits` list of drug B, that will cause an interaction because no pathways are left for drug A to metabolize. We can define a function, `inhibition-interaction`, that compares the `inhibits` properties of the precipitant drug with the `substrate-of` properties of the object drug, and a similar function `induction-interaction` that checks the `induces` property of the precipitant drug.

```
(defun inhibition-interaction (precip obj)
   (null (set-difference (substrate-of obj)
                         (inhibits precip))))

(defun induction-interaction (precip obj)
   (null (set-difference (substrate-of obj)
                         (induces precip))))
```

Three-way or more complex interactions occur when a drug can be metabolized by multiple enzymes (as above) and each of several other drugs inhibits only one or two of them, but the combination eliminates all pathways. This would not be found in the binary drug interaction compilations, and the labor to create a database for it would grow exponentially with the number of enzymes involved. The approach of modeling the underlying mechanisms requires only one entry for each drug, and allows the functions or rules to compute the consequences. For three-way interactions, where there are two precipitant drugs and an object drug, we have,

```
(defun inhibition-interaction-3 (precip-1 precip-2 obj)
   (null (set-difference (substrate-of obj)
                         (append (inhibits precip-1)
                                 (inhibits precip-2)))))

(defun induction-interaction-3 (precip-1 precip-2 obj)
   (null (set-difference (substrate-of obj)
                         (append (induces precip-1)
                                 (induces precip-2)))))
```

Generally for any number of precipitant drugs, we collect all the `inhibits` or `induces` properties (using `mapcar`), and append them together, before taking the set difference with the `substrate-of` list. In the generalized version, the object drug is the first argument, so that the variable number of precipitants can become a `&rest` argument, allowing for any number of them. The `induces` function needs to be applied to all the precipitants and all the resulting lists need to be combined using `append`.

This operation will be useful in many other situations, where we want to map a function over a list to produce a list of lists and then flatten it into a single list. Following Norvig [305, page 19] we define a function, mappend, to combine the mapping and flattening

```
(defun mappend (fn the-list)
  (apply #'append (mapcar fn the-list)))
```

Then using mappend, the interaction test for any number, n, of precipitants can be written very simply,

```
(defun inhibition-interaction-n (obj &rest precips)
    (null (set-difference (substrate-of obj)
                          (mappend #'inhibits precips)))))
```

```
(defun induction-interaction-n (obj &rest precips)
    (null (set-difference (substrate-of obj)
                          (mappend #'induces precips)))))
```

These two functions can be parametrized of course as a single function by providing the attribute accessor function, inhibits or induces as a parameter.

Although this is effective, encoding the logic of these enzymatic interactions in procedures has the disadvantage that it is not easily extensible, when the logic gets more complex, for example, when we incorporate genetic variations, and other effects. So, the drug interactions investigation of Boyce et. al. [40,41] instead formulated this same knowledge using a more powerful formalism, that of First Order Logic (FOL), which will be introduced in Chapter 2 along with examples of how logic formalism can actually be used to formulate the same knowledge as above in a more modular way.

1.2.6 XML

A system of tags that define specific categories of data and specific structures is called a *markup* language. The most ubiquitous application of markup languages is in creating *documents*, notably formatted reports, articles for publication, and books. For example, the LaTeX document preparation system [247] was used in writing and producing this book. The HTML markup language [289] is used in documents that are intended to be viewed by a Web Browser program. However, as we have shown, markup can be used in data description as well. The most widely used markup languages share a common syntax, in which the markup terms are enclosed in angle brackets (<>) and the tag that ends an item or section or element is (usually) identical to the tag that begins it, with a / added at the beginning of the tag text. A standard called SGML, for "Standard Generalized Markup Language," defines a metalanguage, or set of rules for defining a markup language using a standard syntax.[46]

---

[46]The SGML standard is ISO standard 8879 (1986) and may be found at the ISO web site, http://www.iso.ch/cate/d16387.html. A book [133] written by the inventor of SGML provides additional background.

A subset of SGML, known as XML [149], or Extensible Markup Language, is the most popular syntactic standard in use at the present time, for defining markup languages. The idea of XML is to provide more standardized and possibly more sophisticated support for markup, specifically to standardize the syntax for embedding tags into the text of a document to label its parts. This allows for retrieval of the parts as entities. It is an alternative to the tag-value representation we presented in Section 1.2.3.

However, like the symbol/list tag system, XML only defines the syntax for tags and descriptions of structure. It does not define the meaning of any particular tag. Thus, XML by itself does not give any meaning or interpretation to a document. The writer or designer of the tag language associates meaning with particular tags. In order to do any computation on the document text, a computer program must have some information in it about what to do about each particular tag. This is parallel to the role that the class definition plays in being able to write programs that interpret tagged text, as in Section 1.2.3.

One should not be misled into thinking that XML's Document Type Definition (DTD) facility solves this problem of defining the meaning of tags. Having a DTD for classes of documents or data records is important, but a DTD only defines what is the allowed and required structure for a particular kind of document, and is generally only used by programs to do structure checking, to determine if the document is correctly formed or not. The meaning of the tags and the structures containing them is not specified by the DTD, but is something that the programmer must decide or know separately.

XML schemas are a more flexible way to specify the structure and allowed tags of a document, as well as constraints on the content of fields. However, these are still only syntactic and do not specify the meaning of the data contained within a tagged field. Nevertheless, DTDs and schemas are important to assure at least one level of data quality, namely that the documents are correctly constructed. An application program that processes an XML document can be much simpler if it can assume that the tags and data are consistent with the DTD or schema, rather than having to check syntax as the document is processed. Using a DTD or schema allows for running a validation program on a document before an application attempts to process the data.

Here is how the patient data instance in Section 1.2.2 might look when labeled with XML style tags.

```
<PATIENT>
  <NAME>Ira K</NAME>
  <HOSPITAL-ID>99-2345-6</HOSPITAL-ID>
  <AGE>63</AGE>
  . . .
</PATIENT>
```

A computer program could translate the above text into a series of tokens, rather similar to our symbol and list representation, and then search for particular tags within a particular structure. In this example, a program might, on detecting the <PATIENT> tag, create an instance of a patient class, and then interpret the remaining tagged items up to the closing </PATIENT> tag as names of slots and data to be put in the corresponding slots in the newly created patient instance. Updates to existing data structures could be handled the same way. Alternately, the electronic medical record system might be using a relational database system to store and retrieve patient data. In this case, the program that is processing the XML input would create appropriate commands in the database system's data manipulation language, to add a new record to the patient data table, or update an existing record.

The above example is about the simplest possible use of XML imaginable. Each tag has no attributes. It just marks the beginning (and its closing version marks the end) of a *container*, which has some data. Containers can be nested, just like lists.

It is also possible to have attributes associated with a tag, rather than creating separately tagged items aggregated within a container labeled by the tag. Here is an example of such a structure, an excerpt from a record in the Gene Ontology (abbreviated as "GO") [18,19,126], a large hierarchically structured set of data about genes and gene products, including cellular component, biological process, and molecular function information.[47] Each tag has a *prefix* which usually means that it has an intended meaning defined somewhere other than the document containing it.

```
<owl:Class xmlns:owl=" http://www.w3.org/2002/07/owl#"
           rdf:ID="GO_0003954">
  <rdfs:label>NADH dehydrogenase activity</rdfs:label>
  <rdfs:subClassOf rdf:resource="#GO_0005489"/>
  <rdfs:subClassOf rdf:resource="#GO_0016651"/>
</owl:Class>
```

This particular "document" describes a class. Every entry in GO is a class. The class tag includes two attributes. The first, labeled with the tag, xmlns:owl, provides a URL where more information is available about the tags that have the prefix owl, in this case, the *class* tag. The second, labeled rdf:ID, provides a unique identifier within the Gene Ontology, for this document. Then, the parts of the document include a label, which is some human readable text describing the entity, and two data elements providing unique identifiers of the parent processes of which this is a subclass or specialization. The general layout of a container tag with attributes is

```
<tag-name attrib1=value1 attrib2=value2 ...>
...content here
</tag-name>
```

The tags beginning with the prefixes rdf, rdfs, and owl have meanings defined by World Wide Web Consortium (W3C) standards. Resource Description Framework (RDF) provides a set of predefined tags that can be used in a way that (because they are described in a published document adopted by a standards body) can be relied on by different groups exchanging data. The Web Ontology Language (OWL) similarly adds some predefined tags, with meanings that

---

[47]URL http://www.geneontology.org/ is the official web site of the Gene Ontology project.

can be used to represent biological, medical, or other data and knowledge so that it will be interpretable without individual agreements having to be reached by people every time some data needs to be encoded and exchanged. These standards and much more can be found at the W3C web site, `http://www.w3.org/`.

GO is organized into an extensive class hierarchy, so that you can determine from it if a particular class of gene product's function is in some sense a subclass or specialized version of some more general kind of function. In the example above, the process "NADH dehydrogenase activity" is a specialization of each of two more general processes. Only their identifiers are given here so you have to search GO to find the full records for each. A lookup directly in the XML file would involve a huge amount of parsing. However, once the data are converted into structures like lists and symbols, we can use standard search methods for sequence data. Then we will find that `GO_0005489` is the identifier for "electron transporter activity" and `GO_0016651` is the identifier for "oxidoreductase activity, acting on NADH or NADPH." We can in turn look for the full records for these two, to determine if they are subclasses of more general processes, and so on. So, given two GO identifiers, we can search GO to find their relationship, if any. This will be developed more extensively in Chapter 2, Section 2.3.

A full description of XML is beyond our scope here, but to give a flavor of how to generate XML tagged documents from data in a running program and how to transform XML tagged documents to data within a program, we provide a simple XML writer, and a short description of an XML parser available with one of the commercial Common Lisp systems.

It is moderately easy for a computer program to write XML versions of simple objects like instances of the patient class described in Section 1.2.2.2. In fact for any instances of classes that are implemented like the patient data described in Section 1.2.3, we can use a modified version of the `put-object` function developed in that section. The changes that are needed to `put-object` are:

- where we print a tag (a slot name) we need to write an XML opening tag containing the slot name,
- where we are done producing the printed representation of the data item, we need an XML field ending tag (with a slash before the slot name).

We just build the solution with the same basic code as in `put-object`, replacing the output part with a fancier one that writes all the brackets and slashes of XML syntax. This new function is called `put-object-xml` to distinguish it from `put-object`. This version has to write both an opening tag and a closing tag, each in the right place. The opening tag is written before writing out the slot contents, and we put that detail in a separate function, `print-xml-tag`. Similarly, after printing the slot contents, the function `put-object-xml` calls another function, `print-xml-end-tag`, to write the closing tag.

```
(defun put-object-xml (object out-stream &optional (tab 0))
  (let ((tag (class-name (class-of object))))
    (print-xml-tag tag out-stream tab)
```

```
(mapc #'(lambda (slotname)
         (when (slot-boundp object slotname)
           (print-xml-tag slotname out-stream (+ 2 tab))
           (case (slot-type object slotname)
             (:simple (tab-print (slot-value object slotname)
                                 out-stream 0 t))
             (:object (fresh-line out-stream)
                      (put-object-xml
                       (slot-value object slotname)
                       out-stream (+ 4 tab)))
             (:object-list
              (fresh-line out-stream)
              (mapc #'(lambda (obj)
                        (put-object-xml obj out-stream
                                        (+ 4 tab)))
                    (slot-value object slotname))))
           ;; closing tag for each slot regardless of content
           (print-xml-end-tag slotname out-stream (+ 2 tab))
           ))
       (set-difference (slot-names object) (not-saved object)))
  (print-xml-end-tag tag out-stream tab))); terminates object
```

The two tag printing functions use `format` in a straightforward way. First, here is `print-xml-tag` for the opening tag. Given a tag, a stream, and a tab value (an integer), an XML tag is generated and printed after the appropriate number of blank spaces, as specified by the value of `tab`. A newline is always output afterward.

```
(defun print-xml-tag (tag stream tab)
  (format stream "~a<~a>~%"
          (make-string tab :initial-element #\space)
          tag))
```

Next, here is `print-xml-end-tag` for the closing tag. Given a tag, a stream, and a tab value (an integer), an XML closing tag is generated and printed after the appropriate number of blank spaces, as specified by the value of `tab`. Here also, a newline is always output afterward.

```
(defun print-xml-end-tag (tag stream tab)
  (format stream "~a</~a>~%"
          (make-string tab :initial-element #\space)
          tag))
```

These two functions are so similar you could parametrize them and write just one:

```
(defun make-xml-tag (tag stream tab &optional close)
  (format stream (if close "~a</~a>~%"
                     "~a<~a>~%"
          (make-string tab :initial-element #\space)
          tag))
```

Then you would just use this function in place of the other two tag generating functions in `put-object-xml` and use the appropriate value for the `close` parameter in each case.

Here is a convenience function analogous to `put-all-objects`. It opens a file named `filename`, and iteratively calls `put-object-xml` on successive elements of the list `object-list`. If a file named `filename` already exists, a new version is created.

```
(defun put-objects-xml (object-list filename)
  (with-open-file (stream filename
                          :direction :output
                          :if-exists :new-version)
    (dolist (obj object-list)
      (put-object-xml obj stream))))
```

This code does not write out tags with attributes, so it is very limited in terms of the mapping of CLOS objects to XML. Much more information about a CLOS object is available than just the slot names and values. How much of this is important will depend on the application. Insofar as most additional information is class level information rather than instance level information, it is kept with the class definition, not the serialized instances.

Reading and parsing XML into data structures in a computer program is a much bigger programming job. Implementing `get-object` was relatively easy because we used a textual representation of data that the built-in Lisp parser already handles. So, we did not need to write a parser to convert text representations to internal data representations. Implementing an XML version of `get-object` would be an arduous task because we would have to add low level operations to deal with the syntax of XML with its angle brackets and other punctuation. However, the need for this is so widespread that there are now XML parsers for most programming languages, including Common Lisp. Several are open source, and some are incorporated with programming language compilers and systems. Here are some web links for Common Lisp code that implements XML parsers and other XML applications:

* URL `http://homepage.mac.com/svc/xml/readme.html` is the site for a Simple XML Parser for Common Lisp by Sven Van Caekenberghe,
* URL `http://common-lisp.net/project/xmls/` is the site for XMLS, a simple XML parser for Common Lisp by Drew Crampsie,
* URL `http://www.emacswiki.org/cgi-bin/emacs/XmlParser` is the site for an XML parser written in Emacs Lisp,
* URL `http://www.agentsheets.com/lisp/XMLisp/` is the site for the XMLisp project [339], which provides a strategy for integration of XML and CLOS.

There are two more or less standard approaches for an application programming interface (API) to an XML parser. One, the Document Object Model, will be described in some detail here. The other, SAX, is only briefly described.

Document Object Model (DOM) parsing reads the entire XML file and creates an instance of an XML Document Object Model object. A DOM document is a tree-like structure that has components organized according to the tags, attributes, and components, at the lowest syntactic level. An API for DOM provides accessor functions to get at the tags and their associated values, much as we illustrated for our list structured tag architecture. First, your program calls the DOM parser function, then you examine the resulting object to do computations on it or its components. This is simple in that it completely separates the processing of XML syntax from the processing of the resulting data structure. It has the advantage that you can get arbitrary access to any element of the document, in any order rather than having to deal with it serially, component by component. For looking up cross references such as the GO example, this is very useful. A disadvantage is that the entire document must fit into the available memory address space, so for large documents performance could suffer if not enough physical memory is available. A standards document defining the Document Object Model, its interfaces and some language bindings, is available at the W3C web site mentioned earlier.

A Simple API for XML (SAX) parser will read components and call back a function you provide to do whatever you want with each component. This approach is more complex in that your program has to have lots of logic to deal with whatever might come next in the data stream. It has the advantage that you can incorporate such logic as the data elements are read. This can be very useful when different things must happen depending on what data are input. It is particularly useful when dealing with data streams coming through a network connection rather than from reading a file. Unlike DOM, SAX is not a W3C standard. Its official web site as of this writing is http://www.saxproject.org/ and it is also affiliated with Sourceforge (http://sourceforge.net/).

The DOM standard defines a language binding for the java programming language, not surprisingly, since DOM originated as a way to make java programs portable among various web browsers. However, the standard is abstract and it is possible to define language bindings for other languages, including Common Lisp. We will develop here a small example of how to use the XML DOM parser provided with Allegro Common Lisp, a well-supported commercial Common Lisp programming system.[48] This implementation defines a Common Lisp language binding for DOM.

A document consists of a root node, a *document* object, and child nodes. The child nodes can be *elements*, corresponding to units that are enclosed by (and include) beginning and ending tags. A child node can also be a text node, which corresponds to a sequence of characters, a CDATA section node (containing uninterpreted data), or other node type. Nodes can have attributes as well as values and children. To illustrate this, the patient document example above would be represented as a document object, whose document element would have as its node name, PATIENT, and it would (ignoring the ellipses) have three child nodes, corresponding to the three tagged fields.[49] These three child nodes would be accessible as a list of element objects.

---

[48] Allegro CL^{TM} is a product of Franz, Inc., Oakland, CA.

[49] Actually there is a typographical problem which we will explain shortly.

Each element object corresponds to a tagged field. The element object may contain just text or it may in turn contain nested elements, so we have to examine the child node(s) to find its contents. In the patient example, the node name of the first element is NAME and its value is actually contained in the child node list. In this case there is one child node, and it is a text node. That node has name "#text" and its value is in the node value slot, since a text object can't have any child nodes. If there were also some nested tagged elements, the list of child nodes for the NAME field would include text objects and element objects.

If a tag has no attributes associated with it, its attribute list will be empty, otherwise it will contain a list of attribute nodes. An attribute node has in turn a name and a value (but no children).

So, if everything worked as we just described, we could use the DOM API as follows. The function `parse-to-dom` takes a string or an open stream as input, and returns a DOM instance. Let's suppose our patient record were contained in a file called `dom-test-patient.txt` and we are using the interactive Lisp environment to explore. In order to use the Allegro DOM libraries we need to use `require` to insure the code is present in the environment (the module name is `:sax` because both the DOM parser and the SAX parser are included in the same module), and we also arrange to use the package, whose name is `:net.xml.dom`, so that we don't have to retype the package name every time. Then we just open a stream and call the parser.

```
> (require :sax)
T
> (use-package :net.xml.dom)
T
> (with-open-file (s "dom-test-patient.txt")
    (setq doc (parse-to-dom s)))
#<DOM1-DOCUMENT (1) @ #x716ac80a>
```

The result is that the variable `doc` now has as its value a DOM document object. We can then look up things as described earlier. To print the text in each of the fields (elements), we would iterate over the list of elements, and for each, look up the child node, and print its value. Note that all the values are text. There is no concept of the variety of data type representations that we have in printed values of lists, symbols, strings, numbers, etc. If the content of a text node is actually a number or a list, it needs to be further parsed, for example using the Lisp function `read-from-string`. For this example, we will just define a generic function, `print-dom-node`, whose job is to print the text that is contained in a node, if any, or if the node is an element, look up its components and print those, or if it is a top level document, print its contents, the top level element.

```
(defmethod print-dom-node ((node dom1-text))
  (format t "Value: ~A~%" (dom-node-value node)))

(defmethod print-dom-node ((node dom1-element))
  (let ((name (dom-node-name node)))
```

```
    (format t "Element name: ~A~%" name)
    (mapcar #'print-dom-node
            (dom-child-node-list node))
    (format t "End of ~A~%" name)))

  (defmethod print-dom-node ((node dom1-document))
    (format t "DOCUMENT")
    (print-dom-node (dom-document-element node)))
```

When the `print-dom-node` function is called, it will examine the type of its input, determine which method is the most specific applicable method, and use that. Each of the above methods recursively calls `print-dom-node` but in general, the recursive call will use different methods depending on what kind of input is provided. For example, the method for a document object will provide a node of type `element` as input so the method for `element` will then be used, and so on. So, the generic function dispatch takes care of whatever the function finds, as it traverses through the tree of nodes. This function only handles simple structures, but it can easily be generalized to apply to attribute processing, and to other types of nodes. Of course, instead of printing to the display, the action taken could instead be to construct a class instance, and put values in its slots.

Here is the result of applying the function to a file containing the little patient example.

```
> (print-dom-node doc)
DOCUMENT Element name: PATIENT
Value:

Element name: NAME
Value: Ira K
End of NAME
Value:

Element name: HOSPITAL-ID
Value: 99-2345-6
End of HOSPITAL-ID
Value:

Element name: AGE
Value: 63
End of AGE
Value:

End of PATIENT
NIL
```

It looks like it almost went according to our plans, but there is one strange problem. It appears there is some extra output before and after every element in our document. In fact, there is some text in between the tags ending one element and beginning the others. It is *whitespace*, characters we usually ignore, like blanks, tab characters, and new line characters. But the XML standard does not specify that these should be ignored. There could certainly be a document element with some text and then an element with a tag, and then more text. This introduces a considerable complication. One way to avoid it is to take a strictly data-centric view, and demand that such whitespace simply not be present. This makes the data unprintable. Another, perhaps more robust, is to add a filter to the code above, to ignore whitespace. In Chapter 4, Section 4.3.4, we will show how to do this, in the context of querying PubMed for bibliographic data.

You can imagine now how complicated it would be to write the analog of the `get-object` function. Although `put-object` was easy, sorting through the DOM structure to get slot values into the right places, and particularly handling object-valued slots, will be more difficult than it was for our simple representation with parentheses, symbols, and Lisp style text representations. Which way is better? It depends on how much existing context is already committed to XML. It should be noted that a huge amount of effort in biomedical informatics is devoted to dealing with decisions that have already been made about data representation and storage. This is as true in the context of biomedical research data as in clinical and public health practice.

Finally, it should be noted that several other projects in the Lisp community have as their goal to be able to translate between XML representations of data and their realization in programs as class instances. One noteworthy one is the XMLisp project [339]. The idea here is to extend Lisp to support *X-expressions*, in which XML data elements are interpreted in terms of semantics like class definitions I described earlier (Section 1.2.3). Instead of having an intermediate structure like a DOM object, which then takes a lot of work to translate to and from class instances, XMLisp provides a more direct mapping.

XML is intended only to be a "markup" language, a way to support somewhat self-describing or semi-structured data. As pointed out in an excellent book introducing XML [149], XML is *not* a programming language and it is *not* a database system or database schema language. Although you cannot write a program in XML or even a complete data model, the extensions, RDF and OWL, used above in the GO excerpt, provide a substantial set of reserved keywords and associated semantics to define classes, instances, properties, and even rules. However, you will still need a translator that converts this representation into the appropriate constructs in a full programming environment. A program to apply the rules and perform operations on the collection of classes and instances is also necessary. Chapter 2 describes how such rule-based programs work and how they might be integrated with representations of classes and instances.

1.3 Database Concepts and Systems

Although the tag-value system described in Section 1.2.3 works well in some biomedical applications, often the regularity of structure of the data suggests that a more efficient system could be built. The problem of dealing with large and complex but regular data is sufficiently broadly important that an entire subfield of computing, the study of database systems, has developed to address it. The following are definitions of some useful terms in reference to database systems.

Entity   a collection of data treated as a unit, such as a radiologic image, a patient, a radiation beam, a laboratory test panel, an encounter record.

Relationship expresses associations between entities, such as that a particular radiologic image *belongs to* a particular patient. This is *not* the same as a *relation*, sometimes called *table*, which is specific to the Relational Data Model, and is described in Section 1.3.1.

Schema   a description of the kinds of entities in a database, and their relationships. A schema can include many inter-related tables or other structures (sometimes hundreds).

Query   specifies an operation on a database, resulting in a new collection of data, arranged in some way, possibly a new schema.

Data model a general framework for defining schemas and queries. Examples are the relational model and the object-oriented model.

View   the result of a query (in the relational model a view is a table, so it can be the input to another query), but views are not stored in the database system; they are computed.

Many different data models are in use, including the relational model, the object-oriented model, the (hybrid) object-relational model, the entity-attribute-value model, the entity-relationship model, and others. In Section 1.2.2 we gave an illustration of object-oriented modeling with entity-relationship diagrams. A more elaborate treatment may be found in a new edition [32] of a classic text [358].

1.3.1 The Relational Model

When data logically match the idea of many identically structured records with relatively simple structure, or a collection of such structures, the most powerful model for describing, storing, and querying large data sets appears to be the Relational Database Model [69,83,334]. So much of the data used in business, education, government, and other organizations has this regularity that the relational model is the dominant model for database systems. Many products implement relational database systems, and large numbers of applications are built with a relational database back end. In this section we give only a brief sketch of the core ideas of relational databases. A more in-depth treatment can be found in textbooks devoted specifically to database systems [83,105,334].

The core idea of the relational model is that data are organized into a set of *relations*, often called tables, though strictly speaking a relation is not really a table. A relation is a collection (or set) of *tuples*. A tuple contains data about some entity and can be thought of as expressing a statement about it. The tuple consists of a set of values, each of which is associated with a particular attribute of the entity it describes. Thus the relation idea is that this tuple relates the values together for the individual entity. A row in a relational table, as typically shown in database textbooks, displays the contents of a tuple. The columns correspond to the attribute values, and the column headers provide labels for the attributes. There is no significance to the order of the rows or columns, but the columns serve to identify which value in each tuple goes with which attribute.

Each attribute has a type associated with it. All tuples must have attribute values conforming to the specified type for each attribute. An important requirement on attribute values is that only a single value for each attribute in a tuple is allowed. A relation should not have tuples in which there are multiple values for an attribute in a single tuple. For example, it may be that an immunization database has records for each patient of immunizations and the dates on which the vaccine was administered. So one of the attributes could be the administration date. Each tuple can only have a single date. Repeat vaccinations such as "booster shots" each will be recorded as a separate tuple. A relation meeting this requirement is said to be in First Normal Form.

While most databases have attributes that take on relatively primitive values, such as (small) integer, short character sequence (string), floating point number, etc., the relational model in the abstract supports structured types and other complex and user-defined types. A type specification defines what are the allowed values, analogously to a type definition and specification in a programming language. Since a relation specifies a set of allowable tuples that are valid for that relation, a relation itself defines a type, and so it is possible even for an attribute type to be a relation. This means in an informal sense, that an attribute can have an instance of another relation as a value. It must be noted that there is some controversy over this. The idea is related to the extension of the relational model called the *Object-Relational Model*. The Object-Relational Model seems to provide some of the most important ideas of an object-oriented database while preserving the essential integrity of the Relational Model.

The distinction between variables and values that we made in describing a computer program written in LISP also applies here. In a relational language, one can have variables, whose values may be individual tuples, or sets of tuples. The values, tuples and sets of tuples, exist independently of whether they are assigned as values to variables. However, this is not the usual view of variables and values in more conventional programming languages.

Furthermore, each tuple must be unique. It must differ from every other tuple in the relation in at least one attribute value. This makes relations analogous to mathematical sets, in that each tuple has an identity determined by its values, and having the same element twice in a set does not make any sense. To give a medical example, in a relation that contains laboratory tests performed on patients, we find a tuple that says a creatinine level was done on Jones from blood drawn at 9:40 AM on September 24, and the result was 1.04. There should only be one such entry. If the actual database (files or other storage) contains two tuples with the same set of values, it is a duplicate entry and a mistake was made. This is different from data collection in some scientific experiments where the identity of individual data points does not matter, duplicates are allowed, and the purpose of the experiment is in fact to count the number of occurrences of various combinations of values.

An example, a table that might be found in a tumor registry database, is shown in Table 1.6.[50]

---

[50]You may notice that there is a misspelling in this table, where someone has put in "Prostrate" instead of "Prostate." This is deliberate on the part of the author, to illustrate a point about data quality (see Section 1.4).

Table 1.6 A Table for the *Patients* Relation

Name	Hosp_ID	Tumor_site	Age
Joe Pancreas	11–22–33	Pancreas	38
Sam Vacuum	15–64–00	Glioblastoma	60
Sue Me Too	99–88–77	Lt. Breast	45
David Duck	88–77–99	Pancreas	46
Norbert Pivnick	−6	Prostrate	62
...			

Thus, a table is *not* a spreadsheet, which allows free form values in any box representing a particular row and column. In a spreadsheet, the row numbers serve to index the rows and similarly the column numbers index the columns. In a spreadsheet, anything can go in any box, and a single column can have different types of data in different rows. By contrast, tuples cannot be referred to by row and column numbers. They are instead referred to by content. One can for example look for all the tuples in which the *Tumor_site* attribute has the value "Brain," or the *Age* attribute has a value less than 21. These operations are called *restrictions*. They are sometimes also called selections, implemented by the SELECT command of SQL.

In addition to restriction, one can also collect values from a tuple, corresponding only to certain atttributes, rather than all. This operation applied to a set of tuples gives a new set, that is a different relation, a *projection* of the first relation. Both of these operators produce new tables from the existing tables. Such new tables are called *views*.

In a relational database, a schema is a statement that describes a relation or collection of relations. It specifies the name of each attribute, and its associated data type. It may specify many other properties as well. The standard data definition, manipulation, and query language for relational databases is SQL. It provides for creating, updating, and searching tables, and forming new tables (views) from existing ones.

Standard database textbooks cited earlier provide discussion in depth about the implementation of database systems, particularly relational database systems. They also treat such questions as query optimization, the process of translating the user's query into a form that is more efficient for the database system to execute than the original query.

1.3.1.1 A Brief Introduction to SQL

SQL (for "Structured Query Language") is the standard relational database programming language. Despite being called a "query" language, the standard includes a data definition language (DDL), a data manipulation language (DML), and a query language (QL). Other capabilities built into SQL include the ability to embed SQL statements in a conventional programming language like C, or Lisp, and most implementations provide some way to do the converse, that is, to incorporate procedure calls in such languages into SQL statements. It provides statements for defining tables (relations), and statements for populating, updating, and querying them. The following illustrates the essential ideas.

To define a table structure, one uses the CREATE TABLE construct. This is the core of the Data Definition Language. Here is an example. The *Patients* schema consists of a single table, described by the following:

```
CREATE TABLE Patients
              (Name CHAR(20), Hosp_ID CHAR(10),
               Tumor_site CHAR(20), Age INTEGER)
```

The name of the table follows the CREATE TABLE command, then in parentheses, separated by commas, is a sequence of attribute descriptors. Each attribute descriptor is required to have a name and type declaration.

Table 1.6 is an excerpt from an instantiation of this table definition. To make changes to the table, we need a Data Manipulation Language. Such tables are maintained by INSERT, DELETE, and UPDATE commands.

The INSERT operator adds a new tuple to a relation. The name of the relation is needed, along with the names of the attributes in some order, and the values to be associated with each attribute (in the same order).

```
INSERT
INTO Patients (Name, Hosp_ID, Tumor_site, Age)
VALUES ('A. Aardvark','00-12-98','Nasopharynx',57)
```

The Patients relation now has a new value, consisting of the original set of tuples plus one more. The DELETE operation does just the reverse, removing a specified tuple from the relation. In this case, we need some way to identify the tuple(s) to be deleted. This is done with a WHERE clause. This clause includes a logical condition to be tested against each tuple. If the condition is true, the tuple is deleted, but if not, it is left in.

```
DELETE
FROM Patients
WHERE Name = 'I. Kalet'
```

In this example, it is unambiguous what the attribute Name refers to, but in cases where two or more tables are used in an operation, it will be necessary to use "qualified" attribute names. A qualified name consists of a prefix and a dot, followed by the attribute name. The prefix can be the name assigned to the relation by the CREATE TABLE operation, or it can be a local name associated with the relation, as in the following example. This example shows the use of the UPDATE operator, which uses a SET clause to update an attribute value for a tuple identified by a WHERE clause.

```
UPDATE Patients P
SET P.Hosp_ID = '65-21-07'
WHERE P.Name = 'Norbert Pivnick'
```

Table 1.7 A View Table that Results From a Query of the Patients Relation

Name	Age
Joe Pancreas	38
David Duck	46
...	

A *query* consists of a specification of which columns, which table(s), and selection criteria to use. The query commands make up the Query Language. Generally, the result of a query is a new table, called a *view*. Views are not stored in the database, but are temporary creations, that are manipulated in an interactive context or in a program.

The SELECT operator is used for queries. The format is that after the SELECT one provides a sequence of column labels, separated by commas, and qualified by a symbol that is a proxy for a table name (the P. in the following example). As before, the proxy name is not needed if the column name is unambiguous, that is, if only one of the tables referenced contains a column of that name. Next there is a FROM clause, associating table names with the proxy names, then the selection criteria (if any), specified by a WHERE clause. Here is an example on the Patients relation:

```
SELECT P.Name,P.Age
FROM Patients P
WHERE P.Tumor_Site = 'Pancreas'
```

which gives the result shown in Table 1.7.

An extremely important property of the relational model is that queries on relations produce other relations. The input to a query is one or more tables and the output is another table. Thus, a relational database is *closed* under query operations. This makes nested queries possible. The output of one query can be part of the input of another. Suppose we have another table, with treatment records:

```
CREATE TABLE Treatments
            (Name CHAR(20), Treatment_Type CHAR(10), Dose INTEGER)
```

and the table is populated as in Table 1.8.

Then with this example, nested queries are possible. The following query would list all the patients and tumor sites that were treated with neutrons.

Table 1.8   A Table of Treatment Records for Each Patient

Name	Radiation Type	Dose
Sam Vacuum	neutrons	960
Norbert Pivnick	neutrons	840
Sue Me Too	electrons	5040
Joe Pancreas	photons	6000
David Duck	photons	7000
...		

Table 1.9 A View Table Resulting From a Nested Query Involving Both the Patients Relation and the Treatments Relation

Name	*Tumor_Site*
Sam Vacuum	Glioblastoma
Norbert Pivnick	Prostrate
...	

```
SELECT P.Name,P.Tumor_Site
FROM Patients P
WHERE P.Name IN (SELECT T.Name
                 FROM Treatments T
                 WHERE T.Treatment_Type = 'Neutrons')
```

This query would produce the result shown in Table 1.9.

Originally SQL was intended to be a data manipulation "sublanguage," and did not have the full power of a "Turing-equivalent" programming language like Lisp, C, FORTRAN, or java. It provides a very powerful way to define data organization and to search and manipulate it as is most often needed with database systems. The more recent SQL standard includes extensions that make it capable of arbitrary computations, but many problems remain [83, Chapter 4]. Most database management system implementations also include a way to embed SQL expressions in a regular programming language, so that ultimately any computation may be done with the data.

1.3.1.2 Normalization

Although it is required that each tuple in a relation should be unique, it may be that there are many ways to represent the data. Sometimes the attributes in a relation are not really independent, but there are constraints between them. For example, if we were to represent lab tests of a patient in a table with the patient's name, address, attending physician, diagnosis, and other information about the patient, every tuple will have something different about it, making it unique. Nevertheless, there will be significant redundancy in such a table. The patient's address will (usually) be the same in every tuple for that patient, and other attributes will also be determined once the patient is determined. It would seem more efficient to split this information into two tables with links between them. Depending on the relationships among the data, there could be multiple levels of splitting that are possible. These different ways of handling dependencies between attributes are called *normal forms*. If a relation satisfies the basic requirement that no attribute's value in any tuple has multiple values, it is said to be in First Normal Form, or 1NF. Every valid relation is in 1NF.

The laboratory data example, illustrated in Table 1.10, has dependencies between attributes, so it can be reorganized into two separate tables. This process is called *normalization*.

The Name, Birth Year, and Diagnosis all appear to be related. One might think of splitting this into two tables, one with just those fields, and the other with all the lab test data, together with just enough information to tie each lab test back to the right individual. In reality, the

Table 1.10   A Table With Attributes That Apparently Have Dependencies

Name	Birth Year	Diagnosis	Test Name	Result	Date of Test
Joe Pancreas	1967	Pancreatic Ca.	Creatinine	1.7	24-May-2004
David Duck	1954	Prostate Ca.	PSA	14.3	29-May-2004
Joe Pancreas	1967	Pancreatic Ca.	Creatinine	1.3	2-Jun-2004
Joe Pancreas	1967	Pancreatic Ca.	Creatinine	1.6	19-Jun-2004
David Duck	1954	Prostate Ca.	PSA	16.1	23-Jun-2004
David Duck	1954	Prostate Ca.	PSA	0.4	9-Aug-2004
...					

name is not enough, but such tables often have many attributes, a few of which together serve to identify the patient. So, one relation would have the Name, Birth Year, and Diagnosis, and the other would have the Name, Test name, Result, and Date of Test.

This more highly normalized arrangement would avoid the redundant repetition of the Birth Date and Diagnosis information. In the original arrangement, if this data had to be re-entered manually with each new lab test, typing errors would certainly occur, and would be extremely hard to find and correct. One effect of such errors would be that important records would not be retrievable.

1.3.1.3 Constraints and Keys

A reasonable database system would support ways of insuring that the data stored in it are consistent with the schemas, and allow for additional constraints to be specified where needed. Already there are enforceable constraints on elements that can be in each column, because *type* specifications are included in the table definitions. These can be enforced by the data entry facilities. In addition, as noted earlier, no two rows (tuples) in a table can be identical. Each pair of tuples must differ in the value of at least one column, or attribute. Further constraints can be specified in SQL, called *key constraints*, by defining one or more fields in a relation to have the property of being a *key*.

If the relation is at least in 1NF, that is, it satisfies the requirement of each tuple being unique, there is always some combination of attributes that can be used as a key, whose value will uniquely specify a single tuple. The worst case, of course, is that the entire tuple is the key. More often, a single attribute will suffice to identify the record, or a combination of two or three will do it. There may even be more than one choice of attribute that serves as a key. In the following example, the hospital ID (or medical record number), which is the attribute labeled Hosp_ID, is identified as a key by declaring it UNIQUE, so in the Patients table, no two records will have the same value for the Hosp_ID field.

```
CREATE TABLE Patients
     (Name CHAR(20), Hosp_ID CHAR(10),
      Tumor_site CHAR(20), Dose INTEGER, UNIQUE (Hosp_ID))
```

It is possible to choose one key and designate it the *primary key*. This helps the database system optimize queries, for example by generating an index. In order for a set of attributes to serve as a primary key, those attributes must be declared NOT NULL as well as unique in combination.

In order to relate one table to another, one can also designate an attribute or combination to be a *Foreign key*. This attribute is then considered a link to connect information from one table to that in another table. So, for example, for the Treatments table, a better design would be

```
CREATE TABLE Treatments
    (Hosp_ID CHAR(10), Treatment_Type CHAR(10),
     PRIMARY KEY (Hosp_ID),
     FOREIGN KEY (Hosp_ID) REFERENCES Patients)
```

The ability to combine tables to generate new ones, as well as use restrictions and projections makes the relational model very powerful. All the description so far is really a description of the abstract logic of relations and does not imply anything about the actual implementation. How the data are stored in files, actually represented in memory, whether a query is processed directly in some obvious way or transformed first into an equivalent but more efficient form, are all details of which the end user normally would not be aware.

1.3.1.4 Relational Algebra

We already showed the *selection* operation, which picks some subset of rows from a relation. This can be thought of and noted formally as a function on a relation that returns another relation. The Greek letter σ is used to denote a selection operation. The selection operation $\sigma_{\text{Tumor\_site=Pancreas}}(P)$, where the symbol P refers to the Patients table, does part of the job of the SELECT operation example shown earlier. The other part of that operation is to only display two of the columns, not the entire set of columns. That is a *projection* operator.

The *projection* operation picks a subset of columns (fields). So, the SQL SELECT operator can do just a projection, by not specifying a WHERE clause. Projection is denoted by the Greek letter π, so the projection in the same example as above would be $\pi_{\text{Name,Age}}(P2)$, where $P2$ is the result of the selection operator above. Thus the SQL query would be rewritten as $\pi_{\text{Name,Age}}(\sigma_{\text{Tumor\_site=Pancreas}}(P))$.

Often the information we need is a combination of data from two or more tables, rather than a single table. We need operations to combine tables into bigger tables, and then to select and project from those. A *cross product* forms a new composite table from two other tables – for every combination of an element from one with an element of the second, there is a row in the cross product. The cross product is denoted by the symbol $\times$.

This does not usually get us exactly what we need. In the cross product table there may be repeated columns, if the two tables have a column in common, that is, the same kind of data appears in each. We only want to link up the rows in which the data correspond, where the contents from one table in its column are the same as the contents of the other table in the corresponding column. For example, we combine two tables that each have patient names. We want only the cross product records where the name field from the first table is the same as the

Table 1.11   A Table Showing Both Tumor Site and Treatment for Each Patient, the Result of the Natural Join, Patients ⋈ Treatments

Name	Hosp_ID	Tumor_Site	Age	Radiation_Type	Dose
Sam Vacuum	15-64-00	Glioblastoma	60	neutrons	960
Sue Me Too	99-88-77	Lt. Breast	45	electrons	5040
Norbert Pivnick	−6	Prostrate	62	neutrons	840
...					

name field in the second table. So we can do a selection on that basis, and then a projection that just leaves out the now duplicate column.

A *join* is a cross product followed by a selection and/or projection operator. The symbol for a join is ⋈. A common kind of join, that accomplishes what is described in the previous paragraph, *selects* rows in which fields with the same (field) name are equal. In that case, the two equal columns are condensed to a single column; the selection is followed by a projection. So Patients joined with Treatments gives a table where the "Radiation type" and "Dose" fields are added on for each patient for which there is a record in both tables. The result of such a join looks like Table 1.11. This "natural join" can be expressed in relational algebra as Patients ⋈ Treatments. Since no condition for selection or projection is specified, the two columns whose names match are taken to be the ones where equal fields select the rows, and the columns are then condensed to one.

Other operations on tables are possible, such as set union, intersection, and others. Standard database texts such as [83,105,334] develop all these ideas in more depth. A motivation for expressing schemas and queries in relational algebra is that complex queries can be transformed into equivalent queries that may realize huge efficiency gains. This is a subject for database system design. A well-designed system will need a way to take a schema and create an implementation, some efficient way to store the data and maintain appropriate indexing of records for fast search and retrieval. A query interpreter is needed to translate SQL (or some other query language) into basic operators that can be implemented in the system. A query optimizer then looks at the operations requested to see if the query can be reorganized to a more efficient form, perhaps by changing the order of selection operations or by handling rows in groups rather than one by one. Finally an efficient execution engine will perform the optimized query using the index and other facilities available in the system.

1.3.1.5 The Catalog, Metadata, and Object-Oriented Databases

Use of the CLOS Meta-Object Protocol in Section 1.2.3 to serialize data showed that programs can be used as data in order to generalize operations on the data that the programs describe. This is similar to how a database system works. The SQL CREATE TABLE expressions are code in an SQL program, but they can also serve the same meta-object role as the CLOS class definitions. They *describe* the structures of relations. Database implementations would convert the text of such statements into an internal form that can be used to interpret further SQL operations and tuple data. This information is itself stored in a relation (or set of relations) called the *catalog*, sometimes also called the *data dictionary*. The point here is that a relational database is typically built on its own machinery, much as the Common Lisp Object System is built using CLOS classes and methods.

One implication of this is that one could write a program to create a database schema from a class model in a programming language, if the class descriptions were available. It might seem that this way you could add *persistence* or non-volatile storage, to a programming environment like Common Lisp. There are systems that augment Common Lisp to provide persistent objects, essentially an object-oriented database, but they do not use a relational database as a back end storage or implementation scheme. Instead they use the MOP to implement an object-oriented database model more directly. An example of early work on persistence in the context of Lisp systems is Heiko Kirschke's PLOB system [231,232]. A commercially supported implementation of persistence in a Lisp system is the AllegroCache^{TM} system [119] from Franz, Inc., which is an extension of Allegro Common Lisp to provide object-oriented database capabilities. This approach comes at the problem from the opposite end of the database field, where relational database technology and SQL have been extended to incorporate ideas from general-purpose and object-oriented programming. The work on object-oriented databases in the Lisp environment emphasizes extending the programming environment to support persistence for program data, so there is no issue of an object-oriented query language. The Lisp programmer just writes Lisp code in the usual way to manipulate persistent data just the same as transient data.

1.3.2 Indexing and Tree Structures

In a relational table, designating an attribute as a primary key can speed up query operations. The database system achieves this speedup by creating and maintaining an *index* for each primary key. Without an index, getting the tuples that match even a simple query requires that the query processor retrieve and process every row in the referenced table, checking if the value(s) of the attributes in the query match the query. This may require a lot of disk reads and is $O(n)$, or linear, in the size of the table. Having an index can reduce the cost significantly, from $O(n)$ to $O(\log n)$. A query that requires a join is even more burdensome, and there is even more potential performance improvement with the availability of an index.

Two ideas are involved. First, each tuple is assigned some kind of "row number" or identifier that the system can use to retrieve the record directly, without reading sequentially through the table (as for example in a disk file). This is analogous to storing each tuple's contents as an element of an array, and referring to the tuple with an array index. Second, a separate structure that is quickly searchable contains an entry for each possible value of the indexed attribute, together with the row number of the tuple that contains that value. This is possible because in order for an attribute to be a key, it has to have a unique value in each tuple. If there were duplicates, the attribute would not qualify as a key.

For example, an electronic medical record table like the `patients` table would have an index associated with the `Hosp_ID` attribute, so that finding the tuple for a specific patient would be

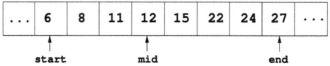

Figure 1.30 A range within an array of numbers, showing the current search range and the mid-point (rounded to an even array index).

very efficient. This is important because EMR systems are mainly used as transaction systems, in the process of dealing with one patient at a time.

1.3.2.1 Binary Search on Arrays

The index may take any of several forms. If the possible values of the indexed attribute are able to be ordered, the index can be sorted (and kept in sorted order). Knowing the size of the index and storing the entries in an array, a "divide and conquer" method will find an entry in $O(\log n)$ time (where n is the size of the data set (or array). The principle here is to retrieve the middle entry and determine if the desired entry is before, equal to, or after the middle. If equal, the entry is found. If before, the process is repeated on the first half of the array, and if after, it repeats on the rest of the array. Figure 1.30 illustrates the array and the pointers to the start of the current search, the midpoint, and the end of the range.

Here is an implementation from Graham [137, page 60]. The finder function does the work, but it has start and end inputs to specify where in the array to confine the search. The higher level function bin-search sets the initial bounds and then uses finder to do the work.

```
(defun bin-search (obj vec)
  (let ((len (length vec)))
    (and (not (zerop len))
         (finder obj vec 0 (- len 1)))))

(defun finder (obj vec start end)
  (let ((range (- end start)))
    (if (zerop range)
        (if (eql obj (aref vec start))
            obj
            nil)
        (let ((mid (+ start (round (/ range 2)))))
          (let ((obj2 (aref vec mid)))
            (if (< obj obj2)
                (finder obj vec start (- mid 1))
                (if (> obj obj2)
                    (finder obj vec (+ mid 1) end)
                    obj)))))))
```

This code can be improved. First, using the `labels` operator would make the `finder` function local, to avoid name conflicts and show more clearly that it is an auxiliary function to the search function (it is renamed `bs-aux` in the following). Second, instead of assuming the items in the array are always numbers and explicitly using the < operator, we can provide the comparison function as an optional input. This requires some reorganizing of the logic, since it will not be known in general what the opposite comparison should be. So, an explicit check if the middle item is the desired one has to come first, then the test if the desired item is before the middle item. If these both fail, you can safely assume that the desired item is *after* the middle item.

Finally, there is one case that Graham's version does not handle. If the desired item is *before* the beginning of the array, the `finder` function will not terminate. This happens because the `start` and `mid` indices become equal due to rounding. An additional check is necessary to detect this and not do the wrong recursive call in this case.

```
(defun binary-search (search-obj input-vec &optional (compare #'<))
  ;; define a local recursive function that tests and splits
  ;; a range specified by start and end
  (labels ((bs-aux (obj vec before start end)
             (format t "~A~%" (subseq vec start (+ end 1)))
             (let ((range (- end start)))
               (if (zerop range)
                   (if (eql obj (aref vec start)) obj nil)
                   (let* ((mid (+ start (round range 2)))
                          (current (aref vec mid)))
                     (cond ((eql obj current) obj)
                           ((and (funcall before obj current)
                                 ;; this check prevents an infinite loop
                                 (not (eql start mid)))
                            (bs-aux obj vec before start (- mid 1)))
                           (t (bs-aux obj vec before (+ mid 1) end))))))))
    ;; apply the function to the full input array
    (let ((size (array-dimension input-vec 0)))
      (and (not (zerop size))
           (bs-aux search-obj input-vec compare
                   0 (- size 1))))))
```

One more level of generality can be added to this operation. The most common case is that the items stored in the array are each instances of a structured data type with several fields. One of these fields contains the index value (even for an index in a relational table, this is the case, as each index entry will have a *pair*, the index value, and the row number of the tuple containing it). In that case, two additional inputs are needed, as in the built-in `find` function and similar sequence functions in Common Lisp. One is a lookup function that obtains the element to be compared with `search-obj`, and the other is an equality test function to be used instead of using `eql`. These two inputs are analogous to the keyword arguments `:key` and `:test` used in the standard sequence functions. It is left as an exercise to add this to the code above.

Although the algorithm is presented here in the context of indexing relational tables, it is useable as an organizational strategy more generally. However, one disadvantage of creating sorted arrays is that it is expensive to add new entries. Although it is easy and fast to find the array location where the new entry should be put, the insertion code must then move the rest of the array in order to free up the storage. In turn this requires that there must be expansion room at the end of the array, or somehow the programming environment must provide for dynamically extendable arrays (this is indeed available in Common Lisp).

1.3.2.2 Binary Search Trees

When data can be indexed by attributes that support ordering or sorting, it is possible to organize them in a way that provides both $O(\log n)$ time performance and ease of updating the data structure. One such structure is the binary tree. A binary tree is made up of *nodes*. A node is a structure that has a value slot for the item represented by that node, and has two links, one to a subtree of nodes that come before it (in order by element value) and one for the subtree of nodes that come after it. As with all well-designed tree structures, there is no difference between the "root" node and any other node. This is the key to being able to design a recursive function that operates on any tree or subtree, always the same way.

The node structure is specified by a `defstruct` form.[51] The `prev` slot is initially empty but can contain a pointer to a node (or subtree) whose contents are before the contents of the specified node. The `after` slot similarly is initially empty but can contain a pointer to a node whose contents come after the contents of the specified node.

```
(defstruct node contents (prev nil) (after nil))
```

To insert a new node in a tree, you start at the root of the tree. If the tree is empty, meaning the current tree input to the insertion function is `nil`, just make a new node with the specified contents. If the current node's contents are the same as the new one, we don't allow duplicates and it is already in the tree, so just return that node as is. The node includes the pointers to any further tree layers, so, if the node is found somewhere in the interior, the entire subtree consisting of that node and its children is returned. This helps distinguish between not finding anything and finding something whose `contents` slot contains `nil`.

If the contents are not the same as the current node, add a new node with the same contents as the node being examined. If the new item belongs before the current contents, the insertion is applied to the `prev` subtree, and the `after` subtree remains the same. If the new item belongs after the current item, the insertion is applied to the `after` subtree and the `prev` subtree is linked unchanged.

```
(defun bst-insert (obj bst before)
  (if (null bst) (make-node :contents obj)
```

---

[51]The code here follows that of Graham [137, pp. 72–76], with some minor changes for clarity. For example, Graham uses the < as a variable whose value is the comparision function, where I decided to use the name `before` to emphasize the generality of these functions.

```
(let ((contents (node-contents bst)))
  (if (eql obj contents) bst
    (if (funcall before obj contents)
        (make-node
          :contents contents
          :prev (bst-insert obj (node-prev bst) before)
          :after (node-after bst))
      (make-node
        :contents contents
        :after (bst-insert obj (node-after bst) before)
        :prev (node-prev bst)))))))
```

This does not change the original "current node" so the original tree still exists. The function bst-insert returns a *new* tree, that *replicates* the chain of nodes up to the insertion point, and *shares* the remainder of the tree. This is similar to other non-destructive functions in Lisp that return new structure, leaving the old structure intact.[52]

Figure 1.31 shows a tree built from some names, organized so the tree represents alphabetic ordering. The symbol tree has this tree's root node as its value.

Once the tree is built, it can be updated by continuing to use the bst-insert function for each new entry. The cost is small. Figure 1.32 shows the result of adding an entry with the name "Kalet" to the tree in Figure 1.31. The binding of the symbol tree is unchanged, and the new tree with the new name inserted is the value of another symbol, new-tree. The nodes in new-tree

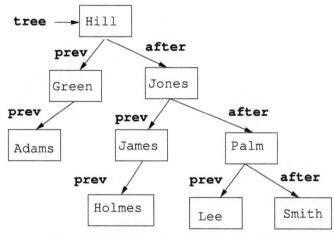

Figure 1.31 A binary search tree built with some patient names sorted in alphabetic order.

---

[52]A destructive version of this same idea appears also in Graham [137] on page 202. One reason for doing destructive operations is to make the code more space efficient and minimize the need for garbage collection.

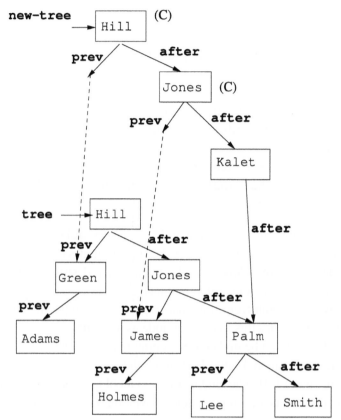

Figure 1.32 The new binary search tree with the node for "Kalet" added.

above the new node named "Kalet" are *copies* of the original nodes, but the remainder, under "Kalet" is the corresponding subtree of the original.

To find an entry is even simpler, much simpler than the divide and conquer strategy for an array, because the tree already represents the contents as split up. If the current node contents match the item you are searching for, you are done. Otherwise check if the object is before the current node. If it is, search the prev subtree, otherwise search the after subtree.

```
(defun bst-find (obj bst before)
  (if (null bst) nil
    (let ((contents (node-contents bst)))
      (if (eql obj contents) bst
        (if (funcall before obj contents)
            (bst-find obj (node-prev bst) before)
          (bst-find obj (node-after bst) before))))))
```

Removing a node is more work, because the link to it must be filled in with the rest of that node's subtree. The reader is referred to Graham [137, page 74] for the details.

1.3.2.3 Hash Tables

The two preceding methods apply to the case where there is an ordering of the set of possible attribute values, and a predicate that expresses the ordering (a comparison operator that determines if one value is before another). If there is no ordering, these methods can't be used.

An alternative is to implement the index as a *hash table*. A hash table consists of entries whose index values are used to compute the entry references, rather than searching for them. This reduces the lookup time even further, to $O(1)$, since the time it takes to compute the row number or retrieve the hash table entry is the same regardless of the order in which the table rows are actually stored, or how many there are. Hash tables are provided in Common Lisp as a part of the standard. Here is an example of use of the built-in data types and functions. Instead of implementing the `patients` table as a relation, it will be a hash table with the `Hosp_ID` field value as the hash key. Each patient's record will be an instance of the class `patient`, defined in Section 1.2.2.2. The scope of the `patient-table` variable is only within the `let` expression, so the hash table will only be accessible through functions that are defined within the `let` expression. However, the binding has indefinite extent, meaning that the hash table will be available to these functions for the entire duration of running the program, since the function definitions are global.

```
(let ((patient-table (make-hash-table)))
  (defun add-patient (pat)
    (setf (gethash (hospital-id pat) patient-table)
          pat))
  (defun find-patient (hosp-id)
    (gethash hosp-id patient-table))
  (defun list-patients-by-names ()
    (maphash #'name patient-table))
  ...<other useful functions to be added here>
)
```

This is an example of how to make information private so that it will not be accidentally affected by other parts of a large complex computer program. It is even safe to have another place in such a program where the symbol `patient-table` is used. That will refer to a different binding of this variable name. Operations on it will not affect the hash table in the above code, because when the `let` expression evaluation completes, the hash table is no longer the value of that variable. It is only referred to internally by the defined functions.[53]

In other programming languages, basic hash table functions would typically not be part of the base language standard but would be available in code libraries that are so commonly used that they are *de facto* standards. However, in a computer science course on data structures and algorithms one learns how to implement hashing algorithms, not only to understand the basis

---

[53] Other examples of this technique may be found in Graham [137, page 108] and Norvig [305, pages 92–95].

for the libraries, but also to be able to implement special purpose variants that may be more suited to specific problems. Many textbooks are available, such as [437] which uses the java programming language.

1.3.3 Time-Oriented Data and Interval Trees

Much of the information found in electronic medical records has a meaning associated with some aspect of time. In many cases, such as the description of laboratory tests in the LOINC coding system, the test requests and reports themselves, and the composite medical image data for a particular image study, a time stamp appears explicitly as a property of the information object. These are, for example, the time that the test was ordered, the time the specimen was obtained, the time the test was completed, and the time it was reported. Medical images typically have acquisition dates and times associated with the images, while the reports will have a date and time that the report was written. There may only be an implicit reference to a time or date for some data. For example, a current medications list will not always have such information as when a medication was started.

Clinical conditions of a patient change over time, and may or may not be recorded in detail in the electronic record. Worse, the beginning, end, and duration of conditions are often stated without much precision, either because the patient cannot report more precise data or because the condition is not precisely specified. Nevertheless it is important to medical decision making. Thus representing and reasoning about time-oriented medical data has been studied for a long time [74, 162, 196, 260, 331, 373, 393]. A recent review and position paper [5] presents ideas and challenges for research on representation and reasoning about time-oriented data, a methodological review [448] treats aspects relating to medical natural language processing, and an even more recent book is devoted to this subject [72]. In this section a relatively simple way to represent and process time-oriented data should serve as an introduction. This material is based on the Ph.D. dissertation research of Daniel Capurro [60] at the University of Washington. The data structures and algorithms here are formally defined using Common Lisp. Dr. Capurro did an implementation in the java programming language.

The presentation here focuses on organizing time-oriented data to facilitate queries involving multiple patients, rather than on reasoning about the time course of disease and therapy for a single patient. Such queries may be designed for research purposes, to identify suitable subject for a clinical trial or retrospective study, but they are also important for such administrative applications as quality assurance. In both cases, these uses are called *secondary* uses of medical data. Much work has been published on the specific problem of querying temporal clinical data for research purposes [60, 73, 234, 330]. The development here presents the basic ideas and a simple implementation. It shows how to create interval labeled data from clinical laboratory test results, how to represent relations between intervals, how to organize interval data into efficient tree structures, and how to search for intervals that match queries. This formalism has several nice features. First, the tree structure is easy to update with additional data. Second, complex queries that would be very difficult to implement in a relational database context can be easily expressed.

Third, the result of a query is in the same form as the input data to be searched, so queries can be nested. This is similar to a very important property of SQL and the relational model, in which the output of a query on a table or tables is another table, so there too, queries can be nested.

Laboratory results (and other clinical data) will typically have a time stamp or some kind of label indicating when the observation was made. In order to construct intervals, it is necessary to order the data linearly in time, and then consider how to abstract intervals from subsequences of the sorted results sequence. The starting point for this development will be a simple representation of results, representing each result as an event at an instant in time. The problem of how to get this information out of a particular electronic medical record system is not treated here. Such processing is specific to each system and is nowhere dealt with in a systematic or principaled way. The general problem of extracting information from arcane systems such as electronic medical record systems in order to apply data analytics or reasoning such as medical practice guidelines or other computations for clinical decision support remains unsolved. It has been referred to as the "curley braces" problem, identified early on in connection with the development of Medical Logic Modules using Arden Syntax [164]. One possible way to address this problem would be to write a parser for HL7 messages, following up the development presented in Section 4.3.2 of Chapter 4. This would then require a way to solicit HL7 messages on demand, or a way to accumulate them in a repository for later lookup, neither of which is practical in the context of current EMR systems.

1.3.3.1 Constructing intervals from results

The discussion here will consider only a simple representation of test results, those that consist of an identifier, a value, a time stamp, and upper and lower limits delineating the "normal" range. From a sequence of these, it should be possible to create intervals that represent periods during which a value remained stable, or increased or decreased, and by what amount or fraction. These are the terms of typical queries, such as "identify patients who experienced acute kidney failure during hospital stay." Acute kidney failure is defined as an absolute increase of serum creatinine of at least 0.3 mg/dL, or a relative increase of at least 50%, from the baseline. The interpretation of collections of events or observations at single points in time, to produce intervals, is the focus of many of the previously cited works. The focus here will be on organizing and searching the resulting tree structure to answer queries. The structure of the `interval` data type will thus be very simple.

```
(defstruct interval
  (start 0.0)
  (end 0.0)
  (description nil))
```

The description will be a symbol or list containing the data that specify what happened in that interval, for example, the list,

```
(creatinine low)
```

More complex examples are provided in Daniel Capurro's dissertation [60, Chapter 6]. The development here only focuses on the time intervals.

1.3.3.2 *Constructing an interval tree from intervals*

Many tree structures have been described in textbooks and other sources, including tries [305, pages 343 and 472] and interval B-trees [14]. Also many different types of interval relationships have been described, as for example, in [11]. However it is sufficient to be able to represent only three relations, of which all the others can be considered special cases. These three are:

*   interval A is *before* interval B,
*   interval A is *after* interval B, or
*   interval A *overlaps* with interval B.

Strictly speaking, it is not necessary even to include the *after* relation, since it can be expressed in terms of *before* by reversing A and B. However, for simplicity we represent it separately to make the tree structure transparent. So each node in the tree will have three branches or subtrees. A node in the tree will be a structure with contents (the interval data itself), the subtree of all intervals that are strictly before that node's interval, the subtree of intervals that overlap with the interval, and the subtree of all nodes that are strictly after the interval.

```
(defstruct interval-node
  contents
  (before nil)
  (overlap nil)
  (after nil))
```

Relations compare intervals, and return t if the relation is satisfied by the interval pair, in the specified order. The three relations are predicate functions, and can easily be explicitly defined by comparisons on the interval end points.

```
(defun before (int1 int2)
  (< (interval-end int1) (interval-start int2)))

(defun after (int1 int2)
  (> (interval-start int1) (interval-end int2)))
```

These two could be combined into one predicate function by noting that if int1 is *before* int2 that is the same as determining that int2 is *after* int1. In other words, the relations are antisymmetric. However, it is easier to read code that uses the two names, rather than having to keep track in your head which one is intended in a particular situation. The overlap function *is* symmetric in that if int1 overlaps int2 then it is also true that int2 overlaps int1. This lumps together all the kinds of overlap (adjoining, contained in, partial overlap) into a single relation.

```
(defun overlap (int1 int2)
  ;; need to decide if intervals are open or closed - what about
  ;; matching end points?
  ;; could also define this one in terms of the others, as
```

```
;; (and (not (before int1 int2)) (not (before int2 int1))) but note
;; that would implement closed intervals for overlap
(and (< (interval-start int2) (interval-end int1))
     (< (interval-start int1) (interval-end int2)))))
```

To construct a tree from a sequence or list of intervals (not sorted, since there is no way to do such a sort), the new interval is compared with the current node and percolated through the subtrees according to the results of each comparison. The interval, int, is checked with each test until one succeeds. Which one succeeds determines how the insertion takes place. For example, if the before test succeeds, the new interval belongs to the before subtree, so a new node is constructed from the current node, its before subtree will be the result of (recursively) inserting the new interval into the original before subtree, and its after subtree will be the original (unchanged) after subtree. This is a non-destructive operation, in that the original tree is still intact, but some new structure is created that incorporates the old structure.

```
(defun interval-insert (int tree)
  (if (null tree) (make-interval-node :contents int)
    (let ((contents (interval-node-contents tree)))
      (cond ((equal int contents) tree)
            ((before int contents)
             (make-interval-node
              :contents contents
              :before (interval-insert int
                        (interval-node-before tree))
              :overlap (interval-node-overlap tree)
              :after (interval-node-after tree)))
            ((overlap int contents)
             (make-interval-node
              :contents contents
              :before (interval-node-before tree)
              :overlap (interval-insert int
                          (interval-node-overlap tree))
              :after (interval-node-after tree)))
            ((after int contents)
             (make-interval-node
              :contents contents
              :before (interval-node-before tree)
              :overlap (interval-node-overlap tree)
              :after (interval-insert int
                        (interval-node-after tree))))
            ))))
```

This construction is exactly like the construction of the binary tree described in Section 1.3.2.2 but with a three-way branching instead of a two-way branching. It has all the same advantages, including ease and efficiency of search as well as ease of adding new nodes.

1.3.3.3 *Queries and searching the interval tree*

Once the intervals are organized into the tree structure, it is easy to perform the basic operations from which complex queries are built. The queries are:

*   to find an interval in the tree and return that node together with its subtrees,
*   to find all the intervals that are before a given interval,
*   to find all the intervals that overlap a given interval, and
*   to find all the intervals that are after a given interval.

The `interval-find` function searches for a specified interval, `int` in an interval tree, `tree` and returns the subtree for that interval if found, or `nil` if the interval is not present. It is very straightforward. If the interval corresponds to the current node, return the node, otherwise compare it with the interval in the current node and search the appropriate subtree.

```
(defun interval-find (int tree)
  (if (null tree) nil
    (let ((contents (interval-node-contents tree)))
      (if (equal int contents) tree
        (interval-find int
                       (cond ((before int contents)
                              (interval-node-before tree))
                             ((overlap int contents)
                              (interval-node-overlap tree))
                             ((after int contents)
                              (interval-node-after tree))))))))
```

If the interval is in the tree, the remaining queries are already answered, as they each are satisfied by simply collecting all the intervals in the corresponding subtree. Writing this code is left as an exercise for the reader.

If the interval is not already in the tree, the search is a little more difficult. First consider the search for all intervals before the given one, assuming the interval is *not* in the tree.

```
(defun find-before (int tree)
  (if (null tree) nil
    (let ((contents (interval-node-contents tree)))
      (cond ((before int contents)
             ;; search the before AND overlaps subtrees
             (append (find-before int (interval-node-before tree))
                     (find-before int (interval-node-overlap tree))))
            ((overlap int contents)
             ;; search all three subtrees
             (append (find-before int (interval-node-before tree))
                     (find-before int (interval-node-overlap tree))
                     (find-before int (interval-node-after tree))))
            (after int contents)
            ;; include the current interval AND
            ;; search the overlaps and after subtrees
```

```
(cons contents
    (append
        (find-before int (interval-node-overlap tree))
        (find-before int (interval-node-after tree)))))))))
```

The idea is that if the interval is before the current one, don't include the current one or the ones that are *after* it, since they will surely not be before the given interval, but do search the other two branches. If there is overlap, the current interval is still not included but all three subtrees need to be searched. If the given interval is after the current interval, the current interval *does* satisfy the criterion so it is put on the beginning of the list resulting from searching the overlaps and after subtrees.

The logic for the `find-overlaps` and `find-after` functions is similar and is again left as an exercise. Queries that specify minimum, maximum, or explicit elapsed times between the query intervals can be implemented by filtering the resulting list with simple comparisons. A higher level query processor then combines all this to construct a new tree from the intervals that satisfy the query, thus making nested queries or queries that combine relations by standard logical operations such as AND, OR straightforward.

Implementing these ideas in a fully functional system would involve not only solving the "curley braces" problem mentioned above, but also providing a suitable user interface so that the end user would not have to write programs (though a well-implemented system would provide that capability), just as relational database systems provide full ability to write SQL in addition to graphical user interfaces. Considerable work has been done on user interfaces, including visualization of results [234].

1.3.4 The Entity-Attribute-Value (EAV) Model

We consider one more flexible approach related to the tagged data idea, the Entity-Attribute-Value (EAV) model. In this model, every piece of data is in the same form, a triple, in which the three elements are: a reference to some object, the name of an attribute of that object, and the value associated with that attribute. The value could possibly be a reference to another object, or it could be a basic datum (a number, a string, or other elementary datum). Some attributes may have multiple values, for example, PSA test results, in which case there will be multiple EAV triples, one for each test result.

Here are some examples, just written as three element lists:

```
(patient-6 creatinine 1.2)
```

```
(radiation-beam-3 gantry-angle 45.0)
```

A value could be another entity, as in.

```
(patient-6 treatment radiation-beam-3)
```

This could be represented in a relational database as a simple three column table. The efficiency is offset by the complexity of queries, some of which may not be possible in SQL.

Nevertheless, such systems are used as the basis for some EMR implementations, notable the MIND system at the University of Washington.

In some hospital IT settings this model is also called the vertical model, because a graphical display of the data when extracted by a simple SQL query runs vertically down the page, row after row, rather than horizontally across the page, column after column. EAV modeling in production databases is commonly seen in clinical data repositories, with such areas as clinical findings, clinical procedures, etc.

Another way to think about the EAV model is that it is a concrete realization of a collection of RDF triples. This suggests that an effective way to query EAV data would be similar to searching through an RDF network using path specifications like Xpath. Exploring this idea further is beyond the scope of this book.

1.3.5 Online Analytical Processing (OLAP)

Indexing of keys in a relation provides efficiency for some kinds of queries, but there are a growing number of applications for which it is central to be able to retrieve and process large numbers of attribute values from tables based on combinations of the values of the other attributes. In effect, one would like to think of the relation as an array of values of one attribute, with the array indices corresponding to the values of the other attributes. Extensions of relational databases that support such queries are not yet standard, but they have been applied to interesting scientific problems. This section provides an introduction to these extensions in the context of an extremely important bioinformatics problem that of analyzing large data sets resulting from dynamic simulation of protein folding and unfolding. The presentation here is written and contributed by Andrew Simms and derives from his research [384] and doctoral dissertation [383]. Except for Figure 1.33 as noted, the figures in this section were created and contributed by Dr. Simms, as well as Table 1.12.

1.3.5.1 Background

The relational database was first described in 1970 by Codd [69] and grew to become the dominant data management paradigm for Online Transaction Processing (OLTP). Much of

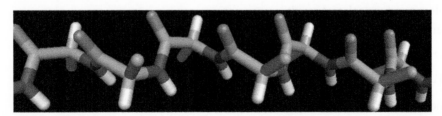

Figure 1.33 An alpha strand using alternations of (−50,−50)/(+50,+50) for the protein backbone dihedral angles ϕ and Ψ; Engh-Huber bond geometry is used throughout. Figure and caption from Wikimedia Commons, URL http://commons.wikimedia.org/wiki/File:Alpha_strand_50_50.png.

Table 1.12   The Result of the Query in Figure 1.35 is a Two Dimensional Table Consisting of 362 Rows and 54 Columns. Here Only the First nine Columns and a Subset of the Φ Angle Bins From −120 to −100 Degrees are Shown

	PRO	ARG	THR	ALA	PHE	SER	SER	GLU	GLN
−120	(null)	452	472	249	116	4	(null)	(null)	(null)
−119	(null)	461	446	237	151	5	(null)	1	(null)
−118	(null)	450	431	246	138	3	(null)	(null)	2
−117	(null)	495	433	264	145	3	(null)	1	1
−116	(null)	463	380	227	129	9	(null)	1	(null)
−115	(null)	466	391	260	153	7	(null)	(null)	3
−114	(null)	473	330	285	148	8	(null)	(null)	3
−113	(null)	455	319	289	176	10	(null)	1	(null)
−112	(null)	475	308	272	185	12	(null)	(null)	3
−111	(null)	436	293	303	176	22	2	(null)	3
−110	(null)	447	307	351	165	24	(null)	(null)	1
−109	(null)	469	297	344	188	20	1	5	6
−108	1	488	277	351	163	19	1	4	4
−107	(null)	468	265	397	207	38	(null)	4	8
−106	(null)	480	292	416	213	42	2	6	3
−105	(null)	438	240	424	222	37	(null)	11	10
−104	(null)	474	305	433	231	44	(null)	18	15
−103	1	417	280	442	260	51	2	14	15
−102	(null)	396	268	466	259	68	3	19	22
−101	(null)	406	310	530	297	89	3	31	28
−100	3	414	268	529	284	95	7	19	30

this success can be attributed to just three features: the Structured Query Language (SQL), transactions, and constraints. SQL is described as "intergalactic dataspeak," [400] and is the most common language used for expressing queries. Transactions enable multiple agents to safely access and modify data, eliminating the potential for data corruption and inconsistency when multiple writers and readers attempt to access the same data. Transactions also protect against hardware failures by insuring that changes are either written in their entirety, or rolled back in their entirety in the event of a failure. Finally, constraints insure data integrity by expressing business logic as declarative statements that are enforced at the server level, removing the need to duplicate these checks across client programs.

Ironically, two of the three features that make relational databases such a powerful solution for OLTP are actually a detriment for analysis of large, static data sets. The overhead of transaction support is a significant performance burden. Constraints, specifically foreign key constraints, introduce significant storage overhead when every fact data row must contain multiple foreign key columns linking to dimensions. Codd recognized the requirements of a database to support analysis are fundamentally different than the requirements for a transactional database and coined the term Online Analytical Processing (OLAP) in 1993 [70]. Codd's report outlines a set of principles that should be supported by an analysis-centric database. Unlike Codd's papers describing the relational model, this document was written under commission for a software company and not published in a peer-reviewed journal. More importantly, it did not contain a mathematical description of the concepts, leaving this as an implementation detail for future developers. Nonetheless, the report defined OLAP databases and specifically called for the support of sparse multidimensional matrices as the fundamental unit of analysis. The report also emphasized the complementary approach of OLAP to other database technologies and highlighted the concept that OLAP could mediate between other data sources to present a consistent model for analysis.

1.3.5.2 OLAP fundamentals and an example

Multi-dimensional matrices are the fundamental unit of analysis in OLAP systems. These matrices are referred to as hypercubes, and the term hypercube is typically shortened to just "cube." Vendor products vary widely in their fundamental storage engine implementation, they all are designed to efficiently locate fact data by filtering on dimensions (slicing), to pre-calculate aggregate functions along dimensional hierarchies (aggregation), and to efficiently accommodate missing data (sparsity). OLAP systems typically do not support transactions, data modification after initial loading, nor constraint frameworks to protect against malformed data. Although some error checking is available during import, OLAP systems tend to assume that data being loaded has already been checked for errors (scrubbed). Thus OLAP systems are not used as the primary data store or "store-of-record" and depend on other systems, typically relational databases, to perform this function. Data are typically imported into OLAP cubes using a process called extraction, transformation, and loading (ETL). A final optional step, called "processing," can be called after ETL to compile imported data into one or more cubes.

As described in Simms and Daggett [384], multidimensional data can be implemented in a relational database by translating the dimensional model into a set of fact and dimension tables. Fact tables use columns to represent fact data (also known as measures) and additional columns that link each fact row back to one or more dimension tables. The links between facts and dimensions are implemented explicitly as primary keys in dimension tables and foreign keys in fact tables. Optimal performance on specific hardware is achieved by designing tables, indexes, and constraints. In an OLAP system, the translation of the dimensional model to fundamental storage varies widely. For example, Oracle, Microsoft, and many others provide customized versions of their relational database software preinstalled and tuned for specific server hardware configurations and are sold together as an information appliance. Other vendors, such as Neteeza, provide a relational database engine stripped of transactional functions and that is tightly integrated with massively parallel proprietary hardware. Another approach to support OLAP is to abandon the relational model entirely. A number of OLAP products, such as SQL Server Analysis Services (SSAS) and Oracle Hyperion, use a proprietary multi-dimensional storage engine, and there is no concept of tables at all. The multi-dimensional store created by these systems can then be queried using the Multi-dimensional Expressions (MDX) query language [282]. Additional details on SSAS and MDX can be found in Gorbach [135], Webb [432], and Whitehorn [439].

An example data set generated by mathematical simulation of protein folding dynamics will serve to illustrate the essential ideas of OLAP. Protein molecules are composed of a linear chain of the same repeating pattern of three single-bonded atoms: a nitrogen (N), a carbon (C^α), and a second carbon (C'), called a peptide unit. The middle carbon atom, C^α, can be attached to one of a set of amino acid residues called a side chain. Interactions with the environment cause the protein to assume a functional three-dimensional structure. This structure is flexible, but subject to a complex set of constraints collectively known as steric hindrance. One method for understanding the flexibility of a protein is to measure the rotation of the peptide unit about the N–C^α and the C–C' bonds. These are the dihedral angles phi (Φ) and psi (Ψ), respectively; and assume values from $-180°$ to $180°$. Figure 1.33 shows an example backbone segment in which the dihedral angles alternate between $-50°$ and $+50°$.

In a folded protein, steric hindrance will limit the possible rotations to very specific values. The more limited the range of values, the more rigid the protein at that specific residue. A molecular dynamics simulation can be used to sample possible conformations of a protein in a variety of conditions, and dihedral angles can be computed from the raw atomic coordinates. The explanation following uses the dihedral angles to illustrate the ideas of dimensions, data cubes, tuples, measures, and the structure of a query.

1.3.5.3 Dimensions

Dimensions are the fundamental data structures used to organize and locate fact data in an OLAP database. Dimensions contain two types of information: attributes and attribute relationships. Attributes are ordered, discrete quantities that are associated with continuous fact data. Similar to the ruler in diagram A of Figure 1.34, a dimension can have multiple attributes. Here the figure shows lengths in 1/8th in. increments, 1/2 in. increments, and whole inch increments. These attributes participate in two attribute relationships; the 1/2 in. increments are exactly equal to 4

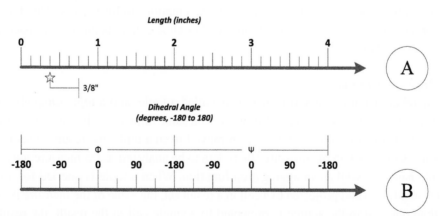

Figure 1.34 OLAP dimensions are similar to a ruler, and individual values, such as 3/8 in., are called members (A). Unlike a ruler, members are not limited to integers and can be organized hierarchically (B).

1/8th in. increments, the whole inch increments are exactly equal to 8 1/8th in. increments. A specific value, such as 3/8th in., is called a member of the dimension. In order to uniquely locate data along the dimension, one attribute must serve as the key. A member of the key attribute uniquely identifies its associated fact data along the dimension, and thus the key is also called the granularity attribute. In Figure 1.34, numbers labeling the 1/8th in. ticks (not shown) form a key. However, members are not limited to integers.

Attributes are bound to one or more specific columns in the source data table, and the component columns of the key (granularity) attribute will uniquely identify a row. Attributes can be related to each other using attribute relationships. Attribute relationships allow attributes to be organized into hierarchies that reflect the organization of underlying data. In diagram B of Figure 1.34, a dimension for organizing dihedral angles (a measurement used to analyze protein structure) is illustrated. It includes an integer numeric value (the angle bin) as well as the name for the angle. This is a simple two-level hierarchy with a range of possible values for dihedral angles associated with each named angle. An important aspect of dimension attributes and associated hierarchies is that these data are ordered. This is a fundamental difference between a multi-dimensional data store and the relational model of unordered sets.

1.3.5.4 Cubes

A cube is an object that contains both dimensions and the data being analyzed, which are called facts or measures. Cubes are typically used to facilitate analysis of specific sets of fact data and are largely self-contained, although there are facilities for linking to other cubes. Cubes contain cells, one cell for the Cartesian product of all the key attributes of the cube dimensions. Inside each cell are the measure data. Cubes are in effect multi-dimensional arrays. However, unlike an array where space is allocated for every possible value from all dimensions, the multi-dimensional storage engine allocates space only when fact data are present in a given cell. In addition, dimension members can include discrete values such as strings, integers, or dates. These values are reduced from their printable representations to bit vectors. This loading and compaction of data, referred to as processing, makes cube data structures very space efficient but effectively read only.

1.3.5.5 Tuples and measures

A set of members from all cube dimensions is called a tuple, and a tuple uniquely identifies one specific cell in the cube. Groups of tuples of the same dimensionality are called tuple sets or simply sets. If one or more members are removed from a tuple, the resulting set of cells is called a slice. Cells contain the measure (fact) data to be analyzed. Each measure must have its own unique name, as well as an aggregation function. When a tuple is queried for a measure (one cell is identified) and projected to a cell in a result set, the value of the measure is returned. If more than one cell in the source is projected to a single cell in the result, the result of the aggregation function is returned.

1.3.5.6 Queries and the Multi-dimensional Expression Language

Multi-dimensional Expressions (MDX) is a query language designed specifically for multi-dimensional data, and is available on several vendors' OLAP platforms. At first glance it appears similar to SQL because the main statement of the language, SELECT, uses some of the same keywords; however, the languages are completely different. The purpose of MDX is to operate on and to produce multi-dimensional result sets, or subcubes, by selecting a set of data from a cube, applying calculations, and returning a result projected to multiple axes.

The primary statement used to retrieve data is the SELECT statement, an example is shown in Figure 1.35. MDX employs a two-pass query process. The first step, slicing, selects the set of data to be analyzed and produces a logical subcube that is used for the rest of the query. Slicing is controlled by the WHERE clause of the query, using a tuple set expression. The second pass, dicing, projects the desired results onto one or more axes. After both passes, the result cube is returned to the caller. MDX makes no distinction between result axes, making it possible to build a variety of result sets. Among the most useful of these options are matrices, which would involve complex UNION and/or PIVOT statements to achieve in SQL.

In addition to projecting existing attribute or fact data onto axes, MDX also supports a variety of scalar and set calculations. The results of these calculations fall into two categories, named sets and calculated members. Named sets facilitate additional calculation by creating additional cells; calculated members provide a mechanism for creating new measures. Measures and calculated members are always evaluated in an aggregation context. As mentioned previously, if a single measure from multiple cells is projected into a result cube in a single cell, the values of the measures will be aggregated according to their defined aggregation function, typically summation. A small variety of other aggregation functions are also available.

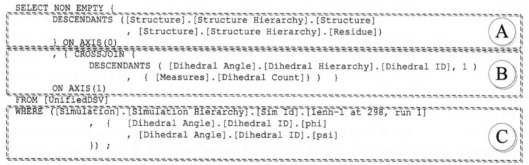

```
SELECT NON EMPTY {
        DESCENDANTS  ([Structure].[Structure Hierarchy].[Structure]
                    , [Structure].[Structure Hierarchy].[Residue])         A
        } ON AXIS(0)
    , { CROSSJOIN (
          DESCENDANTS ( [Dihedral Angle].[Dihedral Hierarchy].[Dihedral ID], 1 )
                    , { [Measures].[Dihedral Count]} )   }                  B
    ON AXIS(1)
FROM [UnifiedDSV]
WHERE ([Simulation].[Simulation Hierarchy].[Sim Id].[1enh-1 at 298, run 1]
        , {   [Dihedral Angle].[Dihedral ID].[phi]
            , [Dihedral Angle].[Dihedral ID].[psi]                         C
        }) ;
```

Figure 1.35 A Sample MDX Query. This query counts the number of times Φ and Ψ angles (measured in degrees) appear in whole numbered bins for each residue in the engrailed homeodomain protein (1enh) in a molecular dynamics simulation. The query consists of column definition (A), a row definition (B), and is filtered (sliced) to look only at a specific simulation and only at Φ and Ψ angles.

1.3.5.7 Dihedral Angles: Query Results

An OLAP cube can be used to store the dihedral angles, and a simple binning function (ROUND (angle,0)) is used to place angle values into whole numbered degree bins matching the dimension illustrated in diagram B of Figure 1.34.

In this example, native state simulation data for the engrailed homeodomain (1enh) including dihedral angle data were loaded into an OLAP cube. The MDX query in Figure 1.35 looks across all residues and shows the number of times the Φ and Ψ angles at the given residue appear in bins from $-180°$ to $+180°$. This query constructs a two-dimensional result by first selecting (slicing) a specific simulation from the Simulation hierarchy and a set of desired angles from Dihedral Angle dimension. Slicing is accomplished via the WHERE clause, shown in part C of Figure 1.35, which lists specific dimensions members to include in the result. Then a set of members representing the residues in order from the protein structure are projected onto columns. In the original model, proteins are organized by structure, chain, and residue. The DESCENDANTS function starts at the structure level and requests all residue children (part A of Figure 1.35). A set of dihedral angle members, which are limited to only Φ and Ψ angle bins in the WHERE clause, are created by using the DESCENDANTS function. The CROSSJOIN function combines the set of dihedral angle bins with the set of dihedral angle measurements to produce a single set with all combinations of bins and dihedral angle measurements (see part B of Figure 1.35). The query calculates the number of dihedral measurements at the intersection of the residue and angle/bin combination. The complete result set consists of 362 rows and 54 columns, and a portion of the result is shown in Table 1.12.

1.3.5.8 Clinical trial feasibility: a second example

Conducting clinical research involving human subjects is extremely expensive. Before recruiting begins, a power calculation will be performed to estimate the number of patients needed to detect effect of a treatment. Given this estimate, the next critical question is how long it will take to recruit the number of patients required by the power calculation who match the inclusion criteria of the trial. Inclusion criteria can include items such as age, sex, or an ICD-9-CM diagnosis code. For example, a research coordinator would like to understand how many women, aged 18–50, with a diagnosis of diabetes, are expected to visit an academic medical center and associated neighborhood clinics on a weekly basis. An OLAP cube with a simple dimensional structure can be assembled from a relational database of medical record data. Dimensions captured include date of birth, sex, billed ICD-9-CM codes, and visit date. The cube was then populated with patient data associated with clinic visits. The query in Figure 1.36 can be used to identify the set of patients matching the inclusion criteria and count how many of them are available on a weekly basis.

The WHERE clause limits the data considered by using the provided inclusion criteria. It specifies a specific quarter (Q2, 2012), a range of birth years (1962–1994), and a range of ICD-9-CM codes corresponding to fracture injuries. Here the measure [Measures].[Patient Count]

```
-- Patients by visit date, stratified by Sex
SELECT   NON EMPTY ( [Sex Dimension].[Sex Abbreviation].CHILDREN
                 ) ON COLUMNS
             , NON EMPTY (( [Admit Date Dimension].[YW].[Week].MEMBERS),
                     [Measures].[Patient Count]
                 ) ON ROWS
FROM [Az ADT]
WHERE ( [Admit Date Dimension].[YQMD].[Quarter].[2012-Q2]
       , ([DOB Dimension].[Year].[1962]:[DOB Dimension].[Year].[1994] )
       , ( [ICD9 Dimension].[ICD9 3-Digit].[800]:[ICD9 Dimension].[ICD9 3-Digit].[829] )
     );
```

Ⓐ

Figure 1.36 Feasibility MDX Query. This query looks at how many patients treated in a practice meet the inclusion criteria for a study. Initially a simple count stratified by sex is used to understand the weekly flow of patients.

utilizes a distinct count of patient identifiers as the aggregate function, which identifies the number of patients at the intersection of dimensions. Results are given in Figure 1.37.

Additional details behind the summary results can be easily exposed by adding members to the axes. For example, to better understand the age distribution of patients, date of birth could be added as columns. However, rather than adding each individual birthdate, a summary is used. In the cube definition, an attribute called YearGroups was defined as copy of the Year attribute, but with the addition of a discretization function. When the cube is processed, the function creates five equal-area member bins. Figure 1.38 shows this calculated attribute added to the query.

	Week	Measure	F	M
	2012-14	Patient Count		3
	2012-15	Patient Count	2	9
A	2012-16	Patient Count	6	12
	2012-17	Patient Count	1	9
	2012-18	Patient Count		2

Figure 1.37 Feasibility results with increasing level of detail. The same base query is run three times, each time adding additional detail called a drill down. Results for the initial query (Figure 1.36) summarize patient count by week and sex only.

```
-- Patients by week, stratified by Sex and Birth Year Groups
SELECT   NON EMPTY (( [DOB Dimension].[YearGroups].MEMBERS ),   ⬅
                  [Sex Dimension].[Sex Abbreviation].CHILDREN
                 )   ON COLUMNS
             , NON EMPTY (( [Admit Date Dimension].[YW].[Week].MEMBERS),
                     [Measures].[Patient Count]
                 ) ON ROWS
FROM [Az ADT]
WHERE ( [Admit Date Dimension].[YQMD].[Quarter].[2012-Q2]
       , ([DOB Dimension].[Year].[1962]:[DOB Dimension].[Year].[1994] )
       , ( [ICD9 Dimension].[ICD9 3-Digit].[800]:[ICD9 Dimension].[ICD9 3-Digit].[829] )
     );
```

Ⓑ

Figure 1.38 Expanded Feasibility MDX Query. The DOB Dimension defines the attribute Year-Groups with 5 equal area bins based on birth year. Adding this attribute to columns exposes more details about the distribution of ages in the set.

	Week	Measure	All	All	1959 - 1973	1959 - 1973	1974 - 1988	1974 - 1988	1989 - 2012	1989 - 2012
			F	M	F	M	F	M	F	M
B	2012-14	Patient Count		3		1		2		
	2012-15	Patient Count	2	9	1	1	1	4		4
	2012-16	Patient Count	6	12	3	3	1	5	2	4
	2012-17	Patient Count	1	9		3	1	1		5
	2012-18	Patient Count		2				1		1

Figure 1.39 Results for the second query (B) reveal more about age distribution by stratifying to bins defined in the cube. Although the cube defines 5 bins, the key word NON EMPTY will cause empty cells to not be included in the result.

```
-- Patients by week, stratified by Sex, Birth Year, ICD9
SELECT  NON EMPTY (( [DOB Dimension].[YearGroups].MEMBERS ),
                  [Sex Dimension].[Sex Abbreviation].CHILDREN
                )    ON COLUMNS
       , NON EMPTY (( [Admit Date Dimension].[YW].[Week].MEMBERS),
                  [ICD9 Dimension].[ICD9-1].[ICD9 4-Digit].MEMBERS,
                  [Measures].[Patient Count]
                ) ON ROWS
FROM [Az ADT]
WHERE (   [Admit Date Dimension].[YQMD].[Quarter].[2012-Q2]
        , ([DOB Dimension].[Year].[1962]:[DOB Dimension].[Year].[1994] )
        , ( [ICD9 Dimension].[ICD9 3-Digit].[800]:[ICD9 Dimension].[ICD9 3-Digit].[829]  )
     );
```

Figure 1.40 A Third Feasibility MDX Query. Additional stratification can also be added on rows; the hierarchy of ICD-9-CM codes is specified such that 4th digit classifications (and children) are summarized.

The results are given in Figure 1.39.

MDX provides many convenient ways to assemble member sets, especially if the underlying data are organized as hierarchies. For example, ICD-9-CM diagnosis codes consist of a 3 digit primary code, and in most cases subclassification codes that add a 4th or 5th digit. The ICD9 Dimension contains a user-defined hierarchy derived from the structure of the codes with levels for 3, 4, and 5 digit codes. Figure 1.40 adds a level specification from this hierarchy for the 4th digit of the code (results filtered to the 16th week of 2012).

The results for codes 820.0–828.0 are given in Figure 1.41. The Patient Count measure will be calculated for a diagnosis coded to the given 4th digit as well as any 5th digits below it (for example, 823.0 will also include 823.00, 823.01, and 823.02).

Multi-dimensional arrays, hyper cubes, or simply cubes are a natural representation for many scientific problems, including some from biomedical informatics as illustrated here. The relational model can accommodate this structure, but it suffers from performance and storage efficiency issues. Dedicated OLAP systems circumvent these issues by assuming data are primarily read only. This assumption allows relational OLAP systems to remove transaction and constraint frameworks, while other OLAP systems implement extremely efficient multi-dimensional storage and bit vector indexes. Both approaches partition data into two forms: dimensions and facts,

	Week	ICD-9	Measure	All F	All M	1959-1973 F	1959-1973 M	1974-1988 F	1974-1988 M	1989-2012 F	1989-2012 M
	2012-16	820.0	Patient Count	1	1	1			1		
	2012-16	820.8	Patient Count		1				1		
	2012-16	821.0	Patient Count	1	2	1			2		
C	2012-16	821.2	Patient Count		1						1
	2012-16	823.0	Patient Count		4		1		1		2
	2012-16	823.2	Patient Count		2		1		1		
	2012-16	823.8	Patient Count		1				1		
	2012-16	826.1	Patient Count		1		1				
	2012-16	827.0	Patient Count		1				1		
	2012-16	828.0	Patient Count		1						1

Figure 1.41 Results of the third query (C) show additional detail at the 4th digit ICD-9-CM code using a hierarchy. Each row will contain the given 4 digit code as well as any 5 digit sub-codes. Note that results were filtered to only show week 16 and codes 820.0 through 828.0.

which are the discrete categories and continuous values being analyzed, respectively. The MDX query language is a powerful and flexible means to query cubes, including the exploitation of hierarchical relationships within dimensions to assemble results of interest.

1.4 Data Quality

All the discussion to this point somewhat implicitly assumes that the data contained in the various long-term storage schemes are of high quality, that there are no misspellings of names of patients, or diseases, or other text strings. In Figure 1.6, the term "Prostate" is misspelled as "Prostrate," which is a position, not a disease or an organ. This typographical error then propagates to views, as in Figures 1.7 and 1.11. This kind of error often happens when the data entry is done by typing into a text field in a computerized form. One problem with this is that the record will not turn up or be counted in a search, for example, for all patients with *prostate* cancer.

To alleviate the data entry problem, large "controlled vocabularies" have been built to standardize the terminology that people use in recording medical information. Another strategy for reducing such errors is to provide structured data entry facilities, where users choose from menus rather than typing in a word or phrase that is part of a standardized vocabulary. The complexity of biological and medical data makes this a challenging design problem. Especially in the context of medical practice, time pressure on the provider actually favors traditional dictation and transcription rather than direct data entry.

Making up names for entities as part of a controlled vocabulary is a very challenging task. One strategy for managing the problem is to design a naming structure. In organic chemistry, for example, the names of small hydrocarbon compounds and compounds related to the hydrocarbons follow a convention that indicates which family of molecules they belong to. The names of alcohols, aldehydes, ketones, and others are formed from the corresponding hydrocarbon names, with a suffix added to indicate the family. So, from the common hydrocarbon, ethane (formula C_2H_6), there is the even more common alcohol, ethanol (formula C_2H_5OH). Constructing names in this way becomes a major effort, managed for chemistry by the IUPAC. Many naming systems have been developed for medicine, including LOINC, described in Section 1.1.4, and others mentioned later in this section. It might be tempting to parse such complex names to infer facts about the named entities, but the results will be unreliable. Even if family membership could be derived from names, properties of such entities vary in eclectic ways within a family. Instead, it is now widespread practice to create terminology systems where the relationships between entities are explicitly represented in some computational formalism rather than encoded in naming conventions. Chapter 2 develops this idea in detail.

In many circumstances, data are already available in computerized form and can be transmitted from system to system. An example that has been available for a long time is the planning and delivery of radiation therapy (see Chapter 2, Section 2.2.8 for more background). For many decades it has been possible to create a plan for radiation treatment of tumors using a computer simulation system. The system displays cross sectional images of the patient and can display the radiation pattern produced by the treatment machinery. The planner, called a "dosimetrist," uses interactive controls on the display to position the radiation beams. When the plan is completed, it can be written on tape, disk, or other media, or transmitted over a computer network to a computer-controlled radiation treatment machine.

In the early days of such facilities, the radiation therapist was required to enter the machine settings from a printed sheet produced by the planning computer. The computerized version of the settings was used by what was called a "Record and Verify" system, to have the machine check if the therapist entered the data correctly. In my opinion (loudly expressed at the time, early to mid-1980s) this is exactly backward. The computerized data and network connection have the potential to save time and improve reliability. The role of the radiation therapist should be to check that the machines are operating correctly, and not for the machines to check the humans. Fortunately, this view prevailed. Current practice is to use a network protocol (DICOM, mentioned earlier) to transfer the data, and to have the operators (dosimetrist and therapist) provide the quality assurance of human oversight, to insure that the machines are set up properly.

Electronic medical record systems store data in eclectic ways, depending on the medical software vendor. One system uses B-trees, another uses an EAV model, and yet another uses an RDB schema. When data are put into such systems, either by HL7 messages or a data entry facility, some conversion is necessary, just as for the medical digital image data described in Section 1.1.5. The DICOM standard goes a long way toward mitigating this problem, at least for digital image data, radiology reports, and radiation treatment data, by providing a common data model, but the actual details of how the data are stored, retrieved, and managed within a particular system is strictly a proprietary matter unique to that system.

For display purposes, when a medical practitioner looks up information on a patient, the presentation is controlled by tables that specify what appears on the screen. These tables can be very large. In a typical table there are entries for each possible laboratory test, or group of tests, for example. The way in which tests are classified is not standard but it is up to each hospital or clinic to create the tables specifying how the data are to be handled. The motivation behind the LOINC naming system described in Section 1.1.4 is to have a common set of test names and definitions.

Despite the widespread use of LOINC, problems can still occur. One of the most challenging issues is to maintain consistency when updates occur. MINDscape [409] is a Web-based system that integrates access to various clinical data sources at the University of Washington Medical Center and related practice sites. Its back end system, MIND, builds tables of test names dynamically, with caching. The first time a test is displayed, the (vendor system supplied) name is cached. It is not subsequently updated when changes occur. Normally these names don't change, but on one occasion a test was displayed with the name "HTLV 1" when the result should have been labeled "HTLV 1+2." The original entry apparently was realized to be incorrect, and was changed, but the update did not get into the cached table.

Configuration tables in electronic medical record systems (that are used to generate displays of results and other information) have evolved over time and are not generally the result of systematic software development. Sometimes vendor-supplied enhancements to such tables will conflict with local enhancements installed by the system users at a particular hospital. Coordination of configurations between different software components and systems presents additional challenges. Much remains to be done.

Another consistency problem is implied by the table examples in Section 1.3. A spelling error in a person's name may be discovered and corrected in one place, but by the time such errors are discovered, often many copies of the information have propagated to other systems. For example, when a CT image data set is transmitted from the Radiology PACS (picture archiving and communications system, a large-scale online storage system for medical image data) to the Radiation Therapy department's radiation treatment planning systems, the data must be matched with other data for that patient. The matching must rely on the patient's name, medical record number, and possibly other information. Such matching algorithms will make errors when there are spelling mismatches. This problem is difficult to solve, because no matter how loud the cry for one vendor to provide all the medical computing applications for an institution, it is clearly a naive wish that ignores the complexity of medical practice. There will always be many vendors, for all the huge variety of medical equipment and systems, including radiology imaging systems, infusion pumps, radiation treatment machines, anesthesia monitoring systems, vital signs monitors, and so on, not to mention the imaginary central repository, the Electronic Medical Record. Even in cases where a single vendor supplies all the systems or components, there are often incompatibilities whose resolution is difficult, and which introduce additional workload on users.

The problem of managing context, for example by simultaneously accessing multiple systems for data on the same patient, has given rise to software products and attempts at standards, such as CCOW (Clinical Context Object Workgroup). This group is a part of the HL7 standardization effort. Although products exist that are compliant with this standard, and that leverage compliance to present an appearance of seamless integration, they are more at the level of providing a cover over what remains a pretty messy situation.

In addition to managing consistency of identifiers for patients, there is a related issue of managing user access credentials and authorization of users of all these different systems. Another popular wish is for "Single Sign On" (SSO). The goal of SSO is that a user presents credentials such as a login ID and password (and perhaps additional authentication such as "SecurID"), at the user's desktop, and then it is passed through or otherwise managed by the SSO layer so that the user does not have to repeatedly authenticate with each application. This requires synchronization of credentials and authorization among many applications, not an easy task. Systems for SSO exist, but again, they tend to take a "sweep the trouble under the rug" approach, rather than really solving the underlying problem.

Finally, looking ahead to the problem of creating large-scale knowledge systems in addition to data systems raises the question of whether the knowledge, facts and ideas, terminology, etc., are consistent with each other. The FMA, mentioned earlier, is a collection of millions of facts about human anatomy. We know that anatomic entities satisfy constraints and are not arbitrarily interconnected. For example, lymphatic vessels have as part of their description the names of other lymphatic vessels which are downstream and upstream. The upstream vessels should *always* be lymphatic vessels, never arteries or veins, and never other kinds of organs or tissues. The downstream vessels are similarly constrained, with two exceptions, the Thoracic Duct and the Right Lymphatic Duct, which connect to veins, as they are the terminal elements, or trunks, of the two lymphatic trees that make up the lymphatic system. Another constraint is that every lymphatic vessel, by following the downstream relation, should reach one of the two mentioned above. This is equivalent to the observation that there are no lymphatic vessels that are not part of the two trees. Such requirements can be formulated as constraints, and a construct like the FMA can be automatically checked by a computer program making queries and tracing the data. This will be further developed in Chapter 4.

Another example of the possibility of automated data quality checking is in systems like EcoCyc [215, 217, 218]. EcoCyc is a comprehensive symbolic model of all the metabolic pathways in the bacterium E.coli. It is one of many such models, collectively referred to as BioCyc. The BioCyc knowledge library at SRI International, Inc [31, 218], is a collection of pathway databases for over 150 different species, primarily focusing on metabolic processes. It has been and continues to be very useful for answering difficult questions about metabolic processes in microorganisms, and has more recently been extended to include important eukaryotic species. URL `http://www.biocyc.org/` is the web site for EcoCyc and all the other metabolic pathway knowledge bases built by the BioCyc project.

EcoCyc has a representation of biological molecules, biochemical reactions, and pathways that consist of linked reactions. The molecule entries include information about how many of each type of atom are in the molecule. The reactions include details of how many of each molecule participate in the reaction as reactants and how many of each of the products are produced. In effect the information usually found in a chemical reaction equation are stored. It is therefore possible to check each reaction to determine if the number of each type of atom is balanced on the left side and the right side of the reaction. This is the requirement of *stoichometry*. It is a useful quality check to identify possible data entry errors when creating such large-scale information resources.

Terminology systems that are the basis for standards like HL7, as well as medical terminology systems such as the System of Nomenclature for Medicine (SNOMED) [355] and the Unified Medical Language System (UMLS) [37,295], also can be examined to determine if they meet reasonable constraints [63]. Many misclassifications or anomalies exist. Such misclassifications can have serious consequences when these systems reach the point where they are relied on in a clinical application.

The HL7 Version 3 Reference Information Model (RIM) has been the subject of much criticism [387,388] because it systematically misclassifies many entities of different types as if they were the same types. Even HL7 Version 2 has problems with inconsistencies in the formal descriptions of messages, and other problems that stem from allowing the use of arbitrary terminology and labeling systems. The implementors (meaning the installers, not the software developers) of HL7 message routing systems must somehow resolve all these discrepancies, which are often undocumented and must be discovered by actually looking at data streams to determine what a system is actually sending.

1.5 Data, Information, and Knowledge

Before leaving the subject of biomedical data, it is worth considering how the terms, "data," "information," and "knowledge" are used. In the next chapters we will show how to encode biomedical knowledge and compute with it. Is knowledge different from data? What about the seemingly broader term, "information?"

In distinguishing among the commonly used terms, "data," "information," and "knowledge," the dictionary is not much help. The Random House Dictionary of the English Language [395] defines "data" as "facts, information, statistics or the like, either historical or derived by calculation or experimentation." It does, however, also offer the definition from philosophy, "any fact assumed to be a matter from direct observation," which seems to exclude the derivation by calculation. "Information," on the other hand, is defined as "knowledge communicated or received concerning a particular fact or circumstance." The words "data" and "facts" are presented as synonyms for "information." The entry for "information" distinguishes between information as unorganized and knowledge as organized. The dictionary defines "knowledge" as "acquaintance with facts…" and suggests that "information" is a synonym.

An additional complication is that the term "information" is indeed used in the sense of communication of data or knowledge, as in *Information Theory*. This is the quantitative study of how data are transmitted and received through communication channels, accounting for noise, redundancy, synchronization, and other matters. Two good elementary introductions to this subject are the reprints of the original pioneering work of Claude Shannon [374] and a very entertaining treatment for the layman by John R. Pierce [329]. A message containing a large amount of data can be transmitted but it may not convey a large amount of information. Shannon's notion of information is that a message contains information when the content is not known in advance. The amount of information the message conveys is related to the amount by which the unknown is reduced. If the only messages that can be conveyed are "yes" and "no," the message is said to convey one *bit* of information. When there are many more possibilities, the message will contain more information. This has some relevance in biology, and is being pursued by some researchers [3]. The related concept of "entropy" is used in medical image registration, pattern matching, and optimization [410,411]. Section 3.4 in Chapter 3 introduces the basic ideas and Section 3.4.5 provides some examples in biology and medicine.

Despite the growing success of systems biology,[54] many still hold the view that biology and medicine are fundamentally different from physics and chemistry. To some extent this view is manifest in discussions about the nature of information [33,35,36,110]. Blois [34] asserts the "need to inquire vertically from one hierarchical level into lower lying ones." This vertical, rather than horizontal or single level, orientation leads Blois to conclude "the structure and explanations of medical science are qualitatively unlike those of the lower level natural sciences."

As outlined in the preface to the first edition, and elaborated in the next chapter, this book takes a strong view in the opposite direction, that there is more similarity between physics and biology than we may have thought, that much is to be gained still from the development of sound computational models of biological processes and medical phenomena. A difference is that in medicine there are more interacting processes than just a few, and the initial conditions are harder to measure. So-called high level concepts like "pneumonia" or "acute renal failure" are vague, as Blois and others point out, but we need to get beyond thinking that these vague notions are somehow fundamental. They are not. Such vague phenomena occur in physical systems as well, but no one would consider them to be fundamental in the same sense as the notion of an electromagnetic field. In biology and medicine (especially) we are still seeking to understand what is fundamental and what is an emergent phenomenon. This problem comes up as a huge challenge in the construction of biomedical vocabularies such as the Unified Medical Language System, when one attempts to attribute precise semantics to the terms and their relationships.

One point that is critical here is that the symbols representing data are meaningless by themselves. Each symbol or datum needs to have an interpretation that means something to us. So, it is reasonable to define the term "information" as "data plus meaning (interpretation)." This is however not sufficient to proceed to the construction of biomedical theories, or general knowledge. The underlying view we take is that the real-world entities represented by our symbols behave in a way that can be accurately described by computational relationships. If

---

[54]Systems Biology is a huge subject, too rapidly growing to provide a single reference or even a few. Specifically for the Systems Biology Markup Language (SBML) there is an official web site, URL `http://www.sbml.org/Main_Page`. One possible starting point for a glimpse of Systems Biology is the Seattle-based Institute for Systems Biology home page at URL `http://www.systemsbiology.org/`. Another possible starting point is an article in Wikipedia, under "Systems Biology."

that is true, once the mapping from entities to symbols and relationships is asserted, we can then use computational processes to determine what the behavior of these entities will be. The predicted data then get mapped back into (identified with) real-world entities that are then observed to either confirm or disprove the hypothesized relationships. Even if the entities are represented by probability distributions and related by structures such as Bayes Nets (this topic is taken up in Chapter 3), the same philosophical approach still applies.

Blois and others are of course right in pointing out that the entities we describe are vague (as distinct from uncertain). One response to the need to deal with vagueness is the idea of "fuzzy logic." Its origin is generally attributed to Zadeh [447]. A book by Xu [443] provides a comprehensive treatment of fuzzy logic in bioinformatics.

For our purposes, we use the convention that "data" refers to matters of direct observation of specific instances, or computed assertions about those specific instances. So, blood test results such as glucose level, red blood cell counts, etc., for a particular blood sample from a patient are data. We will usually use "facts" as a synonym for "data." The data usually are accompanied by precise interpretations or meanings in terms of things in the real world, so it is appropriate to refer to them as "information."

We reserve the term "knowledge" to refer to statements that are true of classes of individuals, or what we might also call "general principles." So, a general principle, or chunk of knowledge, might be that, in general, the drug heparin prevents clotting of blood. Together with some facts about an individual, these chunks of knowledge can be applied to derive new facts, by logic or other calculations. When there are enough chunks of knowledge to provide a comprehensive and internally consistent representation of a biological system or process or entity, we call that collection a "theory."

What about information? The dictionary seems to suggest that unorganized information is data, and organized information is knowledge. This does not distinguish between information about individuals vs. information about general principles, nor does it address information as a way to quantify communication (Information Theory). We will see that similar ideas about looking up information will apply to both data and knowledge in various forms. So, although one might imagine a progression from data to information to knowledge as shown in Figure 1.42, in reality the progression is at least two-dimensional, from unorganized to organized, and from particular instances to general laws and principles, as shown in Figure 1.43. The assignment

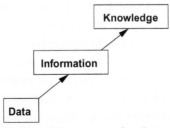

Figure 1.42   A diagrammatic illustration of the progression from data to information to knowledge.

Organized	Structured Data	Biomedical Theories
Unorganized	Raw data (Facts)	Unstructured Knowledge
	Specific	General

Figure 1.43   A two-dimensional view of data, information, and knowledge.

of meaning to data and symbols is yet another dimension, not illustrated in the figures, but in Figure 1.43 it is assumed that the interpretations or meanings are known and assigned.

The general laws and organized information can be used to infer more facts, and possibly to discover further general laws. Data (and their meanings) are the means by which we test theories, and the source of insights and patterns from which we might guess theories. This is the activity of the theoretical side of scientific research.

In clinical practice, too, one can see the same process. For example, we start with some facts from a medical record. Patient X has an adenoid cystic carcinoma (a kind of cancer) in the nasal cavity, behind the nasal vestibule, between the upper left turbinate and the nasal bone. This is information about patient X. We know also that this location is drained by the submandibular lymph nodes and the retropharyngeal lymph nodes. This drainage information is knowledge, because it is true in general about human anatomy, not just for a particular patient. But, it can be applied to the particular patient to suggest that these two nodes are at risk for metastatic disease and should be treated with therapeutic irradiation. So, the combination of data and knowledge leads to more data and a useful clinical insight. The process of medical diagnosis can also be seen as an application of knowledge to data, enabling a stepwise theory refinement. In both diagnosis and treatment planning, instead of trying to construct a theory of biomedical or pathological processes in general, the physician is constructing a theory for a particular patient at a particular time. This is referred to as the hypothetical-deductive approach [378, pp. 70–71]. Some early medical expert systems were designed along these lines.

These distinctions, while useful, are not so rigid as one might surmise. Consider the FMA [353], described in more detail in Chapter 2. It is an *ontology*, a collection of terms naming the entities in the human body and links naming the relationships between them. The intent of the authors of the FMA is to represent as precisely as possible what is known about human anatomy. To a student learning anatomy by browsing the FMA, it is a collection of information, to be understood and perhaps to some extent memorized. To a software designer, the terms and links may be considered data that are subject to search and manipulation by a computer program. To a biological scientist it is a formal theory of human body structure. The difference in labeling the contents of the FMA are really about different uses of the same things. As long as the meaning, structure, and intended use are clear, we will not worry too much about the labels, "data," "information," and "knowledge."

1.6 Summary

Some essential ideas come out of our exploration of how to represent and manipulate biomedical data in a computer:

- Biomedical data are heterogeneous, highly variable in size as well as type and structure, but from molecules to humans to populations, there are some common themes.
- Biomedical data are symbolic and structured, not just numbers and text. Programming languages designed to handle symbols and structures (such as Lisp) help in working with such data.
- Self-describing data (and use of metadata) organization strategies can pay off in simplifying access and storage. These methods also provide flexibility and robustness in the face of change.
- When raw speed is needed, such as for medical images and large sequences, special techniques are appropriate, but abstraction is still important.
- When dealing with generic descriptive data rather than characteristics of individuals, the distinction between data and knowledge is less clear.

In addition, we saw that the Lisp programming language provides strong capabilities for handling biomedical data. We will see in the next chapter that these same capabilities are powerful tools for handling knowledge as well as data. Special features of Lisp that are particularly important are:

- Data, not variables, have types,
- Lisp supports symbols and lists, which can be used as abstract elements in theories and as representations of non-numeric information,
- Functions and other code (such as class definitions) are also data, making possible the use of metadata, and multiple layers of description.

In the next chapter, we make the transition from data to knowledge, along the dimension described, from the particular (data) to the general (knowledge) and then examine how to combine data and knowledge in the solution of biomedical problems.

1.7 Exercises

References to the "book web page" in the Exercises may be found at URL `http://faculty.washington.edu/ikalet/pbi/`. On that page there are links at the bottom to a file containing all the code in the book and to some useful data files.

In the following, provide brief explanations of any code you write, and include a transcript of your test runs that you did to produce answers to the questions.

1. Data manipulation - queries on drugs

    Use the `get-all-objects` function and related code from Section 1.2.3 to read into a global variable the entire set of drug instances whose data appear in the `drug-objects.txt` file on the book web page. You can define your own functions and execute them, or just type expressions at the Lisp prompt to answer the following questions:

(a) How many drugs are there in the file?

(b) How many of them are *prodrugs*? (A prodrug is one in which the form administered to the patient is not the active form but a precursor that is quickly transformed into the active version.)

(c) How many of them are substrates of the enzyme CYP3A4 (an abbreviation for Cytochrome P450 type 3A4)?

(d) Which drugs are substrates of more than one enzyme?

2. Why not just use a database and SQL?

Write out how you would use SQL to define a table of drugs, and SQL queries to answer the questions above. Don't worry about populating an actual database and running it - this is just a theoretical question. If you run into technical difficulties, write a short explanation of why the difficulties came up.

3. A drug interaction checking system

The world-reknowned pharmacologist, Dr. L. Dopa, has hired you to help him design a drug interaction database. You of course point out that the code in Section 1.2.5 can check a suitable database for interactions, but he says that he would like to have a function that just takes a list of drugs and checks each one against each of the others. Write such a function, that will take any number of drug names, retrieve the record for each from a list of drug class instances (using the class definition in Section 1.2.5), and and use the inhibition and induction interactions functions together with some calls to `format` to print out a list of nicely formatted interactions for the drugs in the list.

4. A Documentation exercise

Write a short set of directions for how to use your code from the previous exercise, including what should the user type exactly, at the Lisp top level prompt, assuming the code and database are already in the Lisp environment, and what exactly will the user see as the result of a typical function call.

You can find a readymade test file of drug records at the book web page as noted above, readable by the `get-all-objects` function defined in Section 1.2.3.

5. Using Drug Class Information

For this exercise you will need to look up two articles. The first, by Hanlon et al. [143], provides two tables, one a simple classification system for drugs commonly prescribed in elderly patients, and the other a table relating drugs and drug classes to diseases. This second table indicates that certain drugs (or classes of drugs) should not be given when a patient has the indicated disease. The second, by Kuo et al. [242] claims that the recommendations implicit in these two tables have redundancies and other interactions. They (Kuo et al.) define four types of interactions and they report on how many of each they found in the tables of Hanlon's paper.

On the book web site, there is a link to a computer readable rendition of Table 2 in Hanlon, the drug-disease table (a file named `drug-disease-table.cl`). Add to this a similar rendition of the drug class table, Hanlon's Table 1. Write a program that will compute each type of guideline interaction described in the Kuo paper and apply it to the Hanlon tables. You should be able to reproduce the numbers Kuo reports at the beginning of the results section, except that your result may differ by 1 from Kuo for Type 2 interactions. Can you explain why?

6. Object-oriented design- a discussion question

   Other than the radiation therapy and radiology examples we have discussed, where else in biology, medicine, or public health could object-oriented design be a significant help, for example by making system design and implementation easier, more extensible or other benefits? Give a small, concrete example, with a sketch of actual code. Keep in mind what we have learned about what object oriented programming is about.

7. Meta-objects and XML - more discussion

   Section 1.2.3 presented a strategy for representing instances of classes as text in a file, serialization of objects. Many such strategies have been proposed and implemented. Another one is XMLisp, mentioned in Section 1.2.6. In addition, many other programming languages provide XML parsers that turn XML documents into some kind of program data structures (the so-called Document Object Model), or provide a way for your program to structure the data as it comes in (SAX parsers). Compare and contrast these various ideas. What advantages would you get by using XML instead of the keyword-value representation in Section 1.2.3? What are the disadvantages? Looking carefully at the XMLisp documentation will help you identify some of these advantages and disadvantages.

8. Higher dimensional abstraction

   Medical physicist Di Com has decided to use the serialization code in Section 1.2.3 in her DICOM system implementation to read and store images. Di claims that the code for two-dimensional array version of `read-bin-array` will work just fine for three and higher dimensions, unchanged, so there is no need for different branches in a `case` expression. Is she right? Explain.

9. Reading, Writing, and...

   Implement the function `write-bin-array`, described in Section 1.2.3 as the counterpart of `read-bin-array`. Arrange for it to handle one, two, and three dimensional arrays. Don't worry about writing efficiency.

10. Putting it together

    Modify the function `put-object` to handle the `bin-array` data type, using the `write-bin-array` function of the previous exercise.

11. Size may matter

    You will need access to a copy of the FASTA format file containing the complete Swiss-Prot protein sequence database. It can be downloaded from the ExPASy web site (`http://www.expasy.ch/`). The file you need is `uniprot_sprot.fasta.gz` which has been compressed with the `gzip` program.

    Use the `read-fasta` code and related functions from Chapter 5 (Section 5.1), to search the Swiss-Prot sequence data to find out which is the largest protein sequence in the database and which is the smallest. As far as I can tell, this is not an available function at the ExPASy web site, though it could be that someone has implemented such a query somewhere. Sometimes a researcher (or even a medical practitioner) needs the results of a query that is not a standard function of an existing tool. The only alternative is to write your own code, and the point of this exercise is to show that it is not unthinkable.

    There are over 270,000 proteins in SwissProt. It is probably not reasonable to read all the records into your program and then process them. A more efficient strategy is to read a record at a time, decide what to do with it, and then dispose of the data and read the next record, and so on until you reach the end of the file.

    There is some ambiguity in the above specification and it is a part of this assignment to identify the ambiguities and make decisions about how you want to resolve them. Thus it is possible that different programs will give different (valid) answers. This is a recurring problem in dealing with medical and public health data, not just molecular biology data.

12. Genes and proteins

    At the book web page you will find a link to a file containing the DNA sequence for BRCA1 and a link to a file containing the amino acid sequence. How many bases are there in the DNA sequence? How many in the amino acid sequence? Why is the DNA length a lot more than three times the amino acid length, since each amino acid supposedly corresponds to a three base sequence?
    Hint: this is a biology question, not a programming question

13. Base counts

    Use `naive-read-from-file`, `dna-count-simple` and other functions in Section 1.1.2 with the file `brca1.txt` available at the book web page to compute the percentages of each of the four bases in this fragment of the BRCA1 gene.

    If you've done it right, you'll get a stack overflow error. As you may have seen with the factorial function, on page 33 recursive programs are very elegant, but require a lot of stack space to run. However, a carefully written recursive program can be compiled so that it runs as an iterative process. This avoids the use of stack space and gains some speed too. The advantage of Lisp is that you can write a very elegant (but seemingly inefficient) program, and the compiler translates it into more efficient form for you, instead of you having to do the optimization work.

Try compiling and re-running the code. First, arrange for all the code to be in a convenient file, with extension .cl. At the Lisp prompt, if you are using Allegro CL, :cf will compile a file; :cload will compile and load a file; :load will load a file after it's already been compiled. A compiled Lisp file has the extension .fasl instead of .cl. You can just provide the filename without the extension in all these cases.

If you are using some other Common Lisp implementation, there is a standard way to do the compile and load. The ANSI standard operator, compile-file, will do the compile, and as you may already have learned, the ANSI standard operator, load, will load either source or compiled code. So the compile and load steps are:

```
> (compile-file "my-code")
...output here from the compiler
> (load "my-code")
MY-CODE
>
```

You replace my-code with the name of your file, of course.

14. Beyond biology

   The main ideas here are to represent real-world things that are made up of long chains of subunits as sequences (lists), to be able to automate operations on them, and to be able to process large amounts of data in a small space and time by doing sequential processing. Describe briefly two other applications of these ideas, from your own experience with medicine, health, or other applications of computers that are significant.

CHAPTER 2

Symbolic Biomedical Knowledge

Chapter Outline

Principles of Biomedical Informatics. http://dx.doi.org/10.1016/B978-0-12-416019-4.00002-0

> *A little knowledge is a dangerous thing.*
> *– Alexander Pope*
> ***An Essay on Criticism***

    In Chapter 1, we distinguished between data and knowledge by reserving the term "knowledge" for general principles and assertions about classes or categories, rather than individuals. The power of such general principles comes from their compactness. A few general principles may be sufficient to infer or calculate or derive a huge range of assertions that are entailed by these principles. The derivation of a particular result may involve a lot of computation, but it is done only when needed. Precomputing all possible results (or alternatively, doing experiments to observe, record, and catalog all possible facts) is usually (though not always) out of the question. Even when it is possible, it raises substantial questions about indexing and retrieval, some of which will be addressed in Chapter 4.

    A second important aspect of representing knowledge in the form of general principles is that one may take the view that the principles are descriptive. They are not just a compact summary of the facts of the world, but statements about how the world really works. These two points of view were espoused by two very prominent scientists. Ernst Mach held that the theories (equations and formulas) of physics were powerful and compact summaries of all the known observed data of the physical world, and nothing more.[1] Albert Einstein, on the other hand, took the view that the theories were approximate descriptions of how the world really works.[2] This is a substantial point of discussion among biologists and medical scientists. One example is the study of biomedical pathways (in metabolism, for example). One view of pathways is that they do not exist as such but are only convenient constructs to organize information about biochemical reactions. Others may take the view that such pathways represent fundamental principles of operation of living cells and organisms. Both views have a place in biomedical informatics.

In fact, three views of knowledge representation are possible. One is the view of Mach, mentioned above, which is knowledge representation as information management. If the focus is on organizing what is known (or believed) for efficient retrieval and communication, it may be acceptable to have contradictory assertions in the knowledge repository.[3] The human users of such systems would make judgments about what to do in cases of conflicting information. Information retrieval is dealt with in more depth in Chapter 4. A second view, which will extensively be developed here, is that of knowledge representation ideas and methods as the basis for constructing biomedical theories. In this context, any particular theory must be constructed free from contradictions in order to be able to unambiguously determine the predictions of the theory. When there are competing ideas, we construct multiple theories, and separately compute their predictions. The experimental facts will ultimately decide which theory is right (of course they may both be wrong). Finally, there is the view of knowledge representation as a model of human activity or human intelligence, modeling what people believe or do. This is the point of view of much research in the field of Artificial Intelligence. The goal is to understand the nature of intelligent thinking, not so much to create a model of the world of things. The only area in this book where we address this is in Chapter 3, where we present decision modeling that incorporates preferences and beliefs rather than truths about biomedical entities and relationships.

Thus, in this chapter, rather than settling this philosophical debate, we will develop the ideas, methods, and formalisms from which biomedical theories may be constructed. This in turn provides the foundation on which many applications have been built. Some of these are described in this chapter. Logic is at the base of the rest of the formalisms that are widely used today so it is the first topic to be developed here. Biomedical knowledge also involves cataloging and describing relations between categories, so systems for representing categories and relations come next, in particular, semantic nets, frame formalisms, and description logics.

---

[1] A succinct summary of Mach's ideas may be found in an article in Wikipedia under his name.

[2] See for example, [102], which has been reprinted in several anthologies of Einstein's essays, including [103,104].

[3] Mach would not likely have accepted this view. Like other physicists, he saw that the compactness of the representation depends on the ability to infer or compute things from the basic knowledge rather than enumerate all the facts, thus consistency would be critical.

2.1 Biomedical Theories and Computer Programs

Formally speaking, a theory is a set of statements called "principles," or "axioms," and reasoning methods from which you can infer new statements from the principles or axioms, given specific assertions, or facts, about a particular case. In sufficiently rich theories, one may also infer general statements that are consequences of the axioms. Such derived generalities are called "theorems" in a mathematical context. These in turn can be applied to specific cases just as if they were axioms. The language in which the statements are expressed is characteristic of the topic or domain. In physics, algebraic symbols are used to represent variables or elements of the theory, since their values are numeric and may vary with time and position. A frame of reference is needed to relate the symbols to things we see or otherwise experience. These relations express the meanings of the symbols. The axioms or laws are represented by equations that relate quantities to other quantities. The methods of algebra, geometry, calculus, and higher mathematics can then be used to manipulate the equations using some particular values for the symbols, and arrive at other values.

2.1.1 A World Class Reasoning Example

A remarkable example of the power of mathematics and reasoning is the method by which Eratosthenes determined the circumference of the Earth [302, Volume 1, pp. 206–208]. The diagram in Figure 2.1 shows the geometric and physical picture from which the reasoning follows. The diagram shows the sun's rays striking the Earth (vertical lines) at Syene and Alexandria, in Egypt. The angles a and b are equal, and measured in Alexandria (b) to be about 7°. The argument of Eratosthenes is as follows:

1. Axiom: light from the Sun comes to the Earth in parallel rays.
2. Axiom: the Earth is round (remember, this took place around 200 B.C.E.).
3. Fact: at Noon in the Summer, sundials cast no shadow in Syene (in Egypt).

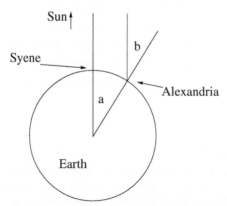

Figure 2.1   Diagram showing how Eratosthenes calculated the circumference of the Earth.

4. Fact: at Noon in the Summer, sundials *do* cast a shadow in Alexandria (also in Egypt), about 7° off of vertical.
5. Fact: the distance from Alexandria to Syene is about 4900 stadia (a stadion is an ancient unit of distance, corresponding roughly to about 160–180 meters).

By Euclidean geometry, the angle subtended by the arc from Alexandria to Syene (as measured from the center of the Earth) is the same as the angle of the Sun from vertical (by a theorem from Euclidean geometry: when a line intersects parallel lines, corresponding angles are equal).

So the distance from Alexandria to Syene represents about 7° of arc length, from which Eratosthenes calculated about 700 stadia per degree. Then, since a full circle is 360°, the circumference of the Earth is simply 700 × 360 or about 252,000 stadia. The size Eratosthenes used for a stadion is not known exactly. Eratosthenes' result was somewhere between 39,690 and 46,620 km. The current best value is about 40,000 km.

What would it take to represent the reasoning of Eratosthenes in a computer program? Certainly the arithmetic is easy. What about the logic that led to the right arithmetic formula? That requires a computer program that can implement symbols representing the propositions, methods for combining the symbols in abstract formulas, and finally a proof method to derive the final formula (deriving theorems or results from axioms and other assertions).

Of course, the Earth's surface is not exactly a perfect circle, and the light from the Sun does not exactly travel to us in parallel rays. This calculation nevertheless is close enough to reality to be useful. Success in science depends not only on accuracy of measurement, but also on making the right approximations and being able to create abstractions that leave out unimportant details. In addition, it is critical to apply the abstractions using sound reasoning in a consistent way. Ignoring sound reasoning requirements can lead not just to results that are somewhat inaccurate but to results that are just plain contradictory and therefore patently false. Lots of good traditional examples of paradoxical and unsound reasoning may be found in [220], also available in [221].

2.1.2 Biological Examples

Physiological systems can be described by mathematical variables and the general knowledge about such systems as equations. The solution of those equations for a particular individual case may be compared to measurements for that case. The measurements are the data, the formulas or equations are the knowledge.

However, as we saw in Chapter 1, much of what we know in biology and medicine is not quantitative, as it is for the subfield of physiology. Even in biochemistry, where we might also have equations relating concentrations of biochemical compounds, more often the form of the equations is unknown, or the rate constants are unknown. Nevertheless, what *is* known is that certain molecules react with certain other molecules in various ways. We know the reactants (sometimes called *substrates*), products, and catalysts, but not the numerical details. Is this useful? It turns out that catalogs of such reactions and indeed whole biochemical pathways are very useful for answering important biological questions. For example, the EcoCyc knowledge

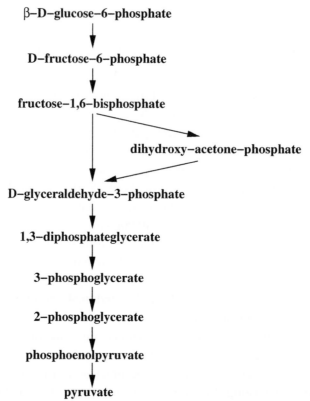

Figure 2.2 A diagrammatic representation of the first part of the glycolysis pathway.

base [215] contains the complete metabolic pathways of the *E. coli* organism, in symbolic form. Figure 2.2 shows a diagram of an example pathway, glycolysis, which transforms glucose to pyruvate.

Such a collection of pathway representations can answer a question like: "can *Escherichia coli* grow in a medium consisting of the following material…?" It is a graph search problem. Are there pathways that start with the given nutrients and end (individually or in combination) with all the necessary compounds for the organism to build proteins, etc., that it needs to survive? In Section 2.2.1.5 we develop this idea in detail.

2.1.3 Symbolic Theories

The variables in some theories thus take on discrete values rather than continuous ones, or sometimes just one of the values "true" and "false." Functions can also have discrete values in their range as well as their domain. If a function's range is the set consisting only of the values "true" and "false," it is called a *predicate function*, or simply, *predicate*. We will still be able to write relations between these variables, functions, and values, and in some sense solve the "equations" that we write as the axioms of our theory. These ideas form the basis of *first order logic*, abbreviated as "FOL."

The discussion of the nasal cavity tumor that may metastasize (in Chapter 1, page 172) can be put in the form of first order logic. The goal is to express the knowledge that tumor cells spread from the primary tumor site through the lymphatic system, and lodge in lymph nodes, so that we can use knowledge of the structure of the lymphatic system to predict the locations of regional metastasis. Then, having a formal theory of tumor spread, the theory can be embodied in a computer program, to automate the application in a practical clinical setting.

We define the `location` function, a function of two variables, such that if `location`(x, y) has the value "true" (which we represent by the symbol `t`), then the object denoted by x is located at the anatomical site y. Similarly the lymphatic-drainage function, called `drains`, has two variables, an anatomic location and a lymphatic network or chain. If `drains`(x, y) has the value `t` that is interpreted to mean that the lymphatic chain y drains the anatomic location x. Finally we define the metastatic risk function, called `risk`, a function of one variable, a lymph node. This means that `risk`(x) has the value `t` if the lymph node is at risk of metastatic disease. Then the first axiom of our theory may be written as a logic formula (the notation is described in more detail in Section 2.2):

$$\forall x, y, z \quad \text{tumor}(x) \wedge \text{location}(x, y) \wedge \text{drains}(y, z) \rightarrow \text{risk}(z). \tag{2.1}$$

In this formula, x, y, and z play the role of variables. Here, however, their values are not numbers but *symbols* representing things in the world, like a particular tumor, an anatomic location (taken from the Foundational Model of Anatomy, or FMA, introduced in Chapter 1 and more fully described later in this chapter), and a lymph node or lymphatic chain (also from the FMA).

The formula represents general knowledge. For an individual patient, we would have assertions about the location of a particular tumor, and the lymphatic drainage of that location. We'll consider a tumor in the nasal cavity, which has the retropharyngeal node group as its lymphatic drainage. For this patient, we have these additional formulas:

$$\text{location}(T4, NC) = \text{t}, \tag{2.2}$$

$$\text{drains}(NC, RP) = \text{t}, \tag{2.3}$$

where $T4$ is the tumor, NC is the nasal cavity, and RP is the retropharyngeal node group. Then we can infer from Formula (2.1) the result

$$\text{risk}(RP) = \text{t}. \tag{2.4}$$

and from that we would infer that we should include these nodes in a radiation treatment plan to insure that these suspected metastatic tumor cells are destroyed.

What about the functions? How are they defined? The `drains` function can be implemented as a query of an anatomy model such as the FMA. This is possible because an ontology like the FMA serves not just as a standardized vocabulary, but also contains relations between the terms, such as `contained-in`, `part-of`, and `lymphatic-drainage`, as well as subsumption relations (one thing is a specialization of another, sometimes called an `is-a` relation). From this perspective the FMA represents *knowledge*, not just data.

In this simple example, it is a lot of notation for something easy to glean from reading the textual version of the knowledge. But even in this example, it is known that tumor cells may spread through lymphatic vessels for some distance beyond the initial drainage sites. Thus, to predict the extent of metastatic (but unseen) disease, one must know or look up many details of connectivity of the lymphatic system. When the number of formulas gets only a little larger and the number of anatomic locations and lymph nodes also gets large, it will pay to formalize the knowledge. Why? Once we have a symbolic form for it, we can have a computer do the work. Even better, if the symbolic language we use for formalizing the data and knowledge is a computer interpretable language (a programming language), we can eliminate one translation step.

The EcoCYC knowledge base mentioned earlier is another example of a symbolic theory. It is a theory in the sense that it represents a collection of known biochemical principles, basic knowledge about biochemical compounds, and a collection of known biochemical pathways. By applying reasoning principles or methods, one can specify a particular problem or condition and then predict the behavior of the *E. coli* organism under that condition.

One can go further and predict possible pathways by searching at a more detailed level, chaining together known biochemical reactions to see if there is a way to, for example, metabolize glucose, that is, manufacture other useful molecules from it, with the available enzymes and reactants. This and other examples are developed later in this chapter.

This idea of "theoretical biomedicine" is controversial. Certainly biomedical theories are taking a different shape than theories in physics or chemistry. In biology and medicine, these theories contain many specific assertions (possibly millions) rather than a few general rules or equations from which large numbers of solutions can be derived. Also in medicine (and biology), although we are writing categorical statements here, every statement we will put in logical form has some uncertainty associated with it. Finally in many circumstances, particularly in medical practice, such theories are very incomplete in coverage, and thus may have limited applicability. Even where the coverage is good, one can question whether much new or interesting can be predicted from such knowledge compilations. We are going to leave these questions to be answered by future research and experience. Despite these limitations, there is a large potential, and growing activity in this area.

Expressing ideas in biomedicine as computer programs gives three benefits. One is that a program gives precision to the theory, so that its properties can be formally examined, deduced, and discovered. It may be possible, for example, to determine through the formalism if the proposed theory is self-consistent or not. A second benefit is that the proofs, derivations, and calculations that in the physical sciences until recently have been done by hand (the algebraic and other mathematical manipulations of the formulas, not the computation of specific numeric results) can be done automatically by the computing machinery. Third, the program can in principle be incorporated into a practical system for use in the laboratory, the clinic, or in other biomedical contexts.

These are lofty goals. The extent to which they are easy or difficult to achieve will be seen in the following sections. The citations and summary at the end of this chapter provide an entry to research aimed at extending the ideas and solving the difficult problems.

Another way to look at the problem of representing medical knowledge is referred to as "Evidence-Based Medicine" or EBM. This name is misleading. It has been described as the idea "that clinical care should be guided by the best scientific evidence" [155, page 75]. This way of describing it to the lay person usually gets the response, "So, how have doctors been doing it until now?"

Clinical practice, done well, has *always* been guided by the best scientific evidence. Until recently, progress in biomedical science has been slow enough for research findings (evidence) to gradually and carefully move from the research literature to the textbooks and to practice. Now, with the explosion in biological data and the rapid increase in sheer volume of scientific publication, the amount of new information is overwhelming our old system of knowledge transfer. When people talk about EBM, what they are really talking about is finding new and more efficient methods of distilling, summarizing, and transmitting (making available) scientific discoveries. This process needs to go more rapidly than before in order to cope with the huge increase in the rate of discovery of new knowledge.

From this perspective, the role of logic, inference, ontologies, and the like, is to provide new organizing tools so that we can record, search, and retrieve chunks of knowledge. So, we will have to keep in mind these two points of view, knowledge-based systems as scientific theories and knowledge-based systems as information management tools. In this chapter and Chapter 3, the emphasis will be on building scientific theories, and in Chapter 4 the emphasis will be on information management.

2.2 Logic and Inference Systems

The supreme task of the physicist is to arrive at those universal elementary laws from which the cosmos can be built up by pure deduction.

– Albert Einstein
from "Principles of Research," address at the celebration of the sixtieth birthday of Max Planck, reprinted in [103]

Biology is not physics.
–Larry Hunter
(personal communication)

Logics are formal languages for representing information and knowledge such that new assertions can be inferred from the initial knowledge, assertions that were not part of the initial information or knowledge. Many different kinds of logic systems can be invented, each with a range of capabilities and limitations. The most widely used and studied, propositional calculus and first order logic, will be described here.

A logical system consists of a collection of statements, or sentences, made up of symbols that have some predefined meanings. In order to make it possible to do automated deduction and analysis, we have to specify what constitutes a sentence and what is not a sentence, just as in ordinary written languages. The allowed symbols constitute the *vocabulary* of our logic. The rules for constructing grammatically correct sentences specify the *syntax*. In order to make any practical use of the logic system, we need to specify also an *interpretation* of the symbols, essentially a mapping of symbols to things in the world of our experience. The interpretation gives meaning to the symbols and sentences (formulas). This mapping is also called the *semantics* of the logic system.

Some of the symbols will have meanings that can be completely specified independently of interpretation of the other symbols. These are the familiar *connectives*, such as *and*, *or*, and *not*, and (when dealing with variables) *quantifiers*, like *for all* and *there exists*. For example, if A and B are sentences that could be either true or false, then $A \wedge B$ is also a sentence, where the symbol $\wedge$ represents the "and" connector. The meaning of $\wedge$ is that if the sentence A is true and B is also true, the compound sentence is true, otherwise the compound sentence is false. This only depends on the truth value of the two terms, A and B, not on their actual real world interpretations or meanings.

In any logic system there are also *primitives*, whose meaning depends on an interpretation. Because of this, a sentence that is true in one real world interpretation may well be false in another. For example, the statement, "Mach is the best in the class," could be either true or false, depending on what the terms "Mach," "best," and "class" refer to. One interpretation is that Mach is the famous 19th century physicist, "class" refers to a mathematics class he enrolled in, and "best" compares his numeric grade in the class to that of the other students. I don't know if this is true; it certainly could be false. On the other hand, in more modern times, this sentence could refer to Mach, a variant of the UNIX operating system, and the class referred to is the class of UNIX variants. "Best" could mean many things, and this too could be either true or false. The overall plan for building logical theories is to start with sentences that are true in the real world interpretation of interest (these are *axioms*), and then by a reliable method of proof, find out what other sentences are true in that interpretation (these are the *entailments*).

Thus a logic consists of:

- a set of primitives,
- ways of building formulas (connectives, quantifiers, punctuation), and
- inference rules (how to derive new formulas).

In the following subsections we will develop a simple logic, Predicate Calculus, with biological, medical, and public health examples. We will then consider First Order Logic, which includes variables and quantifiers. At the end of this section, we will consider some limitations of the formal logic approach, which will then lead to some alternatives.

2.2.1 Predicate Calculus

A simple logic in which to illustrate the ideas is *predicate calculus*, sometimes also called *propositional calculus* or *propositional logic*. We will represent primitives by symbols. There will be a few special symbols, the connectives, whose meaning is independent of any particular real world interpretation. We will then show how to do proofs, to determine what the implications are of a starting set of formulas. The standard connective symbols of predicate calculus are:

$\vee$	or (inclusive)
$\wedge$	and
$\neg$	not
$\rightarrow$	implies

Using these symbols and also symbols for primitives, we can construct sentences in our logic, called "Well Formed Formulas" (sometimes abbreviated as "wff"). Here are the rules for constructing a wff from primitives, connectors, and other wffs (in the following, we use the Greek letters, α, β, and γ as symbols for propositions, in addition to the ordinary Latin alphabet):

- Every primitive symbol is a formula.
- If α is a formula, then $\neg \alpha$ is a formula.
- If α and β are formulas, then $\alpha \wedge \beta$, $\alpha \vee \beta$, and $\alpha \rightarrow \beta$ are all formulas.

In cases where combining formulas with connectors might result in ambiguity as to the meaning (for example because the resulting series of symbols might be grouped in different ways), parentheses are used to make clear the order in which the connectors are applied. For example, $(\neg \alpha) \vee \beta$ is not the same as $\neg(\alpha \vee \beta)$.

The language of logic is compositional. The meaning and value of compound formulas depends on the meaning of the individual terms. When a term appears in several places in a single formula, it is required to have the same interpretation everywhere. For the relation $\alpha \rightarrow \beta$, for example, we can assign truth values that depend only on the truth values of the individual primitives, α and β, regardless of what the symbols stand for.

		β	
		T	F
α	T	T	F
	F	T	T

and similarly for $\alpha \vee \beta$ and $\alpha \wedge \beta$ and $\neg \alpha$.

The point of all this is to take knowledge in biology, medicine, and health, and where possible reformulate it in this kind of language. Then we can automate the reasoning that would otherwise have to be done manually.

An example from the field of public health surveillance will help to illustrate the idea. This scenario is based on an idealized vision for public health surveillance in a book chapter by

Dean [84]. In Dean's example, the epidemiologist has at his/her disposal computer systems that can retrieve information about patterns of activity and other data, displayed geographically, so that the epidemiologist can correlate outbreaks with environmental conditions. The epidemiologist notices an unusual increase in the number of clinic visits which appears to be a result of an increase in the incidence of asthma in a particular area. After looking at several other data sources the epidemiologist identifies a factory that is upwind of the neighborhood, where industrial waste is being moved by bulldozer. The action chosen is to request that the waste be sprayed with water so that it will not so easily be spread by wind currents. There are many components of this scenario, including statistical analysis of time series and geographic data to identify anomalous signals. However, here we just show how some of the reasoning of the epidemiologist can be expressed in logic formulas.

Here are the English language statements, in a simplified form:

- If particulate material is being moved and it is dry it will become airborne.
- If particles are toxic and airborne, then there will be an increase of incidence of asthma in the vicinity.
- If there is an increase of asthma incidence, there will be an increase of clinic visits in the area.

How can this be formalized? We abbreviate the things being described by symbols and use the connectors above to represent the causal relations. Thus:

- $\texttt{moved} \wedge \texttt{dry} \rightarrow \texttt{airborne}$
- $\texttt{airborne} \wedge \texttt{toxic} \rightarrow \texttt{more} - \texttt{asthma}$
- $\texttt{more} - \texttt{asthma} \rightarrow \texttt{more} - \texttt{clinic} - \texttt{visits}$

One can use single letter symbols if only a few are needed, as we used the Greek letters earlier. However, in most biomedical and health applications, the number of things in the world that are coded as symbols can be in the thousands or even hundreds of thousands, so longer names are needed.

The formulas above can be combined to reach a conclusion from the environmental facts, which can explain the observation that matches this conclusion. Intuitively, we can see that if someone is moving (transporting in an open truck) some significant amount of dry and toxic waste, the above statements imply that we will see an increase in clinic visits. In the following sections we will give this inference process a formal framework, and then show how it can be automated in a computer program.

2.2.1.1 Truth and Consequences

The formulas about toxic waste cleverly use symbols that we recognize as words with meanings, but a computer program has no way to assign meaning to symbols. So, in order for a computer program to process the information, we need a more narrow idea of "meaning." When we read the statements we intuitively assign each term to correspond to something in the physical world, a property of the world that can be either true or false. In any particular instance, this connection between the term and its real world counterpart should determine if it is actually a true assertion or a false one. When we choose the connections, we are creating what is called an *interpretation*. The meanings determine whether each formula is true or false in that interpretation. In the domain of logic, the term "interpretation" means simply to assign a value of either true or false to each of the primitives that appear in all the formulas. This must of course be done consistently. Wherever the same primitive appears, the value must be the same for it. In the context of a biomedical theory or knowledge base, however, we get these assignments of true or false values from the meaning of each primitive in the real world.

In a set of formulas (we sometimes call such a set a "knowledge base"), when a primitive appears in different formulas, its meaning should be the same everywhere. Of course, to represent knowledge, rather than arbitrary statements, each of the *statements*, or formulas, should be true in our chosen interpretation. In that case, it could be that additional statements must be true as well (this is the idea of inference described informally above). A formal reasoning process would enable us to determine what else is true if all the formulas (we call them *sentences*) in the set are true.

- If a sentence is true in an interpretation, we say it is *satisfied* in that interpretation.
- A sentence α is said to be *entailed* by a set of sentences S if α is true in every interpretation in which all the sentences in S are true.

Determining entailment is a difficult problem. Generally it involves computing and examining truth tables for every combination of primitive terms that appear in the set of sentences, and that make the sentences true. As an example, we can show that the formula $\alpha \rightarrow \beta$ is equivalent to $(\neg\alpha) \vee \beta$, that is, the second formula is entailed by the first, and vice versa. This can be verified by using truth tables. For each of the four combinations of the values *true* and *false* for α and β, we determine the value of the first formula and the value of the second formula, to find that they are identical in every case. The truth table for $\alpha \rightarrow \beta$ is already available above. When α has the value *true* (T) and β does also, $(\neg\alpha) \vee \beta$ has the value T, because that is the meaning of the "or" connective. $(\neg\alpha)$ may be F but if β is T, then the whole expression is T. Each of the other combinations can be verified the same way. The complexity, or computational burden, of determining entailment in this way is exponential, meaning that the number of steps required increases exponentially with the number of propositions or formulas.

2.2.1.2 *Inference Rules*

Since entailment (using interpretations or truth tables) is so difficult, it is important to find more reasonable strategies for determining it. Such strategies are called *inference procedures*. Inference procedures are purely formal manipulations of formulas from which we derive new formulas. If we can depend on them to produce only formulas that are true, when we start from a set of formulas that are true (assuming of course that the original set is consistent), we say that the inference procedures are *sound*. When the possibilities become extremely numerous, it is a real advantage to have the inference done by machine rather than by an epidemiologist who will be overwhelmed by data.

Usually, when we translate knowledge in ordinary language into logic statements, the formulas are not in the simplest or best form for an automated reasoning system. Also, the automated reasoning system may need to rearrange the formulas in some equivalent but simpler way, as part of the automated reasoning. To help with this, there are some relationships among formulas that are true no matter what the interpretation of the symbols is. These are like algebraic identities and trigonometric identities. Here are some of the useful relations that apply to predicate calculus (again, the Greek letters stand for any term or proposition):

- De Morgan's laws allow us to rewrite formulas that include the $\neg$ (not) operator, either to move the $\neg$ past parentheses or factor it out. The effect of moving the $\neg$ is to change $\vee$ into $\wedge$ and vice versa.

$$\neg(\alpha \wedge \beta) \text{ is equivalent to } (\neg\alpha) \vee (\neg\beta)$$

and

$$\neg(\alpha \vee \beta) \text{ is equivalent to } (\neg\alpha) \wedge (\neg\beta).$$

Also, there are distributive, commutative, and associative laws, that allow us to rearrange formulas that combine $\vee$ and $\wedge$:

- The distributive laws are similar to but not exactly like the rules in algebra for factoring with $+$ and $*$.

$$\alpha \wedge (\beta \vee \gamma) \text{ is equivalent to } (\alpha \wedge \beta) \vee (\alpha \wedge \gamma)$$

and

$$\alpha \vee (\beta \wedge \gamma) \text{ is equivalent to } (\alpha \vee \beta) \wedge (\alpha \vee \gamma).$$

- The commutative laws just state that the order of the terms in $\vee$ and $\wedge$ expressions does not matter.

$$\alpha \vee \beta \text{ is equivalent to } \beta \vee \alpha$$

and

$$\alpha \wedge \beta \text{ is equivalent to } \beta \wedge \alpha.$$

- The associative laws show that we can write expressions with multiple terms all connected by the *same* connective, without grouping by parentheses, since the results are the same for all possible groupings. However, when there are *mixed* connectives, you must include parentheses, otherwise the formula is ambiguous.

$$\alpha \wedge (\beta \wedge \gamma) \text{ is equivalent to } (\alpha \wedge \beta) \wedge \gamma$$

and

$$\alpha \vee (\beta \vee \gamma) \text{ is equivalent to } (\alpha \vee \beta) \vee \gamma.$$

All the equivalence laws described earlier in this section can be used as inference rules, since all of them can be verified to be sound. For example, if your set of formulas includes the formula $\alpha \rightarrow \beta$, you can add to your set the equivalent formula, $(\neg\alpha) \vee \beta$. The following are some important additional sound inference procedures for propositional logic.

- "and" elimination:
  From a conjunction, $\alpha \wedge \beta$, we can infer α and β separately.
- "and" introduction:
  Given two sentences, α and β, we can infer the conjunction, $\alpha \wedge \beta$.
- double negative elimination:
  The double negative, $\neg\neg\alpha$ is equivalent to the formula α.
- Modus Ponens:
  From the two formulas, α and $\alpha \rightarrow \beta$, we can infer β.
- resolution:
  From $\alpha_1 \vee \ldots \vee \alpha_n \vee \gamma$ and $\beta_1 \vee \ldots \vee \beta_m \vee \neg\gamma$.
  we can infer.
  $$\alpha_1 \vee \ldots \vee \alpha_n \vee \beta_1 \vee \ldots \vee \beta_m$$

It might appear that one can write inference rules as sentences *within* the logic system. For example, one might think that Modus Ponens is equivalent to the wff $\alpha \wedge (\alpha \rightarrow \beta) \rightarrow \beta$, but this formula, even though it is always true, is *not* the same as the assertion that we can *add* the *formula* β to the collection.[4] How do we then determine whether a particular inference procedure is sound or not? Indeed there is a connection with the reformulation of inference rules as formulas *within* the system. If such a formula can be shown to be true no matter what the interpretation is, then the inference rule corresponding to it is sound [403].

2.2.1.3 Theorems and Proofs

Entailment is *not* the same as proof. Entailment concerns what is actually true as a consequence of a collection of sentences being true. It is conceivable that there may be truths that follow from a collection of sentences, but there is no way to prove them (in the sense we usually mean by proof, using sound inference procedures, not by checking truth tables). Furthermore, we may invent proof methods that are not reliable, in that they may allow us to find proofs for sentences which are *not* entailed by the given sentences. Unreliable proof methods may lead to false conclusions.

In the following, we are starting with a set of *axioms*, which are formulas that are assumed to be true in the (usually biomedical) interpretation of interest, and which form the basic assertions of our knowledge base (or, put another way, our biomedical theory). The goal is to compute (eventually by an algorithm that can be implemented in a computer program) the entailments of such a set of axioms (usually together with some additional assertions about a particular problem case). Note that an axiom is *not* the same as an assertion that may be true (entailed by the other assertions) but cannot be proved. An axiom is also different from a *definition*, which serves to associate a name with some conditions, entities, or both. Definitions say nothing about the world, but axioms do.

---

[4] A droll treatment of this matter, by Lewis Carroll, reprinted in "The World of Mathematics" [62], can also be found in Hofstadter's "Gödel, Escher, Bach: an Eternal Golden Braid" [159, pages 43–45].

- A *proof* is a sequence of formulas, where the next formula is derived from previous ones using sound inference rules.
- A *theorem* is a formula that can be derived (proved) from a set of *axioms*.
- A *tautology* is a formula that is true regardless of the interpretation. An example is $A \rightarrow (A \lor B)$. This formula is true no matter what the truth values are for A or B. The formula $A \rightarrow B$ is not a tautology. For some combinations of truth values of A and B the formula is false. A tautology is sometimes called a *valid* formula. Sound inference rules can be made to correspond to tautologies, though strictly speaking they are not formulas, but procedures for manipulating collections of formulas. As an example, The Modus Ponens inference rule corresponds to the tautology $\alpha \land (\alpha \rightarrow \beta) \rightarrow \beta$.

It is important to note that a tautology is simply true, independent of the interpretation given to the literals in the formula. All of the Laws mentioned earlier can be expressed as tautologies. An axiom is different. An axiom is a formula that you assert to be true as part of a theory about something in the world, in some particular interpretation. It might or might not actually be true. When our understanding of the world takes the form of a complex theory with many axioms and lots of inferences, it might not be possible to do an experiment or measurement to determine directly whether an axiom is true. In that case, the theorems that follow from that and the other axioms are used to test the theory. If a theorem's statement is in disagreement with the observed world, one or more axioms must be wrong too.

As an example of a proof, we return to the epidemiological example. We start with the set of sentences described previously, together with the three further assertions, moved, dry, and toxic (referring to some large amount of industrial waste).

A: `moved ∧ dry → airborne`
B: `airborne ∧ toxic → more-asthma`
C: `more-asthma → more-clinic-visits`
D: `moved`
E: `dry`
F: `toxic`

What is entailed by these sentences? First, we note that we can use the Modus Ponens inference rule with formulas A, D, and E. The α part is just the combination moved ∧ dry, which comes from D and E. The β part is the assertion airborne. Thus we can add G: airborne. Then we can again apply Modus Ponens, this time to B and the combination of G and F, to get H: more-asthma. One more time, we can apply Modus Ponens with C and H to get I: more-clinic-visits. So there are several entailments that we can prove. This process of going from facts and inferences to new facts and then applying new inferences is called *forward chaining*.

Another way to discover a proof is to start with the theorem to be proved and look for support. This means finding formulas in the knowledge base and an inference rule that together can infer the formula of the theorem. Here, we would look to prove the proposition more-clinic-visits. By C, if we had more-asthma, we could prove more-clinic-visits using Modus

Ponens. So, we look for a way to prove `more-asthma`. That could be done with B if we had both `airborne` and `toxic`. Aha! we do have `toxic` in the original set. Now we just need to prove `airborne`. Fortunately, sentence A provides a way to get `airborne`, if we have *moved* and *dry*. Since we do (D and E), we can then infer that we have everything needed to prove `more-clinic-visits`. This method of starting with the conclusion and finding the support that leads to it, is called *backward chaining*.

Even though a proof might exist, using forward or backward chaining might not succeed in constructing or finding it. Here is a simple example of a set of medical knowledge formulas that have entailments that cannot be deduced by either forward or backward chaining. These sentences form a collection of axioms, a simple theory of bacterial infection and treatment. These statements, involving bacteria and infections, are not entirely biologically accurate. I've artificially simplified the situation in order to illustrate the limitations of Modus Ponens as an inference rule.

* Bacteria are either Gram-positive or Gram-negative (the Gram stain is a procedure that colors bacteria either pink or deep purple, depending on the permeability of their cell membranes).
* There are antibiotics for treating infections from Gram-positive bacteria.
* There are antibiotics for treating infections from Gram-negative bacteria.

We can write these statements in logic notation as follows:

A: `bacteria` $\rightarrow$ (`gram-positive` $\lor$ `gram-negative`)
B: `gram-positive` $\rightarrow$ `treatable`
C: `gram-negative` $\rightarrow$ `treatable`

This set of axioms entails the statement that all bacterial infections are treatable, that is, no matter what organism attacks, there is an antibiotic treatment for it. In logic, with the above predicates, `bacteria` $\rightarrow$ `treatable`. However, this statement cannot be proven by forward chaining or backward chaining using Modus Ponens. The problem is this: in order to use sentence B to conclude that bacteria are treatable, we need to assert that bacteria are gram-positive. But they are not all gram-positive. Only some are. For *those*, we can conclude `treatable`. Similarly, for the ones that are gram-negative we can reach the same conclusion, but there is no way to use the Modus Ponens rule to separately reach either `gram-positive` or `gram-negative`.

To deal with axioms using the "or" connector, we need stronger inference rules. Fortunately, the resolution inference rule, also mentioned above, is sufficient to construct a proof. The resolution rule *does* deal with the "or" connector. In order to apply resolution, we need to convert all our formulas to the forms that appear in the rule. This means that we need to have a collection of formulas called *clauses*. A clause is a formula with only primitives and negations of primitives, connected by $\lor$ (or). Later we will show how to convert any formula to a collection of clauses (this is called "Conjunctive Normal Form," or CNF).

Using the fact that the formula $\alpha \rightarrow \beta$ is equivalent to $(\neg\alpha) \lor \beta$, we can rewrite A as

$$\neg\texttt{bacteria} \lor (\texttt{gram-positive} \lor \texttt{gram-negative})$$

and similarly, B and C as

$$\neg\texttt{gram-positive} \lor \texttt{treatable}$$

and

$$\neg\texttt{gram-negative} \lor \texttt{treatable}$$

Then, using resolution with the first and the second, we can eliminate the term `gram-positive` since it appears in one formula as a primitive and in the other as its negation. The result is

$$\neg\texttt{bacteria} \lor \texttt{gram-negative} \lor \texttt{treatable}$$

Combining this with the third formula (derived from C), we can eliminate the term `gram-negative` because it appears in one formula as a primitive and in the other as its negation. The result is

$$\neg\texttt{bacteria} \lor \texttt{treatable}$$

which we can then transform back to the implication form, and thus our result is derivable from the axioms, using resolution as the main inference method.

So far, when we consider proofs and truth, we are referring to truth in the sense of entailment, independent of any real world interpretation of the symbols we are manipulating. In the above expressions we could use the same symbols to mean completely different things in the real world, so long as their relationships are the same. Using these names in such a way would of course be very confusing, though mathematically sound. In mathematics the focus is on the relationships, not on the meaning of the symbols. The use of simple letters or other symbols that do not have natural language connotations helps keep the focus on the formulas and their entailments in a formal sense.[5]

For any collection of logic sentences, we can give the symbols many different interpretations. In some interpretations a sentence may be true, but it is possible that there is no proof for it. The absence of a proof for an assertion does not mean it is false, but only that we can't prove that it is entailed if the other sentences are true. It may be an additional axiom rather than a theorem. It may be that the starting set of sentences is simply incomplete for the interpretation in which that sentence is true. Taking the epidemiology example above, we may observe a spike in clinic visits, that is, `more-clinic-visits` is `true`, but there is no toxic waste being moved about, so, `moved` is `false`. Many other factors can cause an increase or even a spike in the number of clinic visits. These other factors can make `more-clinic-visits` true in our epidemiologic interpretation. This suggests that the epidemiology logic above is not a complete description of what is going on, and we need more axioms and observations.

---

[5]In school mathematics it is common to teach algebra through so-called "story problems" in which people and their activities need to be translated into symbolic notation and algebra equations, that can then be solved. A humorous look at this from the other direction may be found in an essay by Canadian writer Stephen Leacock, called "A, B and C, the human element in mathematics." One of the exercises at the end of this chapter features these characters and includes references to the essay, for your entertainment.

The distinction between *false* and *not provable* is a very important one. The most famous such distinction dates back to the origin of Euclidean Geometry. Euclid, a Greek mathematician who lived in Alexandria, Egypt, in about the 4th Century BCE, is known for his work, "The Elements." In this work, he developed the axiomatic methods we still use today in mathematics. The work covers geometry and number theory. In his formulation of geometry, along with many definitions, he starts with five axioms that define plane geometry. They are:

1. Any two points can be connected by a straight line.
2. Any straight line segment can be extended indefinitely on either end.
3. Given a straight line segment, a circle can be drawn with one end of the segment as the center and the line segment as the radius.
4. All right angles are equal.
5. Through a point not on a given line, one and only one line can be drawn parallel to the given line.

The first four seemed basic enough to require being taken as axioms, but the fifth seemed too complicated. Euclid thought it should be derivable from the others and the definitions. However, it is not possible. It is independent of the others. If it is replaced by some other statement, one can still prove theorems and have a consistent geometry, but it will not be plane geometry. One alternative is that *no* straight line parallel to the given line can be drawn. This gives Riemannian geometry, which describes positively curved surfaces such as the surface of a sphere. In such a geometry, some of the theorems are different. For example, in Euclidean geometry, the sum of the angles of a triangle always adds up to exactly 180°, but in Riemannian geometry, the sum of the angles of a triangle is always more than 180°, with the amount depending on the size of the triangle.

So, the fact that the fifth axiom could not be proved from the others did not make it false, but rather made it independent of the others. In making up logic statements that describe biological processes or entities, we similarly need to be mindful of whether we are making statements that are axioms, theorems, or definitions. We are not completely free to define things any way we wish or to choose any axioms we wish, since they could create contradictions. In building very large collections of statements, typical of biology and medicine, this is a challenging problem of knowledge management.

Finally, there is one more potential problem. We have been considering the possibility that a sentence may be true but not entailed by the given sentences. Then we can add it as an axiom. Since entailment and proof are different concepts, it is also possible that a collection of sentences in a logic may entail another sentence but there is no proof. Such a system would seem to be incomplete in a different way. It seems that the opposite could also happen, that a sentence is provable from a set of sentences, but in fact it is not entailed by them, so it is false when the sentences are true. Such a collection of sentences would be called inconsistent. We will revisit these difficult problems later.

2.2.1.4 Automated Proof Methods

Since the objective of translating biomedical knowledge into formulas using symbols and logical operators is to enable a computer program to search for a proof of an assertion, or compute the entailments of a collection of knowledge, we now turn to the implementation of automated proof methods. With the public health example and the bacterial example, we have informally illustrated several kinds of proof methods, including forward chaining and backward chaining using Modus Ponens, and a strategy using resolution. We will show here some ways to represent propositions and some actual code for computing proofs.

Backward Chaining

To conveniently apply backward chaining as a proof strategy, our knowledge base will be restricted only to a subset of possible formulas, those known as *Horn clauses*. A Horn clause corresponds to a formula which is either a single assertion, or an implication with only a conjunction of assertions on the left side and a single assertion on the right side. These are exactly the ones for which Modus Ponens is applicable. They correspond to clauses in which there is at most one *positive* primitive. It is possible to have a Horn clause in which there are *no* positive assertions, that is, the right-hand side is the constant value, *false*. This is a constraint on the terms on the left-hand side. Not all of them can be true. In typical applications where we are looking to prove some conclusion from some axioms or hypotheses, such clauses are not relevant or useful. But some knowledge may consist of constraints in this form, and it is important in such circumstances to be able to determine if the facts and other assertions are consistent with these constraints.

In the following, you will see that we will also allow expressions involving or as a connective on the *left-hand* side of an implication. This is a convenience in coding. Such a formula can always be rewritten as an equivalent *pair* of formulas, each of which is a Horn clause. In other words, if either of two conditions can infer a conclusion, we can write each separately. The use of or on the *right-hand* side of the implication is the problem we encountered in the bacterial example, and that is what our backward chaining code using Modus Ponens will not be able to process.[6]

Here is a notation for writing propositional statements that can be processed using a computer program that will do backward chaining. The statements will be in the form of *rules*, or implications, with any form of antecedent and a single term as the consequent. The primitives (also called *facts*) are just symbols, and they can be combined using the and and or connectives. Many different notations that can be processed by computer programs have been implemented. We use the same notation as that of Graham [137, Chapter 15] and Norvig [305, Chapter 11].[7] Although the formulas in logic notation are written and read from left to right, our computerized notation will be somewhat opposite, since in Lisp expressions the operator comes first, then the inputs. When writing an implication or an assertion, the logical connective will be first, and

---

[6]Looking ahead, the same consideration will be true for forward chaining using Modus Ponens.

[7]This representation is slightly different from KIF, a draft standard for writing logic expressions [127].

the primitives or subformulas follow, as in normal Lisp syntax. We represent the implication connective by an operator named <-. It will process the inputs from right to left. For example,

```
(<- conseq expr)
```

means that if `expr` can be proved or is true, then `conseq` is true also. This is a representation of a rule, IF `expr` THEN `conseq`. The `expr` can use the connectives, `and`, `or` and `not`. An example would be

```
(<- airborne (and moved dry))
```

These expressions could be used as is, but it is more efficient for backward chaining if we index the rules by their consequents (the `conseq` part in the above). This is useful because we typically would want to look up a term or fact to see if there is a rule that can produce it. If so, we would check to see if the antecedent part of the rule is true or can be proved. Single facts that are asserted as true can be stored the same way as rules. The only difference between facts and rules is that a fact will have an empty list in the `expr` part. The first step is to look and see if the fact is already asserted. If not, the rules are searched to see if a proof can be constructed.

To start, we will define a global variable whose value will be a hash table to store the rules and assertions. The `conseq` part of each assertion or rule will be a symbol that is used as the *key* to look up entries in the table.

```
(defvar *rules* (make-hash-table) "The so-called knowledge base")
```

What we would like the <- operator to do is put an entry in the hash table, using `conseq` as the key and `expr` as the value.

```
(defun <- (conseq &optional expr)
   (push expr (gethash conseq *rules*)))
```

The program we will write has to manipulate symbols and lists of symbols directly as data. In using the <- function, we would need to put quotes in front of the inputs. This can get to be a nuisance. The rules we write would be more readable knowledge statements if we didn't need the quotes. So, we need some way to make the <- operator *not* evaluate its inputs. The way to do this is to implement it as a *macro* instead of a function. We define macros using the operator `defmacro` (which is itself a macro!). The inputs are not evaluated by the Lisp interpreter. Instead they are put into the body of the macro unchanged, and then the macro body is evaluated to produce an expression, which is then evaluated. To implement <- as a macro, then, we need to have its body construct the list containing the `push` operator, etc., which will then be evaluated. One might think that we could just use `list` as follows (we have spelled out "consequent" and used "antecedent" instead of "fact" to make the roles of these two inputs clearer):

```
(defmacro <- (consequent &optional antecedent)
   (list 'push antecedent (list 'gethash consequent '*rules*)))
```

In other words, the macro <- constructs an expression containing the code and then evaluates the resulting code. So, applying it to the formula asserting `airborne` would look like this:

```
(<- airborne (and moved dry))
```

and that expression, when evaluated, would expand into

```
(push (and moved dry) (gethash airborne *rules*))
```

and then this constructed expression would be evaluated, which looks almost like what we want. However, this is *not* exactly right. Notice that in the macro expansion, the symbol `airborne` and the list beginning with and are put in without quotes. The evaluation of this expression will give an error – the intent was to just use the symbols and expressions as is. Putting quotes in the macro definition will not solve the problem; it will give just the symbols `antecedent` and `consequent` instead of their values.

There is a solution. We use `list` inside to add the quotes. What we need to do is construct a list beginning with the symbol `quote`, in each case. Here is a version that will work,

```
(defmacro <- (consequent &optional antecedent)
   (list 'push (list 'quote antecedent)
         (list 'gethash (list 'quote consequent) '*rules*)))
```

and in the use above, this version will expand into

```
(push '(and moved dry) (gethash 'airborne *rules*))
```

which is what we want. However, this is so awkward that it would be useless if we had to write macros this way. Fortunately, Common Lisp has a special syntax to make it easier to deal with this problem and to write macros in general.

Instead of using the `list` function, you write an expression preceded by ' which is referred to as *backquote*.[8] It is like the ordinary `quote` except that variables and expressions inside can be "unquoted" or evaluated, by preceding them with a comma. Inside a backquoted expression, ' can appear and it will *not* be interpreted as the `quote` operator, but will be left in place, for example, after an expression marked by a comma is evaluated. So, we could define our macro as follows:

```
(defmacro <- (consequent &optional antecedent)
   '(push ',antecedent (gethash ',consequent *rules*)))
```

The body now more closely resembles the expression it will transform into. The actual (unevaluated) argument for `antecedent` will get put into the indicated place instead of the

---

[8]In Scheme, the other currently popular dialect of Lisp, it is called *quasiquote*.

symbol, antecedent, and so on. The quotes in front of the commas will remain, and the right expression will result.[9]

So, following Graham [137, Chapter 15], we implement a macro, <-, that adds a rule or a fact to a hash table of rules (kept as the value of the global variable *rules*, the "knowledge base" in our system). The fact (the consequent part of the rule) is used as the index into the table, and the content of an entry is a list of all the items that can be used to prove the consequent or fact. Note that a simple fact (a rule without an antecedent part) will have the symbol `nil` (the empty list) added to its hash table entry, indicating that no further proof is needed. Also, we have enclosed the call to the `push` macro in a call to the `length` function so that the <- operator can be used interactively. Without it, the call to `push` would return a (possibly) big list, which is not useful to see at top level.

```
(defmacro <- (consequent &optional antecedent)
  "adds antecedent to the hash table entry for consequent, even if
  antecedent is nil"
  '(length (push ',antecedent (gethash ',consequent *rules*)))))
```

Once we have our facts and rules in the hash table, we can manipulate this data using ordinary functions. A proof will involve at each step two possibilities: either a single fact needs to be proved, or an expression needs to be processed. To prove a single fact or consequent, we define (and use) the `prove-simple` function. It looks up the fact in the knowledge base. If there is no entry, there is no way to prove it and the proof fails (the `prove-simple` function returns `nil`). If it was asserted as a fact, a `nil` will be present in its hash table entry (that is, there will be an entry, a list containing just a single `nil` in it, not the same as an empty list or no entry), and the proof succeeds (the `prove-simple` function returns `t`). If there are other antecedents in the hash-table entry, we need to try to prove one of them. If *none* of this succeeds, `prove-simple` returns `nil` for failure.

```
(defun prove-simple (pred)
  "checks if an entry is present, and succeeds if there is a nil for
  simple assertion, or an expression that itself can be proved"
  (multiple-value-bind (ants found) (gethash pred *rules*)
    (cond ((not found) nil)
          ((member nil ants) t)  ;; find won't work here!
          (t (some #'prove ants)))))
```

Here we are using the fact that the `gethash` function returns *two* values. Most functions return a single value. However, sometimes the result of a computation consists of several values. In the case of `gethash`, it returns the hash table entry if one is found, or `nil` if not found, but it also returns an indicator of whether the lookup succeeded. This second value is very important. What if there is an entry for the key (in this case the value of `pred`), but its value is the symbol

---

[9]You can read more about macros and backquote in Graham [137, Chapter 10]. Note that none of the examples in that chapter have the problem we had to solve here, because none of them need to manipulate expressions that should not be evaluated. A deeper discussion of macros may be found in Graham's earlier book [136].

nil? This is a different case from the possibility that nil is returned because there is no entry for that key. So the success indicator, the local variable found in the code above, is provided to distinguish between these two cases.

Rather than return a two element list, *multiple values* are returned. If the second value is not of interest, the caller can just ignore it as if it were not there. If it is of interest, the multiple values can be tied to local symbol bindings. This is what multiple-value-bind does. In the multiple-value-bind expression above, the symbol ants (for "antecedents") will have the first returned value, which will be either the hash table entry or nil, and the symbol found will have the value t or nil depending on whether the lookup succeeded or failed. So the two cases are: no entry is found (ants is nil and found is nil), or an entry is found but its value is a list containing just nil (ants is a list containing nil and found is t). In the second case, the proof succeeds.

If the lookup failed, the proof should fail and prove-simple returns nil. If the results list contains the value nil, the proof succeeds because this means that the predicate was asserted as a simple fact, and prove-simple returns t. To check if the results list contains nil, it will not work to use find, since find will return nil no matter what (if it finds nil in the list, it will return it). Instead, member will return a *list* containing nil if successful.

Otherwise, it just goes through the list of antecedents and tries to prove one of them. In prove-simple the some function applies a new function, prove, that we have not written yet, to each element of a list and returns t if one of those applications succeeds, meaning it returns non-nil. We need a more general function, prove, because some of the antecedents will be expressions involving the logical connectors, and, or, not.

We now define the function prove, which will handle atoms (single terms) with prove-simple, and expressions with a little more work. The antecedent part of a rule may be absent, which means that the consequent is simply asserted to be true. If present, it may be a single term, or it may be an expression beginning with one of the three basic logic connectors, and, or, not. For each of the types of expressions that starts with a logical connector, the prove function needs to recursively check the subexpressions, which could each be a single term or another expression using and, or, not.

```
(defun prove (expr)
  (if (listp expr)
      (case (first expr)
        (and (every #'prove (reverse (rest expr))))
        (or (some #'prove (rest expr)))
        (not (not (prove (second expr)))))
    (prove-simple expr)))
```

Note that expr should never be nil. In the case of an expression beginning with and, every subexpression must be provable in order to have a successful application of the rule, so the every function is used in the application of prove to the elements in the subexpression. For or, if any subexpression succeeds, that is sufficient, so the some function is used. For not, we just determine if a proof of the (single) subexpression fails, in which case we assert that the negation is true.

Allowing a not term in an expression would seem contrary to the distinction we made earlier between "false" and "not provable." However, some kinds of assertions do include negation. An example is the assertion that a child is at risk for measles if exposed and *not* immunized. While we could reformulate the negation as a positive, for example, rename it *unimmunized*, typically medical and public health systems record dates and types of immunizations, treating immunization (not its absence) as a positive status. If we recorded both kinds of assertions, we would have to build a system that can remove assertions when new ones come in. So, when a record that an immunization occurred came in, the unimmunized assertion would have to be removed. Such systems are called non-monotonic. Reasoning in these situations is more complex. For now, we will allow negation and keep in mind that potentially unjustifiable conclusions might be derived. In this example, it's not a problem, since it *is* legitimate to infer unimmunized from (not immunized).

This code implements a systematic search of the knowledge base. It processes each rule found in the knowledge base that matches a goal, in the order in which the matching rules are stored in the hash table, and for rules with multiple terms on the left-hand side, the terms are processed right to left (because of the call to reverse in the and branch of the case expression). The logic programming language, Prolog [68], follows a similar order.

One might think this exhaustive search will find a proof if it exists, or return nil if it does not. Unfortunately, although we are dealing with a finite set of primitives, or symbols, it is still possible to write rules that cause this search to go into an infinite loop. A simple example is the tautology, $p \rightarrow p$, which we would write in our programming system as (<- p p). To prove the proposition p, the program would look up this rule, and determine that if it can prove p, then it can conclude that p is true. Thus the program continues to loop, trying to apply the same rule over and over. For now, we will just have to be careful in writing rules to insure that our logical formulas have no such infinite loops. This same problem occurs in writing programs in the programming language, Prolog. Even when a proof *should* terminate, it may not, depending on the order in which the rules are written. The order of statements in a Prolog program determines an order in which they are used.

We can now rewrite the epidemiological example in terms of our simple backward chaining system. We use the same symbols as before. Here are the rules.

```
> (<- airborne (and moved dry))
> (<- more-asthma (and airborne toxic))
> (<- more-clinic-visits more-asthma)
```

Here are the assertions for the day of the incident.

```
> (<- moved)
> (<- dry)
> (<- toxic)
```

We would like to prove that these conditions lead to an abnormal increase in clinic visits, or in other words, the assertion `more-clinic-visits` is to be proved. So, we just supply it as an input to the `prove` function. Note that the symbol `more-clinic-visits` needs to be quoted here, since `prove` is a function and the interpreter will attempt to evaluate its input.

```
> (prove 'more-clinic-visits)
T
```

The result returned by `prove` is t if the proof is successful, and `nil` if not. We can see how the proof works by tracing the `prove` function.

```
> (trace prove)
(PROVE)
> (prove 'more-clinic-visits)
  0[2]: (PROVE MORE-CLINIC-VISITS)
    1[2]: (PROVE MORE-ASTHMA)
      2[2]: (PROVE (AND AIRBORNE TOXIC))
        3[2]: (PROVE TOXIC)
        3[2]: returned T
        3[2]: (PROVE AIRBORNE)
          4[2]: (PROVE (AND MOVED DRY))
            5[2]: (PROVE DRY)
            5[2]: returned T
            5[2]: (PROVE MOVED)
            5[2]: returned T
          4[2]: returned T
        3[2]: returned T
      2[2]: returned T
    1[2]: returned T
  0[2]: returned T
T
```

For completeness, we will also examine the hash table, `*rules*`. The `maphash` function will apply a function to each element of a hash table, much as `mapcar` iterates over a list. Unlike `mapcar`, it returns `nil`, so the function we use in the call to `maphash` is going to print the information line by line. It will have to take two inputs, the hash key value and the contents associated with that key. We just use an anonymous function that displays the values in the simplest possible way. For convenience we wrap this all in a "debugging" function called `printhash`, defined as follows:

```
(defun printhash (hashtable)
  "prints out the contents of hashtable"
  (maphash #'(lambda (key val)
```

```
            (format t "Key: ~S Value: ~S~%" key val))
       hashtable))
```

Then, we can apply it to the hash table from our public health problem above. It should be easy to see how the prover can look up an item in the hash table, where it will find a list for that item.

```
> (printhash *rules*)
Key: MOVED Value: (NIL)
Key: AIRBORNE Value: ((AND MOVED DRY))
Key: TOXIC Value: (NIL)
Key: MORE-ASTHMA Value: ((AND AIRBORNE TOXIC))
Key: MORE-CLINIC-VISITS Value: (MORE-ASTHMA)
Key: DRY Value: (NIL)
NIL
```

If the list contains `nil`, the term is a fact, assumed to be true. If the list contains another symbol or an expression, the prover has to then attempt to prove that assertion. If there are multiple items in the list, it means that there are multiple proof paths, and the prover tries each until one succeeds or they all fail. When failure occurs, the prover backs up to the previous step and tries the next available proof path. If none are left the proof fails.

Forward Chaining

There are limitations of the backward chaining algorithm. It can be very inefficient. In some scenarios, many propositions would be checked multiple times and the proof could grow exponentially with the number of rules or assertions in the knowledge base. This could possibly be dealt with by *memoizing* the proof, recording something as true once a proof is found, so that its individual proof does not need to be reconstructed each time it is needed in another part of a proof. Even if this is done in the proof code, the entire proof is rediscovered each time a query is made, so it would seem that the caching of results in a more permanent or global way would be useful.

Aside from efficiency, there are situations where it is not convenient to enumerate all the queries one might like to ask, but rather the need is to generate a collection of results, and then organize them. Thus, it is useful to be able to start with data and rules and simply generate the implications. In a practical system, some guidance as to which subsets of rules to consider is often part of the control strategy.

Earlier we excluded clauses with no conclusions (that is, constraints), but we can allow such clauses, and introduce another method, *forward chaining*, that also overcomes the efficiency problem. The idea of forward chaining is that as each new proposition or rule is introduced to the knowledge base, the procedure derives all the consequences and adds them to the knowledge base, so they do not have to be rederived later. A later query to determine if an assertion is supported by the knowledge entered will be a simple lookup, since all the inference has been done. If the assertion is not already present, it cannot be derived from the existing propositions. This also allows for *consistency checking*, determining if adding a new formula might make the knowledge base inconsistent, and alerting the knowledge base maintainers of this. Automated consistency checking is a considerable aid to maintaining quality as the knowledge base is built. Of course, nothing is free. This kind of system uses a lot of storage space, since a lot of possibly uninteresting conclusions will be inferred and never used.

To implement forward chaining efficiently we need a somewhat different representation of the facts and rules. We will want to index the rules by their *antecedent* parts, rather than the consequent parts. In a forward chaining procedure, we look for inferences whose antecedents are available, and we infer whatever consequents we can, rather than looking for a way to prove or derive a particular consequent.

The following is a demonstration of how to write a series of assertions and rules suitable for forward chaining, and then a forward chaining procedure itself. The formulas will be a little simpler than the one we developed for backward chaining. We will write each clause separately, rather than allow embedded OR expressions, and we will not use negation. It is possible to assert a negation by writing a clause with a single term on the left-hand side and *false* on the right-hand side. Such a clause corresponds to an expression in which the consequent part is `nil`.

The clauses will be written with a macro similar to the `<-` macro that we used for backward chaining. We will name it `->` to indicate we are creating structures that will be used for forward reasoning. The idea here is that the antecedent part will be a list of symbols, each representing a single proposition, the consequent part will be a single symbol (possibly `nil`), representing the conclusion or proposition that would be inferred if all the antecedents turned out to be true, and we need a *count*, that tracks how many of the antecedent propositions remain to be proved. When the count reaches 0, we can add the consequent to the list of proved, or true, propositions. If we infer a consequent of `nil`, we have generated a contradiction, and the set of propositions is inconsistent.

So, we start with a global variable to hold the clauses as they are produced. Each clause will be an instance of a structure type, `clause`.

```
(defvar *clauses* nil
  "The cumulative list of clauses, aka knowledge base.")

(defstruct clause
  ants con count)
```

The way our expressions will look is the reverse of the way we wrote expressions using `<-`. The antecedents will come first and the consequent will be second. The pollutant example we used earlier would look like this:

```
> (-> (moved dry) airborne)
```

meaning that if a substance is being moved and is dry, we can infer that it will become airborne. Since we are *only* using conjunction in the antecedent part, we don't need the `and` tag.

The -> macro will be a little more complex than <-. It will construct a `clause` structure, making the `ants` slot be the list of antecedent symbols and the `con` slot the consequent symbol. The initial value of the `count` slot of course is simply the length of the antecedent list.

We need one further structure in order to make our code simple and efficient. It will be useful to track for each symbol (proposition) the clauses in which it appears in the antecedent part. This indexing enables us to look up the next clauses to be affected when a proposition becomes proved. As each proposition becomes asserted, we will look up the clauses in which it appears and decrease the count for each. Since propositions are symbols, we will use a feature of symbols, the *property list*, to store this information with the symbol itself. The property list of a symbol is a list of property name and property value pairs, in which the pairs are indexed by the property names. The Common Lisp `get` function retrieves property values, and `setf` can be used with `get` to set or update a property value for a symbol. So, we make up a property name, `on-clauses`, and use it to store the clauses in which a symbol appears.

```
(defmacro -> (antecedents consequent)
  '(let* ((ants ',antecedents) ;; to avoid multiple eval
          (con ',consequent)
          (clause (make-clause :ants ants
                               :con con
                               :count (length ants))))
     (dolist (pred ants)
       (push clause (get pred 'on-clauses)))
     (push clause *clauses*)
     clause))
```

As each clause is constructed, the `dolist` expression goes through the symbols in the antecedent part and adds the new clause to each symbol's `on-clauses` property. Later we will retrieve this information in order to do the actual forward chaining. The -> macro also adds the new clause to the global `*clauses*` list.

As the forward chaining process proceeds, we need to keep track of what propositions are true, and of those, which we have already used and which are not yet applied in the reasoning process. In the forward chaining code, we will keep a *stack*, a last-in first-out queue, for the propositions that are true but not yet processed, and a results list for the ones that we have finished with. The `push` function adds a new item on the front of the stack (which is really just a list), and the `first` function returns the item on the top of the stack, the last one in.[10]

We need a way to initialize the stack variable, from all the clauses. This can be done very simply. We start with an empty list. For each clause, we check if the count is 0. If it is, that means the consequent (the `con` slot value) is true without further reasoning, and we add it to the stack. After all the clauses are checked, the resulting stack is returned.

---

[10]There is a pop function that actually modifies the stack, like `push`, but we are not going to use it. Instead, we will use `first` and `rest`, as we will be passing the stack value in a recursive function call.

```
(defun init-stack (clauses)
  (let (stack)
    (dolist (clause clauses)
      (if (zerop (clause-count clause))
          (push (clause-con clause) stack)))
    stack))
```

Now we can implement forward chaining. The algorithm is straightforward. First, examine the stack. If it is empty, we are done and the result list has all the propositional entailments of the original set of formulas. If there are items on the stack, take the first item, check all the clauses on which it appears (look that up as the on-clauses property), and decrement the count slot of each. If a count gets to 0, check the consequent (the con slot) of that clause. There are three possibilities:

- it is nil, meaning we have an inconsistency in the knowledge base, or
- it has already been processed as a consequent of some other clause, in which case do nothing, or
- it has not yet been processed, in which case we add it to the new items that will go on the stack to be processed.

Then, just recursively process what is left on the stack (including any new additions), adding the proposition just processed to the results list. This is all done by an inner function, fc-aux, as usual, and the outer function just calls it with the initialized stack and an empty results list to start with. In the following, we have added a few format expressions to print out information about the progress of the proof.

```
(defun forward-chain (clauses)
  (labels ((fc-aux (stack results)
             (if (null stack) results
                 (let ((prop (first stack))
                       (newprops nil))
                   (format t "Current prop: ~A~%" prop)
                   (dolist (clause (get prop 'on-clauses))
                     (format t "Clause: ~S~%" clause)
                     (if (zerop (decf (clause-count clause)))
                         (let ((concl (clause-con clause)))
                           (format t "Concl: ~A~%" concl)
                           (if (null concl) (return-from fc-aux 'fail)
                               (unless (find concl results)
                                 (push concl newprops))))))
                   (fc-aux (append newprops (rest stack))
                           (cons prop results))))))
    (fc-aux (init-stack clauses) nil)))
```

In the case of backward chaining with Modus Ponens, we saw that there is a lot of searching. Because there are so many different ways to accomplish solutions to most problems, it is important to consider the efficiency of the solution. We express the *complexity* of a problem (and similarly an algorithm that solves it) in a notation called "Big O" notation. The complexity question asks, how does the cost of computation scale up as the number of items in the problem increases? If the number of items doubles, does the computation time double also? Perhaps it quadruples, or worse. This relation can usually be expressed in terms of a mathematical function of the number of items, n. An algorithm that takes a time proportional to the problem size is said to be linear, or $O(n)$. If it is quadratic, or grows in proportion to the square of the problem size, we write $O(n^2)$. If the problem grows with some definite power of the problem size, even if the power is large, we say it is doable in polynomial (P) time. If it is worse, for example it grows exponentially, like $O(2^n)$, we say it is exponential, or in any case, non-polynomial (NP). In some cases, a problem may have a solution that is exponential, but it can be proved that the problem also has more efficient, perhaps polynomial, solutions. So, the complexity of an algorithm may or may not be worse than the (intrinsic) complexity of the problem. It is of course never better, by definition.

A common misconception is that recursive algorithms have worse complexity than iterative algorithms. This is simply wrong. Each can be linear, or exponential or anything in between. The recursive functions we have defined so far are mostly linear in terms of time. There is an iterative algorithm for solving the famous Towers of Hanoi problem,[11] but it is exponential. The Towers of Hanoi problem is intrinsically exponential, and no better algorithm is going to be found. One difference that often is true between recursive and iterative algorithms for the same problem is that a naive recursive algorithm is not only linear in time, but also in space, that is, it uses a lot of memory for the recursive call context. However, back in Chapter 1 on page 36, we saw how to write tail-recursive functions that can be compiled into iterative processes taking up very little space. This way you get the space efficiency of an iterative process with the simplicity and elegance of a recursive algorithm.

We saw that the backward chaining code is exponential. How efficient is the forward chaining algorithm? There are two repetitive processes here, the recursive call, and the iteration over clauses. The maximum number of recursive calls in the outer loop is the number of positive terms that appear on the right-hand sides of clauses, since we don't repeat the processing of any term. This can't be more than the number of clauses. The maximum number of iterations each time in the inner loop is less than or equal to the number of clauses, since the maximum length of any symbol's `on-clauses` property is the total number of clauses (for example, if a symbol appeared in every clause). Therefore the computation is definitely no worse than $O(n^2)$ where n is the number of clauses. In general it is much less, and in any case this is far better than exponential.

---

[11] The Towers of Hanoi are three posts, on which are placed concentric disks (with holes in the middle) of decreasing size. The disks start out on one post, and the goal is to move one disk at a time until all are on another post. However, one can never place a larger disk on top of a smaller disk. Discussions of it may be found in [48,440] and in [123] where it is described as part of a legend originating in India.

To test the code, we create some clauses and then call the appropriate functions. We'll run through our well-exercised pollution example. As noted earlier the code for forward-chain includes some format expressions to print out a progress report as the calculations proceed. So the trace that follows shows those as well as the final resulting list of entailments.

```
> (-> (moved dry) airborne)
#S(CLAUSE :ANTS (MOVED DRY):CON AIRBORNE :COUNT 2)
> (-> (airborne toxic) more-asthma)
#S(CLAUSE :ANTS (AIRBORNE TOXIC) :CON MORE-ASTHMA :COUNT 2)
> (-> (more-asthma) more-clinic-visits)
#S(CLAUSE :ANTS (MORE-ASTHMA) :CON MORE-CLINIC-VISITS :COUNT 1)
> (-> nil moved)
#S(CLAUSE :ANTS NIL :CON MOVED :COUNT 0)
> (-> nil dry)
#S(CLAUSE :ANTS NIL :CON DRY :COUNT 0)
> (-> nil toxic)
#S(CLAUSE :ANTS NIL :CON TOXIC :COUNT 0)
> (forward-chain *clauses*)
Current prop: MOVED
Clause: #S(CLAUSE :ANTS (MOVED DRY):CON AIRBORNE :COUNT 2)
Current prop: DRY
Clause: #S(CLAUSE :ANTS (MOVED DRY):CON AIRBORNE :COUNT 1)
Concl: AIRBORNE
Current prop: AIRBORNE
Clause: #S(CLAUSE :ANTS (AIRBORNE TOXIC):CON MORE-ASTHMA :COUNT 2)
Current prop: TOXIC
Clause: #S(CLAUSE :ANTS (AIRBORNE TOXIC):CON MORE-ASTHMA :COUNT 1)
Concl: MORE-ASTHMA
Current prop: MORE-ASTHMA
Clause: #S(CLAUSE :ANTS (MORE-ASTHMA):CON MORE-CLINIC-VISITS :COUNT 1)
Concl: MORE-CLINIC-VISITS
Current prop: MORE-CLINIC-VISITS
(MORE-CLINIC-VISITS MORE-ASTHMA TOXIC AIRBORNE DRY MOVED)
```

As you can see, forward chaining generated all the consequents. In this case, there is no extraneous information, but in a realistic system with hundreds or thousands of propositions, there will be a big difference between exponential and quadratic scale in performance, even with lots of extraneous information. A problem with use of a forward chaining system for this example, though, is that it is difficult to extract the actual proofs, or explanations, for any given conclusion.

An example of a system that incorporates forward chaining as an inference strategy is the IMM/Serve system [283] which was designed to provide advice on management of childhood immunization. In this system, a hybrid design is used to organize knowledge in whatever way best matches each phase of the decision support task. Parameters that specify the time points at which various vaccines are recommended are organized in tables. The logic of identifying status of a particular child's immunizations, and which are due, or overdue or too early, is encoded in rules (our Horn clauses). The complexity of actual scheduling of dates on which to perform vaccinations led the developers to implement that part as procedural code.

Enhancements of this simple code are possible. One that would be very useful is to modify both the -> macro and the `forward-chain` function to support incremental additions of assertions to the knowledge base, or list of clauses. A more complex problem is that of retracting assertions. This is a special case of a more general problem of support for *non-monotonic* reasoning. In some situations it is important to be able to assert facts or rules as *defaults*. This means that if there is no reason to believe otherwise, these assertions are considered to be true, but they can be overridden by assertions later added. Any other assertions that were derived from them may no longer be entailed by the new knowledge base. Systems that can manage these kinds of changes are called *truth maintenance* systems. The drug interaction modeling problem addressed later in this chapter illustrates the use of such a system for managing evidence about drug metabolism, and about how drugs affect metabolic enzyme production.

Resolution

As we saw with the example of bacteria and treatability, when disjunctions appear on the right-hand side of a rule (implication), it is not possible to apply Modus Ponens as an inference method. Fortunately we were able to apply a more general and more powerful inference rule, resolution. In the code so far for backward chaining and forward chaining using only Modus Ponens, we have restricted the form our logic expressions can take to Horn clauses. However, much medical knowledge is not expressible this way. It seems that more powerful methods are needed.

Resolution, together with a suitable strategy for its application, can be shown to be both sound and complete when applied to propositional logic. Using a kind of forward chaining, given a set of propositional sentences, one can simply apply the resolution rule to pairs of clauses in the set until no further new clauses can be generated. If this does not produce the empty clause, the set is consistent. If it produces the empty clause, the set is inconsistent. So, a refutation method will work here. To prove that a formula, F, is entailed by a set of formulas, S,

1. append the negation of F to S,
2. convert all the formulas to clauses in Conjunctive Normal Form (CNF),
3. systematically consider every possible pair, and derive any new clauses using the resolution rule,
4. add the new clauses to the set.

The last two steps are repeated until either the empty clause results from an application, which means the set is inconsistent, and the formula F is entailed (since its negation is inconsistent with the set), or no new clauses are derivable, in which case the proof fails, and the formula F is not entailed by S.

This procedure is guaranteed to terminate because there are only a finite number of literals or terms in the original set and the formula F. Only a finite set of clauses can be formed from a finite number of literals. This is because repeating a literal in a clause does not produce a new clause, and including a literal and its negation makes that part of the clause a tautology, which can be removed. So, the longest clauses are the ones in which every literal or its negation appears once. The order also does not matter, since the ∨ operator is commutative. Thus the resolution procedure terminates on every finite set of clauses, whether the new formula is entailed or not.

The method will handle any set of formulas whatsoever. This is because any formula(s) can be transformed into an equivalent set of formulas in clause form (CNF). Any formula that is entailed by the original set is also entailed by the set of clauses, and vice versa. The steps for producing clauses from arbitrary formulas are:

1. eliminate all implications ($\rightarrow$) by using the fact that the formula $\alpha \rightarrow \beta$ is equivalent to $(\neg\alpha) \lor \beta$,
2. move all the negations inward until negations only appear with individual terms, using the double negative replacement rule and De Morgan's laws,
3. use the distributive and commutative laws to rearrange the "and" and "or" connectors, until you have a series of disjunctions connected by conjunctions.

Altogether, this may sound pretty inefficient, and it is. Many variant strategies for applying the resolution rule exist that can be more efficient. They are described in the standard AI texts.

While this may sound like our ultimate solution, propositional logic is very limited in what can be expressed. It is impossible to write general statements that can then be specialized to particular individuals. For example, the claim that substances that are dry and being moved about become airborne is a general one, to which we would like to add that a specific substance is being transported, rather than always reasoning in general. Not all substances that are transported in dry form are toxic, but propositional logic has no way to distinguish them without having lots of terms, which then need rules for each. We will shortly present a generalization, called First Order Logic, which adds variables and quantifiers, to be able to express much more. Unfortunately, the price we will pay is that reasoning is much harder, and not guaranteed to succeed.

2.2.1.5 Example: Biochemical Reachability Logic

Before moving onto First Order Logic, a more elaborate example, searching for biochemical pathways, shows that interesting problems can be represented with only propositional logic, even though it is very simple.

At some level, life is a large collection of cooperating biochemical processes. These processes are regulated by the activity of genes and proteins. In this section, we will show how to do some symbolic biochemical process modeling, just using basic knowledge about biological molecules and biochemical reactions. The specific process we will be examining is metabolism, the process by which cells transform energy sources in the form of sugar molecules into more useable forms such as ATP. Although in reality all kinds of biochemical reactions are going on simultaneously in the cell, it is possible to think of these transformations as sequential, a series of reactions, starting with a molecule such as glucose, going through one reaction after another until we end up with another molecule, pyruvate, and some by-products. The products of one reaction become the reactants or inputs to the next. Such a sequence of molecular reactions is called a *pathway*. Although pathways don't exactly exist as such in cells, the pathway representation provides an explanation of how things can happen. If there is no pathway from a given start point to a given end point, we know for sure it cannot happen. If a pathway exists, it may or may not be energetically favorable. The reactions may require too much energy, or other conditions such as the pH may make the reactions too slow in one direction to have any impact. We will ignore these details for now.

The problem we will solve here is: suppose you have a bunch of molecules in a cell, and you also have a list of possible (known) biochemical reactions. Is it possible from these starting molecules to generate some other molecule, and how?

Pathway generation is one way to answer the question of whether a process is possible. One can represent biochemical reactions in predicate logic, and use the backward chaining theorem prover to answer questions about whether molecules can be synthesized in various environments.

Essentially the reasoning behind the biochemical reactions is propositional logic. So, it should be possible to reformulate the problem as a logic problem and use proof methods. The right proof methods will be able to combine multiple branches, if necessary. They are not limited to a single chain of inference steps (chemical reactions).

Here is an implementation of some of the reactions and molecules involved in glucose metabolism, as rules in propositional logic. Reactions are rules, where each product is a consequent of some reactant or combination of reactants (ignoring enzymes for now), and the initial environment is defined by a set of assertions. The interpretation is that a symbol has the value *true* if the corresponding molecule is present in the cellular environment. First we list all the reactions, without distinguishing between "products" and "other products." Where a reaction produces multiple products, we have added a rule for each of the products. Where a reaction has several reactants, they are in the rule antecedent part combined with the and operator. Thus our representation consists of strict Horn clauses, and we can use the backward chaining (or forward chaining) proof methods described here. Here are all the reactions.

```
(<- f1p fru)
(<- glh f1p)
(<- dap f1p)
(<- g3p glh)
(<- g6p (and glu atp))
(<- f6p g6p)
(<- fbp (and f6p atp))
(<- dap fbp)
(<- g3p fbp)
(<- g3p dap)
```

```
(<- bpg (and P NAD + g3p))
(<- 3pg (and bpg adp))
(<- 2pg 3pg)
(<- pep 2pg)
(<- pyr (and pep atp))
(<- aca (and pyr NAD + coa))
(<- ict cit)
(<- akg (and ict NAD+))
(<- sca (and akg NAD + coa))
(<- suc (and sca gdp P))
(<- coa (and sca gdp P))
(<- fum (and suc FAD))
(<- mal (and fum H20))
(<- oxa (and mal NAD+))
(<- adp (and glu atp))
(<- adp (and f6p atp))
(<- NADH (and P NAD + g3p))
(<- H+ (and P NAD + g3p))
(<- atp (and bpg adp))
(<- H20 2pg)
(<- adp (and pep atp))
(<- NADH (and pyr NAD + coa))
(<- H+ (and pyr NAD + coa))
(<- CO2 (and pyr NAD + coa))
(<- NADH (and ict NAD+))
(<- H+ (and ict NAD+))
(<- CO2 (and ict NAD+))
(<- NADH (and akg NAD + coa))
(<- H+ (and akg NAD + coa))
(<- CO2 (and akg NAD + coa))
(<- gtp (and sca gdp P))
(<- FADH2 (and suc FAD))
(<- NADH (and mal NAD+))
(<- H+ (and mal NAD+))
```

Next, a series of assertions define which molecules are in the initial environment. We have left out the enzymes, since they are not used in the rules.

```
(<- atp)
(<- adp)
(<- gdp)
```

```
(<- gtp)
(<- fru)
(<- glu)
(<- NAD+)
(<- FAD)
(<- P)
```

Can we produce pyruvate (pyr) from the above system? It is a simple call to the prove function.

```
> (prove 'pyr)
T
```

The result is not news, but the fact that we got the result confirms that this formulation appears to be valid. To see in more detail what is happening, we can use the trace operator. Tracing the prove function shows a pathway by which pyruvate can be made from the above environment.

```
> (trace prove)
(PROVE)
> (prove 'pyr)
  0[2]: (PROVE PYR)
    1[2]: (PROVE (AND PEP ATP))
      2[2]: (PROVE ATP)
      2[2]: returned T
      2[2]: (PROVE PEP)
        3[2]: (PROVE 2PG)
          4[2]: (PROVE 3PG)
            5[2]: (PROVE (AND BPG ADP))
              6[2]: (PROVE ADP)
              6[2]: returned T
              6[2]: (PROVE BPG)
                7[2]: (PROVE (AND P NAD + G3P))
                  8[2]: (PROVE G3P)
                    9[2]: (PROVE DAP)
                      10[2]: (PROVE FBP)
                        11[2]: (PROVE (AND F6P ATP))
                          12[2]: (PROVE ATP)
                          12[2]: returned T
                          12[2]: (PROVE F6P)
                            13[2]: (PROVE G6P)
                              14[2]: (PROVE (AND GLU ATP))
                                15[2]: (PROVE ATP)
                                15[2]: returned T
```

```
              15[2]: (PROVE GLU)
              15[2]: returned T
            14[2]: returned T
          13[2]: returned T
        12[2]: returned T
      11[2]: returned T
    10[2]: returned T
   9[2]: returned T
  8[2]: returned T
  8[2]: (PROVE NAD+)
  8[2]: returned T
  8[2]: (PROVE P)
  8[2]: returned T
 7[2]: returned T
6[2]: returned T
 5[2]: returned T
  4[2]: returned T
   3[2]: returned T
    2[2]: returned T
     1[2]: returned T
       0[2]: returned T
   T
```

This seemingly elegant scheme has several limitations. First of all, it only solves the reachability problem, it does not actually return the pathway. Such is mathematics. Many elegant proofs are non-constructive. However, there is another problem: it does not find all the pathways, but terminates when the first one is found. These can all be fixed by extending the prove implementation to return the actual proof in some standardized way and optionally iterating over all proofs supported by the knowledge base. Rewriting the above in Prolog would certainly be a straightforward way to accomplish this.

Another way to view this problem is that a pathway can be considered as the result of a search through a network in which the nodes are subsets of biochemical molecules and the arcs connecting the nodes are the biochemical reactions. This approach will be developed in Section 5.2.6 of Chapter 5. As will be seen, the backward chaining algorithm here has a big advantage over the search formulation, in that it can handle reactions with multiple reactants, and backward chaining will seek reactions to produce all of them. Simple path search will not be able to pick up the branch paths that are needed to support the proof. So, possibly our search for a path for malate may succeed, where it will fail with simple path search. Naively, we should see it fail if we do not have the initial presence of citrate and Coenzyme A.

```
> (prove 'mal)
  0[2]: (PROVE MAL)
```

```
   1[2]: (PROVE (AND FUM H2O))
    2[2]: (PROVE H2O)
      3[2]: (PROVE 2PG)
        4[2]: (PROVE 3PG)
...
    2[2]: (PROVE FUM)
      3[2]: (PROVE (AND SUC FAD))
        4[2]: (PROVE FAD)
        4[2]: returned T
        4[2]: (PROVE SUC)
          5[2]: (PROVE (AND SCA GDP P))
            6[2]: (PROVE P)
            6[2]: returned T
            6[2]: (PROVE GDP)
            6[2]: returned T
            6[2]: (PROVE SCA)
              7[2]: (PROVE (AND AKG NAD + COA))
                8[2]: (PROVE COA)
                  9[2]: (PROVE (AND SCA GDP P))
                    10[2]: (PROVE P)
                    10[2]: returned T
                    10[2]: (PROVE GDP)
                    10[2]: returned T
                    10[2]: (PROVE SCA)
...
```

The proof does fail, but not in the way we expected. In an effort to establish the presence of succinyl Coenzyme A (sca) the proof ran into a circularity in the rules.[12] Instead of failing because there was no coa in the environment from which to make sca, it tries to synthesize coa from another possible reaction, which leads back to sca.

So indeed Coenzyme A needs to be present and it is produced and consumed as the cycle proceeds. Several kinds of things are both produced and consumed over all, including the ATP and ADP molecules, the NAD family, etc.

It is possible to make the proof succeed by adding coa and cit to the environment. If the assertion for the simple presence of coa is reached before the rule that attempts to generate it, the proof will avoid an infinite loop.

```
> (<- coa)
2
```

---

[12] There is one other odd thing about the proof. The prove function attempts to prove the presence of water, because we didn't include water in the initial environment. Fortunately it is indeed the by-product of a series of possible reactions.

```
> (prove 'mal)
NIL
> (<- cit)
1 > (prove 'mal)
T
```

This works because the rules and facts are searched in order of last in, first out (LIFO). However, if coa had been asserted *before* the rule producing it from sca, the infinite loop would still persist. This is exactly the problem that one runs into more generally in writing Prolog programs, where the order of assertions makes a huge difference in how the rules are used by the control strategy. Nevertheless, these problems with our simple model can be overcome, opening a wide range of possibilities in modeling the logic of biochemistry.

2.2.2 Unsound Inference: Abduction and Induction

> *There is something fascinating about science. One gets such wholesale returns of conjecture out of such a trifling investment of fact.*
>
> *– Mark Twain*
> **Life on the Mississippi, 1874**

So far, we have looked at how to do reasoning with a pair of sentences, where one is an assertion of some fact and the other is an implication, that is, from P and $P \rightarrow Q$ we can infer Q. This is a sound reasoning method. But there are several other possible combinations. What about the other way around? From Q and $P \rightarrow Q$ can we infer P? This is known as "abduction," or abductive reasoning. It is definitely *not* sound, but it turns out to be very useful in medicine and biology, and in many other areas as well. The idea here is that we observe an effect, Q, and we know that it can be produced by a cause, P. Is seeing the effect sufficient to infer that the cause is present? No, certainly not. Other things might cause the same effect. For example, it may be that $N \rightarrow Q$ is also true. So, then, which cause is it? It is possible that Q is present without any cause even though it is also always present when P is.

Many medical diagnosis problems take this shape. A patient is experiencing congestion, which could have been caused by an allergy to something in the environment, or by a viral infection, or by high blood pressure. These three possible causes can be hypothesized as *explanations* for the congestion. This idea of using implications and inference rules as a route to explanations is called *abductive* reasoning. It helps form hypotheses. Further observations can help narrow the list of hypotheses that are consistent with the observations. For example, one of the hypotheses may be used with another implication and the sound inference of Modus Ponens to infer something else. An example might be, a viral infection causes elevated body temperature. Thus, observing whether the patient has a fever can possibly rule out the virus hypothesis. This process (the hypothetical-deductive approach described in Chapter 1 on page 172) is very common in medical diagnosis.

Similarly, we can consider whether it is any use to infer the $P \rightarrow Q$ relation when we observe both P and Q? This inference method is also unsound. There are many situations where two phenomena are observed together, but one cannot necessarily conclude that they always will be. Indeed many correlations are coincidental, and the two things are unrelated. But as for abduction, which can be a guide to hypothesis formation, this form of reasoning can be a guide to theory construction. It is known as *induction*, or inductive reasoning, or sometimes in a more general form, *machine learning* or *data mining*. Carefully constructed scientific experiments are designed to observe whether relations between variables *always* hold, independently of other factors in the environment. When a strong correlation is seen, it is likely that some kind of mechanistic relationship exists. In molecular biology, using automated methods to search for correlations, suggesting causal relationships, is widely used.

However, strong correlation results from data mining or from a series of well-designed experiments, *by themselves*, cannot establish a causal relationship or even suggest a direction for it. Observing P and Q together could equally well reflect the possible reverse relation, $Q \rightarrow P$, rather than $P \rightarrow Q$. It is simply not possible to generate theories by processing data. The theories or explanations must rely on the imagination of the scientist and on other knowledge which sets a context for new theories or hypotheses.

2.2.3 First Order Logic (FOL)

First order logic (FOL) extends predicate calculus by introducing the idea of variables and functions. Variables are symbols that can take on values representing individuals, just as in ordinary algebra. Then, in addition to writing formulas expressing assertions about individuals, we can write more general formulas expressing relations among classes of individuals.

A function is a relation between two sets of individuals, in which specifying the first of the individuals is sufficient to determine the second one. The one that is to be specified is the input, or *argument* of the function. Simple examples of functions include: the *title* function, where the input is an individual book, and the title function has a value that is the text naming the title of the book (assuming every book has a single title…). The relation between a child and his/her parent does not define a function, since any particular child typically has two parents. Similarly relations like *sister*, *brother*, *cousin* do not represent functions. The *mass* relation defines a function since every object has a definite mass.[13]

We sometimes write functions in the way they are usually denoted in mathematics, but typically with multi character names, since so many are needed in representing even a small amount of biological knowledge. More often we will use Lisp notation, since we are going to implement functions in real code. In the following, book17 is a particular book, and car54 is a particular car. The title of that book is a string of text, but the color of the car car54 is represented by a symbol.

```
title(book17) = "Principles of Biomedical Informatics"
color(car54) = silver
```

---

[13]That is, at any particular instant, though my mass seems to fluctuate on a daily basis depending on my indulgence in salty foods.

```
color(?x) = ?y
...
```

In the third example, ?x and ?y are *variables*. In the formula, their values are unspecified or unknown. In Lisp computer programs that manipulate such expressions, it is common to use symbol names beginning with ? to indicate that the symbol is being used as a variable rather than an individual (also called a *constant*). In some programming languages, variables are distinguished from constants by other typographical conventions. In those languages the typographical conventions are processed by the compiler that translates the source program into a runnable form. However, here the symbols are going to be manipulated by the programs we write, not a compiler program. So we choose a naming convention that our programs will use. In a conventional program, variables will be assigned values, and they really represent just storage locations for the values. Here, what will happen is that the variables will have bindings associated with them, but not in the usual way of assigning a symbol a value.

One can define functions of more than one argument, or input variable, as well. In every case, in order to be a function, it must have a unique value for every combination of values of its *arguments*, or input variables. For example, we could define a function, *length*, that refers to the number of pages in a book. This function would have two inputs, a title and an edition, and could be denoted as length (?booktitle,?edition), so, for example, if we refer to the title "Medical Informatics: Computer Applications in Health Care" as medinfo, we have the value

```
length(medinfo, 2) = 715
```

Finally, functions whose values can only be one of the two values, *true* (we denote this in Lisp by the symbol t) and *false* (denoted by the symbol nil), are called *predicates*. Although parent(?x) is not a function, the parent relation with two inputs can be a predicate function. Thus, parent(?x ?y) can have the value t if the actual value of ?x is in fact a parent of the individual that is the value of ?y.

```
parent(ira alan) = t
parent(alan morrigan) = t
parent(ira susan) = nil
```

To summarize, as with predicate calculus, first order logic is made up from primitives and connectors, and semantics for the connectors. Primitives in first order logic can be:

- constant symbols (individuals) as in predicate calculus,
- function symbols,
- predicate symbols.

A function symbol names a function, whose value depends on the inputs to the function. In a formula using the function, the inputs can be named individuals, or they can be variables.

Predicate symbols name special functions whose values can be only the values "true" and "false."

First Order Logic also provides connectors called "quantifiers." Quantifiers allow iteration over individuals that are referred to by variables. There are two kinds of quantifiers. The universal quantifier ∀, reads "for all," meaning that the formula in which the quantified variable appears is true for every possible value of the variable. The existential quantifier ∃, reads "there exists," meaning that there is at least one value for the quantified variable such that the formula in which it appears is true.

A *sentence* is a formula with no unquantified variables. The same (quantified) variable can appear in many different sentences, and in each case it may take on values independently of values it takes on in the other formulas. What this means is that we can "reuse" variable names when writing many formulas. However, *within a single formula or sentence*, the variable's meaning and value must be the same everywhere in that sentence. This usage is exactly like that in writing mathematical formulas.

The ∀ quantifier is used to construct a sentence in which something (or a combination of things) is true for every instance or object to which the predicates or functions apply. It is almost always used with the → connective. An example is the statement that all mammalian cells have nuclei. We would write it in FOL notation as

$$\forall x : \text{mammal-cell}(x) \rightarrow \text{has-nucleus}(x).$$

This assertion is actually not universally true (red blood cells don't have nuclei) but the formula does express the meaning of the natural language sentence.

The ∃ quantifier is almost always used with the ∧ connective, asserting that there is some individual or instance satisfying the stated predicates or formulas. There could be more than one, but the quantifier just asserts that there is at least one. An example is the statement that there is at least one antibiotic that is effective against *streptococcus* bacteria. In FOL notation, we would write

$$\exists x : \text{antibiotic}(x) \wedge \text{treats-strep}(x).$$

2.2.3.1 The Art of Writing Logic Formulas

Using quantifiers, we can construct a much wider range of formulas than those of predicate calculus. However, it is often difficult to decide what the predicates and the variables should be. Some choices are awkward for doing the kind of reasoning we would like to be able to do. Others seem to work naturally. The following examples attempt to illustrate some of this.

First, we consider a family of antibiotics called *fluoroquinolones*. A popular textbook on pharmacology [148] states, "The fluoroquinolones are potent bacterial agents against *E. coli* and various species of *Salmonella, Shigella, Enterobacter, Campylobacter* and *Neisseria*." The names, *E. coli, Salmonella, Shigella*, etc. are categories of common types of pathogenic bacteria. The name "treats" will refer to a function of two inputs, x and y, that will have the value "true" if the value of x is a potent bacterial agent against the bacterial species that is the value of y. So, the term, treats(x, y), means "x is an effective treatment for y." It looks reasonable to write separate sentences for each of the categories of bacteria. The abbreviation fluor(x) will mean that x is a fluoroquinolone.

$$\forall x : \text{fluor}(x) \rightarrow \text{treats}(x, \text{ecoli})$$
$$\forall x : \text{fluor}(x) \rightarrow \text{treats}(x, \text{salm})$$
$$\forall x : \text{fluor}(x) \rightarrow \text{treats}(x, \text{shig})$$
$$\forall x : \text{fluor}(x) \rightarrow \text{treats}(x, \text{entb})$$
$$\forall x : \text{fluor}(x) \rightarrow \text{treats}(x, \text{campb})$$
$$\forall x : \text{fluor}(x) \rightarrow \text{treats}(x, \text{neiss})$$

Next, we note that the generic name, Levofloxacin, refers to one of several kinds of drugs that are fluoroquinolones. A simple way to express this would be fluor(Levf) though we will see shortly that this way of defining the relation is not so useful.

Many drugs are available from major drug manufacturers as *brand name* drugs. Most brand name drugs have "generic" equivalents, essentially the same drug but manufactured by a different company than the original patent holder. Levaquin is a brand name drug corresponding to the generic, Levofloxacin. We can use the abbreviated name Levq for Levaquin, but writing Levq = Levf is not correct. It is true that the generic name applies to the brand name drug as well as those versions that have no brand name. But there are other brand names that correspond to the generic Levofloxacin, so a correct way is:

$$\forall x : \text{Levq}(x) \rightarrow \text{Levf}(x)$$

and we should rewrite the previous formula as

$$\forall x : \text{Levf}(x) \rightarrow \text{fluor}(x).$$

It is also known that Levofloxacin can cause tendonitis or ruptured tendons. For this, we can just write

$$\forall x : \text{Levf}(x) \rightarrow \text{tend}(x).$$

To be fancy and embed this in a richer environment, one would write formulas referring to the individual who is experiencing tendonitis because he/she took Levofloxacin.

For the next example, we consider the following excerpt from a radiologist's report of a CT scan of a patient's chest. The patient (we will call him "patient K") had a big kidney tumor removed, and later turned out to have a lung nodule that might be metastatic kidney cancer.

> *There is a 3 mm nodule in the right lower lobe (of the lung of patient K),*
> *corresponding to a 1.5 mm nodule seen on the previous scan.*

Here are some generally true statements about CT scan findings and treatment of kidney cancer:

- If a nodule is getting bigger, it may be either a malignancy or a reactive node.
- Malignancies need treatment.
- Reactive nodules don't need treatment.
- If a patient does not need treatment he will be OK.
- If a patient has a nodule that needs treatment and an effective treatment is available for that nodule the patient will be OK.

One more thing is known about patient K. Interleukin-2 (IL2), a human protein involved with the immune system, is an effective treatment for nodules that have been previously found in patient K.

How do we express all this in FOL, and what can be inferred? It looks like this patient will be OK, but can it be proved from the general knowledge and specific findings? Here is a possible translation. We will call the nodule N. The meanings of the names for the predicate functions should be apparent. We will encode the general statements first. To be complete, we should include the predicate nodule(x) in most of these formulas, but it does not add any insight, so I am omitting it.

$$\forall x : \text{bigger}(x) \rightarrow \text{malignant}(x) \lor \text{reactive}(x), \tag{2.5}$$

$$\forall x : \text{malignant}(x) \rightarrow \text{need-treat}(x), \tag{2.6}$$

$$\forall x : \text{reactive}(x) \rightarrow \text{no-treat}(x). \tag{2.7}$$

So far, this all looks right. However, the last two sentences are a little tricky. A straightforward translation would be:

$$\forall x, y : \text{contained-in}(x, y) \land \text{no-treat}(x) \rightarrow \text{OK}(y), \tag{2.8}$$

$$\forall x, y, z : \text{contained-in}(x, y) \land \text{need-treat}(x) \land \text{effective}(z, x, y) \rightarrow \text{OK}(y). \tag{2.9}$$

This will be sufficient for our purposes. Now we add the specific sentences about the nodule and patient K. One is a constant term, a predicate expressing a property of nodule N. Another is the statement that nodule N is contained in patient K. The last is a sentence about nodules in general, but patient K and IL2 treatment in particular. We can of course substitute N in this formula too.

$$\text{bigger}(N), \tag{2.10}$$

$$\text{contained-in}(N, K), \tag{2.11}$$

$$\forall x : \text{effective}(\text{IL2}, x, K). \tag{2.12}$$

Recalling experience in reasoning in propositional logic, it looks like a generalized form of resolution will be needed here, because we have a sentence which is not a Horn clause. Also, a way is needed to relate the general statements with variables to the specific statements about individuals.

2.2.3.2 Additional Inference Rules for FOL

Statements that have universally quantified variables capture general and powerful knowledge about some biomedical or health idea. We would like to apply them together with statements about specific individuals. So, it should be possible to take any general statement and from it get a statement about a particular individual to whom the predicates and other functions apply, by simply substituting the individual for the quantified variable everywhere in the sentence. This is called *Universal specialization*. So, from $\forall x : p(x)$ we can get $p(A)$, where x is a variable, and A is some particular entity, represented here by a constant.[14]

We may also have a sentence in which we would like to do variable substitution that will make expressions match up. Although we can put any individual in place of a variable (provided the predicate functions are applicable to it), some choices will be useful and others not. In order to apply inference rules we had previously with propositional logic, we need to have the same term appear in two different formulas. To make terms match, the variables must match, or the variable substitutions with individuals (constants) should make the predicates match. Finding the substitutions that make such terms match up is called *Unification*.

In the metastatic cancer example, we have an assertion about a specific nodule, N, and a general rule about the predicate, `bigger`. A good strategy seems to be to substitute the specific nodule N in the rule for `bigger`, giving

$$\text{bigger(N)} \rightarrow \text{malignant(N)} \vee \text{reactive(N)}. \tag{2.13}$$

We can go further with this and put nodule N as a specific instance in all the rules where nodules appear in predicate functions. We can also substitute IL2 for the variable referring to treatment and K for the patient, in the rules that apply.

Now it looks like we could do some reasoning, but before that, we need to convert the formulas to clauses, and then we can apply resolution to prove that indeed the patient will be OK.

2.2.3.3 Clauses and Conjunctive Normal Form

The general procedure for converting an arbitrary clause to CNF in First Order Logic is essentially the same as in propositional logic, with some additional steps to handle the quantifiers and variables. The goal is to finish up with expressions which are disjunctions of either primitives or negations of primitives, and to have only universally quantified variables or constants. We will need three additional ideas, rules involving quantifier movement within formulas, being able to rename variables, and being able to replace existentially quantified variables by constants or functions.

---

[14]A note of caution here is in order: p is a function, which, as part of its definition, has a specific *domain*, the set of individuals for which the function p is defined. This means that universal specialization is not quite universal, but is limited to those entities in the domain of p.

Quantifier Order

In propositional logic, we saw that there are algebraic rules for transforming composite expressions, with the various connectives in different orders and combinations. Now that we have introduced quantifiers, it is important to establish the rules for changing the order of quantifiers with each other and the connectives.

Changing the order of universal quantifiers does not change the meaning of a sentence. Here is an example: if a physician is the family physician of an individual, we can infer that she will provide medical diagnosis or treatment for that individual at some point.

$$\forall x \forall y : \text{family-physician}(x, y) \rightarrow \text{treats}(x, y).$$

The formula is symmetric with regard to the quantifiers, so the following means exactly the same thing.

$$\forall y \forall x : \text{family-physician}(x, y) \rightarrow \text{treats}(x, y).$$

The same is true of existential quantifiers. The statement that there is at least one antibiotic that works on some bacterial species can be rendered as follows:

$$\exists x \exists y : \text{antibiotic}(x) \wedge \text{kills}(x, y).$$

Again, since the formula is symmetric with regard to the quantifiers, interchanging them gives a sentence with the same meaning.

$$\exists y \exists x : \text{antibiotic}(x) \wedge \text{kills}(x, y).$$

However, changing the order of a universal with an existential *does* change the meaning. Such formulas are not symmetric, and the predicate functions that appear in such sentences are usually not symmetric. Here is a sentence that says that every person has a family physician (this is definitely not true, but an ideal perhaps we can aspire to achieve).

$$\forall x \exists y : \text{family-physician-of}(x, y).$$

On the other hand, the following sentence says something very different, that there is someone who is the family physician of everyone. This is certainly *not* something we would consider as an ideal or expect to ever see as a reality.

$$\exists y \forall x : \text{family-physician-of}(x, y).$$

Similarly with formulas involving implication, we get very different meanings when we interchange universal with existential quantifiers. This sentence says that every species of bacteria has some nutrient medium on which it grows. It is certainly true, since otherwise that bacterial species would not exist.

$$\forall x \exists y : \text{bacteria}(x) \rightarrow \text{nutrient-medium}(y) \wedge \text{grows-on}(x, y).$$

Interchanging the quantifiers gives a very different statement, that there is a universal nutrient medium on which all bacteria grow, which, to my knowledge, is *not* true [301, page 114].

$$\exists y \forall x : \text{bacteria}(x) \rightarrow \text{nutrient-medium}(y) \wedge \text{grows-on}(x, y).$$

So, in FOL we will have some laws or rules similar to the ones we presented earlier for propositional logic, but we need to be very careful about applying them in formulas with quantifiers.

Quantifier Movement

Another simplification step we will want to follow in transforming formulas is to move all quantifiers to the left-hand side of a formula, so that the rest of the parts can be manipulated with other transformation rules. In general, you can move quantifiers to the left in a formula, without changing its meaning, subject to the constraint we already showed above. For example, you can move a quantifier past an implication connective, as in the following sentence, which says, "Antibiotics treat everything."

$$\forall x : \text{antibiotic}(x) \rightarrow \forall y : \text{can-treat}(x, y).$$

The sentence resulting from moving the y quantifier to the left means exactly the same thing (and is just as untrue).

$$\forall x \forall y : \text{antibiotic}(x) \rightarrow \text{can-treat}(x, y).$$

This is true even if the quantifiers are different, as long as you do not move one past the other. The following sentence says that every bacterial species has some nutrient medium on which it grows, just as did the sentence in the previous section.

$$\forall x : \text{bacteria}(x) \rightarrow \exists y : \text{nutrient-medium}(y) \wedge \text{grows-on}(x, y).$$

You can also move such quantifiers to the left past an and connective or an or connective, but you can *not* move quantifiers past a negation. Consider the following sentence, which says that no disease is worse than a fatal disease.

$$\forall x : \text{fatal-disease}(x) \rightarrow \neg \exists y : \text{worse-than}(y, x).$$

This may be debatable, of course, depending on the meaning of "worse than," but it certainly is not the same as saying that for every fatal disease, there *is something* that is not worse than it.

$$\forall x \exists y : \text{fatal-disease}(x) \rightarrow \neg \text{worse-than}(y, x).$$

Nevertheless, it is possible to move all the quantifiers to the left of a formula. In the case of negation, when a quantifier is moved past a negation, in order to preserve the meaning you change an existential quantifier to a universal and vice versa, that is, with negation the quantifiers have a kind of complementary relationship.

$$\forall x : \text{fatal-disease}(x) \rightarrow \neg \exists y : \text{worse-than}(y, x),$$
$$\forall x : \text{fatal-disease}(x) \rightarrow \forall y : \neg \text{worse-than}(y, x),$$
$$\forall x \forall y : \text{fatal-disease}(x) \rightarrow \neg \text{worse-than}(y, x).$$

Conversion to CNF

The step-by-step process of conversion of an arbitrary formula to CNF is as follows:

- Eliminate all implications ($\rightarrow$) by using the fact that the formula $\alpha \rightarrow \beta$ is equivalent to $(\neg\alpha) \vee \beta$.
- Move all the negations inward until negations only appear with individual terms, using the double negative replacement rule and De Morgan's laws. Note that in FOL, moving negations past quantifiers requires some care, as we have just seen.
- Standardize the variables. In a formula with several quantified parts, each quantified variable is independent of the others, and should be given a different name.
- Eliminate existential quantifiers. An existential quantifier says that some individual exists, but does not identify it. For each one, assign a unique (constant) name. For example, the assertion, $\exists x : \text{tumor}(x)$, could mean there is some tumor present, but we don't know where. We would replace x with a constant, M, and rewrite the formula as $\text{tumor}(M)$ without the quantifier. If an individual depends on another variable, perhaps universally quantified, we replace the existentially quantified variable with a function (unspecified but named), that depends on the other variable. For example, in the sentence, "For every bacteria, there is a medium on which it will grow," which we wrote as

$$\forall x \exists y : \text{bacteria}(x) \rightarrow \text{nutrient-medium}(y) \wedge \text{grows-on}(x, y)$$

  the variable y depends on the variable x, so instead of a constant for y we put a function, $m(x)$, and we rewrite the formula as

$$\forall x : \text{bacteria}(x) \rightarrow \text{nutrient-medium}(m(x)) \wedge \text{grows-on}(x, m(x)).$$

  The use of a function accounts for the dependency, namely that different bacteria may need different media.

  These constants and functions are called Skolem constants and Skolem functions, named after their inventor, Thoralf Skolem, and the process is called *skolemization*.
- Convert to *prenex* form. This means move all the universal quantifiers to the far left, so the scope of each is the entire formula. At this point we have a bunch of quantifiers and a bunch of terms. The part without the quantifiers is called the *matrix*.
- Use the distributive and commutative laws to rearrange the matrix so that it consists of a series of disjunctions connected by conjunctions.
- Eliminate the universal quantifiers. Since every variable is universally quantified, we don't need to keep the quantifiers. We just adopt the convention that any variable that appears is universally quantified.
- Rewrite the formula as a set of separate formulas, each of which is a disjunction, rather than connecting them all by "and" connectors.

We can apply these procedures to convert all the formulas about nodules and patient K into clauses. Then we can use resolution in a systematic way, using refutation, to prove that indeed patient K will be OK. First, here are the clauses from the general formulas, after substituting the individuals N, K, and IL2.

$$\neg\text{bigger}(N) \vee \text{malignant}(N) \vee \text{reactive}(N), \tag{2.14}$$

$$\neg\text{malignant}(N) \vee \text{need-treat}(N), \tag{2.15}$$

$$\neg\text{reactive}(N) \vee \text{no-treat}(N), \tag{2.16}$$

$$\neg\text{contained-in}(N, K) \vee \neg\text{no-treat}(N) \vee \text{OK}(K), \tag{2.17}$$

$$\neg\text{contained-in}(N, K) \vee \neg\text{need-treat}(N) \vee \neg\text{effective}(\text{IL2}, N, K) \vee \text{OK}(K). \tag{2.18}$$

The following are the three specific formulas about the nodule N and patient K. In the third I have also substituted N for x.

$$\text{bigger}(N), \tag{2.19}$$

$$\text{contained-in}(N, K), \tag{2.20}$$

$$\text{effective}(\text{IL2}, N, K). \tag{2.21}$$

Now, we add one more clause, since we will do a proof by contradiction. We will add the clause

$$\neg\text{OK}(K) \tag{2.22}$$

and then derive the empty clause from the resulting set. The process will be to start with one of the clauses, use resolution to derive a new clause, and apply each clause in turn to the result of the previous step. Effectively this is a kind of forward chaining strategy. Eventually (if our theorem is indeed true, that the sentences entail that patient K will be OK) we will reach the empty clause. This strategy is called SLD resolution.

Here is the sequence. We start with clause 2.18. Resolve it with 2.20, giving

$$\neg\text{need-treat}(N) \vee \neg\text{effective}(\text{IL2}, N, K) \vee \text{OK}(K). \tag{2.23}$$

This clause then can be resolved with clause 2.22 to give

$$\neg\text{need-treat}(N) \vee \neg\text{effective}(\text{IL2}, N, K). \tag{2.24}$$

Next we resolve this with clause 2.21 to produce

$$\neg\text{need-treat}(N). \tag{2.25}$$

Now we need to use a clause that contains need-treat(N) in order to make further progress. The only choice is clause 2.15 which gives

$$\neg\text{malignant}(N). \tag{2.26}$$

By now you can see that for the next step we need a clause with malignant(N), which is clause 2.14, and this next step then gives

$$\neg\text{bigger}(N) \vee \text{reactive}(N). \tag{2.27}$$

We have two possible ways to go from here. We can use 2.16 or 2.19. Let's use the smaller of the two, clause 2.19, giving

$$\text{reactive}(N). \tag{2.28}$$

Now we use clause 2.16, since it is the only choice available. This brings us to

$$\text{no-treat}(N). \tag{2.29}$$

It may seem that we are going in circles, but in reality we are systematically eliminating all the possibilities. So, now we use clause 2.17, which we have not used yet, and is our only option. Now we have

$$\neg\text{contained-in}(N, K) \vee \text{OK}(K). \tag{2.30}$$

The end should be in sight. We can now use clause 2.20 to eliminate the first term, giving

$$\text{OK}(K) \tag{2.31}$$

which, together with clause 2.22, gives the empty clause, and our proof is completed. Variants of this process can be automated, and theorem proving programs that use resolution have been written. However, the complexity here is exponential. Because of this, for biomedical knowledge systems, other methods have been sought that trade off some expressive power to gain more tractable proof methods.

Because in general the formulas of FOL can involve functions and variables, one can have infinite chains of reasoning. Certainly, neither exhaustive forward chaining nor backward chaining is guaranteed to terminate. It is known that in fact no strategy will be guaranteed to terminate if the proof does not exist. This is the worst-case scenario. However, in many cases, the typical problem to be solved may be better behaved.

We have described here and in the propositional logic section a refutation proof method, since this is known to be complete. However, sometimes we know the form of a query but not the answer to verify. It would be nice to have an "answer extraction" method, some way to automatically generate the variable substitutions that satisfy the proof. To do this, instead of appending the negation of the goal, $G(x)$, you would append the tautology, $G(x) \vee \neg G(x)$, and do the same derivation. At the end of the proof, instead of the empty formula, you will have the formula $G(A)$ where A is the substitution that made the proof go through (after some unification step or steps). Then A will be an answer to the query. Finally, it is sometimes the case that a predicate can actually be evaluated to be either true or false, for example in an arithmetic statement. Evaluation of such terms is done by procedural attachment, which is invocation of actual function calls rather than just matching of predicates, thus simplifying the resulting expression.

2.2.4 Rule-Based Programming for FOL

This section describes how backward chaining application of Modus Ponens with Horn clauses can be implemented for First Order Logic. It is a more general version of the code in Section 2.2.1 in the backward chaining subsection. The additional capability needed here is to be able to do unification, or pattern matching. We will need also a way to write predicate functions and represent variables. The code here is reproduced (with permission) from Graham [137, Chapter 15]. It implements a kind of very simplified logic programming in the style of the programming language Prolog.

The form of our rules and assertions will be as before, with an operator, <-, similar to the one defined on page 213, that takes our predicate expressions and puts them in a hash table for later use. The only difference will be that the version here will have to take apart predicate *function* expressions, where before we just had symbols.

A predicate will consist of a list beginning with a symbol naming the predicate, and zero or more inputs, which can be constants (symbols or other terms), or variables. Since symbols are used as constants here, we need some way to indicate that a symbol is a variable instead of a constant. Following Graham, we represent variables by symbols that begin with ? (question mark). So the expression.

```
(tuna charlie)
```

encodes the assertion "charlie is a tuna." Similarly, the expression,

```
(tuna ?x)
```

encodes the assertion that something (we don't know what, yet) is a tuna. Then we can write rules, or implications, in the same style as before. So, to encode the sentence "Cats like to eat fish" we write,

```
(<- (likes-to-eat ?x ?y) (and (cat ?x) (fish ?y)))
```

This assertion reads from right to left as before, translating to, "If ?y is a fish and ?x is a cat, then ?x likes to eat ?y." In this limited rendition of FOL, we will assume that all variables are universally quantified. More expressive languages have also been defined to incorporate existential quantifiers and other kinds of operators too, but we won't deal with that here.

As before, the hash table key is going to be the predicate name in the consequent (left-hand side) of the rule. So, instead of just using the consequent, we need to use the first element of the consequent as the key and the rest of the consequent will go into the table with the antecedent, as the value. We'll need this information in order to match the variables with each other and with constants.

```
(defmacro <- (con &optional ant)
    "adds ant to the hash table entry for con, even if ant is nil"
    `(length (push (cons (rest ',con) ',ant)
                  (gethash (first ',con) *rules*)))))
```

Next we consider what to add to the prove function. In this case, every expression to prove will be a list, beginning with a predicate symbol, or one of the three connectors, and, or, not, as before. But now we have variables, and will have to keep track of possible temporary assignments of values to the variables, called *bindings*. The bindings result from the unification, or matching, operation, which we will describe shortly. So, in this version of prove there is an additional parameter, binds, which is a list of bindings. This bindings list needs to be passed to each of the specialized proof functions as well.

```
(defun prove (expr &optional binds)
  (case (first expr)
    (and (prove-and (reverse (rest expr)) binds))
    (or  (prove-or (rest expr) binds))
    (not (prove-not (first (rest expr)) binds))
    (t   (prove-simple (first expr) (rest expr) binds))))
```

As before, we have a prove-simple function when there is just a single predicate expression. For the other cases, we used the built-in Common Lisp functions, every, some, and not, but here we need specialized functions for these logical operators, because they need to handle the bindings, and possibly other complexities. In the and clause, as before, we use the reverse of the list of predicates to prove, implementing a right to left control strategy as in Prolog.

First, we consider the prove-simple function. This time it takes three inputs, the name of the predicate, a list of the arguments for that predicate, and a set of bindings. It will look up the predicate in the hash table of rules and assertions. The table contains a list of expressions, which could include constants, or antecedent parts of rules, associated with variables. As before, we test each entry in this list to see if it matches with the arguments, given the current bindings. If a match occurs, that could generate new bindings, or it could be that now we have another predicate we need to prove. In the first case, the new bindings are added to the existing ones, and in the second case, a recursive call to the prove function is needed.

```
(defun prove-simple (pred args binds)
  (mapcan #'(lambda (r)
              (multiple-value-bind (b2 yes)
                  (match args (first r) binds)
                (when yes
                  (if (rest r) (prove (rest r) b2)
                    (list b2)))))
          (mapcar #'change-vars
                  (gethash pred *rules*))))
```

The Common Lisp function mapcan is a useful combination of the action of nconc and mapcar. It applies a function to a list, giving a new list for each *item* in the original list, and then flattens the whole thing into a single list as nconc would.[15]

Because the variable names in the stored rule data might conflict with the current variable names in the args expressions, before using any of the expressions from the table, the variables should have their names changed to temporary unique names. This is the job of the change-vars function.

```
(defun change-vars (r)
  (sublis (mapcar #'(lambda (v) (cons v (gensym "?")))
               (vars-in r))
       r))
```

The vars-in function just goes through an expression and looks for symbols beginning with?, as tested by the var? function, returning a list of the variables. Then mapcar uses gensym to make an association list of old names and new ones. Then the sublis function replaces each variable with its corresponding new one.

```
(defun var? (x)
  (and (symbolp x)
       (eql (char (symbol-name x) 0) #\?)))

(defun vars-in (expr)
  (if (atom expr)
      (if (var? expr) (list expr))
    (union (vars-in (first expr))
           (vars-in (rest expr)))))
```

Now we are ready to consider the matching process. Two expressions match if:

- they are identical, that is, they are of the same symbol, so gentamycin matches with gentamycin, or they are the same variable, so for example, ?x matches with ?x without any bindings, or
- there is a binding for one of the two in the binds list, and the binding matches with the other (checked by a recursive call to the match function), or
- one is a variable (with no binding, since the previous lookup did not succeed), in which case the other expression becomes the binding, and that is added to the current bindings, or
- both are conses, in which case try to match the two first elements in each cons, and if that succeeds, try to match the remainders with each other.

---

[15]The function nconc is like append except that nconc is allowed to destructively manipulate its inputs to gain efficiency. It is OK to use nconc here because the lists it operates on are newly constructed and modifying them will not affect anything else in the environment.

In each case, the list of bindings (whatever is put in together with any new ones) is what the match function returns, along with a success or failure indicator. Here are some examples:

```
> (match '(calico ?x) '(calico pudge))
((?X. PUDGE))
T
> (match '(likes-to-eat ?x ?y) '(likes-to-eat pudge ?z))
((?Y. ?Z) (?X. PUDGE))
T
> (match 'pudge 'charlie)
NIL
> (match '?104 '?104)
NIL
T
> (match '?x '?x)
NIL
T
```

Note that the simple matches of the same variable did not generate any bindings but the second value returned is t indicating success. Thus, it is important in the prove-simple function to use both values. Similarly the code for match itself needs to use both values returned in the recursive call to match in the last part of its implementation. In the earlier branches, the function returns whatever the recursive call returns, so both values are simply passed on.

```
(defun match (x y &optional binds)
  (cond
    ((eql x y) (values binds t))
    ((assoc x binds) (match (binding x binds) y binds))
    ((assoc y binds) (match x (binding y binds) binds))
    ((var? x) (values (cons (cons x y) binds) t))
    ((var? y) (values (cons (cons y x) binds) t))
    (t
      (when (and (consp x) (consp y))
        (multiple-value-bind (b2 yes)
            (match (first x) (first y) binds)
          (and yes (match (rest x) (rest y) b2)))))))
```

The binding function looks up the binding of a variable, in the current list of bindings. If it finds one, it recursively looks to see if the result of the binding has a further binding, and so on, since a variable can be bound to another variable, which in turn may have a further binding.

```
(defun binding (x binds)
  (let ((b (assoc x binds)))
```

```
(if b
    (or (binding (rest b) binds)
        (rest b)))))
```

Here are the specialized functions that handle conjunctions, disjunctions, and negations. Remember that these only appear in the right-hand side, not the consequent side, so we are still dealing only with Horn clauses. The `prove-and` function takes the clauses in right to left order, applying `prove` to each, and all the bindings are then put together as a big list (using `mapcan` to combine both these jobs). If the proof of any of the clauses fails, `prove-and` returns `nil` (because `prove` will return `nil`).

```
(defun prove-and (clauses binds)
  (if (null clauses)
      (list binds)
      (mapcan #'(lambda (b)
                  (prove (first clauses) b))
              (prove-and (rest clauses) binds))))
```

The `prove-or` function tries to prove each clause, and returns the bindings for any that succeed, not just the first one.

```
(defun prove-or (clauses binds)
  (mapcan #'(lambda (c) (prove c binds))
          clauses))
```

The `prove-not` function is really simple. It tries to prove the input clause, and if the proof fails, the current bindings are returned. Note that here, as in the propositional case, proving $\neg A$ means that there is no proof for A. It does not mean that A is definitely false.

```
(defun prove-not (clause binds)
  (unless (prove clause binds)
    (list binds)))
```

To illustrate how the inference code works, here is a simple example, with some assertions about some common knowledge from the animal world. We start with the following set of sentences, encode them in our "mock Prolog," and then perform some queries.

1. $\forall x \forall y : \text{cat}(x) \land \text{fish}(y) \rightarrow \text{likes-to-eat}(x, y)$
2. $\forall x : \text{calico}(x) \rightarrow \text{cat}(x)$
3. $\forall x : \text{tuna}(x) \rightarrow \text{fish}(x)$
4. tuna(Charlie)
5. calico(Pudge)

Here is the encoding of these FOL sentences in the Graham inference formalism. Although there is not much reasoning to do here, when we trace the proof, it will become apparent how many steps there are.

```
(<- (likes ?x ?y) (and (cat ?x) (fish ?y)))
(<- (cat ?x) (calico ?x))
(<- (fish ?x) (tuna ?x))
(<- (tuna charlie))
(<- (tuna herb))
(<- (human ira))
(<- (calico pudge))
```

The query interface provided by Graham, `with-answer`, calls `prove` to get all the successful bindings of any variables in the query, and then executes its body once for each set of bindings.

```
(defmacro with-answer (query &body body)
  (let ((binds (gensym)))
    '(dolist (,binds (prove ',query))
       (let ,(mapcar #'(lambda (v)
                          '(,v (binding ',v,binds)))
                      (vars-in query))
         ,@body))))
```

Here is an example of how this operator is used, assuming we have the rules defined above for cats and fish. The `prove` function tries to find a match for the expression `(cat ?x)` and the only binding of the variable `?x` that works is pudge. That value is put in the `format` expression and the expression is evaluated, resulting in a single printout.

```
> (with-answer (cat ?x) (format t "~ A is a cat.~%" ?x))
PUDGE is a cat.
NIL
```

Another query could be to ask what Pudge likes to eat. This should give two answers, Herb and Charlie.

```
> (with-answer (likes-to-eat pudge ?x)
       (format t "Pudge likes to eat ~A.~ %" ?x))
Pudge likes to eat HERB.
Pudge likes to eat CHARLIE.
NIL
```

The reasoning proceeds by matching the `likes-to-eat` predicate with the query expression, which succeeds by binding the first variable in the original rule (suitably renamed) to pudge, and then a search proceeds to find a binding for the second variable, which will become the binding for ?x in the query expression. We can trace the `prove` function, to see how the bindings are produced.

```
> (trace prove)
(PROVE)
```

```
> (with-answer (likes-to-eat pudge ?x)
        (format t "Pudge likes to eat ~A.~%" ?x))
 0[2]: (PROVE (LIKES-TO-EAT PUDGE ?X))
   1[2]: (PROVE (AND (CAT #:?119) (FISH #:?120))
               ((?X . #:?120) (#:?119 . PUDGE)))
     2[2]: (PROVE (CAT #:?119) ((?X . #:?120) (#:?119 . PUDGE)))
       3[2]: (PROVE (CALICO #:?123)
               ((#:?123 . PUDGE) (?X . #:?120) (#:?119 . PUDGE)))
       3[2]: returned
               (((#:?123 . PUDGE) (?X . #:?120) (#:?119 . PUDGE)))
     2[2]: returned (((#:?123 . PUDGE) (?X . #:?120) (#:?119 . PUDGE)))
     2[2]: (PROVE (FISH #:?120)
               ((#:?123 . PUDGE) (?X . #:?120) (#:?119 . PUDGE)))
       3[2]: (PROVE (TUNA #:?126)
               ((#:?120 . #:?126) (#:?123 . PUDGE) (?X . #:?120)
               (#:?119 . PUDGE)))
       3[2]: returned
               (((#:?126 . HERB) (#:?120 . #:?126) (#:?123 . PUDGE)
               (?X . #:?120) (#:?119 . PUDGE))
               ((#:?126 . CHARLIE) (#:?120 . #:?126) (#:?123 . PUDGE)
               (?X . #:?120) (#:?119 . PUDGE)))
     2[2]: returned
               (((#:?126 . HERB) (#:?120 . #:?126) (#:?123 . PUDGE)
               (?X . #:?120) (#:?119 . PUDGE))
               ((#:?126 . CHARLIE) (#:?120 . #:?126) (#:?123 . PUDGE)
               (?X . #:?120) (#:?119 . PUDGE)))
   1[2]: returned
               (((#:?126 . HERB) (#:?120 . #:?126) (#:?123 . PUDGE)
               (?X . #:?120) (#:?119 . PUDGE))
               ((#:?126 . CHARLIE) (#:?120 . #:?126) (#:?123 . PUDGE)
               (?X . #:?120) (#:?119 . PUDGE)))
 0[2]: returned
               (((#:?126 . HERB) (#:?120 . #:?126) (#:?123 . PUDGE)
               (?X . #:?120) (#:?119 . PUDGE))
               ((#:?126 . CHARLIE) (#:?120 . #:?126) (#:?123 . PUDGE)
               (?X . #:?120) (#:?119 . PUDGE)))
Pudge likes to eat HERB.
Pudge likes to eat CHARLIE.
NIL
```

The first step is to generate two new goals from the likes-to-eat rule, to be able to prove that Pudge is a cat, and to find a binding for the fish predicate. The rule that a calico is a cat, and the fact that Pudge is a calico, then is used to establish that pudge satisfies the first requirement. Nothing directly satisfies the fish predicate, but the rule that a tuna is a fish generates another goal, to find a tuna. This succeeds twice, once for Herb and once for Charlie, so prove returns two lists of bindings.

If we asked the more general query, who likes to eat what, there would be two variables for which to find bindings, which works just the same way, but with a little more work. All the valid combinations will produce bindings and `with-answer` will run its body with each combination. Of course here we are not doing much with the results, but one could write arbitrary code that uses the results, even to run more proofs.

To build a practical rule-based system in the context of (for example) medical data, one more problem must be solved. The rules that form the knowledge can be created and used in the context of convenient and user-friendly reasoning systems that implement one or more of the methods we have described. Such systems provide the advantage of nice user interfaces and knowledge management facilities. Integration into an arbitrary programming environment is more difficult. Each provides some particular interface depending on the favorite environment of its creator. One popular example is JESS (Java Expert System Shell). This system is at URL `http://herzberg.ca.sandia.gov/`. It also uses Lisp S-expression style syntax, and provides integration with `java`. Also, a much more elaborate system like the backward chaining code described here is in Norvig [305]. That code has been integrated with the Common Lisp Object System [118] so that objects can have slot values that are asserted as facts in the rule base, and of course inferences can result in updating objects or even creating new ones.

However, being able to write programs that combine conventional code with inference is not sufficient to solve the medical decision support problem. The modularity of rules suggests that one could define standards for writing chunks of medical knowledge this way, and making such knowledge repositories widely available for incorporation in electronic medical record systems. There is no standard interface between the data in the electronic medical record system and the assertions in the knowledge base. This problem is typically solved only by local developers for their own specific system, making knowledge interchange very difficult. In general it remains an unsolved problem in medical informatics.

2.2.5 Example: Predicting Drug Interactions by Logical Inference

This section presents the first of several realistic applications of rule-based inference to important medical and biological problems. A drug interaction system is presented here that can reason about metabolic interaction mechanisms, as does the procedural code in Chapter 1, Section 1.2.5. However, in this formulation, the knowledge is declarative rather than procedural, and has the advantage that it is easier to add more rules and assertions than it would be to modify function code. It retains the power to predict interactions that may not have been entered in a catalog, an important step toward creating deep theories of biochemical and physiological processes important in medicine.

2.2.5.1 Reasoning About Pharmacokinetics

Some aspects of knowledge about pharmacokinetics can be expressed in first order logic (FOL), as described in this chapter. This opens the possibility of constructing a reasoning system for drug interaction prediction, which in turn means that some potential drug interactions can be flagged even if there have been no compiled case reports or clinical studies. Similarly, there is a potential to reduce the false positive output of a drug alert system if there is some way for the system to reason about the clinical significance of a potential interaction. Richard Boyce and colleagues have experimented with this approach with some success [40,41]. This section introduces the basic ideas and illustrates with examples taken from that work.

It is possible to identify and model pharmacokinetic origins of drug interactions. These interactions are consequences of mechanisms which are known for many drugs. A typical course in a pharmaceutics program would catalog and document a wide range of these mechanisms. Here we just provide a few examples.

For orally administered drugs whose primary route of absorption into the bloodstream is through the intestine, the serum level (concentration in the bloodstream) can be strongly affected by the condition of the intestine. Low intestinal motility (as caused by some narcotics) will increase the time the drug spends in the intestine and may increase the drug's serum level. Conversely, if there is damage to the gut, or the person is having intestinal problems such as diarrhea, absorption could be dramatically reduced. These can be expressed as FOL formulas. In the first formula, x is a drug and y is a patient. The predicate low-mot(y) is true if patient y is experiencing constipation or related problems that increase the time the drug spends in his/her gut.

$$\forall(x, y) : \text{oral-adm}(x, y) \wedge \text{low-mot}(y) \Rightarrow \text{inc-absorb}(x, y) \tag{2.32}$$

In the next formula, the opposite condition leads to reduced absorption.

$$\forall(x, y) : \text{oral-adm}(x, y) \wedge \text{high-mot}(y) \Rightarrow \text{red-absorb}(x, y) \tag{2.33}$$

These could be included in a reasoning system that infers low-mot(y) and other terms from further information about the patient and other drugs that are part of the patient's treatment regimen.

The next two formulas express the inhibition and induction effects mentioned earlier. In these cases, we have added one more condition to the formula, namely that the enzyme z is the primary clearance enzyme of drug y. This means that the metabolism of y is mostly through the action of z, not other possibly active enzymes.

$$\forall(x, y, z) : \text{inhib}(x, z) \wedge \text{prim-clear-enz}(z, y) \Rightarrow \text{red-clear}(x, y) \tag{2.34}$$
$$\forall(x, y, z) : \text{induce}(x, z) \wedge \text{prim-clear-enz}(z, y) \Rightarrow \text{inc-clear}(x, y) \tag{2.35}$$

A change in pH can affect the rate of absorption as well. Some drugs are administered in a form that is not directly effective, but is transformed in the patient's body to the active form. Such drugs are called "prodrugs." If a drug, y, inhibits the hydrolyzation of an acid-hydrolyzable prodrug, x, then a reduction in x's absorption follows. The both-adm(x, y) means that both x

and *y* are being taken at the same time in the same patient.

$$\forall(x, y) : \text{prodrug}(x) \wedge \text{oral-adm}(x) \wedge \text{acid-hyd}(x) \wedge \text{both-adm}(x, y)$$

$$\wedge \text{ inhibit-hyd}(y) \Rightarrow \text{red-absorb}(x)$$

$$(2.36)$$

A drug interaction resource that utilizes these mechanisms would take a very different shape than a simple table of pairwise interaction entries. Instead, for each drug one would catalog (in symbolic form so a program could use it) the above information, namely, what are its effects on various body systems, what enzymes does it inhibit or induce, and what enzymes metabolize it. Then one could use a simple reasoning system to correlate entries and check if two drugs interact according to the data on each. This could be done procedurally, or with rules and forward or backward chaining as previously described.

Obviously, there is a large reduction in data management using this approach. The compilation of pairwise interaction data for *n* drugs scales as $O(n^2)$ but the reasoning system allows us to reduce it to linear, $O(n)$. For a thousand drugs, this is a thousand-fold reduction in effort to build the knowledge base.

The fact that one drug affects the enzyme or enzymes that metabolize another drug is not sufficient to be certain that an interaction will be significant in a person. But it is the beginning of the process of building a more complex theory that could reliably make such predictions. In the following, we sketch out this rudimentary theory and leave it to the reader to further develop it. In particular, we will only model the CYP enzyme system, and not consider absorption or circulation.

A Rule-based Drug Interaction System

Using the backward chaining logical inference system described in Section 2.2.4, Boyce put knowledge about metabolic drug interaction mechanisms into a small computational theory. The following shows the rules pertaining to inhibition of clearance; a similar set of rules can be implemented for metabolic induction.

```
(<- (metabolic-inhibit-interact ?precip ?object ?enz)
    (and (inhibits-primary-clearance-enzyme ?precip ?object ?enz)
         (narrow-ther-index ?object yes)))

(<- (inhibits-primary-clearance-enzyme ?precip ?object ?enz)
    (and (inhibits-partial-clearance ?precip ?object ?enz)
         (major-pathway ?object ?enz)))

(<- (inhibits-partial-clearance ?precip ?object ?enz)
    (and (inhibits-effectively ?precip ?enz)
         (substrate-of ?object ?enz)
         (primary-clearance-mechanism ?object metabolic)))
```

```
(<- (inhibits-effectively ?drug ?enz)
    (and (inhibits ?drug ?enz)
         (or (inhibit-strength ?drug ?enz strong)
             (inhibit-strength ?drug ?enz moderate)))))
```

Next we need to create a drug knowledge base (KB) containing the necessary drug facts for inference with the rules. Collecting facts from the primary literature is an arduous task, as is sifting through the drug labeling information available from the FDA. Fortunately, a very compact pocket reference on clinically significant drug interactions is available [145]. This reference also includes facts on each drug's potential for inhibition or induction of CYP450 enzymes.

In our original experiments we augmented our drug KB with information from a Continuing Education Module containing pharmacokinetic information on drugs commonly prescribed to elderly epileptic patients [71]. In addition to facts on potential CYP450 modulation, this reference lists the relative importance of each drug's clearance enzymes. We included several drugs not found in Hansten and Horn [145]. In total, the resulting drug KB contains facts useful for mechanism-based inference for 267 currently prescribed drugs. These can all be written as assertions in the style of the inference code in Section 2.2.4. For example,

```
(<- (substrate-of fluvoxamine cyp1a2))
(<- (substrate-of fluvoxamine cyp2d6))
(<- (inhibits fluvoxamine cyp1a2))
(<- (inhibits fluvoxamine cyp2c9))
(<- (inhibits fluvoxamine cyp2c19))
(<- (inhibits fluvoxamine cyp3a4))
(<- (substrate-of warfarin cyp1a2))
(<- (substrate-of warfarin cyp2c19))
(<- (substrate-of warfarin cyp3a4))
```

Of course to make the rules applicable, we need other facts about the drugs. These are available as well. The entire collection of assertions can be generated by representing the information about each drug in a frame, or an instance of the CLOS class, drug. The following are queries on this knowledge base for any drugs that inhibit or induce the primary clearance enzyme for another drug that has a narrow therapeutic range (NTI), whose clearance is primarily metabolism.

```
(with-answer
    (metabolic-induce-interact ?precip ?object ?enz)
    (format t
     "Drug ~A induces ~A, a primary clearance enzyme, of drug ~A~%"
            ?precip ?enz ?object))
(with-answer
    (metabolic-inhibit-interact ?precip ?object ?enz)
    (format t
     "Drug ~A inhibits ~A, a primary clearance enzyme, of drug ~A~%"
            ?precip ?enz ?object))
```

Table 2.1 Predicted Inhibition Interactions that were not Found in any of Four Online References at the Time of our Study [40]. In Each Row, the Precipitant Drug Inhibits the Primary Clearance Enzyme (PCE) of the Object Drug Based on Information in the Drug KB

Precipitant	PCE	Object
Amiodarone	CYP2C9	Phenobarbitol
Disulfiram	CYP2C9	Phenobarbitol
Fluorouracil	CYP2C9	Phenobarbitol
Fluconazole	CYP2C9	Phenobarbitol
Gemfibrozil	CYP3A4	Carbamazepine
Gemfibrozil	CYP2C9	Phenobarbitol
Gemfibrozil	CYP2C9	Phenytoin
Leflunomide	CYP2C9	Phenobarbitol
Miconazole	CYP3A4	Carbamazepine
Sulfamethizole	CYP2C9	Phenobarbitol
Sulfamethoxazole	CYP2C9	Phenobarbitol
Sulfinpyrazone	CYP2C9	Phenobarbitol
Sulphaphenazole	CYP2C9	Phenobarbitol
Zafirlukast	CYP3A4	Carbamazepine
Zafirlukast	CYP2C9	Phenytoin
Zafirlukast	CYP2C9	Phenobarbitol

The first of the two references to Boyce's work [40] reports the results. The queries returned a total of 90 predicted drug-drug interactions. These included 12 induction interactions involving three precipitant and five object drugs, and 78 inhibition interactions involving 35 precipitant and four object drugs. It would be interesting to see if these also turned up in any of the popular online drug reference databases. On checking for these predicted interactions with four such references,[16] Boyce found that 16 of the 90 predicted interactions could not be found in any of the four. These 16 predicted interactions are listed in Table 2.1.

So, this seemingly simple set of rules, together with some readily available facts, generated some interesting new possibilities. The next step is to examine these to see if they are verifiable or false positives. When we did this, we found several complications, that led to developing a more sophisticated model. First, there was considerable variation in the methodology associated with the evidence relevant to some of the assertions that went into reasoning. Second, of course some information was missing. Third, it was clear in discussions among us that the knowledge base is dynamic and the design needs to allow for that.

Phenobarbitol is an example where strength of evidence might be a problem. Phenobarbitol has been shown to be metabolized by CYP2C9 *in vitro* but there is little *in vivo* evidence to support this assertion. Unfortunately, *in vitro* evidence on drug interactions does not always map to predictions of *in vivo* occurrences [175,257]. Also, in the knowledge base, we asserted disulfiram as a CYP2C9 inhibitor based on Hansten [145]. While some human studies indicate

---

[16]The ones we checked were: First Data Bank's Micromedex, WebMD's Medscape, Discovery Health's http://health.discovery.com/, and Cerner Multum's Drugs.com.

that disulfiram may be a CYP2C9 inhibitor, it has not been found to inhibit S-warfarin or tolbutamide, both of which are known to be CYP2C9 substrates. So, information from other sources may call this assertion into doubt.

Examination of the evidence for other drug properties in this pilot system revealed that important drug mechanism knowledge is sometimes missing. For example, all 11 of the interactions involving phenobarbital in Table 2.1 are predicted to occur by inhibition of phenobarbital's primary metabolic clearance pathway which we listed as CYP2C9 based on one source [71]. We could only *indirectly* support the hypothesis that phenobarbital is a CYP2C9 substrate with two studies from the early 1980s that identified an apparent metabolic interaction between sodium valproate and phenobarbital [212,213] and an *in vitro* study conducted years later showing that sodium valproate is a CYP2C9 inhibitor [438]. This pattern of inference is weak since it assumes that the interaction could only occur by means of CYP2C9 inhibition while pharmacology research has exposed other means by which apparent metabolic interactions can occur, including inhibition of transport proteins. We could find no studies, such as an *in vitro* assay, designed specifically to examine whether phenobarbital is metabolized by CYP2C9. Without this missing information, there remains considerable uncertainty behind this assertion.

It is important to note that these predicted interactions not found in any drug reference were not necessarily false predictions. It is not possible to test every possible drug combination in a clinical trial and the effects of drug interactions can be very hard to recognize, so that some drug interactions escape notice in the scientific literature until years after a drug comes to market. The system's predictions were based on pharmacokinetic principles that are considered valid indicators of potential interactions in FDA guidelines [421]. The clinical relevance of these predictions was based on the assumption that any change in the exposure of a patient to a NTI drug is of clinical interest. Thus, it is possible that some of these predictions are valid interactions that have not been studied.

In the next sections, we show how the three problems can be addressed. The second and third items, missing information and the requirement to be able to dynamically update a knowledge base, will be addressed first. The way Boyce's project dealt with these is to reconstruct the drug knowledge base and the reasoning system by using a *Truth Maintenance System* (TMS), which supports making and retracting assumptions as well as updating the results of reasoning when the premises change. This also lays the foundation for addressing the first problem, the varying degrees of certainty in the support for drug metabolism and enzymatic inhibition. The TMS assertions would be generated from an evidence base which includes an ontology of evidence and a means for generating assertions from evidence specifications. We will describe that in Section 2.2.6.

2.2.6 Truth Maintenance Systems

One way to handle missing knowledge when it is important for reasoning is to assume some truth state for the knowledge until proven otherwise. This is a form of *default reasoning* whose various forms include inheritance in semantic networks, circumscription, and default logic

[360]. Implementing default reasoning in a system that performs logical inference requires that the system be *non-monotonic*. Conceptually, this means that the system can retract or reinstate inferences as the belief state of assertions change. For example, a conservative knowledge-based system might, by default, assume that all patients potentially possess the poorly functioning form of each polymorphic enzyme. Such a system would overpredict the occurrence of drug interactions, generating a lot of likely false positives. Then, when a clinician decides that their patient carries the normal functioning form of the enzyme, the system would not only remove the assumption made previously, it would also retract all inferences, and thus all predictions about drug interactions, that depends on the patient having the poorly functioning enzyme. Any inferences that predict interactions by other mechanisms would remain valid.

One type of non-monotonic logic system is a Justification-based Truth Maintenance System (JTMS) [112]. Typically, a JTMS works in conjunction with a rule engine to manage assumptions and their effects on inference. The JTMS adds two new capabilities to the rule-based inference previously discussed. First, it supports the idea of *assumptions* that can be made and retracted, and used in the inference process. Second, the system keeps track of *justifications*, by which it can *undo* only those reasoning steps that become invalid when an assumption is retracted, rather than having to repeat all the reasoning steps from the beginning.

As we saw in Chapter 2, rules expressed as logic formulas have an *antecedent*, or IF portion, and a *consequent*, or the THEN portion. The IF part must be true for the THEN part to be true. Many rule engines, including the drug interaction system, model domain theories in terms of Horn clauses (defined in Section 2.2.1). These are rules in which the antecedent is a conjunction of zero or more positive terms, and the consequent consists of a single positive term. In predicate calculus, the consequent of a Horn clause is always true if all terms in its antecedent are true. This is not the case in a system using a JTMS. Rather, a consequent can depend on other assertions in addition to the ones in its antecedent. The set of assertions a consequent depends on is called its *justifications*. In order for a consequent to be true in a JTMS, all of its justifications must be true. Justifications can include *assumptions*. Assumptions are assertions that can be :IN or :OUT by assignment; they do not require any supporting justifications. The JTMS labels an assumption node :IN, or *enabled*, when the rule engine *asserts* belief in it, and :OUT when the rule engine *retracts* that belief.

The JTMS represents every assertion in the rule engine's working memory as a node possessing a *label* reflecting its current belief state. Every rule in the rule engine specifies a set of justifications its consequent depends on for belief. The JTMS labels a consequent in a rule as :IN when all of its justifications are :IN. If any of a consequent's justifications is labeled :OUT, then the consequent is labeled :OUT. In this way, the JTMS maintains a dependency network of assertions and justifications. A change in belief in any assertion or assumption node recursively propagates through the dependency network, changing the belief state of any other node that contains the changed node in its set of justifications.

The examples here use Forbus and De Kleer's ANSI Common Lisp rule engine (JTRE) and JTMS from [112], as does the experimental system of Boyce. A rule used by Boyce et al., for testing their drug interaction knowledge representation system will serve to illustrate how the management of justifications works. This rule is the analog of one of the rules in the earlier basic drug interaction model, but has the additional feature of using justifications.

```
(rule
  ((:IN (inhibits ?x ?y))
   (:IN (substrate-of ?z ?y)))
  (rassert!
   (inhibit-metabolic-clearance ?x ?z ?y)
   (nil
     (inhibits ?x ?y)
     (substrate-of ?z ?y)
     (inhibitory-concentration ?x ?y)
  ))))
```

This rule declares that if a drug ?x inhibits an enzyme ?y, and drug ?z is a substrate of ?y, then drug ?x inhibits the metabolic clearance of drug ?z. It has an additional condition that ?x must be present at a sufficient concentration to effectively inhibit ?y. Unlike the simpler rule-based system, this condition is not part of the antecedent list, but is part of a *justification* list.

The syntactic structure of the rule is that it is a list, beginning with rule, declaring that this is a rule. The next two expressions are a list of antecedents, as in the simpler rule formalism, and an assertion expression, a little different from the consequent part of the simpler rule formalism.

The antecedents each have a tag, :IN, and an expression to be satisfied. Expressions are in the same form as before, including variables as symbols beginning with ?, that can be used with a unifier like the match function described in Section 2.2.4. However, in the JTMS, assertions are tagged either :IN or :OUT, so an assertion can be present, but tagged as :OUT. In order for this rule to be applied, both of the assertions must unify with assertions that are :IN.

For example, the following expression,

```
(assert!
  '(inhibits 'CMYN 'CYP3A4)
    'dikb-assertion),
```

asserts that clarithromycin (CMYN) inhibits CYP3A4. The result is that the JTMS creates a node for the assertion and assigns it an :IN label. Similarly, to assert that CYP3A4 metabolizes carbamazepine (CARB), we evaluate the expression,

```
(assert!
  '(substrate-of 'CARB 'CYP3A4)
    'dikb-assertion),
```

which says that carbamazepine is a substrate of CYP3A4. These two relations could also be asserted as the result of applying a rule, but the purpose of this example is in fact to illustrate how that happens.

The consequent section in this rule is the expression beginning with `rassert!`. The first part is just the assertion that ?x inhibits the metabolic clearance of ?z by ?y. Next is a list containing an *informant*, which is a symbol, and a series of justifications for the consequent. The *informant* is a variable that further explains the inference; in this case it is set to `nil`. The justifications represent assertions or assumptions that must be `:IN` in order for the consequent assertion to be `:IN`. When the rule interpreter determines that the antecedents are both `:IN`, it will create a node for the consequent if there is not one already. It will not mark it as `:IN` unless all the justifications are also marked `:IN`. If so, the JTMS labels the node `:IN`, otherwise if the node was just created, it is marked `:OUT`, or if it already exists it is left unchanged.

The justification mechanism provides two capabilities. One is that justifications can include default knowledge as JTMS assumptions. The second is that such assumptions can be retracted, and the effect of retraction can be computed efficiently by examining the rules in which the assumption appears as a justification.

The drug interaction theory as described here thus assumes by default that all precipitants reach a concentration sufficient to cause a clinically significant effect on the enzymes they are known to inhibit. In the example above, the knowledge base would initially create an enabled assumption (marked `:IN`) declaring CMYN to be at a concentration sufficient to inhibit CYP3A4 using `assume!`.

```
(assume!
 '(inhibitory-concentration 'CMYN 'CYP3A4)
  'default-inference-assumption)
```

In this expression, the symbol `'default-inference-assumption` is an informant, identifying this assumption as a default assumption. With this and the assertions above, the node

```
(inhibit-metabolic-clearance 'CMYN 'CARB 'CYP3A4)
```

will be marked `:IN`, and will be linked to the three justifications. However, if further data causes the belief state of our default assumption to change to false, then the program can retract belief, using `retract!`.

```
(retract!
 '(inhibitory-concentration 'CMYN 'CYP3A4)
  'default-inference-assumption)
```

The effect of changing this belief is that the node

```
(inhibitory-concentration 'CMYN 'CYP3A4)
```

will be marked as `:OUT`. In addition, the system then examines all nodes whose justification requires this node. Consequently, it will also mark the node

```
(inhibit-metabolic-clearance 'CMYN 'CARB 'CYP3A4)
```

as : OUT. It should be clear that *any other assertions or inferences* that depend directly, or indirectly, on either of these assertions will now also be labeled : OUT.

A Drug Interactions Theory with Justifications

To complete this new drug interaction theory, we just need a few more rules, and then we need a way to create and manage what facts are asserted or assumed about the particular drugs in the drug interaction knowledge base. The next rule relates the inhibition of clearance to increased drug exposure, provided the object drug and precipitant drug are not the same ones.

```
(rule
 ((:IN
   (inhibit-metabolic-clearance ?x ?z ?y)
                 :TEST (not (equal ?x ?z))
  ))
 (rassert!
  (increase-drug-exposure ?x ?z ?y)
  (nil
   ;;justifications
   (inhibit-metabolic-clearance ?x ?z ?y)
   (primary-clearance-mechanism
                     ?z 'METABOLISM)
  )))
```

The next two rules express the fact that there are two different conditions under which a drug may be classified as at risk for clinically significant effects due to changes in concentration. A drug may have a narrow therapeutic range, outside of which there can be serious consequences, or a drug may be considered a *sensitive substrate*. The FDA defines a sensitive substrate as a substrate that exhibits a 5-fold or greater increase in exposure with the addition of an inhibitor. These rules are necessary because typically, systems based on forward or backward chaining are designed to have only conjunctive antecedents.

```
(rule
 ((:IN (narrow-therapeutic-range ?z)))
 (rassert!
  (nti-or-sensitive-substrate ?z)
  ;;justifications
  (nil
   (narrow-therapeutic-range ?z))
  ))
(rule
 ((:IN (sensitive-substrate ?z)))
 (rassert!
```

```
(nti-or-sensitive-substrate ?z)
;;justifications
(nil
  (sensitive-substrate ?z))
))
```

The final rule here specifies conditions that are necessary to have a high likelihood of a clinically significant inhibition interaction.

```
(rule
  ((:IN (increase-drug-exposure ?x ?z ?y))
   (:IN (primary-clearance-enzyme ?z ?y))
   (:IN (nti-or-sensitive-substrate ?z)))
  (rassert!
   (metabolic-inhibition-interaction ?x ?z ?y)
   ;;justifications
   (nil
     (increase-drug-exposure ?x ?z ?y)
     (primary-clearance-enzyme ?z ?y)
     (nti-or-sensitive-substrate ?z)
     )))
```

Now all that remains is to build the knowledge base of assertions with which to apply the rules. To do this, we need an evidence management strategy, which involves deciding what assertions will be :IN. The architecture will be as shown in Figure 2.3.

The evidence is a collection of assertions about drugs and metabolism. Depending on some criteria to be developed, some assertions from the evidence base are selected to be :IN. The rule interpreter and JTMS generate inferences (respond to queries, as in the previous rule-based

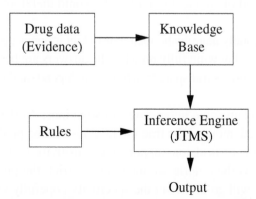

Figure 2.3 An architecture for an evidence-based drug interaction inference system.

system). If the set of base assertions is changed, the JTMS will retract assertions that are no longer supported, and a revised set of inferences will be produced.

Knowledge base maintainers would place evidence for or against each assertion about a drug's mechanistic properties in an *evidence base* that is kept current through an editorial board approach. Maintainers would attach metadata describing the source type of each piece of evidence in the evidence base and users of the system can define specific belief criteria for each assertion in the evidence base using combinations the evidence metadata.

The system will have a separate *knowledge base* that contains only those assertions in the evidence base that meet user-specified belief criteria. The reasoning system will use assertions in this knowledge base, so it will only make drug interaction predictions using those facts considered current by the system's maintainers and believable by users. The system uses a JTMS to handle both default reasoning and the effects on inference of changes in the knowledge base as new evidence causes assertions in the evidence base to meet, or fail to meet, belief criteria.

Reasoning Based on Evidence Types

Now we address the question of how to decide what assertions are : IN and which are : OUT, and under what conditions they might change. We noted earlier that the strength of evidence for or against various properties of drugs and enzymes varies with the different drugs. Since curation is a common component of building knowledge and data resources, it seems very important to decide what kinds of information sources are acceptable and which are not. However, there is a problem, whose solution opens some interesting possibilities for exploration. The problem is this: even among experts at the same institution, in the same group, we found that there are strong differences of opinion, regarding what kinds of supporting evidence are required under what circumstances.

So, rather than attempt to get a consensus, it seemed reasonable to just record with each assertion what evidence types support it. Then we built a system that could accept from a user a set of evidence categories, and pick out just those assertions that met the criteria. This way, each user could decide, and of course different users would therefore get different predictions from the system.

This may seem like a totally anarchistic idea, but in fact it opens up the possibility of asking questions in different ways, knowing fully on what the answers are based. This seems far better than using a system that is not as transparent, where you depend on the curator's reputation and have no flexibility.

Moreover, it also opens up some interesting investigations. As the criteria for inclusion of assertions are more relaxed, more of the true interactions will be predicted, although the false positives will too. In terms of decision support, the sensitivity will go up but the specificity will go down. Similarly, as the criteria are made more strict, the predictions are going to be fewer, and the sensitivity will go down but the specificity hopefully will go up. So, being able to adjust the criteria means you can take a pile of evidence and determine the receiver operating characteristic (ROC) curve for predictive accuracy vs. level or strength of evidence by running computing experiments. This would certainly be a very interesting project.

The way to do this is to create an evidence classification system. We need to be able to store information about properties of drugs, not just the values of the properties. So, it makes sense to define an `assertion` class, with slots in which we can record the information about the assertion. Referring back to the `drug` class, we can elaborate it by putting instances of the assertion class in some slots, instead of just values. Of course other slots are categorically known, and there is no need for anything but a value. For the `drug` class, the slots that are categorical are: `generic-name`, `drug-class`, `form`, `prodrug`, etc. However, the slots containing the values from which we will derive the assertions in the knowledge base are *evidential* slots, such as `substrate-of`, `inhibits`, `induces`, `level-of-first-pass`, etc.

The `evidence-model` component of the system stores instances of objects that the system reasons about. The purpose of the evidence model is to manage knowledge about the properties of the objects it stores and communicate their current state of belief to the knowledge base. Both the set of assertions in the knowledge base and their belief state changes as the evidence model accumulates evidence for and against object property assertions. To satisfy its purpose, the evidence model:

1. stores evidence for and against each property, and evidence metadata,
2. tests the evidence for each property against user-defined criteria for belief,
3. tells the JTMS to retract and assert *enabled assumptions* about the properties of its objects.

So, we need two new class definitions, one for assertions and one for evidence. Here is a workable version of the `assertion` class.

```
(defclass assertion ()
  ((object :documentation "the object's name")
   (slot :documentation "the name of the slot in which this
             assertion goes in the drug class")
   (value :documentation "a value to go in the slot")
   (evidence-for :documentation "a list of evidence types")
   (evidence-against :documentation "a list of evidence types")
   (evidence-rating :documentation "current classification state
                    of this assertion")
   ...))
```

An assertion instance captures the information that would be needed to generate assertion expressions as used earlier in the rules, but with individuals instead of variables. The `slot` entry is a symbol like `inhibits` or `substrate-of`.

Here is a workable version of the `evidence` class.

```
(defclass evidence ()
  ((evidence-type :documentation "a metadata label describing the
```

```
                         type of evidence")
      (doc-pointer :documentation "a pointer to the source of
                    the evidence")
      (quote :documentation "a short textual summary")
      (reviewer :documentation "name of the person who entered it")
      ...))
```

Now, we have not yet said what the evidence types are. The types of evidence that can support drug mechanism facts include, among others, labeling statements, results from *in vitro* studies, expert interpretation of case reports, and various pharmacokinetic trials involving volunteers. Here is a partial list of the types that were included in Boyce's system.

- **RCT** a randomized, controlled, pharmacokinetic study
- **Non-random**

  - a cohort study
  - a case-control study
  - a non-randomized trial with concurrent or historical controls
  - a retrospective study looking at clinical records over time
  - a fixed order study with non-randomized healthy volunteers

- **Case Reports**

  - a single case report
  - a case series

- **FDA Guidances** a statement in an FDA guidance to industry
- ***in vitro***

  - *in vitro* evidence from human tissue, microsomal
  - *in vitro* evidence from human tissue, recombinant

- **Drug Labeling**

  - Non-cited information in a drug label
  - Non-cited *in vitro* information in a drug label
  - Non-cited *in vivo* information in a drug label

Note that some of these types have subtypes. In fact it is possible to create a rich hierarchy of evidence types, a small ontology of evidence, and establish not only class-subclass relationships among the types, but possibly other useful information.

We have a small number of assertion types, corresponding to the predicate function names. When using the system one would choose which evidence types were sufficient to include assertions of different assertion types. To include a report or study that claims that a drug is a substrate of an enzyme, you might accept any of the above types, but for an assertion that a drug inhibits production of an enzyme, you might accept only RCT studies, or perhaps RCT studies and case reports. Once you choose the level of evidence for each type of assertion, the system would then identify the assertions that satisfy the criteria, and include them in the knowledge base. Then you run your query.

Expert users of such a system can define combinations of evidence that they believe will lend different degrees of certainty to the assertion types that the system uses to predict interactions. Different combinations of evidence types might confer different levels of certainty in an assertion and these can be rank ordered to produce "levels of evidence" (LOEs).

The system distinguishes between assertion *instances* and assertion *types*. An assertion instance is a specific fact about a particular object such as a drug or protein. For example, (substrate-of 'CARB 'CYP3A4) is an instance of the generic (substrate-of ?x ?y) assertion type. Users define one or more LOEs for each generic assertion type by creating logical statements listing the level's required evidence types and their multiplicity. The LOEs for an assertion type apply to any instance of that type. Users can also place evidence types that they feel have similar levels of validity into a group called a *ranking category*. They can then use the ranking category just like other evidence types to define LOEs.

For every assertion type users select one LOE as the set of *belief criteria*. The evidence model will tell the system to use a particular assertion instance in inferences if, and only if, there exists a body of evidence for the assertion that satisfies the *belief criteria* for the assertion's type and there is no evidence against the object property. Here are some example LOEs.

- LOE-1 ::= **RCT+** | **FDA Guidance+**
- LOE-2 ::= LOE-1 | **Drug Label+**
- LOE-3 ::= LOE-2 | **Drug Label+** | (*in vitro+* and **Non-random+**)
- LOE-4 ::= LOE-2 | *in vitro+* | **Non-random+**

The symbol ::= means the term to the left is defined as any expression that satisfies the description on the right. The vertical bar, |, means "or" and + means that one or more occurrences of the symbol to the left are allowed. So the first item reads "LOE-1 is defined as one or more **RCT** OR one or more **FDA Guidance** evidence types."

It should be noted that the levels of evidence we are discussing here address the types of observations or measurements reported, not the *quality* of the work. Boyce and colleagues separately established quality criteria that *every* item of evidence needed to satisfy for inclusion at all. So, a poor quality randomized controlled trial would not be included in the evidence base, while a high-quality case report would.

The system that Boyce tested only contains assertions about a family of cholesterol lowering drugs called "statins" and a few other drugs. In addition to six statins, the system currently includes knowledge about 11 other drugs that are sometimes co-prescribed with statins, have been the subject of numerous pharmacokinetic studies (both *in vivo* and *in vitro*), and are thought to be substrates or inhibitors of the same enzyme responsible for metabolizing many of the statins (CYP3A4). For a more complete description of this work, the reader is referred to our second publication [41]. An enormous amount of work was involved in scouring and reviewing the literature to create the evidence base, as well as the engineering of a system with a relatively easy to use front end.

2.2.6.1 Reasoning About Pharmacodynamics

The genesis of the projects described in this section was a simple and naive example using pharmacodynamics to predict drug interactions. Its purpose was to illustrate the use of symbolic representation and computation. This idea was simply that the pharmacologic effects of one drug may dangerously add to or interfere with the actions of another drug. The same knowledge representations and procedures as for pharmacokinetics can be applied here, though the situation is a little more complicated. In order to do this computation we need an `effects` slot in the drug class, that lists intended pharmacologic effects, and we would also use the slots for side effects and contraindications. Here is an example of a drug entry with such information:

```
(make-instance 'drug
    :generic-name 'aspirin
    :drug-class 'nsaids
    :effects '(analgesia blood-thinner
                          platelet-inhibition)
    :side-effects '(heart-attack)
    :contraindications '(gastritis bleeding-tendency
                                   platelet-inhibition nausea)
        ...)
```

The `contraindications` slot is a list of conditions or effects of other drugs that indicate aspirin is not recommended or to be used with caution. So, for example, if a patient is taking aspirin, a potential problem would exist with adding ranitidine, because one of ranitidine's effects appears in the aspirin record's contraindications list.

```
(make-instance 'drug
    :generic-name 'ranitidine
    :drug-class 'antihistamines
    :side-effects '(gastritis gastroduodenal-ulcer)
    :contraindications '(cardiac-arrhythmia
                                   liver-toxicity)
        ...)
```

The procedural code will not distinguish between the intended effects and the side effects, but will consider all of them. However, a distinction and more complicated logic could easily be incorporated.

As for metabolic interactions, the logic is simple. The basic algorithm is to retrieve the entries for each of the two drugs, and match the `effects` and `side-effects` of each against the `contraindications` of the other. Each list is a mathematical set (of abstract elements, not numbers or strings!). The built-in function `intersection` in Common Lisp takes two sets (lists) and returns the elements that are in both sets.

```
(defun interacts (drug-a drug-b)
    (or
```

```
(intersection (union (effects drug-a)
                     (side-effects drug-a))
              (contraindications drug-b))
(intersection (union (effects drug-b)
                     (side-effects drug-b))
              (contraindications drug-a))))
```

The access functions to get the effects and the contraindications for each entry are implemented similarly to the `induces`, `inhibits`, and `substrate-of`.

While this example is still unrealistically simple, it already has a potential here also to predict interactions that could be clinically significant but for which there is no published clinical data.

The enzymatic inhibition procedural code seems very similar to the procedural code here, except that code uses the `set-difference` function. One might think that the two functions could be combined and parametrized, but the logic is slightly different here. In the case of enzymatic metabolic interaction, the requirement is that *all* the important enzymes are inhibited (or induced). Here, there is an interaction if *any* effect appears in the contraindications list. Pharmacokinetics interactions are thus logically different from pharmacodynamic interactions.

The above is a functional representation of the assertion that drugs interact if any of the effects of one is a contraindication of the other. In Prolog, this same information can be represented in assertions and rules. The representation for knowledge about aspirin is thus rendered:

```
effect(aspirin, analgesia).
effect(aspirin, anti-inflammatory).
...
contraindication(aspirin, gastritis).
contraindication(aspirin, bleeding-tendency).
...
```

and for ranitidine,

```
effect(ranitidine,gastritis).
effect(ranitidine,gastroduodenal-ulcer).
contraindication(ranitidine,cardiac-arrhythmia).
contraindication(ranitidine,liver-toxicity).
```

Then we can represent the interaction concept as a pair of rules with variables:

```
interacts(X,Y,Z):- effect(X,Z),contraindication(Y,Z).
interacts(X,Y,Z):- effect(Y,Z),contraindication(X,Z).
```

Then one could query the system as follows:

```
?- interacts(aspirin,ranitidine,M).
M = gastritis
?- interacts(X,ranitidine,Y).
X = aspirin
```

```
Y = gastritis
... (possibly other answers)
```

Much more can be done, by leveraging the existence of ontologies, or classification systems, for drugs. Aspirin is one of a class of many drugs that share common effects and contraindications, so lookup by class is potentially more efficient and a more compact knowledge representation scheme, as in [243]. These authors report a scheme that aggregates drugs into "rollup groups" and "families." Now, with comprehensive multi-level drug ontologies, a range of granularities for organizing this information are possible.

The important point here is that the usual drug classification systems are in fact based on the effects of drugs. Thus, in contrast to pharmacokinetic reasoning, here potentially one *can* use class-based knowledge to reason about pharmacodynamics. This cannot be done directly or easily with First Order Logic. However, alternative approaches can facilitate the organization of knowledge by class rather than individual. Alternatives to FOL will be considered later in this chapter.

While much remains to be done, the application of symbolic representations to drug knowledge provides an intermediate level of knowledge modeling between the lowest levels and the highest. For some drugs there are detailed pharmacokinetic models using differential equations, but for most, the rate constants are unknown, and other details are yet to be discovered. In any case, there is a clear role for qualitative models based on symbolic reasoning systems. As with biological structure, the problem of modeling pharmacologic phenomena calls for rather complex approaches, and not just routine application of well-known programming techniques. Thus once again, the nature of biomedical informatics as the study of biomedical information and its computational properties is distinguished from pure computer science and from the subjects of biology, medicine, and health, while drawing on and building on all these areas.

2.2.7 Electronic Medical Records and Guidelines

Much early work on medical expert systems took on the broad problem of general medical diagnosis, for example, the INTERNIST project [284], or a significant more focused topic, such as blood infections (the MYCIN project [55,377]), pulmonary function [6,7], and other such projects. With the advent of practical electronic medical record (EMR) systems, the focus shifted to providing decision support tools and systems to operate within the EMR context, and to creating standardized formats for knowledge representation so that large libraries of modular units of medical knowledge could be built, contributed by many different sources. A *guideline* is such a modular unit of knowledge. Many different representations of guidelines were developed, and are still under development today. An early example of a standard for guideline representation is the Arden Syntax (Medical Logic Modules) [164,165]. More recent examples include GLIF [321], Asbru [373], and PRO*forma* [405]. Review articles [174,362] provide examples of these representations as well as the execution models used in actual EMR systems. This section adds only a few brief comments to this very active and rapidly changing area of medical informatics.

A guideline, in its simplest form, consists of a single logical rule, together with some details of the context in which the rule should be applied, data needed to apply the rule, and one or more actions to be taken, if the rule applies to the clinical situation that triggered it. Thus, unlike reasoning about biochemical pathways or even drug interactions, guidelines represent shallow reasoning rather than long proofs or derivations. Some challenges include: writing enough guidelines to cover all possibly important clinical circumstances, being able to efficiently index large guideline repositories to find relevant ones, and being able to integrate the application of guidelines into a complex EMR system in a way that makes sense in the clinical workflow. Guidelines can be applied retrospectively to historical clinical records to assess quality of care, but more important is the possibility of applying them in real time to generate alerts in time to deal with impending or emergent adverse events such as stroke.

Although the development of propositional logic and First Order Logic in this chapter presents the effect of application of a rule simply as adding a new proposition or formula to the working data (as in Forward chaining), the more general idea of a "Production system," as described in standard Artificial Intelligence texts, allows for any action to the consequent part of a production (what we have called a "rule"). So, for example, the consequent part of a rule could include an action such as "turn a display box on a user interface screen to flashing red, with text as follows…" or, "block the filling of a drug prescription." These conditions would then have to be cleared by some user action in the system.

Guidelines are often created by manual processes from written documents. It is also possible in principle to develop standards of care by applying machine learning algorithms to large repositories of clinical data, particularly if the data are labeled with outcomes. However, a picture of current practice can be generated even without outcome labeling. Such projects are of course rendered difficult by the requirements of privacy and other regulatory considerations.

2.2.8 Example: Cancer Radiotherapy Planning

Cancer remains the second highest cause of death in the U.S. and other developed countries. While much is known about cancer etiology and prevention, it is still very important to improve our understanding of cancer prognosis and therapy. No great breakthroughs have happened in cancer treatment, but many very significant improvements have been discovered and developed. In particular, the use of ionizing radiation to treat cancer has made great technical strides in the last few decades. Some of this may be attributed to the development of computational modeling techniques that draw on the foundations of biomedical informatics.

Cancer is in reality hundreds of diseases, all sharing the common property that the cells which have become tumor cells do not follow normal growth and death cycles. Tumors grow in an uncontrolled fashion compared to normal tissue. The great challenge of therapy is that tumor cells come from a person's own tissues; they are not foreign bodies like bacteria or parasites. The tumors that arise from different types of organs and tissues have different responses to drugs and radiation, and different growth patterns. The anatomic location, size, and shape will strongly affect the prospect of surgical removal. It is not possible here to present an in-depth discussion of cancer biology. A very readable textbook developed for an undergraduate college course [274] will provide much more background on this subject.

Cancer has a strong genetic basis, and this is certainly an area where the application of modern molecular genetics and epidemiology has yielded important insights. This application of informatics in cancer won't be covered here, but further reading includes the book just mentioned, and for more depth, the encyclopedic references on cancer [89] and on radiation therapy [142] in particular.

The three main modes of cancer treatment are surgery, chemotherapy, and radiation. Surgery has benefited from the availability of digital medical imaging technology, particularly the computer-assisted X-ray tomographic imaging (CT) scanner and more recently the magnetic resonance imaging (MRI) scanner. Chemotherapy for cancer is managed typically by following protocols, many of which are experimental. These protocols are a series of steps and decision points, which are developed by expert oncologists. One of the most important applications of informatics to this process has been the development of knowledge bases and protocol management tools, notably the ONCOCIN project [379]. Radiation therapy is the most technical of these modes of treatment, and has involved computer modeling for many decades [27,80,396,419,424,436]. This section develops the medical informatics aspects of planning radiotherapy, and illustrates how the foundational ideas of knowledge representation, reasoning, information access, and decision making can be applied to facilitate the planning of more accurate and thus more effective radiation treatments.

Radiation therapy directed at malignant tumors involves aiming and collimating a radiation beam at the tumor in such a way as to deposit a large amount of energy in the tumor and as little energy as possible to surrounding (healthy) tissue. Figure 2.4 shows a typical radiation therapy machine. Radiation, such as X-rays (high energy photons), electrons, neutrons, and other particle beams, can kill tumor cells by causing ionization of atoms in or around the cell, and in turn, this can result in direct molecular bond breakage and damage to the DNA of the cell, or it can produce free radicals, active chemical species, which then attack the DNA. In either case, the goal of radiation therapy is mainly one of solving an energy delivery problem, to direct a required amount of radiation into the tumor as uniformly as possible consistent with avoiding overdose of the healthy surrounding tissue. An excellent introduction to the biological effects of ionizing radiation, accessible to the non-medical reader, despite its title, is Eric Hall's book, *Radiobiology for the Radiologist* [141]. A more technical treatise on the physics aspects can be found in the popular textbook by Khan [228].

A modern radiation therapy machine consists of a high energy linear accelerator, producing X-ray (photon) beams or electron beams in the range of 4 to 25 million electron volts (MeV). The machine has a lot of flexibility to aim the beam and shape it in arbitrary ways. The radiation oncologist would like to take advantage of this flexibility by designing a plan that achieves a curative dose in the target region while minimizing the dose to surrounding tissues. This is possible because the amount of "beam on" time for a given daily treatment is only a few minutes. Therefore it is practical to plan and deliver a daily treatment consisting

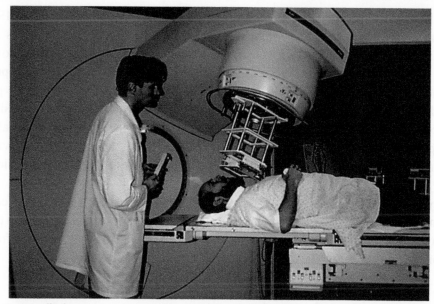

Figure 2.4 A radiation treatment machine, with a member of the UW staff positioned as a patient would be, and a therapist to operate the machine. Once the patient and treatment machine are properly positioned, the therapist leaves the treatment room, then turns on the radiation beam, while monitoring the patient via closed-circuit television. This procedure is repeated for each treatment beam direction.

of multiple short irradiations, each with its own aperture and direction. The idea is that combining many (or even a few) irradiation steps will bring the dose in the overlap region (the target) to the desired level, while no one irradiation step will deliver an unacceptable dose to the surrounding tissue. Thus the radiation therapy planning problem is not to choose the best single direction from which to irradiate, but to create a combination of irradiations that achieves the goal.

Radiation therapy is probably the earliest example of application of computer programming to the solution of clinical treatment modeling and decision problems, dating from the 1950s as noted earlier. A concise review in the context of the development of conformal radiation therapy [116] describes this history in some detail. The first few decades of computer modeling focused on the physics of radiation absorption in human tissue relative to the geometry of radiation beams produced by then current machinery, and on the design of interactive systems that could be used for manual computer-aided design. These systems are called Radiation Therapy Planning (RTP) systems. The physics problem is this: given a direction, aperture, and other settings for a radiation machine to irradiate a particular patient whose body shape is known, how can the radiation dose delivered to any point inside the patient's body be predicted (computed)? Many formulas and algorithms to answer this question have been proposed, implemented, validated, and used in the clinical practice of radiation therapy [153, 195, 228].

With the introduction of Computed Tomography (CT) and Magnetic Resonance (MR) imaging in the 1980s and 1990s, the utility of these computer-aided design programs took an enormous jump. Medical physicists were enthusiastic about the possibility of improving the accuracy of the calculation of internal radiation dose, by using CT image data as a proxy for the radiation absorption properties of the internal tissues, rather than assuming the patient's body had uniform density. The radiation oncologists' attention was drawn to the possibility of more accurately localizing the tumor and estimating its local spread, by using the visual display and computer drawing tools. The next generation of RTP systems provided improved visualization of the beam geometry also, so that the planner could directly apply his/her intuition about anatomy and the capabilities of the machinery.

However, the knowledge of what radiation modalities to choose, and how best to configure them remained largely outside the domain of formal computation. Some of the steps in planning radiation therapy can in fact be formalized in a small-scale computational theory, or fragments of a theory. This section will develop an example of how that can be done.

2.2.8.1 Radiation Therapy Planning (RTP) Systems

The basic steps in radiation treatment planning are:

1. Gather clinical and physical data
2. Decide general approach
3. Select radiation type(s)
4. Delineate the target volume(s)
5. Configure radiation beams
6. Check acceptability and feasibility

The process of designing good radiation plans requires the use of an RTP system. The last two steps may be repeated many times, using the interactive capabilities of the RTP system. A typical screen display from the author's Prism RTP system [203] is shown in Figure 2.5. Various control panels and subpanels provide interactive control of the positioning of radiation beams with respect to the patient, as well as the viewpoints and types of visualizations.

Two factors make radiation therapy effective and practical. Understanding these makes clear the role of computer simulation as a step in the design of radiation treatment.

The first factor concerns the physics of radiation beams. For high energy photons, the maximum dose, or energy deposition, is not at the surface, but can be as much as several centimeters below the skin surface. This is illustrated by the graph of dose vs. depth in Figure 2.6. The quantity on the vertical axis labeled "Depth Dose" is a ratio, the amount of measured dose at the indicated depth, divided by the measured dose at the depth of the maximum dose. The energy deposited is transferred from the X-ray (photon) beam indirectly. The photons ionize atoms in their path, and the resulting electrons travel some distance, on the order of 1–5 cm, before stopping. The electrons have many further local collisions with atoms, causing further ionization. As the photon beam penetrates, the number of electrons set in motion increases rapidly, more

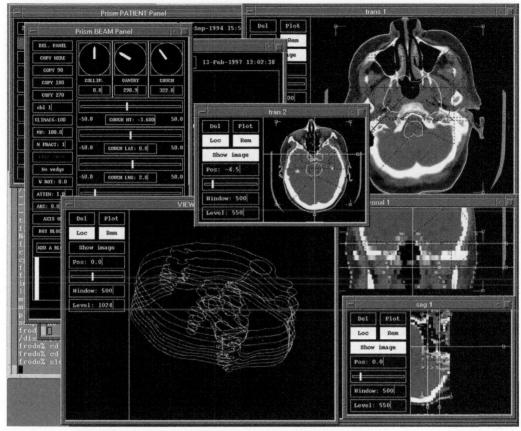

Figure 2.5   An example RTP system display: the Prism system.

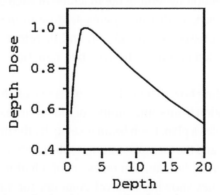

Figure 2.6   Dose from a single radiation beam vs. depth in tissue.

rapidly than the attenuation of the photon beam. Eventually the number of new electrons from photon collisions is equal to the number of electrons stopping because they have reached the end of their range. Then, the effect of attenuation of the photon beam becomes more significant, and the number of electrons and consequently the dose deposition starts to decrease. The depth

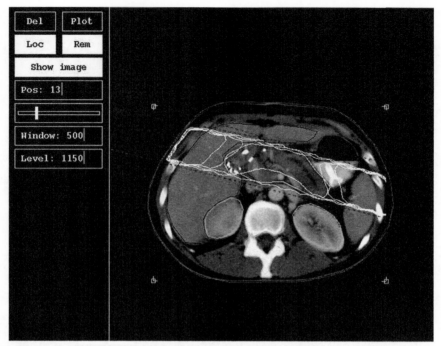

Figure 2.7 A simple radiation treatment plan cross-sectional view.

of maximum dose generally increases with increasing photon energy (the maximum photon energy the treatment machine can produce is designed into it).

The second factor concerns the geometric capabilities of the radiation treatment machine. It is possible to deliver the radiation by aiming the machine in each of several directions, turning it on for a short time from each, so that the aggregate dose to the tumor is high, but the dose deposited in the surrounding tissues is spread out, and therefore lower. Figure 2.7 illustrates a two-dimensional cross section of a plan showing two radiation "beams" overlapping to give this effect.

Thus the problem of deciding how to treat the patient is one of choosing directions, apertures, and relative amounts of radiation from some number of radiation beams. A collection of radiation beams thus specified constitutes a plan. Each beam is set up by the therapist, the machine delivers a preset amount of radiation, and the therapist returns to the treatment room to move the machine to the next position. A typical daily treatment takes about a half hour.

In Figure 2.7 there are lines showing isolevel contours for various radiation dose levels. These come from interpolation of a grid or array of computed values of radiation dose that would result from the specified arrangement of radiation beams. The formulas used to predict the dose at any point within a patient from any specified radiation beam are partly based on principles of radiation physics and partly empirical, interpolating measured data in standard test conditions. The refinement and testing of such formulas is a large and active area of research in medical physics. A survey of the principles can be found in textbooks such as Khan [228], and the particular formulas used in the Prism RTP system are described in exhaustive detail in a technical report [211].

Although methods are well known for computing the physical dose received anywhere in the body from a given beam configuration, there is no established method for solving the inverse problem of computing a set of beam parameters, given a desired dose distribution. In typical clinical practice a dosimetrist (a person with special training and experience) uses an RTP system to display the geometry and dose distribution. The dosimetrist looks for regions of inadequate dose to the target and excessive doses to sensitive structures, and applies modifications that will correct the problem and thereby improve the plan. Therefore the "configure radiation beams" step above is an interactive generate-test loop.

The importance of having powerful three-dimensional radiation treatment planning software tools has been evident for some time [130,131]. Much work has focused on enhancing the delineation of anatomy [52,417], and providing visualization tools [328].

Early systems ran as teletype applications on dialup time sharing systems, then on dedicated minicomputers with frame buffer displays. As the computer systems and graphics hardware became faster and cheaper, the medical physicists writing RTP programs became more ambitious about the kind of interactive graphic displays they implemented [130,132,115,277,348,347, 376]. Through most of this period, from the late 1970s through the 1990s, most of the emphasis was on arcane, clever, and practical features and software tools, rather than on innovations in software architecture or on automating the logic of the planning process itself. With few exceptions [182,177,183,197,198,208,201] this focus remains unchanged today. The size of such systems ranges from 80,000 to over 500,000 lines of source code. Numerical calculations and interactive graphics have a large role in these systems. The programs are commonly written in programming languages like C or C++ and use graphics libraries like OpenGL [311].

Throughout this period of development to the present day, much effort has gone into the search for numerical optimization algorithms and suitable cost functions for use with those algorithms. This subject is covered elsewhere [67,433,434]. The author and colleagues also did investigations of the possibility of using rule-based systems for automating the planning process [206,207,314,315]. The idea was to represent the heuristic knowledge and experience of the physicist and dosimetrist, the qualitative physics of radiation beams, and the anatomy of typical patients, in order to reason about what plan geometries might work best. These experiments did not lead to an effective automated system, but the work laid the foundation for solving another important problem in the planning process, the definition of the area to irradiate, also known as the target, to which we turn next.

2.2.8.2 Locating the Target

The elegance, power, and effectiveness of radiation therapy is that the radiation machinery can be aimed and adjusted to precisely direct the radiation to any desired location, unlike chemotherapy, where drugs are infused into the patient's bloodstream, in hope that the drugs will be selectively

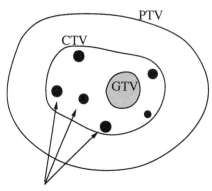

Involved lymph nodes

Figure 2.8 A diagrammatic illustration of the idea of GTV, CTV, and PTV, provided by Dr. Mary Austin-Seymour.

absorbed by the tumor cells. Thus, a basic principle of radiation therapy is that one must know where the target is. In simple terms, you cannot aim precisely at something whose location is unknown. This is a problem involving anatomy, tumor biology, and spatial reasoning.

Radiation oncologists and others, through the International Commission on Radiation Units (ICRU), have created a series of definitions for the entities involved in defining the target for radiation therapy [172,173]. By defining these concepts, they can more precisely describe how treatments should be designed, and ultimately this will also add precision to clinical trials.

In delineating the target of radiation therapy for a particular patient, the radiation oncologist must consider what is palpable and/or can be seen in medical image studies such as CT, MRI, or PET, as well as the more traditional radiographic X-ray images. This is the starting point. The volume that includes only the visible or palpable tumor mass is called the Gross Tumor Volume (GTV). Figure 2.8 shows the GTV as a gray area in a cross sectional view. Although it can usually be seen on images (CT and MR), it is normally not easy to automatically identify with existing image processing techniques. The GTV is usually drawn by hand using interactive drawing software on one or more sets of cross sectional image studies. As with organ delineation, the GTV can be represented in the RTP program as a set of planar contours, with each contour being a set of vertices located in the volume including the patient. This is a tedious task, requiring hand drawing, plane by plane. Unlike high contrast organs like bones and lung, tumors are difficult to detect automatically because their tissue density and other properties are close to the surrounding tissues, and the boundaries are ill defined.

It is well known that for many tumors, there is local and regional spread, to surrounding cavities, lymph nodes, and other structures. This depends on the size of the tumor, the type of cells, and most critically the surrounding anatomy. The volume that includes the local and regional spread is known as the Clinical Target Volume (CTV). It usually includes the GTV plus the lymph node regions around it. These migrant tumor cells cannot be made visible with any imaging techniques known today. Even the nodes themselves are often hard to identify in the images. The task of delineating these nodal regions, which is also usually done by the clinicians, is quite time consuming. Clinicians often elect to perform less aggressive, non-conforming treatment, because they do not have the time to draw the outlines of the nodal regions and CTV, even if they are confident about which node groups are likely to have disease to treat. Involved lymph nodes are illustrated as black dots in Figure 2.8, and the CTV is shown as a contour enclosing them.

Pathology findings following surgery have been reported in the literature, usually to address specific questions such as whether or not to include a specific lymph node group in the target volume. Generally tumor spread largely follows anatomic structure and biologic principles, though this is only manifest in the treatment planning process by hand drawing the CTV. As tumors become larger and time elapses, the probability of more distant node groups being positive for microscopic spread increases. Some data are available that quantify this process, and steps have been taken toward a predictive model of tumor metastasis [209]. Section 5.5.3 develops this idea formally in a small working implementation.

Once the CTV is determined, one must allow for the fact that radiation therapy delivery presents a significant physical alignment problem for the radiation therapist. A typical course of treatment includes 30 daily irradiations, each taking only a few minutes. The patient comes into the clinic each day, and the actual setup geometry can vary because the tumor may slightly shift inside the patient's body, the machine is difficult to align consistently to the external landmarks on the patient, and even with the best efforts at immobilizing the patient, the tumor and/or the body will shift during the few minutes of treatment. Thus, the ICRU defined a third volume, the Planning Target Volume, or PTV, which is enlarged from the CTV to allow for these variations (see Figure 2.8). If the treatment plan is designed for the PTV, one would be reasonably confident that the CTV would receive adequate radiation dose, despite the variations. The data available on setup variability and tumor motion make it possible to compute the PTV from the CTV [21,239]. Evaluation of the accuracy of such a computational model of the PTV is challenging [225,227]. A PTV model implementation is developed in Section 2.2.8.

The ICRU also defined a treatment volume, which represents the volume that was actually irradiated. This volume will depend, of course, on the plan of irradiation that is chosen for actual treatment. Its boundaries will also depend on the critera chosen. The volume irradiated to 50 Gray will be larger than the volume irradiated to 60 Gray, and so on. The treatment volume idea is not practical for radiation therapy planning. Instead it is common practice to compute so-called "dose-volume histograms" [262,264], which represent information about how much dose a particular treatment plan will deliver to the volumes already defined, such as critical organs (uninvolved tissues such as lung, liver, kidney) and the various target volumes. We won't address either of these problems here.

2.2.8.3 Computing the Radiotherapy Planning Target Volume

The automated generation of a "Planning Target Volume" (PTV) from the Clinical Target Volume (CTV) takes account of internal and external motions during and between treatments, based on qualitative facts about the tumor's anatomical location, its histology, and the treatment plan itself. Like the CTV model, it illustrates how a classification system for tumor locations can be used to organize knowledge about their physical behavior. It also ties together numerical and symbolic modeling in a relatively simple example, and illustrates problems inherent in evaluating the accuracy of such computations. The model and code described here is based on the work of Sharon Kromhout-Schiro [21,225,226,239,240].

Classification of Tumor Sites

Knowledge about the etiology, pathology, and treatment of tumors depends so much on the location of the tumor (and ultimately the metastases) that it is used generally in cancer text-books as an organizational theme. Also, for the purpose of formalizing knowledge about the computation of the PTV, as well as for automating other aspects of radiotherapy planning, it is useful to have a classification system for tumors according to anatomic site. As an example of the importance of site, for tumors in the head, neck and brain, the patient can be held quite rigidly with a custom molded plastic mask, while for lung tumors, body position is much less reproducible. Similarly, tumor motion is highly dependent on site, as some tumors are rigidly fixed to the surrounding anatomy while others move as much as a centimeter or even two. When two pieces of knowledge apply to a specific tumor site, one for that site, and one for the larger region (for example, nasopharynx and its region, head and neck), the more specific one should take precedence over the more general one, when there is a conflict.

We will focus on the head and neck, which for radiation oncologists includes everything above the shoulders except for the brain and spinal cord. This is a particularly complex anatomic region, where precision in aiming radiation is very important.

Head and neck cancers are a heterogeneous collection of malignancies arising in multiple anatomic subsites, each with unique clinical, pathological, and prognostic features. The vast majority of these cancers are squamous cell carcinomas that originate in the normal mucosal lining located throughout the head and neck. As these squamous cancers enlarge, they tend to spread from primary location to nearby regional lymph nodes in a predictable sequence. Curative treatment of these cancers, therefore, has historically required attention to the draining regional lymph nodes that could harbor microscopic disease.

Tumor anatomic sites can be represented in a hierarchical structure, a simple directed acyclic graph. Each node in the graph is a tumor site. The nodes can be represented by instances of a CLOS class. The slots should be the `name` of the site (which we represent by a symbol), the immediate larger region to which it belongs (the `part-of` slot), and the subregions or subsites it contains (the `parts` slot). In case the site has no parts, that is, it is the most fine-grained location in common use, its `parts` slot will be empty (it will have the value `nil`).

Here is the representation of the class whose instances will represent the tumor site hierarchy:

```
(defclass tumor-site ()
  ((name :initarg :name :accessor name)
   (part-of :initarg :part-of :accessor part-of)
   (parts :initarg :parts :accessor parts
```

```
                  :documentation "A list of symbols naming
                                  daughter nodes"))
        (:documentation "Each instance represents a single anatomic
                          site or larger anatomic region. The tree
                          structure implied by the parts and part-of
                          slots describes anatomic relationships of
                          tumor sites and groups of tumor sites.")
        (:default-initargs: name nil: part-of nil: parts nil))
```

It is useful to be able to refer to each instance of the `tumor-site` class by name, rather than having to search a global list. One convenient way to do this is to arrange for each instance to become the value of the symbol in the `name` slot, when the instance is created. This also allows for the values in the `part-of` and `parts` slots to be symbols rather than instances, which simplifies the representation of the tree in a file, while keeping traversal of the tree straightforward. To make this happen automatically we just provide a simple augmentation of instance initialization, with the following (see [137, Chapter 11] for a description of how this works):

```
(defmethod initialize-instance :after ((site tumor-site)
                                        &rest initargs)
  "This method makes the site instance available by name"
  (set (name site) site))
```

The `set` function rather than the `setf` macro is used because we want to set the value of the symbol that is in the `name` slot, rather than update the contents of the `name` slot. The `&rest` argument is not used, but is required in order to be consistent with the definition of the `initialize-instance` generic function. Note that this obviates the need for a global list to search for a site (for example, by name). If you know the name, you have immediate access, by just evaluating the symbol.

Here is some of the tumor site hierarchy in terms of constructor functions:

```
(make-instance 'tumor-site
  :name 'INFERIOR-WALL
  :part-of 'NASOPHARYNX
  :parts NIL
  )
(make-instance 'tumor-site
  :name 'NASOPHARYNX
  :part-of 'PHARYNX
  :parts '(POSTERIOR-SUPERIOR-WALL LATERAL-WALL INFERIOR-WALL)
  )
(make-instance 'tumor-site
  :name 'PHARYNX
```

```
:part-of 'HEAD-AND-NECK
:parts '(HYPOPHARYNX OROPHARYNX NASOPHARYNX)
)
```

The full site hierarchy for the head and neck can be constructed by hand from information in an old and venerable cancer treatment reference manual [357], which we used, or from more recent textbooks [89, 142]. There are altogether 78 sites in the head and neck. This hierarchy is implemented exactly as shown, as a file with `make-instance` expressions in it. This only works because of the small size and simplicity of the needed objects, and because the information is static. So, there is no need to use `get-object` or `put-object`. However, a nice printed representation of this site hierarchy can be produced by a simple recursive traversal of the `part-of` relation. The function `print-tree` prints the name of a site (node) and all its subnodes recursively, to the indicated stream (the default is `*standard-output*`), indenting by the number of spaces indicated by `indent`, at each level of subnodes.

```
(defun print-tree (node &key (indent 0) (stream t))
  (when node
    (tab-print (name node) stream indent t)
    (mapc #'(lambda (x) (print-tree (symbol-value x)
                                    :indent (+ indent 3)
                                    :stream stream))
          (parts node)))))
```

Later in this chapter, in Section 2.3 and following, we will see how object-oriented modeling with class hierarchies can be extended to include the idea of *metaclasses*, which we used in Section 1.2.3 to illustrate how computer programs can be considered as metadata. This application does not require the use of metaclasses. In fact, this tumor anatomy model does not even use subclasses in the sense of frames or object-oriented programming. The more specific tumor sites are not subsets or subclasses of the broader categories. The only essential relation here is the `part-of` relation. Note that these are *tumor classification* categories, *not* anatomical entities according to the UW Foundational Model of Anatomy (FMA). The hierarchy in this model is that of the radiation oncologist's organization of knowledge about tumors. So, `part-of` here refers to the fact that tumors originating in the same organs or within a regional locale share some common properties with regard to treatment planning. However, it raises an interesting question: can an alignment of the oncologist's understanding of tumors with the anatomist's understanding of structure provide new insights that could be used for designing better therapy? The construction of these formal models is an essential step in the process of answering such questions.

The site hierarchy allows the organization of general knowledge and specific knowledge to be used efficiently. The control strategy can be to use all applicable rules, or the most site-specific available rule, when rules conflict. Rather than having to repeat general information for each specific site, a simple recursive search from the specific tumor site through all the contained sites, following the `part-of` relation, implemented by the `within` predicate function, collects all applicable rules. This principle underlies the use of many different ontologies that form hierarchies, including subsumption hierarchies as well as those formed by `part-of` relations, as we will see later in this chapter.

First we must convert the slot values in our ontology to assertions in the inference system. This can be done initially by iterating over the collection of site instances, or individually as each site is created with `make-instance`. Specifically, we further augment the `initialize-instance` method above to do this. We need a way to look up values in slots and construct assertions in the inference system from them. The following function will do this.[17]

```
(defun assert-value (pred obj &optional val)
   "converts an object value pair to an assertion named pred"
   (if val (eval '(<- (,pred ,obj ,val)))
      (eval '(<- (,pred ,obj)))))
```

We then use this function when needed. For the `initialize-instance` augmentation, we add to the earlier version a call to `assert-value`.

```
(defmethod initialize-instance :after ((site tumor-site)
                                        &rest initargs)
   "This method makes the site instance available by name
    and registers a rule for the part-of slot."
   (set (name site) site)
   (assert-value 'part-of (name site) (part-of site)))
```

When we write rules to be applicable to various tumor sites, we will want them to be applicable to any *subsites*, that is, sites that are part of (or within) the category that the rule names. The `within` function can be defined in the context of the Graham inference code by two rules. Location X is within location Y if they are the same (each site is considered to be "within" itself), or if X is a part of another location and that location is within the location Y. This second relation makes the `within` relation effectively the transitive closure of the `part-of` relation.

```
(<- (within ?x ?x))
```

```
(<- (within ?x ?y) (and (part-of ?x ?z)
                        (within ?z ?y)))
```

Note that in the second clause, the order of expressions is very important. If the `within` expression comes before the `part-of` expression, the backward chaining algorithm will go into an infinite loop (the order of expression handling with the Graham code is left to right, the opposite of Prolog, so for Prolog, the same consideration applies in reverse). The reason is that it will

---

[17]It's unfortunate that it calls `eval` and uses backquote in this odd way, but it is the only way I could get it to work with the `<-` macro. The code needs to construct an expression and then assert it using `<-`, which was really only intended for literals and variable symbols, not expressions to be evaluated.

keep trying to evaluate (within ?x ?y) by applying the rule recursively and never get to the part-of expression. This is exactly the same ordering issue as mentioned earlier in the biochemical metabolism example.

Here is an example of how it would work:

```
> (with-answer (within middle-ear head-and-neck)
      (format t "The middle ear is within the head and neck. ~%"))
The middle ear is within the head and neck.
> (with-answer (within eye nasopharynx)
      (format t "The eye is within the nasopharynx. ~%"))
NIL
```

In the first case, the query succeeds and the format expression is evaluated, producing the printout seen. In the second case, the query fails, the format expression is not evaluated, and nothing is printed. Of course we will see that the queries needed to determine the PTV will have variables, and we will use the resulting bindings, if any.

Representing Knowledge About Margins

Margins are distance values from which to compute an expanded volume representation from the GTV or the CTV to make the contours of the planning target volume (PTV). There are several different types of margin allowances to account for different kinds of variability in tumor location. In some circumstances the appropriate margin may be different for each of the three directions, x, y, z. For example, it may be possible to immobilize the patient well against left-right motion, but the movement up and down due to breathing cannot be controlled. To accommodate this, the margin is represented as a three element list of numbers rather than a single number.

The way this information is used in an RTP system is that the user draws the CTV (currently this is done by hand, based on clinical judgment, but if the CTV project is successful, that step will be largely automated). Then in practice today, the user selects the amount of the margin to apply, and the RTP system code takes all the contours of the CTV and generates from them new contours by radial or cylindrical expansion. In this section we will show how to record and apply knowledge about how much the margin should be depending on the clinical properties of the tumor and the treatment setup that is being planned.

Figure 2.9 shows a simple cross sectional view of a patient model, where the smallest contour in the center is a drawn CTV, the next one around it is a computed PTV based on the use of a mask immobilization device, and the larger concentric contour is the computed PTV for planning without the use of the mask. The other contours are other anatomic structures, which may need to be excluded from the target volume. The complete PTV tool implemented in the Prism system can do this contour subtraction where needed.

One factor is the mechanical motion of the patient during the treatment, due to the patient's breathing, or to the patient not being completely immobilized. In treating head and neck tumors, for example, it is usual to make a custom-fitted mask from a plastic mesh that softens and

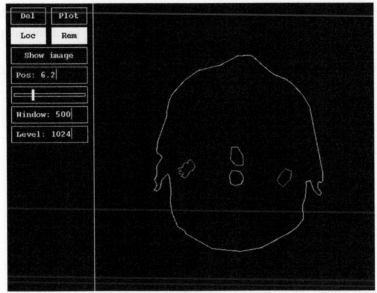

Figure 2.9 A cross sectional view of a CTV contour drawn by hand and PTV contours generated by the rule-based PTV code.

stretches when heated with warm water, then sets once formed to the patient's face shape. The amount of motion allowance in this case is independent of tumor site, and can be expressed in a rule as follows:

```
(<- (pt-movement ?x (0.1 0.1 0.1))
    (and (location ?x ?y)
         (within ?y head-and-neck)
         (immob-dev ?x mask)))
```

This rule states that an allowance of 0.1 cm should be allowed in each direction for movement of the patient during treatment, when a mask is used and the tumor site is in the head or neck.

Another kind of margin allowance is the variation in positioning the patient on the treatment couch, and in aligning the patient with the radiation beam. This is also affected by the kind of immobilization device used, since the device can be marked and laser alignment lights in the treatment room used to position the patient before treatment starts each day. A rule representing day-to-day variability in positioning the patient with respect to the radiation beam might look like this:

```
(<- (setup-error ?x (0.5 0.5 0.5))
    (and (location ?x ?y)
         (within ?y head-and-neck)
         (immob-dev ?x mask)))
```

This rule states that a margin for patient positioning of 0.5 cm in each of the three directions should be allowed, in general for head and neck tumors, if a mask type of patient immobilization device is used.

If no immobilization device is used (there may be practical reasons why a particular patient may be unsuitable for this technique), the margins will have to be larger. Thus we have rules for this case, similar to above.

```
(<- (pt-movement ?x (0.3 0.3 0.3))
    (and (location ?x ?y)
         (within ?y head-and-neck)
         (immob-dev ?x none)))

(<- (setup-error ?x (0.8 0.8 0.8))
    (and (location ?x ?y)
         (within ?y head-and-neck)
         (immob-dev ?x none)))
```

For tumors in the lung, the margins are slightly different. A plastic rigid shell will not allow for breathing, so a customized foam filled box-like device is used instead, called an "alpha cradle." This is not as rigid as the mask used for head and neck treatments.

```
(<- (setup-error ?x (0.6 0.6 0.6))
    (and (location ?x ?y)
         (within ?y lung)
         (immob-dev ?x alpha-cradle)))

(<- (pt-movement ?x (0.2 0.2 0.2))
    (and (location ?x ?y)
         (within ?y lung)
         (immob-dev ?x alpha-cradle)))
```

Finally, there is a third type of variability that requires aiming at a larger volume and therefore contributes to determining the PTV. During treatment, some tumors can move around internally to the patient's body surface, especially those that are in body cavities, viscera (the gut), or low density regions like lung. We refer to these as tumor-movement margins. For example, nasopharyngeal tumors are usually fixed and rigid, but lung tumors can move slightly internally during normal breathing. In the lung, the extent and direction of motion depends on the detailed location within the lung, so a region has to be specified.

```
(<- (tumor-movement ?x (0.0 0.0 0.0))
    (and (location ?x ?y)
         (within ?y nasopharynx)))

(<- (tumor-movement ?x (0.8 0.0 0.0))
    (and (location ?x ?y)
```

```
                    (within ?y lung)
                    (region ?x mediastinum)))
    (<- (tumor-movement ?x (0.5 0.5 1.0))
         (and (location ?x ?y)
              (within ?y lung)
              (region ?x lower-lobe)))
```

These rules come from observations utilizing portal imaging techniques. In general it is difficult to obtain the data from which to write these rules. The work cited earlier gives some indication of sources for the data.

The `within` predicate was defined earlier. The `location` predicate would be asserted in a program using this knowledge base, as would other predicates, such as `immob-dev`. To solve a particular case, a program has to simply assert the particular facts for that case, run a query using the `with-answer` macro to get each of the margin values, and then apply the margin values to expand the clinical target volume contours, resulting in a set of contours for the planning target volume. The function `target-volume` does this.

```
(defconstant *chi-sq-factor* 1.88)

(defun target-volume (tumor immob)
   (assert-value 'location tumor (site tumor))
   (assert-value 'immob-dev tumor immob)
   (let* ((setup-m (with-answer (setup-error ?y ?x)
                      (if (eql ?y tumor) (return ?x))))
          (tumor-m (with-answer (tumor-movement ?y ?x)
                      (if (eql ?y tumor) (return ?x))))
          (pt-m (with-answer (pt-movement ?y ?x)
                      (if (eql ?y tumor) (return ?x))))
          ;; prob-m is a list of the x, y and z margins
          (prob-m (mapcar #'(lambda (m) (* m *chi-sq-factor*))
                      (rms setup-m tumor-m pt-m))))
      (make-instance 'target
        :contours (expand-volume tumor prob-m))))
```

First, `with-answer` is used to retrieve the most specific applicable margin values for each type of margin, then the three types are combined as independent random variables assuming a normal distribution. This is done by multiplying a root-mean-square sum (computed by the rms function) by the value of χ^2 for 2 degrees of freedom. Here is the function, `rms`, that takes three lists of margin values (one for each of the three types of margins) and does the arithmetic for each of the x, y, and z directions, returning a list of the resulting values.

```
(defun rms (list-1 list-2 list-3)
   (mapcar #'(lambda (a b c)
```

```
          (sqrt (+ (* a a) (* b b) (* c c)))))
  list-1 list-2 list-3))
```

The prob-m values will then be used in a function, expand-volume, to create the expanded contours. The overall margin values (x, y, and z) are added to the tumor volume contour points to produce the target volume contour points. The tumor volume contours are lists of x, y coordinate pairs, associated with a particular z value. Each is projected outward cylindrically. The details of constructing a target instance, and doing the geometry of calculating the convex hull of a contour and expanding it are beyond the scope of this discussion, but the entire Prism code is available for the interested reader.

2.2.9 Example: Rule-Based Nursing Knowledge

Although the focus of medical news is on physicians, their discoveries and procedures, nursing science and practice are essential and critical to health care. Fortunately, in recent years more recognition of this important role has been forthcoming. In some institutions, nurses bring critical knowledge of patient care to hospital administration, including serving as high level officials in the administration.

Nurses have sustained a protracted low profile even as innovations and soaring costs have catapulted health care into the public media limelight. Whether this low public profile is by design or by nature, nurses' sustained and protracted political silence on healthcare issues leaves a hole in society's understanding of this large and vital profession. Professional nurses have none the less been deemed essential to the health and function of our society. Nurses are consistently rated as the most trusted professionals in America, yet few of us can really articulate precisely what it is that nurses do—what functions they perform or precisely what need they fill.[18]

Most of us will be seen by a registered nurse only if we are either very sick, or in probable danger of becoming very sick. The role of a physician is arguably to diagnose, prescribe, and treat or cure ailments, illness, and disease. The role of a registered nurse conversely is to promote the highest level of health and function possible in a particular individual patient at a particular moment in time. The health and function of a surgical patient, for example, can vary dramatically from hour-to-hour, even minute-to-minute as the patient emerges from anesthesia, transitioning from an unconscious and entirely dependent intubated and mechanically ventilated state, with fluids and medications flowing at precise rates through intravenous lines, and with catheters and tubes draining their bladder and surgical sites. The post-surgical patient often rapidly recovers from this dependent and vulnerable state to increasing consciousness and independence. This can be a rapid process for young, healthy people, but may be a longer and more dangerous process for the weak, the frail, and the elderly.

---

[18]This and the following paragraphs describing the roles and functions of nurses were kindly provided by Catherine D'Ambrosio, Ph.D., whose research work is described later in this section.

Professional nurses (RNs) collaborate closely with physicians and are responsible for provision of each individual patient's nursing care. This professional nursing care is not necessarily determined by the patient's allopathically diagnosed disease states, nor by their prescribed medications and treatments. The individual's professional nursing care needs can vary quite widely even among patients of the same age, race, gender, socio-economic status, constellation of medically diagnosed conditions, and prescribed medications.

For example (to illustrate how individuals medical condition does not dictate their nursing care needs), two hypothetical 75-year-old gentlemen (we will call them Chuck and Larry) share a hospital room following a prostate resection earlier that day. Both have had hypertension, Adult Onset Diabetes Mellitus, and coronary artery disease for 20 years. Both are on the exact same doses of the exact same eight medications. Both are retired accountants, married, and each has two children. Both hail from the same urban neighborhood and play golf together every week.

Despite the medical and even social similarity of these two men, their nursing care needs can vary quite widely. Chuck (for example) is having some post-operative bleeding and currently requires a blood transfusion. His pain is difficult to control. The RN has already had to administer a medication to block the uptake of his narcotic pain medication because of respiratory suppression and oxygen desaturation. Addressing his pain is further complicated by the acute onset of post-operative delirium causing Chuck to hallucinate, pull at his tubes and catheters, and climb out of bed. He has successfully gripped his foley (bladder) catheter and pulled it out of his bladder with the balloon fully inflated. At the same time that he pulled his foley, he also pulled his intravenous line. His skin is papery and his veins are spidery. He is combative (from the delirium and the pain). He will need a large gauge intravenous catheter for the blood transfusion. His new intravenous line may therefore be difficult to insert.

Chuck's bowel is not yet active, so he cannot eat or drink. He is experiencing and complaining of nausea. He still has (from his surgery) a naso-gastric tube, draining 60–90 ml of green-bile colored gastric output. Even if Chuck's GI system were able to tolerate food, his condition is such at this point that he would not be capable of independently feeding himself. Chuck requires weight-bearing assistance from two to three nurses to get to the chair or the bedside commode. Chuck's temperature has just spiked to 38.8 °C, his heart rate has gone from approximately 80 beats per minute to 130 beats per minute with a newly irregular rhythm. His respiratory rate is slightly elevated at 22 breaths per minute with oxygen saturations in the mid-80s. Chuck has oxygen ordered, but is refusing to keep his oxygen mask on his face. The nurse deduces that Chuck may well be septic so will immediately contact Chuck's physician, who may or may not agree with the nurse's assessment. Any delay in obtaining cultures and starting antibiotics could result in disastrous consequences for Chuck.

Chuck urgently needs blood cultures drawn from two different sites and a urine culture to rule out sepsis. Immediately following the collection of these specimens, Chuck needs to be started on a newly prescribed broad spectrum intravenous antibiotic. Very likely the arrival of the antibiotic will coincide with the arrival of his blood. Since Chuck has never previously received this antibiotic, he cannot receive the antibiotic while he is receiving his blood transfusion. The reason is because in the event that Chuck has an adverse or allergic reaction to either the blood or the antibiotic, it will be impossible to determine which is causing the reaction.

Larry (meanwhile) is up in his chair at his bedside, eating Jell-o™, and visiting with his wife and children. His pain is adequately controlled with oral narcotic analgesics. Larry's intravenous line has been capped. The nurse has asked Larry to use his call light to call her for assistance when he is ready to get back into bed. Larry is bearing weight without the need for any assistance. Larry would like to eat a hamburger and get a shower this evening, but has agreed to wait until tomorrow morning.

Despite their medical similarity, the nursing care needs of these two patients are dramatically different. With some consistently outstanding nursing care and some physiologic luck, Chuck may be fully recovered within a month. If Chuck's professional nursing care is in any way compromised or substandard, his life could well be permanently decimated to a point where he is discharged permanently to a nursing home where he will remain dependent upon assistance for all of his activities of daily living (eating, bathing, dressing/grooming, toileting, continence, transfers, and mobility) for the rest of his life.

2.2.9.1 Nursing Informatics

Although much work has been done in applying knowledge representation and reasoning to medical diagnosis, much less has been done with nursing knowledge. Early work on application of knowledge representation ideas to nursing includes COMMES [79,107] and others [47,64, 158,194,236,325,337].

All these early examples of nursing expert systems notwithstanding, most textbooks in the area of "nursing informatics" today focus on something more like computer literacy for nurses, with particular attention to the organization and use of electronic medical record systems.

2.2.9.2 A Nursing Expert System

In nursing practice, the work is not about diagnosing a medical problem, but analyzing a situation and determining the relevant care plan or action to deal with the evident conditions. In this section we describe an example of how this could be done. However, the setting is not one of acute care as described above, but that of planning for continuing care of patients with various levels of dementia, and how best to provide assistance for their daily living functions. This is also an important role of skilled nursing care.

Catherine D'Ambrosio, who completed a Ph.D. in nursing at the University of Washington in 2003 [82], had many years of experience in private practice, providing home nursing care for patients with dementia and other chronic illnesses. She came to her doctoral studies with the idea that individualized care plans for her patients should be possible to generate automatically with a computer program.

Having already experimented with templates and controlled vocabularies, D'Ambrosio decided that these did not lead to a solution. She proposed that nursing knowledge in this area could be formalized in production rules. She deliberately selected the most basic low-level nursing care decisions she could find, departing from other nursing decision support systems that had focused on calculating the appropriate nursing diagnosis. The main reason for this departure from calculating the nursing diagnosis is the absence of evidence connecting the nursing diagnosis to bedside nurses' decisions regarding the immediate nursing care needs of individual patients. D'Ambrosio's review of nursing informatics and nursing diagnosis literature found that nursing diagnoses have been accumulated without sufficient explanatory theoretical bases from which scientific solutions can be generated. They (nursing diagnoses) lack conceptual clarity, reliability, or operational validity necessary to test any proposition [82, 230].

Her project used the simple mock Prolog inference code from Graham [137, Chapter 15], described in detail here in Section 2.2.4, to implement nursing knowledge about prompted voiding for patients with dementia [82]. Using the rules with the nurse's observations and data from the patient chart, backward chaining could generate a list of patient-specific procedures for the nurse to follow. To test her representation, she created several artificial test cases and had a panel of experts rate the system's recommendations for sufficiency, accuracy, and appropriateness. The following description of a few of the inputs and rules, as examples of how nursing knowledge might be encoded, was also kindly provided by Dr. D'Ambrosio.

Prompted voiding (PV) is a specialized time and labor-intensive (therefore expensive) behavioral intervention that reliably and consistently optimizes continence function even among physically debilitated and cognitively impaired elders with existing incontinence. In order to calculate and determine whether (or not) PV is an appropriate and feasible intervention for this particular patient at this time, the production system requires three calculated values (on that individual patient). The symbols representing these three calculated values (required for a positive PV recommendation) are:

```
P-RQ
BL-CD
TH
```

These inputs (to the PV calculation) are themselves calculated values that are obtained by using a combination of facts that are either entered by the Registered Nurse as facts about that particular patient, or logically calculated when patient-specific facts are combined with rules in the inference engine.

Here are some example rules (the notation is that of the propositional logic programming system described in Section 2.2.1.4):

```
(<- PV-OK (and P-RQ BL-CD TH))
(<- P-RQ (and (not PRMPT-NOT-REQ) PRMPT-LVL PRMPT-RESP-OK)
(<- BL-CD (and FREQ-OK BL-CAP-OK PVR-OK INITIATE-VOID)
(<- TH (and WBN MECH-DEV STR)
```

The meaning of each of the symbols is as follows:

- PV-OK: Prompted Voiding (PV) is recommended for this patient at this time;
- P-RQ: the patient requires and responds appropriately to prompts;
- (NOT PRMPT-NOT-REQ): the patient requires verbal or physical prompting to maintain continence;
- PRMPT-LVL: the granularity of prompting required is logistically feasible for reliable provision of the PV intervention;
- PRMPT-RESP-OK: the patient responds appropriately to toileting prompts;
- BL-CD: the condition of the patient's bladder will not impair the probability of successfully responding to the PV intervention;
- FREQ-OK: the current frequency of positive response to the PV trial is not equal to or greater than the pre-PV incontinence frequency rates;
- BL-CAP-OK: the patient's bladder can hold more than 200 ml;
- PVR-OK: the patients post void residual is not greater than 100 ml;
- INITIATE-VOID: the patient is able to control and initiate voiding voluntarily;
- TH: the type of human assistance required is appropriate to this patient's current needs and is logistically feasible for reliable provision of PV;
- WBN: the patient's weight-bearing needs do not exceed what is logistically feasible to reliably provide PV;
- MECH-DEV: the type of mechanical assistance devices is appropriate to this patient's current needs and capabilities;
- STR: the patient's current strength is sufficient to benefit from the PV intervention.

In order to determine whether an individual patient is likely to benefit from PV and that the nurses will be logistically and reliably able to provide the PV intervention, the approximately 136 required inputs are used to calculate the input values that are used in the PV calculation.

In addition to the recommendation to provide PV or to not to provide PV, the calculated values are used to make nursing care recommendations regarding the prompt-granularity, number of nurses required, type of human assistance required, and the appropriateness of the current mechanical assist devices to safely assist the patient with transfers and locomotion.

The entire system includes 124 such rules, and considerable input/output code. D'Ambrosio's dissertation describes how these rules are used in backward chaining to answer queries to enable the generation of a care plan for toileting of the target patient population. She compared the recommendations of the system for three hypothetical cases with the recommendations of nursing experts. Results varied in details, but the impact of the work was to illuminate the possibility of providing a sound low-level logical basis for nursing decisions.

In addition, this project illustrated several common challenges in practical rule-based system design. The toileting decisions analyzed by D'Ambrosio involve non-monotonic reasoning. This was not simple to represent in a basic forward chaining or backward chaining system.

The granularity required in the rules, how much can be lumped or compiled, and how much to leave as low level details, was also difficult to decide.

Nevertheless, the project was sufficiently intriguing and promising that Dr. D'Ambrosio, at this writing, is working on building an actual physical robot prototype that will incorporate navigation capabilities as well as lift mechanisms, so that for challenging cases, a kind of "robot nurse assistant" can be used to assist nursing home health care professionals. It will be interesting to see how this work develops, especially in light of the success of robotic vacuum cleaner devices and other household applications of artificial intelligence ideas.

2.2.10   Limitations of FOL

Even with resolution, as expressive and powerful as First Order Logic seems, it has some limitations and disadvantages. One question you might ask is whether the process of searching for a proof will always finish (with either success or failure), or whether it will get stuck in a loop or endless search path. Even if it will always finish, how long do you have to wait? Will it finish in some reasonable or at least predictable time? Finally, can we express all the knowledge we have available so that there is adequate coverage to answer lots of interesting queries, or are there important ideas that can't be (or at least, not easily be) put in a FOL formula. We will address each of these in turn.

2.2.10.1   Completeness and Decidability

We indicated that the motivation for using proof methods and sound inference is that entailment is in general hard to compute. However, it is possible for things to be even worse, for example, that a set of logic statements has entailments (that are true) but there really are no proofs for these entailments. Fortunately, there is good news about propositional logic and first order logic. In these two cases, any formula that is entailed by a set of formulas also has a proof: if a set of sentences S entails another sentence R, then a proof of R exists. When a system has this property, that everything entailed is provable, we say the system is *complete*. So propositional logic and FOL are both complete.

It's one thing to guarantee that a proof exists. It is quite another matter to be able to find it. Proof methods and algorithms are also characterized by the possibility that the search for the proof may not terminate. A proof method or strategy that always can find a proof if it exists is also called *complete*. Forward chaining and backward chaining using Modus Ponens are *not* complete proof methods, but resolution (with suitable strategies or algorithms) is a complete proof method. More precisely, since termination is *not* guaranteed if the proof does not exist, resolution is said to be *refutation complete*, since a common method of applying resolution is to add the negation of what is to be proved, and derive a contradiction from the resulting set of formulas.

Now for some interesting bad news: arithmetic is *not* complete. There are arithmetic formulas (statements about numbers) that are true but for which a proof *cannot* exist. If such a proof existed it would make the system inconsistent. This surprising result is known as Gödel's Theorem.

It is described in an English translation [129] of Gödel's 1931 paper and in Hofstadter's book, "Gödel, Escher, Bach: an Eternal Golden Braid" [159]. A very brief sketch also may be found in [359, Chapter 9]. The things that arithmetic adds to first order logic are numbers, operators on numbers that produce other numbers, the principle of mathematical induction, and the fact that the formulas and sets of formulas in arithmetic are an enumerable set. This last idea is that every formula can be assigned a unique number, and every number corresponds to a particular formula. A key to this is the prime factorization theorem, which says that every number can be represented in a unique way as a product of powers of prime numbers.

Proofs being sequences of formulas, they can also be assigned unique numbers. Once you do that, it is imaginable that statements about the existence of numbers satisfying certain arithmetic properties (formulas in arithmetic) can be made equivalent to statements about the existence of proofs. That (with a significant amount of construction) enables the formulation of an arithmetic statement which, if true, cannot have a proof. If you find this mind-boggling, so did a lot of mathematicians, when Gödel first reported it.

There is a second consideration. As we have seen it is possible to write a program to automatically generate proofs in propositional logic, with assurance that they always come to a conclusion. When this is possible, we say that the logic system is *decidable*, meaning that it is always possible to determine whether a given formula is entailed by a set of formulas. Propositional logic, with a finite set of formulas, is decidable. Even a procedure as arduous as enumerating systematically the entire truth tables for a set of formulas will terminate. For FOL, however, the news is bad. For a given set of axioms, a procedure exists that can find a proof if it exists. However, no automated procedure (reasoner) can exist that is guaranteed to terminate when the proof does not exist. Since no automated reasoner can be both sound and complete, we say that FOL is *undecidable*.[19] Arithmetic is both incomplete *and* undecidable.

What is the significance of these things for biology, medicine, and public health? If we hope to have automated reasoning systems to describe biological processes or solve reasoning problems, it would be highly desirable to build systems that are complete and decidable, but the biology itself may simply be too complex. Thus we will be challenged to design systems that operate on practical problems, despite their ultimate theoretical limitations. In medicine similarly it would be disturbing to have decision support systems in a clinical environment that are knowledge rich but not guaranteed to complete their work, even in terms of returning a result of "failure." How long should the clinician wait for the system before deciding it is pursuing an unprovable proposition and will not finish?

2.2.10.2 Tractability

Yet another consideration is the efficiency and tractability of automated reasoning systems. A guarantee that the system will eventually provide an answer to a query will not be of much use if in general it takes several months, or even several hours. Proof methods that grow exponentially

---

[19] Strictly speaking, since it *is* possible to find the proof if it exists, FOL is called *semi-decidable*.

with the size of the problem are not likely to be useful in a medical practice context. However, it is not sufficient to just do the complexity analysis. In some rare cases, an algorithm with worse scalability than another will outperform the more scalable, if the typical computation involves very few items or steps, and the algorithm with worse scalability has much higher efficiency per step. An example of this is the problem of determining the intersection of a line with a planar polygon [205]. This is important because it is the most computationally intensive part of calculation of radiation dose in a patient for radiation therapy planning (for more about radiation therapy planning, see Section 2.2.8).

2.2.10.3 Expressiveness

As expressive as First Order Logic is, many important queries cannot easily be expressed in this formalism. Moreover, many areas of biomedical knowledge do not seem a natural fit to the logic formalism. For example, one might have a set of logic formulas describing parts of the human body. The following formulas assert that the heart is an organ, as are the lung, and the kidney. Here we are considering the predicates `heart`, `lung`, and `kidney` to define the corresponding classes. The variables can have values which refer to particular organs in a particular person, such as my heart or yours.

```
(<- (organ ?x) (heart ?x))
(<- (organ ?x) (lung ?x))
(<- (organ ?x) (kidney ?x))
```

We can also express `part-of` relations in the same way.

```
(<- (part-of ?x ?y) (and (right-atrium ?x) (heart ?y)))
```

Another example statement expressing the `part-of` relation is the assertion that the heart is itself a part of the cardiovascular system. In FOL this looks like

$$\forall x, y : \text{heart}(x) \wedge \text{cardiovascular-system}(y) \rightarrow \text{part-of}(x, y)$$

This can also be expressed in the style of the Graham inference code.

```
(<- (part-of ?x ?y) (and (heart ?x) (cardiovascular-system ?y)))
```

Similarly, other `part-of` relations can be represented by Horn clauses, as, for example:

```
(<- (has-parts ?x (right-atrium left-atrium
                   right-ventricle left-ventricle ...))
    (heart ?x))
```

It might be more useful to assert the `has-part` relationship for each part,

```
(<- (has-part ?x right-atrium)
    (heart ?x))
(<- (has-part ?x left-atrium)
```

```
        (heart ?x))
 (<- (has-part ?x right-ventricle)
        (heart ?x))
 (<- (has-part ?x left-ventricle)
        (heart ?x))
 ...
```

One such relation will be required for every combination of an anatomic structure and each of its parts. This won't scale well to real biomedical knowledge systems, which tend to be very large and complex, perhaps as large as a million statements. Some better organizing ideas are needed. Not only is this an extremely verbose way to manage this information, it does not allow for queries about the kinds, or predicates, themselves. One cannot, for example, ask in such a knowledge base, "What are the parts of the heart?" or "What kinds of organs are there?" One might think that this can be queried with Graham's inference system, as

```
 > (with-answer (part-of ?x heart) ...)
```

but heart is not an individual; it is a class or predicate. A more sophisticated query might be

```
 > (with-answer (and (part-of ?x ?y) (heart ?y)) ...)
```

but this would search the knowledge base for individual hearts and their parts. Depending on what (if any) individual hearts and their parts are in the knowledge base, this query might return 0 or many answers for ?x. Even with answers, it is not possible to query for a given individual, what type of part it is.

Such questions are simply not expressible directly in the kind of first order logic systems we have described, as they are queries about predicates, not about individuals. In addition, we alluded earlier to the problem of non-monotonic logic, to handle situations where new information comes into the system, and we need to be able to remove assertions that are no longer valid. In some cases, it is reasonable to assume certain things as true unless retracted or replaced. This is called *default reasoning*. Systems that handle defaults are also non-monotonic. Such systems are beyond the scope of this book, but are discussed in some of the standard texts on artificial intelligence [46,305,360,408].

One way to handle such queries is to create a new representation language in which the categories, individuals, and relations (predicates) are all on an equal footing. In this new language there are a limited number of operators (analogous to predicates in FOL) which connect categories, individuals, and relations with themselves and each other. The queries that were higher order in the old representation will become first order in this new language. This approach and the code to support it is developed by Norvig in [305, Chapter 14].

An alternative is to develop an entirely different approach, a *frame* formalism, which relies on structured descriptions. This has the virtue of being a good match to large bodies of biomedical symbolic knowledge. Norvig's development also shows how to construct a frame language by providing translations into his relational language. Our approach will be more along the lines of the informal description of frames in Brachman and Levesque [46, Chapter 8], and will have some similarities to frame systems in actual use, for example, Protégé [306], LOOM [265,266,422], and Ontolingua [235].

2.3 Frames, Semantic Nets, and Ontologies

In large data stores such as GENBANK, the sequences are often thought of as the primary data, and are the subject of much computation (sequence comparisons, sequence searching for small patterns like motifs). However, there is a lot of knowledge in the annotations, the text information that accompanies each sequence. This is generally not in an easily computable form, but it is computationally powerful material. For example, information about which protein(s) a gene encodes, and what the function(s) and pathways in which it participates is useful for reasoning about cellular metabolism. Projects like the Gene Ontology [18] are a step toward making these pieces of knowledge the basic elements of a computational scheme. Another example is the EcoCyc ontology of metabolic pathways in *E. coli*, and its generalization to many other organisms [214,215]. The information represented here is symbolic, to be manipulated as such. When an enzymatic reaction is specified in a system like EcoCyc, the symbols and structures represent objects such as a molecule, a particular kind of molecule, a reaction (a relation between molecules as reactants and products). As Karp [217] points out, such structures enable computations that are not possible by searching text or XML data, or using controlled vocabularies.

Similarly, knowledge about laboratory tests and their clinical interpretation is not part of the data or the coding system. This needs to be represented in a different way, which captures the meaning and relationships of the entities being measured, diseases, diagnostic significance, etc. The Unified Medical Language System [258] is an NLM sponsored project that addresses this goal.

Much effort in bioinformatics is now being expended to extract this symbolic knowledge from the text of published articles, and from data resources that have this knowledge embedded as text. Rather than address the issues of parsing and interpreting this text corpus, we will focus on building a representation language for it. We need a language that will incorporate the idea of classification of entities, particularly hierarchical classification systems that are usually referred to as "ontologies." After a brief introduction to frame systems and the use of classes in object-oriented programming languages as a more limited alternative, we will see what kinds of computations can be done on these ontologies.

2.3.1 A Simple Frame System

In this section we develop a really simple frame implementation to illustrate essential ideas. While frames can be used to represent data, the emphasis here is on frame systems as knowledge representation languages. They are an example of a variety of knowledge representation systems called "slot and filler" systems, since the basic structural idea is to group together names of attributes and their values (of course the values can be other frames).

What kinds of information do we want to represent? For any entity in the world, whether an individual thing (a laboratory test for a particular patient), or a category representing many individuals all of the same kind (organ, heart, protein, etc.), we will need to encode a name, perhaps a unique identifier, relationships to other entities such as class–subclass relationships and others, properties of the category as a whole, and properties of specific individuals belonging to the class.

For example, for the category, "heart" we would like to represent its relation to the rest of the body, its parts, and a text description. A complete representation would also include a specification of the blood vessels (arteries) that supply the heart, the veins that return blood from the heart muscle, its connection to the rest of the circulatory system, its lymphatic drainage (what lymph vessels collect the fluid around the heart), and what nerves are connected to the heart muscle. For a specific heart, we would want to be able to record some of its individual properties, such as heart beat rate (pulse), ejection fraction (how much of its volume is it actually pumping out), and how big is it. A more complex attribute of an individual heart would be the results of measuring its electrical activity (an electrocardiogram, or EKG).

The basic unit of representation of entities in the world is the *frame*. Our frame system will use keyword-value pairs and other kinds of tagging as in the patient example in Section 1.2.1 (page 88). However, here we introduce some additional structure, to allow for abstraction of the computations we will perform. A frame can have *slots*, which are accessible by their names, and which may contain associated values, as well as other properties. Each frame needs to have a unique *identifier* (within the context of the knowledge base, not a UID in the global sense). Frames may have some slots that are understood to have special or designated semantics. We will start with a simple representation and augment it as needed.

The main ideas that will be illustrated by this small system are:

- the representation and meaning of the relationship between classes and subclasses,
- the representation and meaning of the relationship between classes and individuals (instances),
- the representation and use of domain specific relations,
- the meaning and use of metaclasses, and
- extensions to support richer semantics for attributes of classes and instances.

In a first try at building a simple system, a frame will consist of a list, in which the first element is a symbol, the name of the frame, and the remaining elements are "slots," which initially we will represent as lists containing keyword-value pairs. In a slot, the keyword serves as the name of the slot, and the value associated with that name can be (in general) any valid Lisp data type. A slot value can also be a reference, that is, the name or identifier of another frame, or the actual frame object itself, or a list of such objects.

Frames can describe individuals or classes, and a frame may describe an entity which is at the same time both an individual and a class. By "individual" we mean a unique single entity, just as in logic. However, we can also talk of categories as individuals. A term can refer to a unique specific category, or class, and although there may be many examples of things that belong to that class (as members), the class itself may be considered a specific thing. This is exactly what we need in order to deal with the problem of reasoning about predicates or categories or relations. We are *reifying* the categories and relations represented previously by predicates, making them into things, or individuals, in our frame system.

We take the body parts as examples. Many organisms have hearts, organs whose organ type is *heart*. We use the term "heart" to name the class whose members are hearts. There is one such class, and it is distinct from the class, *kidney*, which is an entirely different kind of organ. So, the heart class is an entity, a category, with members that are individual hearts of specific beings, like me and you. Having that category, sometimes called *concept*, we can then describe attributes associated with it, that is, things that are true about the heart in general. The class, heart, could be described by the frame:

```
(heart
    (part-of cardiovascular-system)
    (contained-in middle-mediastinal-space)
    (has-parts (right-atrium left-atrium
                right-ventricle left-ventricle
                mitral-valve tricuspid-valve ...))
    (arterial-supply (right-coronary-artery
                      left-coronary-artery))
    (nerve-supply ...)
    ...)
```

Each slot represents a relation between the frame and the value in the slot. The heart is a part of a larger entity called the "cardiovascular system." It has parts of several kinds. It also has an arterial supply. Each of these attributes has a value or values, which are themselves classes, naming kinds of anatomic objects, usually themselves organs, organ systems, or organ parts. So, this frame structure is a way to record knowledge about the heart as a component of a description of the standard anatomy of (for example) a human being.

2.3.1.1 Classes, Instances, and Subclasses

So far, we have a representation of the heart idea, but not of any particular person's heart. A particular heart will have attributes that apply only to that instance, such as whose heart it is, how fast it is beating, what volume of blood it contains (at different parts of its operation), etc. It will also have specific parts, arterial supply, and so on, each of which will similarly be particular to the individual whose heart this is. We might then represent an individual person's heart, as follows:

```
(heart-of-joe
(instance-of heart)
(belongs-to joe)
(beat-rate 72)
(volume 34)
(systolic-pressure 125)
...)
```

We also included a slot named `instance-of` to indicate what *kind* of entity this frame describes. So, Joe's heart is an individual heart, and we say that the frame `heart-of-joe` describes an *instance* of the class `heart`, (which is described by the `heart` frame).

The slots that appear in `heart-of-joe` appear to have come "out of the blue." They seem to have no apparent relation to the `heart` frame, of which this is an instance. This suggests that we did not provide a complete description of hearts in the `heart` frame. When a class such as `heart` is defined, it is usually because all hearts share some common properties. Some of these properties have the same values for all hearts, and are properly part of the class description. Such slots belong to the `heart` frame. Some properties are present in all hearts but the values associated with them may vary from individual to individual, such as the heart beat rate. Those slots should appear in all the instances, but it would be economic to be able to specify as a part of the definition of `heart` (in the `heart` class frame) that these slots should appear in each frame representing an individual heart.

Slots that are listed in a class frame, but intended to be slots with values in instances of that class, are called *template slots*. In the case of the `heart` frame, we should list `belongs-to`, `beat-rate`, `systolic-pressure`, etc., in the class frame, without values, and tag them as template slots.

```
(heart
    (part-of cardiovascular-system)
    (contained-in middle-mediastinal-space)
    (has-parts (right-atrium left-atrium
                right-ventricle left-ventricle
                mitral-valve tricuspid-valve ...))
    (arterial-supply (right-coronary-artery
                      left-coronary-artery))
    (nerve-supply ...)
    (template-slots (belongs-to beat-rate volume
                     systolic-pressure ...))
    ...)
```

The list of names in `template-slots` does not describe slots in the heart frame itself. It specifies what slots will appear in the instances. When we create an instance of this class, the names in `template-slots` will become the names of slots in that instance. Those slots can then have specific values. So, we distinguish between two types of slots. The slots that appear in a frame, and describe the specific entity that frame represents, are called *own slots*, because they are the frame's own slots, not slots that will appear in some other frame. The slots that are specified in a frame but intended to appear (each with values) in instances of the class that frame describes are called *template slots*. Each template slot will have a name, and could have lots of other information about it, to be used when an instance is created. For now, in our system, the template slots list will just include names.

This might be clearer if we explicitly label the own slots with the tag `own-slots`, and put them in a separate list. This list will contain the own slots, implemented as name-value pairs. The list of template slots will just have the names of the template slots, labeled by the name `template-slots`. Then we will have the following structure:

```
(heart
  (own-slots
    (part-of cardiovascular-system)
    (contained-in middle-mediastinal-space)
    (has-parts (right-atrium left-atrium
                right-ventricle left-ventricle
                mitral-valve tricuspid-valve ...))
    (arterial-supply (right-coronary-artery
                      left-coronary-artery))
    (nerve-supply ...))
  (template-slots (belongs-to beat-rate volume
                   systolic-pressure ...))
  ...)
```

We also need to be able to express relationships, for example, of which class is a particular frame an instance. In the first version of Joe's heart, we had a slot named `instance-of`, but it is not listed in the template slots. Every individual frame is an instance of some class. If not otherwise specified, a frame can be thought of as an instance of the class, `thing`, even if it is not specifically an instance of something more specific. So, the `instance-of` slot should be present in every frame, and separate from the `own-slots` and `template-slots`.

Similarly, every frame should have a name (preferably unique) and a unique identifier (we arrange for this to be automatic). In our system, every frame will have a `name` slot and an `id` slot, again, separate from the own slots and template slots.

These ideas are very similar to the ideas of classes and instances in an object-oriented programming language. The principal difference here is that ordinary object-oriented programming languages provide only a means for assigning property values to instances, but not to classes. The frame idea provides a direct way to assign properties to classes too. Classes are themselves entities (individuals) that have descriptions and data associated with them. This is important for representing biological and medical knowledge. Biological knowledge bases (usually called *ontologies*), such as the Gene Ontology, EcoCyc, the FMA, and others, are collections of information about *classes* of biological entities, not data from individuals (as you would find in an electronic medical record). An entry in the Gene Ontology that describes a gene product, such as "NADH dehydrogenase," is describing a type or class, of which there could be many instances in each particular cell in some organism. The information about this type of enzyme is true of all such instances. Similarly, the terms in medical terminology systems such as Galen [336], and the UMLS [37,295] refer to classes of phenomena and entities. For example, a frame called "pneumonia" refers to the disease phenomenon in general, not to particular occurrences in individual patients. For implementing an electronic medical record system, a conventional object-oriented programming system (or alternately a relational database) would be sufficient to describe the data about the patients, facilities, health care providers, etc. However, once you want to incorporate knowledge about the properties of the various classes of things in the database rather than just properties of instances, something more like a frame system will be needed.

In addition to class-instance relationships, we can have frames that describe specialized versions or *subclasses* of the things described by a class. So, for example, the class human-heart is a subclass of heart. A human heart is also a heart and has all the properties of a heart, but it may have some additional properties that are not present in all hearts. In fact, in the above description, the contents of the has-parts slot are specific to human (and some other) hearts. Some hearts have only two or three chambers. So, a better frame description would not have these values in the heart frame, but would move them to the more specific heart subclasses.

These relationships are similar to the relationships among sets, subsets, and elements of sets. The relation between a set and a subset is analogous to the relation subclass-of. The relation of a set to a member of the set is analogous to the relation instance-of. It is very important to realize that subclass-of and instance-of refer to *different* relationships. In our formalism it is not permitted for two entities to be related to each other by both relationships. Entity A cannot be both a subclass of entity B and an instance of entity B. This is analogous to the requirement in the mathematical theory of sets, that a set, A, cannot be both a subset of another set, B, and also a member of B. This won't be enforced by the implementation, but if it is not followed, problems can arise in searching for information and reasoning about the contents of the frames.

2.3.1.2 A More Structured Implementation

Having all these nested lists is getting complex. A better strategy would be to use the Common Lisp Object System to create a frame class (in CLOS), and have our frames all be instances of this CLOS class.[20] So far we see that a class frame describes a category, or set of things, and can also be considered a thing by itself. The idea that a class can have its *own* specific properties is the key to everything we will do in this section. It is not possible in typical object-oriented

---

[20]You might wonder if we should just make frames be CLOS classes and instances. This could be adequate for many applications but CLOS does not provide all the capabilities of a frame system.

programming languages to do this. In languages like C++, classes are not data that can be used by a program. The class definitions in the program source code are used by a compiler to generate the right behavior for instances. Ordinary object-oriented programming focuses on creating and manipulating instances, while here we are treating the classes themselves as data elements of our system.

In order to define a frame structure, we need to decide what properties all frames could have. A frame will have own slots, template slots, and information about which frame (if any) it is an instance of. If a frame describes an entity that is a member of a class, that is the `instance-of` relation. Frames that are classes can also be *subclasses* of other class frames. We will need to put in each frame a list of the classes of which this frame is a subclass. We call these the *superclasses*. The relations `subclass` and `superclass` are inverses of each other. If a frame, A, represents a subclass of another frame, B, we also say that frame B is a `superclass` of frame A.

So, our frame structure needs to include a place for the own slots, template slots, an `instance-of` slot, and a `superclasses` slot. In addition we will need a place for a name and a frame ID. These frame attributes will be initialized by a constructor function, as part of the protocol for making a frame from user specified data.

Finally, we will provide a place to store all the frames for later reference, but it would be convenient if each class frame could keep track of the instances of that class. So, we will also provide a place to put pointers to the instances for a frame, in case we use it as a class frame and make instances of it.

To implement these ideas, we include in the CLOS frame class some special slots with reserved names, whose meaning will be implemented in code. The `instance-of` slot in a frame, F, contains the `id` of the frame describing the class of the object represented by F. In other words, if an object, F, has the symbol `frame-2861`, in its `instance-of` (CLOS) slot, that means that F is a member of the class represented by the frame whose `id` is `frame-2861`. Similarly, the `superclasses` slot and the `instances` slot contain lists of frame IDs, not the actual frames themselves.

Here is a CLOS class that implements this, by providing CLOS slots for each of these pieces of information.

```
(defclass frame ()
  ((id :reader id
       :initform (gentemp "frame-"))
   (name :accessor name
         :initarg :name
         :initform nil)
   (instance-of :accessor instance-of
                :initarg :instance-of
                :initform nil)
```

```
(superclasses :accessor superclasses
                    :initarg :superclasses
                    :initform nil)
(template-slots :accessor template-slots
                    :initarg :template-slots
                    :initform nil)
(own-slots :accessor own-slots
                :initarg :own-slots
                :initform nil)
(instances :accessor instances
                :initform nil)
))
```

The class-subclass relation is implemented in a frame by storing a class frame's *superclasses* because those are known at the point where the frame is defined. The subclasses are not yet known. This is the customary way of organizing classes in object-oriented programming. When we create a specialization of a defined concept after it is defined, the specialized class *inherits* all the properties of the more general class, without disturbing the behavior or definition of anything else in the system. If it were done the other way, proceeding from specific to more general, managing consistency would be quite a bit more difficult.

To make a frame like the heart frame, we will need to put data into each of the slots above. The template-slots slot, for now, can just be a list of the names of the template slots for the heart class. When we make a frame that is an instance of heart, the *instance* will have own slots with names that were specified as template slots in the heart *class*. Instead of having a CLOS slot for each own slot, we just keep the own slots on a list in a single CLOS slot called own-slots. To keep it as simple as possible, the list of own slots will just be a simple association list. The own-slots slot will contain a list of name-value pairs, one pair for each own slot of the frame.

The content of the name slot has no restrictions other than being unique, so the name could be a string or a symbol or even some other data type. Which is best will depend on the context of use of the frame system. The name will be an input to the constructor function that creates the frame. The id will be a unique symbol generated by the initform in the above class definition. The gentemp function will create a symbol with the prefix frame- and a number unique in the current working environment of the program.

Before delving into the details, we need to decide how to keep all the frames in a readily accessible place, so we can look up particular frames, and resolve cross references. A simple way to do this is to arrange for every frame, on creation, to be added to a global list. We use a named global variable, *frames*, for this.

```
(defvar *frames* nil "The global list of all frames")
```

The constructor functions that make frames as well as other functions that you would use to manipulate them should not explicitly reference this variable. In order to allow for later redesign of this simple storage scheme, without changing the code that uses it, we provide a small abstraction layer, with functions to add frames, remove frames, and search for frames by name and ID.

Looking up a frame by ID or name is easy. We use find with a key function to get the data from the pertinent slot. Since the name might be a string, we use equal, rather than the default eql, as the comparison.

```
(defun find-frame-by-id (frm-id)
  (find frm-id *frames*  :key #'id))

(defun find-frame-by-name (frm-name)
  (find frm-name *frames*  :key #'name :test #'equal))
```

Similarly, the functions add-frame and remove-frame provide a way to preserve accessibility of newly created frames, and remove frames from accessibility when (for example) they are superceded or made obsolete.

```
(defun add-frame (frm)
  (push fr *frames*))

(defun remove-frame (frm)
  (setf *frames* (remove frm *frames*)))
```

Note that the test for remove-frame is the default, eql, since we want to remove the actual frame, not one that looks like it. Alternately we can have remove-frame use a frame ID or name to look up the frame to be removed. The name is not guaranteed to be unique, so this could be dangerous.

2.3.1.3 Saving Frame Collections to Files

Since frames are instances of a CLOS class, and the slot values of these instances are ordinary Lisp data, not themselves CLOS objects, it is simple to create a collection of frames, then save it to a file, and later restore it. We just use the serialization functions, put-all-objects and get-all-objects from Chapter 1, Section 1.2.3. Of course, the underlying representation could change, so once again we provide a small abstraction layer. The save-frame-kb function writes out all the frames in the current environment to a file specified by filename, and the function restore-frame-kb reads them back into working memory.

```
(defun save-frame-kb (filename)
  (put-all-objects *frames* filename))

(defun restore-frame-kb (filename)
  (setq *frames* (get-all-objects filename)))
```

Looking ahead, this won't work with the extension described later (Section 2.3.2), where slots can have attached functions. It may be possible to limit what is stored to represent a function, so that this will still work. Alternately, one could extend `put-object` and `get-object` to write out representations of function objects as well as other forms of data.

2.3.1.4 Slot Accessors and Other Abstraction Devices

Now that we have a sketch of the frame representation, it would be useful to write functions that can access the contents of slots, create new frames, and do other operations. While it is possible to write explicit code every time we need to find a slot and retrieve its contents, it is better to provide a little abstraction layer. The CLOS structure just defined for frames already provides generic functions (created by `defclass`) for obtaining the ID, name, class, superclasses, template slots, own slots, and instances. What we still need is some structure and an abstraction layer to get at the information in the own slots.

In this section, we implement an abstraction layer for the structure of an own slot. A simple approach is to just use the keyword-value pair idea mentioned earlier. A slot is as illustrated in the simple list version of our frames, a list whose first element is a symbol naming the slot, and whose second element is the value stored in the slot. We will augment this later when we add more capabilities to own slots, but for now, the following code is sufficient to define the slot abstraction layer. The `slot-name` function just returns the name of a slot. The `contents` function returns the value stored in the slot. In this simple case there is no way to distinguish between `nil` as a value and `nil` indicating no value. Finally, we also provide a nice way to use the `setf` macro to update the data in a named slot.

```
(defun slot-name (slot) (first slot))

(defun contents (slot) (second slot))

(defun (setf contents) (newval slot)
  (setf (second slot) newval))
```

Now, having a way to refer to the name of a slot, we can provide the next abstraction layer. In this layer, there are functions to retrieve a slot by name, get a list of all the names of slots in a particular frame, and a next level of accessors for reading and writing slot data. Later, enhancements to this code that support attaching procedures to slots will go in this layer, not the lowest layer. This makes it easier later to change the underlying representation of slots without changing most of the implementation.

The `get-slot` function returns a particular named slot including its contents. The `slot-list` function returns a list of the symbols naming each of the slots.[21] It answers the question, "What slots does this frame have in it?" Finally the `slot-data` function and its associated `setf` function provide the higher level accessor layer we will need for later enhancements.

---

[21]This is analogous to the `slot-names` function defined in Chapter 1 on page 105, which does the same operation for CLOS classes.

```lisp
(defun get-slot (fr name)
  (find name (own-slots fr) :key #'slot-name))

(defun slot-list (fr)
  (mapcar #'slot-name (own-slots fr)))

(defun slot-data (fr slot-name)
  (contents (get-slot fr slot-name)))

(defun (setf slot-data) (newval fr slot-name)
  (setf (contents (get-slot fr slot-name))
    newval))
```

The `frame-type` function returns the name of the class of which its input is an instance, or nil if it is not an instance of a class frame. This is not just the content of the `instance-of` slot, because that slot contains a `frame-id`, not a name. So we need to look up the frame and get its name.

```lisp
(defun frame-type (fr)
  (let ((parent (instance-of fr)))
    (if parent (name (find-frame-by-id parent)))))
```

While this is not much of an abstraction layer, it provides enough insulation so that we can change the underlying representation if necessary while preserving upper layer code. In particular, we can *extend* the representation (as will be seen in later sections), without changing the code that uses the abstraction layer. In referring to frames, we may refer to them by their names or by their identifiers, depending on the context.

2.3.1.5 Inheritance

If we restrict a class to have no more than one superclass, there will be a single value in the `superclasses` slot. This is called *single inheritance*. For simplicity, some object-oriented programming environments incorporate this restriction. However, many biomedical entities we would like to represent belong to overlapping classes, and so the classes that represent them will have multiple superclasses. We can implement this by simply allowing a list, or more than one value, in the `superclasses` slot.

On the other hand, we restrict each instance to be a direct member of one and only one class. This is not much of a restriction since that class can be a subclass of more than one superclass.

The subclass-superclass relation is transitive, in that if A is a subclass of B, and B in turn is a subclass of C, then A is also a subclass of C. We don't need to replicate all these transitive relations because we can compute them. By recursively collecting superclasses, we can create a list of *all* the superclasses of any particular class. Here is a function to do that.

```lisp
(defun all-superclasses (fr)
```

```
   "returns the frame-ids of all the frames that are superclasses
of
   frame fr all the way up the class hierarchy"
   (let ((direct-sup-ids (superclasses fr)))
     (apply #'append
            direct-sup-ids
            (mapcar #'(lambda (frm-id)
                         (all-superclasses
                          (find-frame-by-id frm-id)))
                    direct-sup-ids)))))
```

An instance should have own slots that correspond to *all* the template slots from all its superclasses. One way to create a list of such slots is to use the `all-superclasses` function to get a list of frame identifiers, and then append together the template slots from all of the superclass frames. The following function, `all-template-slots`, implements this idea.

```
(defun all-template-slots (fr)
  (remove-duplicates
   (apply #'append
          (template-slots fr)
          (mapcar #'(lambda (id)
                      (template-slots (find-frame-by-id id)))
           (all-superclasses fr)))))
```

This use of information from superclasses as well as the direct parent class is called *inheritance*. It is so named, because each specialized class inherits properties from its superclasses. This is only one aspect of inheritance. Other aspects of superclasses may also be passed through to their subclasses. In this particular case, the inheritance protocol is very simple. If a slot name appears in any of the template slots of a superclass, it will become an own slot in the instance. In a more powerful system, one could arrange for richer slot descriptions, with default initial values, type declarations, etc. Then this function would not use `remove-duplicates`, but the constructor function (which we have not yet considered) would have to implement a more complex protocol, where perhaps some slot specifications would override others, or the specifications would combine or augment each other.

We don't include `instance-of` in the template slots list because we can include that slot automatically in the process of using the class frame information to make an instance of the class. This procedural encoding of the semantics of the class-instance relationship will be in the constructor function to make new frames, although the relation itself will be encoded explicitly in the frame instance. Other such relationship semantics can also be procedurally encoded. This is the beginning of the design of the frame system protocol, namely, what rules and constraints will all the frames have, and how does one create, access and relate the frames and their contents.

We might want to require that every frame must have a name, supplied to the constructor function (which we will develop shortly). We will also have to define the protocol for how a frame gets to have own slots and template slots. One possibility is that you should be able to create a frame with any collection of such slots. Another is that own slots come only from a frame being defined as an instance of some class frame, so that the own slots are exactly those specified as template slots in the class frame. We will take the latter approach, which is more restrictive but simpler to implement.

Since it is possible for a frame to have both template slots and own slots, one could construct a frame defining a class, which would have instance frames. It could also be that the class frame is itself an instance of another class frame (not a subclass). This is in our frame system how a class frame will have own slots. These form a somewhat different hierarchy than the class-subclass hierarchy that is most familiar from object-oriented programming. A class whose instances are themselves classes is called a *metaclass*. This class-metaclass relationship can also be used to form a hierarchy of classes and metaclasses (and meta-metaclasses and so on).

The idea of a metaclass and the possibility that a class can have its own properties aside from the specific data that go with instances is very powerful. It means that a system of classes, instances, and operations with them could potentially be built on the same ideas it implements. This requires a Meta-Object Protocol (MOP), defining the building blocks on which the frame system is built. Building blocks may be extensible, by using the ideas of subclasses and generic operations. In fact the Common Lisp Object System is built in exactly this way, that is, most of the entities in CLOS are themselves CLOS meta-objects, and can be specialized and inspected. The Meta-Object Protocol (MOP) for CLOS is not yet part of the ANSI standard for Common Lisp, but it is implemented in most available Common Lisp systems, and is described in [229] mentioned previously.

The Protégé ontology editing system [306] also implements a frame-based knowledge representation, and has some resemblance to CLOS and MOP. In the Protégé system there are system classes that you can specialize to construct your own classes. Protégé supports arbitrary levels of class-metaclass relationships, while CLOS only supports one. Protégé, along with its various plugins, is available at URL `http://protege.stanford.edu/`.

2.3.1.6 Constructing Frames

Now, we can provide a constructor function for frames. A constructor takes a number of inputs and constructs a frame from them. The inputs will depend on whether we are constructing a class frame or an instance frame (or something that is both). A name for the new frame is always required. If it is an instance frame, we will need the name of the class of which it will be an instance. If it is to be a class frame, we need a list of superclasses of which the new class is a subclass (this could be empty), and a list of slot names that will be the template slots for this class. It won't be necessary at this point to look up the template slots for the superclasses. They will be used later by the constructor function when instances of this new class are created. It is not required that every class define template slots, so the `template-slots` parameter could be `nil` or omitted.

For convenience, the parameters `class`, `superclasses`, and `template-slots` will be keyword arguments. The constructor function will then have parameters as follows:

```
(defun make-frame (name &key class superclasses template-slots
                        &rest slot-inits)
    ...)
```

For an instance frame we will need as inputs the initial values of the slots. The `slot-inits` parameter in the constructor function will be used to specify initial values for the "own" slots for the frame. This parameter is an alternating sequence consisting of slot names and the values to put initially in those slots, similar to the way that `make-instance` works in CLOS. Since each value is "tagged" by a slot name, they (slot name and value) can be in any order. They are also optional, so you can provide initial values for some slots and not others, or no initial values. In our constructor function, this list should be converted to a list of pairs, rather than a flat sequence. The `make-pairs` function does that.

```
(defun make-pairs (x)
   (if (oddp (length x)) (error "list length not even")
      (if x (cons (list (first x) (second x))
                  (make-pairs (rest (rest x)))))))
```

We will then use this list of pairs as an association list from which we can create the initial value for the `own-slots` attribute of the new frame. For each slot name, a pair is either retrieved from the list or a new pair with a name, and a value of `nil`, is created. This is the job of `initialize-slot`.

```
(defun initialize-slot (slotname inits)
   (or (assoc slotname inits)
       (list slotname nil)))
```

Now we can write the function `make-frame`. It will construct and return a frame with specified name, and add it to the global storage of frames. If `class` is provided it is used as the name of a class frame of which this frame will be an instance, and thus will have own slots built from the template slots of the parent class. If this is to be a class frame that is a subclass of other frames, `superclasses` should be a list of names of superclasses, if any, and `template-slots` should be a list of names of template slots, or `nil` if this frame will have no direct template slots. Note also that there will be no own slots if this frame is not an instance of some class frame.[22]

```
(defun make-frame (name &key class superclasses template-slots
                        &rest slot-inits)
```

---

[22]This same restriction is also present in Protégé-2000, a widely used frame-based ontology modeling tool.

```
(let* ((parent (if class (find-frame-by-name class)))
       (super-ids (mapcar #'(lambda (x)
                              (id (find-frame-by-name x)))
                    superclasses))
       (fr (make-instance 'frame
              :name name
              :instance-of (if parent (id parent))
              :superclasses super-ids
              :template-slots template-slots
              :own-slots
              (if parent
                  (mapcar #'(lambda (name)
                              (initialize-slot name
                                 (make-pairs slot-inits)))
                        (all-template-slots parent)))
           )))
  (if parent (push fr (instances parent)))
  (add-frame fr)
  fr))
```

Now we can apply the `make-frame` function to create some frames to describe organs, hearts, and other anatomical objects. The following examples are not direct excerpts from the FMA, but very simplified examples of how the frame system works. The `heart` class that we wrote out earlier as a list structure would be created as follows in our frame system:

```
(make-frame 'heart
            :class 'organ
            :template-slots '(belongs-to beat-rate volume
                              systolic-pressure)
            'part-of 'cardiovascular-system
            'contained-in 'middle-mediastinal-space
            'has-parts '(right-atrium left-atrium
                         right-ventricle left-ventricle
                         mitral-valve tricuspid-valve)
            'arterial-supply '(right-coronary-artery
                               left-coronary-artery)
            ...
            )
```

In order to have own slots in this frame, which we are initializing in the above expression with the keyword arguments, the heart must be an instance of a (meta) class that specifies those as template slots. We've specified that the heart is a subclass of the organ class, so we need to define it (actually in terms of execution, the `organ` frame has to be constructed before any subclasses or instances of it). Here is the construction of the organ frame.

```
(make-frame 'organ
            :template-slots '(part-of contained-in
                              has-parts arterial-supply))
```

In constructing the organ frame, we did not specify that it is an instance or subclass of anything else. This is only a small piece of an eventual complete anatomic model. The class hierarchy would start with some general foundational entity definitions, creating the frames that form the root classes, from which everything else would be built up by specialization. Finally, the FMA does not have in it any instances corresponding to individual people, such as my heart or yours. However, it is possible to create such instances, as follows:

```
(make-frame "Joe's heart"
            :class 'heart
            'belongs-to "Joe"
            'beat-rate 72
            'volume 34
            'systolic-pressure 125)
```

To faithfully model human anatomy requires a much richer model, in which slots themselves are structured objects, as well as slot values. The FMA also makes extensive use of the class-metaclass relation, and propagation of slots and values through such relations. This is discussed in some more detail in Section 2.3.4. Here we just address some very simple examples.

2.3.1.7 More About Metaclasses

To illustrate in more detail the class-metaclass relation and its use, we present a somewhat artificial example using information about drugs. The use of metaclasses should be motivated by the knowledge to be modeled. Associating facts with classes in our implementation does not make them directly appear in instances of those classes, or in subclasses either. It is possible to build into the frame system some machinery for automatically propagating information from classes to instances and from classes to subclasses. Protégé implements such mechanisms and as alluded to, the FMA utilizes them. For now, we will look at an alternative approach, that of using procedural semantics to obtain the information as needed, rather than having to maintain consistency among many different frames, where explicit (related and redundant) copies of information are stored.

The example we will use here is that of organizing drugs into a small hierarchy to store information important for certain situations. The problem motivating this example is that many drugs should be either not used at all or used with caution in elderly people with various diseases present. This is described in a paper by Hanlon et al. [143]. The diseases for which drugs may be contraindicated may be associated with major classes of drugs, or somewhat more narrow classes, or with only particular drugs. Rather than replicating (propagating) information, we will associate disease information with drug classes and subclasses as well as instances. Then for a particular query, we will need a procedure to traverse the hierarchy and collect all the relevant information.

Our model is as follows:

- each class of drugs will have an own slot that has a list of diseases,
- it would be desirable to organize the drug classes into a class hierarchy, so that we can apply more specific information where available but also look up more general information where applicable,
- ultimately we want to create instances of a drug class, representing particular drugs that are members of that class, which may have their own properties as well.

In order for the drug *classes* to have own slots, they will need to be instances of a drug metaclass. So, first we create that. We will call it `drug-class` to indicate that its *instances* will be drug classes. This metaclass is created using `make-frame`, specifying that instances (the drug classes) will each have an own slot named `diseases` to record the disease conditions for which this drug or class of drugs should not be prescribed.[23]

```
(make-frame 'drug-class :template-slots '(diseases))
```

In our model the drug classes will be instances of this (meta) class. Each class will have an own slot called `diseases` and we can fill in the information for that class of drugs when creating the frame for it. Here are some specific drug classes.

```
(make-frame 'cns-agents :class 'drug-class)

(make-frame 'narcotics :class 'drug-class
            :superclasses '(cns-agents)
            'diseases '(bph constipation))

(make-frame 'sedatives :class 'drug-class
            :superclasses '(cns-agents)
            'diseases '(copd))

(make-frame 'barbiturates :class 'drug-class
            :superclasses '(sedatives))

(make-frame 'benzodiazepines :class 'drug-class
            :superclasses '(sedatives)
            'diseases '(dementia syncope))

(make-frame 'antidepressants :class 'drug-class
            :superclasses '(cns-agents))
```

---

[23]The information here on drugs is based on Tables 1 and 2 in [143]. However, it should not be relied on, and is only used to illustrate the metaclass idea.

```
(make-frame 'tricyclic-antidepressants :class 'drug-class
            :superclasses '(antidepressants)
            'diseases '(arrythmia heart-block
                        postural-hypertension))
```

This takes care of being able to assign disease lists to drug classes and subclasses, since they are all instances of the `drug-class` metaclass, and thus all have an own slot called `diseases`, to which we can assign a value. In the drug table some classes also have specific instances listed, so we create frames for them too.

```
(make-frame 'codeine :class 'narcotics)

(make-frame 'diazepam :class 'benzodiazepines)

(make-frame 'methylphenidate :class 'antidepressants
            'diseases '(insomnia))
```

Some of these drugs also have specific diseases for which these drugs are contraindicated. How will we record that information? One approach would be to create further subclasses instead of instances, but a class with a single member does not seem like a good model. If we want a drug instance to also have a `diseases` slot, we need to define template slots for a top level class we will call `drug`. The `diseases` template slot in this class will appear in all the subclasses, so an instance of any drug class will have an own slot called `diseases`.

```
(make-frame 'drug :template-slots '(diseases))
```

and then we specify `cns-agents` and other next level drug classes as subclasses of `drug`. Since template slots will be effective for subclasses, we can build a class hierarchy as before, the only difference being that the formerly top level drug classes are now subclasses of `drug`. In this example, only `cns-agents` was "top level," so it is the only one that needs to be redefined.

```
(make-frame 'cns-agents :class 'drug-class
            :superclasses '(drug))
```

Now the `methylphenidate` frame, which is an instance of a drug class, also has a `diseases` slot and we can put specific information there.[24]

```
(make-frame 'methylphenidate :class 'antidepressants
            'diseases '(insomnia))
```

Now the original problem was, suppose we wish to prescribe a drug. We want to know about the diseases to watch out for, so we need to look at the `diseases` slot in that instance, its class, and all the superclasses. Here is a function that uses the `all-superclasses` function to collect together all the relevant diseases at different levels.

---

[24]The generic drug methylphenidate is better known by the registered trademarked brand name, Ritalin.

```
(defun disease-lookup (drug-inst)
  (let ((drug-type (find-frame-by-id (slot-data drug-inst
                                               'instance-of))))
    (append (slot-data drug-inst 'diseases)
            (slot-data drug-type 'diseases)
            (mapcar #'(lambda (x) (slot-data
                                    (find-frame-by-id x)
                                    'diseases))
                    (all-superclasses drug-type)))))
```

Procedural reasoning of this sort may not be scalable, when dealing with very large collections of frames (the FMA has over 140,000). However, there are no entirely satisfactory or appealing alternatives. The requirements of modeling biological and medical knowledge push the limits of every knowledge representation scheme we will present. Clearly much research needs to be done before we have comprehensive theories in significant subdomains of biology, medicine, and health.

2.3.1.8 Subsumption and Other Relations Between Frames

An important class of questions to pose in a frame system is whether a class described by one frame is subsumed by (or a subclass of) a class described by another frame. This is computed essentially by following the values in the superclasses slots in successive frames. If there is a path through the superclass relations, the first class is indeed a subclass of the second. A complication of this is that there could be multiple superclass relationships so we need to trace them all. One way to do this is to search through the result of a call to all-superclasses, but that is not the most efficient method.

To answer the question more efficiently of whether a frame is ultimately a subclass of another frame by a series of class-subclass relationships, we use the superclasses function in a recursive search following all the superclass relations at each level. This is less onerous than retrieving all the superclasses, if the subsumption relation is true, since the search stops when the second frame is found. The predicate function, subsumed-by, answers this question.

```
(defun subsumed-by (fr1 fr2)
  (let ((id-list (superclasses fr1)))
    (cond ((null id-list) nil)
          ((find (frame-id fr2) id-list) t)
          (some #'(lambda (x)
                    (subsumed-by (find-frame-by-id x) fr2))
                id-list)))))
```

When the subsumption relation is false, this will take as much time as the simpler method. Of course, it might be that the relation is the opposite of what is first queried, so one must call the function twice, once with the arguments reversed. This code can then be used with frame representations of biomedical knowledge such as the Gene Ontology to answer questions about relationships among Gene Ontology types or subclasses.

For questions having to do with specific types of instances, instances of a particular class, we can get all the instances given the class name. It is provided by the `instances` slot in the frame class. This slot is maintained automatically by the `make-frame` function.

Up to now we have only examined the relation between a class and its superclasses. The `part-of` relation is also important, since it represents, for example in the Gene Ontology, a compositional relation. In cells, a function may consist of a series of subfunctions, or a cellular component may itself have components. In our representation of GO, the `superclasses` and `part-of` relations are structured identically, so a simple modification of the above code can provide functions to traverse the `part-of` hierarchy. The analog of the `superclasses` function is the `part-of` function.

```
(defun part-of (fr)
  (slot-data fr 'part-of))
```

The analog of `subsumed-by` is the `is-part-of` function. It has almost the same code, except for using `part-of` instead of `superclasses`.

```
(defun is-part-of (fr1 fr2)
  (let ((id-list (part-of fr1)))
    (cond ((null id-list) nil)
          ((find (frame-id fr2) id-list) t)
          (some #'(lambda (x)
                    (is-part-of (find-frame-by-id x) fr2))
                id-list)))))
```

This is a very simplistic representation of "part of." In fact, in biology and medicine, the term "part of" has been misused and ill specified quite a bit. There are many kinds of "part of" relationships, some of which are overlapping and some disjoint. An entity may have constituent parts, for example, the atoms that make up a molecule, or other kinds of components (a molecule may have a specified structure, like an amino acid, so that we can refer to the side chain as a part). Providing formal semantics for the varieties of "part of" relations, and representing them in frames and other reasoning systems, is also the subject of much research [8, 96, 345, 367, 385].

The `is-part-of` relation, as vaguely described here, has the property of being *transitive*, like the `subsumed-by` relation. It is easy to generalize this for a possibly richer ontology with other relations like `superclasses` and `part-of` that correspond to a transitive relation that uses them. We just have to pass the function that finds the links as an input parameter. The essential requirement is that the link function represents a relationship of which we can meaningfully compute the transitive closure.

```
(defun connected-upward (fr1 fr2 link-fn)
  (let ((id-list (funcall link-fn fr1)))
```

```
(cond ((null id-list) nil)
      ((find (frame-id fr2) id-list) t)
      ((some #'(lambda (x)
                  (connected-upward
                    (find-frame-by-id x) fr2 link-fn))
               id-list)))))
```

To traverse the `superclasses` hierarchy, use `superclasses`, as in

> `(connected-upward fr1 fr2 #'superclasses)`

To find out if something is a part of something else, just use the `part-of` function instead of the `superclasses` function. It is essentially a search for a path from one frame to another in a directed frame graph defined by the `link-fn` relationship.

Now it is also possible to see that queries about classes can be easily answered as well. Our earlier query, "what kinds of organs are there?," is just a query on the organ class to find out what subclasses are defined.

Implementing these query and search functions, it should now be apparent that one of the big differences between an ontology and a terminology system is that a terminology system only represents a classification hierarchy, and even at that, often not a consistent one. An ontology built with frames associates a rich collection of information with each term (thinking of a term as a frame name), and a rich set of relationships among the frames. One of the characteristics of a well-designed ontology is that the relationships are precise and consistent, to allow for the kind of procedural reasoning illustrated here. Using the same links for different and possibly inconsistent meanings can lead to nonsensical conclusions, making such reasoning impossible. The HL7 Reference Information Model (RIM) has been criticized from this perspective [387,388].

The UMLS Semantic Network serves as a terminology system, but is somewhat richer, including many link types, analogous to the slots and slot values described here. A semantic network provides a more graphical representation of the same information as a frame system. However, casual assignment of links in the UMLS leads one to some rather strange conclusions. Table 1 of [386] provides some provocative examples, such as "Gene or Genome `location_of` Fungus," and "Food `causes` Experimental Model of Disease." These problems do not come from defects or limitations in the frame or semantic net formalisms. They are a consequence of lack of precision in the meaning and use of the terms themselves. The content is too casual. In human natural language discourse, such mismatches would be ignored or discarded, or at least serve to clarify the actual intent of the terms and links that (mistakenly) led to these statements. But, the enterprise is not just about building repositories of dictionary content, it also is intended to enable automated reasoning. So, much work remains to be done on these "first pass" ontologies.

In classifying or organizing experimental data it is often useful to identify and collect together data for a collection of related genes or gene products. In GO terms, this means pick a term in the GO hierarchy and find all its lineal descendants. In effect we are collecting together the frames that consist of a subtree, starting from a given frame, but searching for subclasses instead of superclasses. It is the reverse of the previous operations.

A straightforward but horribly inefficient way to do this is to go linearly through the entire Gene Ontology and test each entry with the chosen entry, using the subsumed-by function defined above.

```
(defun get-all-children (fr frame-kb)
  "searches the entire frame knowledge base to collect all the
   subclass-of descendants of frame fr."
  (let ((children nil))
    (dolist (entry frame-kb children)
      (if (subsumed-by entry fr)
          (push entry children)))))
```

Similarly, to find all the parts and subparts corresponding to a GO term, use the is-part-of function.

```
(defun get-all-parts (fr frame-kb)
  "searches for items in the part-of subtree"
  (let ((parts nil))
    (dolist (entry frame-kb parts)
      (if (is-part-of entry fr)
          (push entry parts)))))
```

These two can also be combined into a single search function parametrized by the link function to use, either subsumed-by or is-part-of. This generalized version might be called get-subtree, since it extracts the subtree below a particular node in the hierarchy defined by the link relationship.

This is horribly inefficient because the organization of the Gene Ontology (and our frame system in general) is unidirectional, that is, an entry has explicit parents, but not explicit children. In object-oriented programming systems this is the way class hierarchies are organized. A class definition always includes a list of its superclasses, but does not always include a list of its subclasses. This works just fine for programming with classes, because in terms of software evolution, one usually adds specializations, but not generalizations. Adding a more specialized class does not change the behavior of any existing code. On the other hand, modifying a superclass may change the behavior of all the subclasses.

For dealing with class hierarchies as representations of biological entities, it is important to be able to do the traversal in terms of descendants in a more efficient way. One way to make the search much faster is to include links in both directions in each entry. Another is to go once through the entire table and generate an inverted hierarchy, that is, compute the links in the opposite direction once and store them separately. Strategies such as these are provided in object-oriented databases, in order to make such searches and retrieval operations reasonably fast.

2.3.2 Extensions to Frame Systems

For efficiency, we could consider redoing the global storage as a hash table or tables, instead of a list. This would drastically speed up the lookup of frames by ID or by name, for very large collections of frames.

What else can be done? A richer representation of slots would allow for type declarations and other slot constraints. We could represent negation, disjunction, or quantification (which all would end up being conditions on slot values). Here we consider only one of these many possible extensions, the implementation of action functions that get invoked by slot access.

Attached procedures can enforce type checking and other kinds of constraints on slot values, and implement other kinds of logical constraints not directly expressible as slot values or ordinary functions that query a frame collection.

2.3.2.1 Attached Procedures

While we will want to be able to answer questions concerning inheritance, and other relations implemented in slots, it will also be useful to provide a way to specify computations that should be performed when slots are accessed. When a slot value is needed, it may be present (because it was provided when the frame was created), or it may be computed at the time it is accessed. It is something like storing a formula in a spreadsheet, but is a more general facility, in that the procedure can take any action, not just compute a value.

An example of an important use of an "if-needed" or "if-added" function would be to record in a log file when the access occurred and by whom. A medical context for such an access logging function would be an electronic medical record system, where it is required by law (for example, by the HIPAA Security Final Rule of 2003 [88]) to keep records of all access to patient data, specifically data that are categorized as "protected health information" (PHI).

Similarly, one might want to add to a frame an action that takes place when a slot value is added or modified. An example would be: when a medication order is put into the electronic medical record, a safety check could execute, which would generate an alert if the new medication could cause a hazardous interaction with other medications or be contraindicated by a patient's condition.

Yet another possibility is to implement forward or backward chaining using procedural knowledge, by storing procedures as *productions* to be run as attached procedures. An "if-needed" procedure could implement a rule that specifies what else is needed in order to produce the requested slot value. Then, looking up those slot values would trigger further search, until all the prerequisites are satisfied, and the value is computed and returned (and perhaps cached as well). That would be a kind of backward chaining. Similarly, as values are stored, this can trigger the computation of other values, implementing a kind of forward chaining. The big difference between this and the more usual procedural programming is that the procedures are stored, retrieved, and executed, and can be changed as execution proceeds. Maintenance of the procedural knowledge is easier since it is modular, contained in local places without much dependency on the rest of the environment.

One possible way to provide this capability is to adopt a convention that if the slot value is a function it should be called instead of returned as the value. This does not allow for a function to be stored as an actual slot value to be returned rather than called. Also, a slot might need both an if-needed and an if-added procedure, but this simple convention does not allow storing both, or distinguishing one from the other.

A better way is to design some more structure into the slot itself. A simple way to do this is to require a value to be present (it could be `nil`) and have the attached function(s) appear after the value, labeled with suitable keywords. In the slot, if an attached function is present, it would appear after a keyword, `:if-needed` or `:if-added`, which follows the slot value. Indeed, there could be more than one such keyword and function. This allows for any combination of a value, an if-needed function and an if-added function. This arrangement has one limitation: a value must be present as a place-holder (`nil` is OK, or one could reserve a special indicator value that means "no value stored").

Since the functions will be called by internal (frame system) code, their inputs need to be uniform and specified. An `:if-needed` function is required to have two inputs, the frame in which it appears, and the name of the slot containing it. An `:if-added` function takes three parameters, of which the third is the value that was provided in the call to (`setf slot-data`).

Thus, we extend the slot structure by including `:if-added` and `:if-needed` keywords that tag functions to be run. They will appear after the slot name and slot value. If there is no slot value, a `nil` must be there as a place-holder. The following illustrates the structure of such a slot, where the `ssn` slot has an `:if-needed` function and the `birth-date` slot has an `:if-added` function.

```
((age 64)
 (occupation professor)
 (ssn nil
   :if-needed (lambda (fr slot) "SSN not available"))
 (birth-date (4 27 1944)
   :if-added (lambda (fr slot val)
               (schedule-party (slot-data fr 'name) val))))
```

In this example, a lambda expression serves as the `if-needed` function, but a symbol naming a function, or even a compiled or interpreted function object would do also.[25]

---

[25] You may notice that the lambda expression does not have a `#'` prefix. This is needed in source code that will be read in by an interpreter or compiler, but here we are just looking at an actual expression that appears in a list, a piece of data. It is more likely in a real system that what will appear instead of the literal lambda list is a function object created from it.

Getting the functions to execute at the proper time can be done in a completely general way by just modifying the accessor functions `slot-data` and `(setf slot-data)` to execute any function that is present and labeled by the appropriate keyword. In this arrangement the slot value itself can be a function, but it will be returned, not executed. This also allows for a slot to have either or both types of attached functions.

```
(defun slot-data (fr slot)
  "returns the contents of slot slot-name in frame fr, after
   executing any :if-needed function that might be present"
  (let* ((the-slot (get-slot fr slot))
         (if-needed-fn (second (member :if-needed the-slot))))
    (if if-needed-fn (funcall if-needed-fn fr slot))
    (second (get-slot fr slot))))

(defun (setf slot-data) (newval fr slot)
  "updates the contents of slot slot-name in frame fr, after
   executing any :if-added function that might be present"
  (let* ((the-slot (get-slot fr slot))
         (if-added-fn (second (member :if-added the-slot))))
    (if if-added-fn (funcall if-added-fn fr slot newval))
    (setf (second (get-slot fr slot)) newval)))
```

In the `slot-data` function, when looking up the slot value after the `:if-needed` function is run, we again call `get-slot` rather than using the value of `the-slot` because the `:if-needed` function might change the slot value. We want the updated value, not the original one. Also, in `(setf slot-data)` the `setf` operator needs the place name for the slot value, not the local variable in the `let` expression.

Of course we need to provide a way to put these functions in slots from a program when needed. The functions for putting functions in slots just look for keywords and update their associated values. If an attached function is already present, the new one replaces it. If there is none yet, the keyword and function are added to the end of the list that makes up the slot. Since the body is the same for all keywords, we can provide one function, parametrized by the keyword.

```
(defun set-attached-fn (fr slot fn key)
  "puts function fn in the slot named slot in frame fr, using key
   as a tag,:if-needed or:if-added or possibly other types of
   attached functions"
  (let* ((the-slot (get-slot fr slot))
         (length (length the-slot))
         (location (position key the-slot)))
    (if location (setf (elt the-slot (+ location 1)) fn)
      (setf (rest (last the-slot)) (list key fn)))))
```

It is also important to provide a way to remove attached functions from a slot as the knowledge base grows and changes. One way to do this is to literally remove the keyword and the function it labels. This is tricky, as we would have to do some complex list surgery. The implementation above supports the much simpler process of simply putting a `nil` value in place of the function. If `nil` is retrieved, in the `slot-data` and `(setf slot-data)` functions above, the call to `funcall` will not be executed, just as if there were no entry at all.

```
(defun remove-attached-fn (fr slot key)
  "replaces any attached function for key to nil"
  (set-attached-fn fr slot nil key))
```

An alternative to representing slots as lists with keyword tags (association lists) is to implement slots themselves as frames, instances of a "slot" class. The Protégé frame system and the Common Lisp Object System implementation both are designed this way.

2.3.2.2 Rules and Inference

Although subsumption and other kinds of graph traversals work well as procedures, it certainly would be advantageous to be able to add rules and assertions and inference procedures to the frame formalism. Most frame systems support this, but there is no standard way to integrate frames and rules. One thing that is moderately easy is to provide a way to assert facts in a knowledge base (KB) when values are put in slots. This is easy to do with the attached function mechanism described here. You would define a function that takes a slot name and a value, and calls the `<-` macro to add a fact to the KB. Such a function could then be added to a slot as an `:if-added` function, so that the KB automatically updated when the slot was updated. Alternately, a program could do this explicitly.

As noted previously, Norvig [305, Chapters 11 and 12] has done a much more complete implementation of the functionality of the Prolog programming language, embedded in Common Lisp. An extension of this is available for Allegro Common Lisp, which includes integration with CLOS in a somewhat different way. More information about this is given in Appendix A.

2.3.3 Frame System Implementations, Interfaces, and APIs

In the previous section the goal was to show how a frame system could be implemented, in the simplest possible incarnation. However, many good frame systems already exist, and are available for use, either by downloading and building a copy of the code or through a web browser front end to an installation maintained by the creators of the system. These systems provide much richer semantics than the simple system described here. Some support integration of first order logic with the frame system in a more direct way than just embedding a mock Prolog implementation. For example, Ontolingua [109] is based on *description logic* formalisms, LOOM [265, 266, 422], and OWL, the Web Ontology Language, which we take up in the next section.

Ontolingua (`http://www.ksl.stanford.edu/software/ontolingua/`) is based on KIF, and is currently only available as a web accessible application, hosted at Stanford (URL above). LOOM, a Description Logic based Knowledge Representation system, and Ontosaurus, a web-based front end for LOOM knowledge bases, are projects of University of Southern California's Information Sciences Institute. Although not very active at this time, documentation may be found at URL `http://www.isi.edu/isd/LOOM/LOOM-HOME.html`. A book by Gómez-Pérez [134] has useful information about these and other knowledge representation systems in use and/or available for use.

2.3.4 The UW Foundational Model of Anatomy

The FMA is a formal model of human anatomy. According to its creators [351–353], the FMA is intended to be a sound, accurate, and consistent formal theory of human anatomy. The FMA had its beginnings in the Digital Anatomist project at the University of Washington, a project to create a large organized collection of anatomical images and textual information to enhance the teaching of anatomy to medical students. The idea was to link a hypertext (originally implemented with Macintosh Hypercard) and an image repository. The images included annotations, labeling, and overlays of contours or other drawings delineating objects in the images. Figure 1.6 shows an example of an annotated image, an anterior view of the heart, from the Digital Anatomist Thorax atlas.

Later, the facilities became network accessible, and many offshoot projects developed. The Digital Anatomist terminology system was a semantic network that expressed many different kinds of relationships among the concepts named by the terms. This system gained an identity independent of the Digital Anatomist image atlases as an effort to enhance the anatomical content of the UMLS [353,349]. The FMA is maintained by the University of Washington Structural Informatics Group, and is one of many projects applying knowledge representation ideas to structural biomedical information. URL `http://sig.biostr.washington.edu/` is the home page of the University of Washington Structural Informatics Group. At this site there are links to the Foundational Model of Anatomy for download, to a web browser application accessing the FMA, and to other project related to structural informatics.

2.3.4.1 The Components of the FMA

The FMA represents four distinct kinds of knowledge in relation to anatomy. As is typical of most ontological modeling efforts, the FMA includes a *taxonomy* of anatomical entities. This is implemented as a class hierarchy, the Anatomical Taxonomy, or **AT**. Second, anatomy is not simply about classification of entity types, but about relationships between entities. In the case of anatomy, these are *structural* relationships, so the second component of the FMA is a large collection of structural relationships. These structural relationships express such ideas as containment, constituent parts, connectivity, etc. The FMA describes *canonical* anatomy, the relationships that are generally true of human anatomic structure, rather than a particular set of attributes of a particular human being. This part of the FMA is called the Anatomical Structural

Abstraction, or **ASA**. Together with the **AT**, these relationships provide the basis for anatomical reasoning in support of applications.

The FMA also is designed to accommodate things that change with time, in order to describe embryological development. This is the motivation for the modeling of anatomical transformation. This part of the FMA, called the Anatomical Transformation Abstraction, or **ATA**, is much more a work in progress than the **AT** and the **ASA**.

A fourth component of the FMA is the metaknowledge that consists of the rules and principles that the other components are required to follow. These are anatomical *axioms*.

The FMA is realized as a collection of frames (see Chapter 2, Section 2.3). It is built and maintained using the Protégé ontology editor [290, 306]. The FMA can also be translated into a description logic language [25]. It is therefore a computable representation, so we can write programs to search for frames, and trace the various relationships, such as `superclasses` and `part-of`, as we did for the basic frame system in Chapter 2. This is the main idea we will address in this section.

2.3.4.2 Representing Anatomical Relations in the FMA

The FMA includes entries for every element of human anatomy from the body, progressing in levels of detail down to the cellular and subcellular levels. Most important, the FMA represents *relations* between entities, not only in terms of a `superclasses`, or subsumption, hierarchy (class - subclass relationships), but also other relationships such as composition (various `part-of` relations), spatial relations and connectivity (for the blood vessels and the lymphatic systems, upstream and downstream connectivity).[26] In addition to the general relationships already mentioned, there are specialized relationships that apply only to certain subclasses of anatomical entities. The upstream and downstream relations hold between entities of the same type, but there are also relations between different types. For example, *arterial supply*, *venous drainage*, and *lymphatic drainage* are relationships between types of vessels and types of organs. In all, there are nearly 200 relationship types in the FMA.

The FMA is organized as a class hierarchy, as seen in the Protégé screen capture in Figure 2.10. On the left one can see a part of the class-subclass hierarchy, showing that the class *Lung* is a subclass of *Lobular organ*, and continuing up the class hierarchy, the remaining superclasses are, in order, Parenchymatous organ, Solid organ, Organ, Anatomical structure, etc., until finally we reach the overall general Protégé class, `:THING`. The `:THING` class serves as a template for all Protégé classes and instances. It corresponds to TOP in a typical description logic formalism.

Because the FMA is intended to be an artifact embodying a theory of anatomy, it is designed to conform to some key principles. One of these is that the type hierarchy as well as the other relationships must accurately represent the actual experimentally verifiable structure of the human body and its parts. Another is to distinguish, consistent with more general ontology

---

[26]A traditional example of the `connected-to` relation applied to the skeletal system may be found at http://kids.niehs.nih.gov/games/songs/childrens/bonesmp3.htm. This song derives from Chapter 37 in the book of Ezekiel, from the Bible.

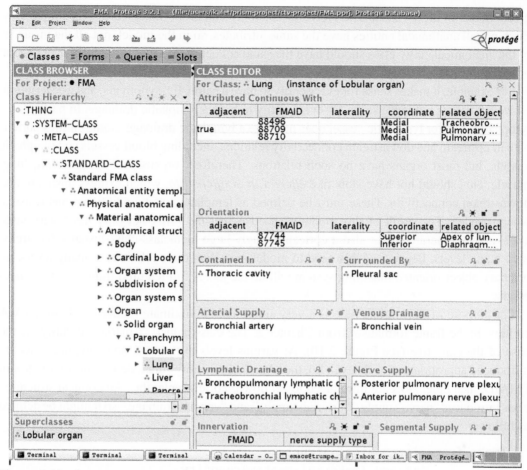

Figure 2.10 **A screen capture of a part of the FMA class hierarchy, displayed in the Protégé class browser window.**

projects such as the Basic Formal Ontology (BFO) [138], between continuants, things that exist at particular moments in time, are bounded in space, and continue to exist indefinitely, and occurrents, which may not be localised in space but are bounded in time. The latter are the things we call "processes," "events," or "changes." The references cited at the beginning of this section provide a formal description of the organizing ideas of the FMA.

2.3.4.3 Metaclasses in the FMA

In the FMA, every anatomical entity is modeled as a *class*, because the intent is to model canonical anatomy. Building class hierarchies is commonplace in the biomedical ontologies world. What is not commonplace is the rich model of relationships in the FMA. These are among the many pieces of information that are associated with the FMA classes. A simple model would be to define a metaclass, anatomical entity, with template slots that would be given values in all the instances. However, anatomy cannot be modeled in this simple way. First, not all anatomical entities have the same attributes, so in the model, not all the instances of the anatomical entity class should have the same collection of *slots*, aside from the question of variations in their values.

For example, it makes sense for every entity to have a name, ID, synonyms, and a description. However, it does not make sense for every entity to have a lymphatic drainage. In particular, a lymphatic chain or lymphatic vessel does not have a lymphatic drainage. Some entities partici-pate in upstream and downstream connectivity relations, including blood vessels and lymphatic vessels, but most organs have no such relations. Therefore, organs such as the heart, lung, muscle, etc., should not have slots like *efferent to* or *afferent to*, which express upstream and downstream connectivity. These must be defined as template slots by additional metaclasses, where needed. The result of this modeling need is that the FMA not only has a very large class-subclass hierarchy, but also a correspondingly large metaclass-class-instance hierarchy, with many levels. Because of this need to model metaclass-class relations at many levels, no ordinary object-oriented modeling system will do the job. A full featured frame system is nec-essary.

For purpose of illustration, here is a very simplified approximate rendering of some FMA entities, in the frame formalism from Chapter 2. The class named "Anatomical entity" is the base of the class tree (see Figure 2.10). At various levels in the tree, new template slots are introduced appropriate to each level. So, the "Anatomical entity" class defines the template slot FMAID, then other slots are introduced lower down. For example, the "Anatomical structure" class introduces "nerve-supply," "arterial-supply," and "venous drainage," so it might be defined as follows:

```
(make-frame 'anatomical-structure
            :class 'material-anatomic-entity
            :superclasses '(material-anatomic-entity)
            :template-slots '(arterial-supply venous-drainage
                              nerve-supply ...)
            :preferred-name "Anatomical structure"
            ...other own slot inits)
```

Similarly, the class, "Lymphatic chain," being a subclass (some levels removed), will inherit these template slots, and add a few more.

```
(make-frame 'lymphatic-chain
            :class 'segment-of-lymphatic-tree-organ
            :superclasses '(segment-of-lymphatic-tree-organ)
            :template-slots '(afferent-to efferent-to ...)
            :FMAID 62865
            :preferred-name "Lymphatic chain"
            ...)
```

Then, all the subcategories of "Lymphatic chain" will also have `afferent-to` and `efferent-to` slots. These are the upstream and downstream relationships, respectively. If chain B appears in the `efferent-to` slot of chain A, that means that chain B is downstream of chain A, and lymphatic fluid flows (usually) from A to B.

Now you might have noticed something shady about the above definitions. I have specified that the `anatomic-structure` frame is both an instance *and* a subclass of `material-anatomic-entity`, and similarly, `lymphatic-chain` is both an instance and a subclass of `segment-of-lymphatic-tree-organ`. This is not forbidden by the frame system, but if you think of class frames as defining mathematical sets, and instance frames as members of those sets, it would be a violation of an axiom of set theory. A thing cannot be both a member of a set and a subset of the same set.[27]

However, this construction is used throughout the FMA because it provisionally solves a very difficult modeling problem. The frames of the FMA are not in fact descriptions of sets. They are formal descriptions of entities in the (human) body. The usual object-oriented model describes canonical objects as classes, with concrete objects being instances. So, as mentioned in Chapter 2, the `heart` class describes the heart as an abstraction, and Joe's heart then becomes an instance of this abstraction. The template slots of the heart class are there to specify what properties instances have, and which may vary from instance to instance. However, we *also* want to model the properties of the heart in general. In an ordinary object-oriented programming environment, these properties are realized as separate data structures, or as procedures. In our frame system, however, we can give a class its own properties by making it an instance of a metaclass. So, the metaclass idea is essential here.

The `heart` class will have slots that describe the kinds of parts a heart has, and other things about hearts. The `heart` class has own slots in which to store these values. These come from the fact that the `heart` class is an instance of the metaclass that provides the slots as template slots. This would not be a problem, except for one thing. As we build the class-subclass hierarchy describing anatomy, we would like to introduce additional properties at different points. In the above example, there are slots in the subclasses that are introduced there. Fine and well, but what I showed are *template slots*. Each frame at every level needs *values* and they are associated with *own* slots, because the values are particular to each frame, not to all frames that have these slots.

One way to have the necessary slots appear as own slots in the classes in the class hierarchy is to make all these classes be instances of the appropriate metaclass. In CLOS, incidentally, this is not the default. When you make a class that is a subclass, the new class does not automatically have the same metaclass as its superclass. But, one could build such a default into a frame system if desired (and possibly even extend CLOS to do this). Even with this, however, to build a complete model will require many thousands of metaclasses, because of the rich structure that needs to be represented. An article by Natalya Noy [307] provides a deeper discussion of this

---

[27] Such things lead to the famous Russell's paradox, about which there is a nice article in Wikipedia.

and other challenging modeling issues in the FMA, as well as how maintaining parallel subclass and instance relations between two classes provisionally solves this problem.

Another complication of the FMA is that relationships are modeled as entities also. The values that appear in slots that link to other entities (like *branch* or *tributary*) could be names of the related entities, but then all the possible kinds of relations would need to be separately identified and defined as slot names. This is unmanageable for large ontologies. Instead, the FMA models relationships as values for slots, so that a generic slot can hold multiple kinds of variants of a generic relationship. The most poignant example is the `part-of` relationship. This and the metaclass modeling problem that is related to it are also described in more detail in [307].

2.3.4.4 The `part-of` Relationship

In some ontologies that describe structure, such as GO, there is a single relation, `part-of`, which is used generically to designate an entity as a component of another entity, in some sense not precisely specified. In the FMA, `part-of` relations are not generic. Rather, there are multiple ways in which anatomical entities are composed of parts. One kind of relation is the spatial subdivision. Some organs are extended in space, and traverse the vicinity of a number of other organs. A much used example is the esophagus. Other examples are the large intestine and the spinal cord. The large intestine is made up of *regional* parts, the cecum, ascending colon, transverse colon, descending colon, etc. Each of these is connected to and contiguous with the next, but there is no distinctive anatomical boundary between these parts. Rather, they are defined by designated or chosen boundaries rather like the boundaries of cities, states, etc. On the other hand, the large intestine, like many cavitated organs, has a wall and a lumen (the interior space), which are its *constitutional* parts. They are separate parts by virtue of having very different characteristics. The subdivisions that are regional parts will also have the same kinds of constitutional parts as their parents, since they are constructed the same way and are regional parts by virtue of defining an arbitrary spatial boundary.[28]

The logic of parthood and its representation has been the subject of much theoretical work [8,96,345,367,385]. This work is important because one cannot build automated reasoning programs unless the semantics of these relationships is clear and consistent. For example, it may be that we need to know which arteries are the arterial blood supply to some region of the large intestine. If we query the system just for parts, we will get a mix including the constitutional parts, for which it is not sensible to define an arterial supply. However, it does make sense to define the arterial supply separately for each of the different regional parts of the large intestine, because they are in fact distinguishable. There is no single blood vessel that supplies the entire organ.

Similar to parthood, other relations in the FMA have been defined with some semantic precision, which ideally will support automated inference. In particular, the connectivity relations involving blood vessels and lymphatics can be traced with some clinical utility. It is possible

---

[28]This is a real application of the old saying, "No matter how you slice it, it is still baloney." Nevertheless, the skin of the baloney is separate from the interior.

to define constraints (axioms) that these relationships must satisfy. The FMA could then be automatically checked to see if all the entries do indeed satisfy the constraints. In order to do this, we need to have a way to query the FMA from a computer program. This will be taken up in Chapter 4.

2.3.5 Representing Drug Information in Frames

The class scheme described for drug information in Chapter 1 works fine if you just want a big catalog of drugs, but some information pertains to whole classes of drugs, and it seems useful to create an entire class hierarchy. We already saw earlier that this calls for something more than an ordinary object-oriented programming language. We would like a class hierarchy, *and* we would like to store information with each class and subclass, not only with instances. Thus, a frame system such as that described in Section 2.3 should provide a better substrate for drug information. Building on the examples later in that section, we can define a more elaborate drug-class metaclass.

```
(make-frame 'drug-class
            :template-slots '(indications contraindications))
```

We include here only template slots for properties that we think will be available for drug *classes* rather than individual drugs. For individual drugs, we will need a lot more information, so the individual drugs will be instances of the drug classes, but the drug classes will all be subclasses of a generic class, drug, as before, which has the template slots needed as own slots by all the individual drugs.

```
(make-frame 'drug
            :template-slots
            '(indications contraindications dosage form
               side-effects substrate-of inhibits induces ...))
```

Then all the particular drug classes are subclasses of this one, and therefore their instances (individual drugs) all will have the own slots specified as template slots above.

```
(make-frame 'cns-agents
            :class 'drug-class
            :superclasses '(drug))

(make-frame 'narcotics :class 'drug-class
            :superclasses '(cns-agents)
            :indications '(analgesia anesthesia sedation)
            :contraindications '(bph constipation))

(make-frame 'sedatives :class 'drug-class
            :superclasses '(cns-agents)
            :indications '(anxiety panic seizures)
            :contraindications '(copd))
```

```
(make-frame 'barbiturates :class 'drug-class
            :superclasses '(sedatives)
            :indications '(intracranial-pressure analgesia)
            :contraindications nil)
;;; etc.
```

Individual drugs or agents may have additional values for attributes that are not true of the entire class. Here for your amusement we provide an entry for a relatively obscure drug class and a very popular drug in that class [Chapter 27, 148].

```
(make-frame 'heavy-metals :class 'drug-class
            :superclasses '(drug)
            ...)

(make-frame 'gold :class 'heavy-metals
            :indications '(hand-pruritis)
            ...)
```

Now it is possible to generalize the `disease-lookup` function from Section 2.3.1 to handle recursive collection of data for any consistently named slot in this two way hierarchy. We just provide the slot name as an input and use it instead of an explicit name.

```
(defun recursive-lookup (drug-inst slotname)
  (let ((drug-type (find-frame-by-id (slot-data drug-inst
                                                 'instance-of))))
     (append (slot-data drug-inst slotname)
             (slot-data drug-type slotname)
             (mapcar #'(lambda (x) (slot-data
                                     (find-frame-by-id x)
                                     slotname))
                     (all-superclasses drug-type)))))
```

If you are only interested in storing and retrieving drug interaction records, that is, creating an online version of a reference like [146], it is even simpler. A simple (sparse) table can represent a collection of assertions that drug A interacts with drug B. The tables on which existing systems rely (such as [146]) enumerate known, documented pairwise interactions. Such a table is difficult to maintain and voluminous, particularly when new drugs are added. Moreover, none of the development here deals at all with the problems of consistent nomenclature, dosage forms, or routes of administration.

A table of drug pairs and information about their interactions is easy to construct. Simple list structure and symbols as tags will work. No arcane programming language syntax is required. As we have seen in so many examples earlier in the book, such structures form data that can be processed directly by a Lisp function. An entry from Hansten and Horn [146] could be represented in essence as follows:

```
(carbamazepine (aka (carbatrol tegretol))
               (interacts-with fluvoxamine)
               (abstract "...lots of text")
               )
```

This only captures the essential information, but is enough to be able to search a collection of such records. Each record for drug 1 is retrieved to see if drug 2 is in its `interacts-with` field. There may be multiple records for each drug, so that in order to ascertain if two drugs interact, many records may need to be examined. Another approach is to have a single entry for each drug; the `interacts-with` field then would contain a list of all drugs that interact with the drug.

A realistic system augments the above representation with many more fields than alternate names and interactions. Typical systems include textual information similar to that available from, for example, MicroMedex [233], or other compendia on drugs. Thus we could write:

```
(carbamazepine (aka (carbatrol tegretol))
               (classes analgesic
                        anticonvulsant
                        ...)
               (dosage-forms
                 (adult "...lots of text")
                 (pediatric "...different text"))
               (contraindications "...")
               (precautions "...")
               (interactions "...")
               )
```

So far, the problem is mainly one of information organization and retrieval. There are many solutions, among them a web interface to a database. Thus, such a system would emphasize information lookup, presentation, and interpretation by the person using it. The main programming job here would be that of indexing and search for appropriate records, followed by formatting the results.

Much effort has been put into drug ontologies. These are motivated by some of the same needs as the other medical terminologies, to establish a consistent and systematic nomenclature that everyone would use, so that there is no misunderstanding, and to enable classification and categorization of information. However, a cautionary note is in order. One must be very careful about reasoning based on drug class hierarchies. The assignment even of the kind of slots that exist at various levels of a drug class hierarchy is problematic, let alone the actual values. It is *not* true, for example, that all members of a given therapeutic class of drugs are metabolized the same way, so no assertion about metabolism can be put at that level. For example, some fluoroquinolones are known to be CYP1A2 inhibitors (ciprofloxacin and enoxacin), but others are known *not* to inhibit CYP1A2 (levofloxacin, lomefloxacin, and ofloxacin) [144].

Hansten [144] also mentions several other cautions concerning common pitfalls in clinical decision making. Here are just a few:

- relying excessively on personal clinical experience, which does not generalize or transfer from one case to another or even from many cases to a new one,
- failure to consider the effects of dose,
- failure to appreciate the time course of drug interactions,
- failure to consider the effects of sequence of administration,
- failure to realize the danger of stopping the precipitant drug,
- failure to consider the pharmacogenetics of the patient.

The last caution is now the subject of a large project, the Pharmacogenetic Research Network and Knowledge Base consortium (PharmGKB) [310, 356], which will accumulate and organize emerging knowledge about genetic variation, functional variation, and its relation to clinical drug responses. The PharmGKB project is primarily aimed at supporting research, but the knowledge base will have tremendous clinical use in the future. It is not hard to see how the representations above can be expanded to include genetic variations and rules or functions to compute the implications. The project is using Protégé [290] to build the knowledge base, but is also using a relational schema to store the same information. An XML schema provides a way to import data from other sources. Protégé provides an application programming interface to ontologies created and maintained with it.

2.3.6 Frames and Object-Oriented Programming Languages

So far, we have looked at how to organize information in terms of class and instance structures and how to automate the management of some of these relationships. However, by now you should have an uneasy feeling that you have seen this somewhere else. Indeed, most of the ideas here can be found in the implementation of some (so-called) object-oriented programming languages. This should not be surprising since the two areas of computer science originated around the same time, and certainly have explicit connections. The idea of structuring a computer program around descriptions of real world entities and their updates as computation progressed originated with the SIMULA programming language [81] in 1966. Some time later Marvin Minsky proposed the frame idea as a way to more easily express knowledge in a form that supported logical inference [286]. Many divergent programming language developments followed.

The Smalltalk programming language implemented the idea that every entity in the programming environment was an instance of a class, even numbers like 7. It recast the idea of operators and function call as *message passing*. So, each class of object has a message interface, and a method for performing the operation requested by the message. The code implementing the method is what we ordinarily think of as a function, but it is associated with the type of (one of) its arguments. There can be other methods of the same name associated with other types or classes. Thus methods became closely associated with classes. The most remarkable thing about

Smalltalk and probably its biggest impact, however, was that it was implemented with a bitmap display and multiwindow graphics environment. The Smalltalk user interface style caught on and eventually became the standard user interface style for all of modern day computing. One can hardly point to a desktop working environment today that does not provide this.

The C++ programming language was built on the framework of the C programming language, and essentially added to C facilities for defining classes, generic functions, and basic operator overloading (extending ordinary operators so they work like generic functions). However, it is a very mixed environment, where some things work as in C, with variables, constants, allocation of storage according to the data type (indeed, associating data types with variable names).

The Common Lisp Object System (CLOS) has some similarities to our little frame system, but of course has much richer facilities. CLOS provides facilities for defining classes, constructing instances, implementing methods for generic functions, and attached functions for slots. Most significantly, it provides a Meta-Object Protocol (MOP) [229], so that the properties of the ordinary objects of a computer program can be determined by the program itself. This is made possible by the fact that in CLOS, classes, generic functions, methods, slot definitions, and other things are not just semantic structures in a program, to be processed by a compiler. They are themselves entities in the program, and can be manipulated by the program. They are meta-objects, instances of metaclasses, which are themselves classes. This means that one can even write a program that extends the system by defining specialized metaclasses, and then creating classes that are instances of these new metaclasses. The main difference between a completely general frame system and CLOS is that in CLOS there is only one level of meta-object abstraction, while in a general frame system there can be arbitrary levels, that is, you can create a model that has classes, meta-classes, meta-meta-classes, and so on. Similarly, instances of classes can themselves be classes, and the instances of those can also be classes, etc.

When do biomedical knowledge-based systems need metaobjects? When is the focus on classes and when on instances? An extreme example is the Foundational Model of Anatomy (FMA), a frame system representing knowledge of human anatomical entities and relationships. In the FMA, being able to describe classes and assign values to "own" slots in those classes is essential. So, classes in the FMA need to be instances of some (possibly more than one) meta-class. The FMA also requires support for an inheritance hierarchy, since much of anatomy can be described in terms of general categories of entities like "organ" and specializations, particular organ types. Relationships in the FMA also have their own properties and are implemented as classes and instances of metaclasses. It would be extremely difficult to implement the FMA as a collection of class definitions in an ordinary object-oriented programming language.

Another example of a large biological knowledge base is EcoCyc [215]. EcoCyc contains information about biological molecules involved in metabolism, biochemical reactions, and pathways composed as sequences of reactions. It is implemented using Ocelot, a frame system descended from the Generic Frame Protocol [219]. Ocelot resembles more closely an object-oriented programming language like CLOS, but provides persistent storage as would an object-oriented database system. In EcoCyc, no use is made of metaobjects, but the object-oriented database capabilities are essential to building a useful system.

It is possible to create (and even somewhat automate) a mapping from a frame-based model of data and knowledge to an implementation as a relational database schema. Maya Hao Li has investigated this in the context of providing support for building and maintaining customized laboratory data management systems [252,253].

2.3.7 Example: The Gene Ontology

The Gene Ontology project [18], according to its creators, "seeks to provide a set of structured vocabularies...that can be used to describe gene products in any organism." An appropriate representation can be a powerful tool for analysis of experiments, and for other applications. To do this, it needs to be put in a form with which inferences and classification reasoning can be done. The purposes proposed by the authors of the Gene Ontology are pragmatic: to facilitate communication between people and organizations, and to improve interoperability between systems. They specifically rejected the idea that GO would provide a formal theory of gene products at the cellular level. As a result, GO developed without a strict concern for logical soundness or consistency. More recently, there is a growing recognition in the bioinformatics community that the information in GO is being used as a collection of elements of a formal theory of cell structure and function, and that the issues of consistency and soundness are important after all.

The Gene Ontology is a collection of information on gene products (proteins). It is organized into three hierarchical structures, which may (loosely) be called an ontology. The three subdivisions concern molecular function, cellular location, and biological process. Known proteins can be assigned a GO ID, associating each with an entry in GO. The GO hierarchy and the annotations within each related entry can then be used to extract useful information about that protein. This is more than just an annotation that might be added to the protein's record. It provides multiple levels of classification.

So, properly organized, GO can be used to answer questions like.

```
> (is? "nadh dehydrogenase" a "transporter")
T
> (is? "nadh dehydrogenase" an "atpase")
NIL
```

Why is this important? An example application is in the annotation of proteins for function [57]. We have a new protein, and we find in GENBANK some entries that are near or good matches, though not identical. We might infer that the new protein's function is the same as one of the near matches. If all the matched proteins have Gene Ontology IDs, we can look up their functions. However, if they are at different levels of the hierarchy with one above the other (one subsumes the other), we would use the more specific one as the annotation of the new protein.

In order to do such queries, the textual information in GO must be converted to a symbolic form. Each GO entry contains an identifier (unique within the GO knowledge base), and several pieces of information identified by tags. It is a good fit to the frame idea described in Chapter 2, Section 2.3.1. Here we use a simple rendition, more like the first try at a frame representation in Section 2.3.1 than the later full system. In this representation, as an example, the GO entry for "nadh dehydrogenase" then looks like this:

```
(0003954
    (NAME "NADH dehydrogenase")
    (SYNONYM "cytochrome c reductase")
    (DEFINITION "Catalysis of the reaction: NADH + H+ + acceptor
                 = NAD+ + reduced acceptor. definition_")
    (isa 0015933)
    (isa 0016659))
```

The entry contains links to the classes of GO concepts of which it is a specialization, identifed by GO identifiers 0015933 and 0016659. These in turn have isa links. The isa link here corresponds to the idea of superclass in the full frame system. The number at the beginning of the frame is the GO ID. Entry 0015933 is

```
(0015933
    (NAME "flavin-containing electron transporter")
    (DEFINITION "An oxidoreductase which contains either
       flavin-adenine dinucleotide or flavin mononucleotide
       as a prosthetic group, utilizes either NADH or NADPH
       and transfers electrons to other electron transfer
       proteins. definition_")
    (isa 0005489))
```

We can continue to follow the isa links until we find

```
(0005215
    (NAME "transporter")
    (DEFINITION "Enables the directed movement of substances
       (such as macromolecules, small molecules, ions) into,
       out of, or within a cell. definition_")
    (isa 0003674))
```

and thus the answer to the query is "yes."

A small program to do this would look up the first name in the GO list, extract the GO IDs of its "parent" classes, labeled by the ISA tags, and recursively follow each of those, until it either found the second name in an entry, or reached the top of the GO hierarchy without success. In the first case, the search would be successful and return t and in the second case, nil. This is exactly the subsumption calculation that is described in Section 2.3.1.

So, we need a way to search the GO entries until we find a match to the name we are starting with. We start with all the GO entries arranged in a big list with each entry structured as a frame, as in the above examples. This list is the value of a variable named *frame-kb* as in Section 2.3.1. However, the subsumed-by function uses the frames themselves, and it would be more convenient to look up frames by name. Looking up entries in GO by name is just a linear search with string matching.

```
(defun find-frame-by-name (name)
  (find name *frame-kb* :key #'name :test #'string-equal))
```

Note that the name function here is the one that is defined as a slot accessor in the frame system of Section 2.3.1.

Then we can query the Gene Ontology frame list (or any similar frame KB) with.

```
(defmacro is? (name1 connector name2)
  `(subsumed-by (find-frame-by-name ,name1)
                (find-frame-by-name ,name2)))
```

The above code is inefficient, but it does logically what we need.[29] Moreover, it is more efficient than pattern matching on strings, which is more typical of how one would approach this problem if programming in Perl. Symbol manipulation, lookup, and comparision are very fast, essentially implemented by single machine instructions. An example of how to do this in yet another frame system has been made available by Jeff Shrager [381].

The subsumed-by calculation is so important and central to bioinformatics, that in most cases, the knowledge base of interest is re-expressed using a description logic formalism such as described in Section 2.4. This has been done with GO. Here is a small excerpt of the OWL representation of GO, showing just a few terms.

```
< owl:Class xmlns:owl=" http://www.w3.org/2002/07/owl#" rdf:ID="GO_0000011">
  < rdfs:label > vacuole inheritance</rdfs:label>
<!-- vacuole organization and biogenesis -->
  < rdfs:subClassOf rdf:resource="#GO_0007033"/>
<!-- organelle inheritance -->
  < rdfs:subClassOf rdf:resource="#GO_0048308"/>
 </owl:Class>
 < owl:Class xmlns:owl=" http://www.w3.org/2002/07/owl#" rdf:ID="GO_0000012">
  < rdfs:label > single strand break repair</rdfs:label>
<!-- DNA repair -->
  < rdfs:subClassOf rdf:resource="#GO_0006281"/>
 </owl:Class>
 < owl:Class xmlns:owl=" http://www.w3.org/2002/07/owl#" rdf:ID="GO_0000014">
  < rdfs:label > single-stranded DNA specific
                    endodeoxyribonuclease activity</rdfs:label>
  <!-- endodeoxyribonuclease activity -->
  < rdfs:subClassOf rdf:resource="#GO_0004520"/>
```

---

[29]The second variable input to is?, the connector variable, is ignored, and is just there to allow typing the word "a" or "an" to make it look impressive when using the function.

```
    </owl:Class>
    <owl:Class xmlns:owl=" http://www.w3.org/2002/07/owl#" rdf:ID="GO_0000015">
      <rdfs:label > phosphopyruvate hydratase complex</rdfs:label>
  <!-- protein complex -->
    < rdfs:subClassOf rdf:resource="#GO_0043234"/>
    < rdfs:subClassOf>
      < owl:Restriction>
        < owl:onProperty>
          < owl:ObjectProperty rdf:about="#part_of"/>
        </owl:onProperty>
      < owl:someValuesFrom rdf:resource="#GO_0005829"/>
      </owl:Restriction>
  <!-- cytosol -->
    </rdfs:subClassOf>
    </owl:Class>
```

The first class (GO ID 0000011) illustrates multiple inheritance, as it is explicitly a subclass of two other GO classes. The next two each have only one parent class. Finally, in this excerpt, GO ID 0000015 is a class that has a single explicit parent class but also has a parent class defined by a restriction on the "part of" property. In essence this subclass clause says that GO ID 0000015 is a subclass of the class consisting of all things whose "part of" property has the value "GO ID 0005829." The verbosity here is not a characteristic of description logic formalism; it comes from the syntax of XML and OWL.

As mentioned earlier, obtaining entire subtrees is important in order to classify and organize data for which the data elements have GO ID tags. An interesting example of an application is the analysis of microarray data based on grouping of genes by function as they appear in a hierarchy in GO [382].

The Foundational Model of Anatomy (FMA) [351,352] has extensive coverage of cellular and subcellular structure. It seems reasonable that the FMA and GO might have considerable overlap, and it would be valuable to generate a reliable mapping between them. This was investigated [20] and has helped to identify directions for refinement of all the related ontologies.

2.4 Description Logics

> *If it looks like a duck, walks like a duck, and quacks like a duck,*
> *it's probably a duck.*
> **– attributed to Senator Joseph McCarthy**
> **1952**

While frame systems provide a lot of expressivity for biomedical knowledge, they also have some challenges. As we have seen, some relations are not easily expressible in frames, and require special custom procedures. Frames and object-oriented programming languages are well designed for constructing objects (instances) and manipulating them. However, much biological and medical knowledge is about the categories themselves and their relations. This is the realm of metaclasses in frame systems. We want to reason about the slots a frame contains, as well as the values in the slots. Alternately the slots we are interested in are "own" slots of class frames. A more direct way of expressing these things would be desirable.

In frame systems and in object-oriented programming, it is necessary to explicitly specify the relationships that make up various hierarchies, such as the hierarchy specified by the contents of the `superclasses` slot. However, an alternate formulation is possible, in which we provide *descriptions* of the various entities in the system, in a way that allows for *computation* of the subsumption relationships from the descriptions. Such systems are called *description logics*. They include formal semantics like FOL (and unlike frame representations where we created ad hoc procedures for computing important results from a collection of frames), but the expressiveness of description logics in general is less than that of FOL. In exchange for the more restricted expressiveness, the benefit is that inference is significantly more tractable. Choosing exactly which kinds of capabilities to include is a matter of balancing this trade-off, including enough to be able to faithfully model biomedical knowledge without making computational reasoning too costly.

The basic idea of a description logic (abbreviated DL) is that entities (individuals and classes) are defined by their *descriptions*, that is, what properties they have. A description is a formula (like a formula in FOL) made up of terms and connectors or operators. There are many description logics, each with their own sets of connectors from which to build descriptions. The entities represented by the terms in a DL formula are *Concepts*, *Roles*, and *Individuals*. They are related by connectors and qualifiers similar to those in FOL, but the collection of connectors may vary from one DL to another. Some DL implementations even leave out individuals and represent only concepts and roles (FaCT, described later in this section). The concepts and roles are kept together (both abstractly and in typical implementations) in a container called the *TBox*. The TBox consists of all the *terminological* descriptions. DL systems that support descriptions about individuals have another container called the *ABox*. The ABox consists of all these *assertional* descriptions, so called because they assert facts about individuals.

A *primitive* concept is one for which there is no composite definition. It has a name, that can be used in other concept descriptions. A defined concept is one that is described in terms of others. A description can define an anonymous concept, and can be used in other descriptions, to make up complex formulas.

A *role* specifies a property, or relation between (or among) concepts. Essentially it describes all those things that satisfy the relation. An example would be `gram-stain`, referring to the bacterial colorization process mentioned earlier. The expression

```
(and bacterium (all gram-stain purple))
```

would describe the gram-positive bacteria. The description

```
(all lab-result normal)
```

would describe the class of individuals each of whose laboratory (blood) tests are all normal. Here `lab-result` is a role, a relation between an individual and a value, the result of a blood test (in this case there are only two values, `normal` and `abnormal`). The value itself names a concept, that is, it refers to a class. Someone who had some normal tests and some abnormal tests would not satisfy the description above. In functional terms, the function, `lab-result`, takes two inputs, a person and a test-name and maps that pair to a value. More often, relations are binary, but it is not required.

Individuals correspond to instance frames or instances in an object-oriented programming environment. These formal entities are abstractions representing individual things in the real world. A concept names or describes a set of individuals, so it corresponds to a class definition in a frame system. So, the author of this book is an individual, described by the phrase "the author of this book," while the phrase "book authors" describes a concept, or class, referring to a set of individuals. The phrase "book authors" describes a subclass of the more general class named "authors." This can be written as a symbolic expression, assuming the term `author` is a primitive concept (it has no composite definition) and the term `work` also is a primitive role, that is, it relates a person and something that person created. Similarly, the term `book` refers to a primitive concept. As used here, the role `work` relates the concept `book` to a person.

```
(and author (at-least 1 work book))
```

This expression describes all individuals who satisfy the concept `author` and the concept of an individual whose work includes at least one book. Allowing for individuals, and referring to this book as `pbi`, we can refer to the author of this book as the individual satisfying the following formula.

```
(and author (at-least 1 work pbi))
```

Since an individual book is also a concept, the role `work` can apply to it as well as books in general. The expression

```
(at-least 1 work pbi)
```

describes anyone who wrote `pbi`. This description is satisfied by only one individual. In the case of a multi-author book (instead of `pbi`) this kind of formula would describe the set of all the authors of that book.

You might think that we should represent somehow that if a person has a book as a work, they are an author, but in fact, printers and editors produce books too. We could use a different term than `work` here, to mean just the writing part. However, editors do writing too, for the books they edit. Is an editor an author?

Class-subclass relationships between the entities can logically follow from an examination of these descriptions. For example, in the above, one can infer that a book author is an author, because the concept `author` is included as a required concept in describing a book author. Similarly, much more complex relationships can be expressed and inferred.

You might think the inference above is derivable from the name "book author." In biology and medicine, naming conventions often go to extremes to make such analysis sound, but it is doomed to fail. Names constructed in this way will be too long for ordinary use. It is an intent of the design of formal coding systems or terminology systems, to encode precise meaning in the formal construction of a name. However, relying on the more informal natural language of biochemistry and medicine leads to very difficult and unreliable computations, involving the same challenges found in other applications of natural language processing (NLP). So, since the intent is to have a computable rendition of the things the terms stand for, with precise semantics, we may as well just get on with it, and build a real formal system, not something resembling human language. A description logic is a formalism for doing this in a way that can easily be manipulated by a computer program.

Several groups have created class hierarchies describing molecules that are of biological importance, such as sugars, fats, amino acids, nucleic acids (components of DNA and RNA). In a frame language, we would define class frames for each and slots to hold their properties. So, a molecule class might have properties like `has-part` whose values would be the elements making up that molecule (its basic chemical formula), `molecular-weight`, `size`, perhaps others. These slots would have values, perhaps with some restrictions. In Chapter 5, Section 5.1 we define class structures for amino acids and show how to compute a variety of useful things from sequences of amino acids. Here we are more concerned with declarative logical definitions than with procedural numeric calculations. So, we could define the concept *molecule* as something made up of elements.

```
(defconcept molecule (all has-part element))
```

This assumes that `element` is a primitive concept, whose allowed values are the known chemical elements that bind in molecules, and `has-part` is a primitive role. We could define a large molecule as

```
(defconcept large-molecule (and molecule (exists size large)))
```

where `large` is also a primitive concept and `size` is a primitive role. We could have also defined `large-molecule` in terms of its `has-part` property.

```
(defconcept large-molecule (and (all has-part element)
                                (exists size large)))
```

which states that a `large-molecule` has parts that are elements and a `size` that is `large`. In this case, it could be deduced that a `large-molecule` is a type of molecule, since it has all the properties required to be a molecule. Thus, looking at properties, one could infer class-subclass relations indirectly as well as directly. It is also possible to rewrite expressions containing defined terms by substituting their definitions, as part of a reasoning process.

With suitable formal notation that can be parsed by a computer program, these descriptions can be organized into hierarchical relations that are analogous to the class hierarchies defined by frames in a frame system. The difference is that the program can compute the hierarchy from the descriptions, rather than compute the descriptions from the hierarchy as we did in collecting the information in slots of superclass frames, following `superclasses` links to trace subsumption.

In medicine, we have names for many abnormal conditions that humans and other living beings sometimes experience, called "diseases." Examples are influenza, pneumonia, cancer, heart disease, diabetes. Large systems for naming and classifying diseases have been built and are now the subject of considerable revision using the knowledge representation methods in this chapter. Many diseases indeed have descriptions and can properly be classified in a hierarchy. For example, the term "cancer" really refers to many different types of cancer, mostly named according to the organ or region in which the cancer originates, such as breast cancer, prostate cancer, lung cancer. However, cancers also are classified according to the type of tissue or cell, for example, sarcoma, squamous cell carcinoma, lymphoma. Cancers are in turn a subclass of diseases called neoplasms, not all of which are malignant. So, we can describe the organ-based cancers by specifying that they are cancers and their `has-origin` property has a particular value, referring to a class of organ. So, breast cancer would be

```
(and cancer (all has-origin breast))
```

This style of description seems more manageable than the equivalent FOL version,

$$\forall x : \text{disease}(x, \text{cancer}) \wedge \text{has-origin}(x, \text{breast}) \rightarrow \text{breast-cancer}(x)$$

In a frame system, if one frame has all the slots of another frame, one cannot say for sure that it defines a subclass of the other frame. In a frame system, such relations need to be explicitly specified. There are certainly examples where two classes exist, where one has all the properties of another but is not a subclass. One could describe a bear as having a color, fur, two eyes, two ears, and perhaps some more properties, and a fox as having that same list of properties, but the class `fox` is not a subclass of the class `bear`, nor is the reverse true. They are both subclasses of the class of things that have a color, fur, two eyes, two ears, etc. In a description logic, the description itself can be sufficient to define this class, of which `fox` and `bear` are subclasses. Giving it a name may be convenient for writing compact formulas, but it is not strictly necessary. Note that the anonymous class described here subsumes the fox and bear classes, but is not equivalent to them, nor is it equivalent to the union of them.

It is possible to incorporate DL semantics into a frame system, by designating sets of slots as being necessary and sufficient to define a class. Then a DL classifier can operate on the frames, and organize them (or reassign them) into a class hierarchy, or adjust their class-subclass relationships.

2.4.1 Description Logics and Notation Systems

The simplest description logics include only a few basic formula types. These are concepts, roles, intersection (combination using `and`), and value restriction, analogous to universal quantification, relating a role and a concept. The language that uses only these forms is known as FL_0. In many systems, two special concepts are also included, known as "Top" and "Bottom." The "Top" concept is all inclusive, meaning "anything" (sometimes it is called "Thing"). The "Bottom" concept is the empty set, or "nothing." So, Top subsumes everything, and all other concepts describe subsets of Top. Bottom, on the other hand, is subsumed by everything. It is analogous to the empty set, since the empty set is a subset of any set. In more expressive languages, additional formulas are allowed, including negation of terms, negation of descriptions in general, the union operator (corresponding to logical or), existential restriction relating a role and a concept, role inclusion, concept inclusion, inverse relationships for roles, and conditions on numbers with roles and concepts.

In this section and following, we will show some examples of description logic formulas and reasoning with them. The system we will use is FaCT [161], a DL system that implements a language called $\mathcal{ALC_{R+}}$, as well as a more expressive system, $\mathcal{SHIQ}$. These names refer to the kinds of expressions that can be built using the system. $\mathcal{ALC_{R+}}$ includes concepts, roles, intersection, value restriction, the Top and Bottom concepts, general negation, union, existential restriction, and transitive roles. $\mathcal{SHIQ}$ adds support for inverse roles and qualified number restriction. The standard notations for some of these terms are shown in Table 2.2. In the table, C denotes a concept, R denotes a role.

As noted earlier, FaCT only implements a TBox, but not an ABox, so it only provides reasoning about classes and relations. According to the FaCT Reference Manual, version 1.6 (1998), FaCT implements named concepts, named roles, intersection, value restriction, the top and bottom concepts, negation, union, and existential restriction. It also supports rules as implication or subsumption axioms between concepts. Table 2.2 only includes the syntax of expressions described in the FaCT Reference Manual.

The reasoning capabilities of FaCT include support for queries about the classification hierarchy of a set of concepts, and relationships between concepts. It is also possible to retrieve concepts and roles by name. To illustrate how this works, we will use some excerpts from a medical terminology system that is provided with the FaCT distribution.

2.4.2 Example: GALEN

In this section we provide some illustrations of the use of a DL for a well-known biomedical terminology system, GALEN [335,336]. URL http://www.opengalen.org/ is the web site of the OpenGALEN Foundation. It provides open access to the products of the GALEN project.

Table 2.2 Some Elements of Description Logic Languages, Standard Notations and Syntax in FaCT

Name	Standard Syntax	FaCT Syntax
Intersection	$C_1 \cap \cdots \cap C_n$	(and $C_1 \cdots C_n$)
Value restriction	$\forall R.\, C$	(all R C)
Top	$\top$	*TOP*
Bottom	$\bot$	*BOTTOM*
Negation	$\neg C$	(not C)
Union	$C_1 \cup \cdots \cup C_n$	(or $C_1 \cdots C_n$)
Existential restriction	$\exists R.\, C$	(some R C)

The FaCT software distribution includes part of GALEN transcribed into a FaCT representation. We draw our illustrations from that. It is contained in the file `galen.tbox` in the FaCT distribution. We can retrieve it using the `load-tkb` function from FaCT. This function reads all the expressions in the specified file, and (by default, though you can also override it) prints a symbol for each item read in, as a kind of progress report. The letter P is printed for a primitive concept, C for a non-primitive concept, R for a role, R+ for a transitive role, A for an attribute, and I for an implication. We have elided most of this printout but left a few lines to illustrate what you would see.

```
> (load-tkb "../../tboxes/lisp/galen.tbox")
RRRR+RR+RR+RR+RR+RRAR+RRARAR+RR+RRAR+R+R+RRRR+RRRR+RR+RR+RRRR
RRRARR+RARRRRRARARRR+RRARRARRRRARARARAARR+RR+RRRRARRARRRRRRR
R+RR+RARRRARRRAAR+RRRRRARARAARRR+RR+RAAR+RAAR+RAARRARRRARRRAR
...
CCCCCCCCCCCCCCCCCCCCCCCCCCCCCCCCCCCCCCCCCCCCPPPPPPPPPPPPPPPPP
PPPPPPPPPPPPPPPPPPPPPPPPPPPPPPPPPPPPPPPPPPPPCCCCCCCCCPPPPPPPPP
PPPPPPPPPPPPPPPPPPPPPPPPPPPPPPPPPPPPPPPPPPPPCCPPPPPPPPPPPPPPP
...
CCCCCCCCCCCCCCCCCCCCCCCCCCCCCCCCCCCCCCCCCCPIIIIIIIIIIIIIIIIII
...
IIIIIIIIIIIIIIIIIIIIIIIIIIIIIIIIIIIIIIIIIIIIIIIIIIIIIIIIIIIII
```

Now, we can use FaCT functions to query the TBox contents. However, we need to know the names of the entities in the knowledge base, as well as the relations. The names in GALEN are case sensitive, so are represented by text within vertical bars. To start with, we look for a concept called "Heart." It seems there is such a thing, and it is a primitive. The FaCT function `what-is?` returns information about a concept, namely its type, which will be one of the four symbols, CONCEPT, PRIMITIVE, ROLE, or FEATURE.

```
> (setq heart (get-concept '|Heart|))
c[|Heart|]
> (what-is? heart)
FACT::PRIMITIVE
```

It would seem reasonable that there would be a concept called "Organ" as in the FMA, and indeed there is but it seems that it is omitted from the subset included with FaCT.

```
> (setq organ (get-concept '|Organ|))
NIL
```

The interesting part of this excerpt is what is known about the heart. First we ask what are the superclasses of the heart, or what anatomical type is the heart. The FaCT function `direct-supers` does essentially what the function `superclasses` does in our frame system (Section 2.3.1).

```
> (setq heart-parents (direct-supers '|Heart|))
!!NOTE!! Haemoglobin is a SYNONYM for Hemoglobin
!!NOTE!! BrachiocephalicVein is a SYNONYM for BrachiocephalVein
!!NOTE!! Myocardium is a SYNONYM for CardiacMuscle
...
!!NOTE!! GastricHemorrhage is a SYNONYM for HemorrhageFromStomach
(c[|LinearBodyStructure|] c[|ActuallyHollowBodyStructure|]
 c[|InternalOrgan|] c[|TubularSolidStructure|])
```

As the search progresses, FaCT prints out a series of messages about synonyms, but eventually returns four concepts, each of which subsumes the concept Heart. Next we ask about the subclasses of the heart. In the frame system, we would have to search the entire frame system to find frames whose superclasses slot (the direct superclasses) contained the ID of the heart frame. This suggests that the frame implementation should be enhanced by adding some caching of relations among frames. Here, it is a matter of looking in the TBox to see what is below the heart concept. It turns out to be nothing.

```
> (direct-subs '|Heart|)
(c[:BOTTOM])
```

Although much of the focus of work on biomedical ontologies is on the taxonomy of terms, the class-subclass relationships, this by itself is not very useful. The really interesting information is contained in other slots, or in descriptions that involve various roles. However it is not obvious what is the best way to make these queries. Suppose we would like to get a list of the parts of the heart. We note that isStructuralComponentOf is a relation or role in GALEN that seems to express the right relationship. So, we can ask about the things that have this role in regard to the heart. The equivalences function only works with named concepts, and so also with direct-subs.

```
> (equivalences '(some |isStructuralComponentOf| |Heart|))
NIL
> (direct-subs '(some |isStructuralComponentOf| |Heart|))
NIL
```

Instead we check if in fact there is any way that this description can be satisfied. We find that it is satisfiable.

```
> (satisfiable '(some |isStructuralComponentOf| |Heart|))
T
```

The way to find out where it sits in the concept hierarchy is to classify it, using classify-concept. This function returns three things, the concepts that directly subsume it (its direct superclasses), the concepts it directly subsumes (its direct subclasses), and all the equivalent concepts.

```
> (classify-concept '(some |isStructuralComponentOf| |Heart|))
(c[:TOP])
(c[|HeartValve|] c[|ChordaeTendinae|] c[|LeftSideOfHeart|]
 c[|RightSideOfHeart|])
NIL
```

Now we have the structural components of the heart, according to GALEN. They are: Heart-Valve, ChordaeTendinae, LeftSideOfHeart, and RightSideOfHeart. Of course this is very incomplete, but at least you can see how classification serves as a query mechanism for more than just subtypes. These concepts do not represent subclasses of the heart, and are not subsumed by the heart concept, but are subsumed by the description of what we are looking for.

So, we continue with another query. The excerpt from GALEN has a concept for the Aorta, the major artery coming out of the heart, which serves as the trunk of the arterial system. It is reasonable to ask what arteries the Aorta branches into.

```
> (classify-concept '(some |isBranchOf| |Aorta|))
(c[:TOP])
(c[:BOTTOM])
NIL
```

It seems that this is not where the information is. However, the Aorta is divided into parts called solid divisions, and there is where we will find the branch information.

```
> (classify-concept '(some |isSolidDivisionOf| |Aorta|))
(c[:TOP])
(c[|AbdominalAorta|] c[|ThoracicAorta|])
NIL
```

These two subdivisions have isBranchOf information.

```
> (classify-concept '(some |isBranchOf| |AbdominalAorta|))
(c[:TOP])
(c[|CeliacTrunk|] c[|RightInferiorPhrenicArtery|]
c[|LeftInferiorPhrenicArtery|] c[|SuperiorMesentericArtery|]
c[|InferiorMesentericArtery|] c[|RenalArtery|] c[|IliacArtery|])
NIL
```

Just to be sure, we check if these divisions have further divisions.

```
> (classify-concept '(some |isSolidDivisionOf| |AbdominalAorta|))
(c[:TOP])
(c[:BOTTOM])
NIL
```

Now we can proceed to further trace the arterial connectivity.

```
> (classify-concept '(some |isBranchOf| |CeliacTrunk|))
(c[:TOP])
(c[|SplenicArtery|] c[|CommonHepaticArtery|]
c[|LeftGastricArtery|])
NIL
```

So, description logics can support interesting and useful queries. However, the problem is that to make this useful (for example, if an artery is blocked, to determine what parts of the body will be denied their blood supply and hence oxygen and nutrients) one needs to write recursive queries, just as in the frame system. Worse yet, as with frame systems, it is crucial to know how the terms are used, and for the terms to be used in a consistent and systematic way, so that queries can be automated.

An important application of DL formalism, like FOL, but more tractable, is the ability to do consistency checking of large-scale ontologies. One example is the translation into a DL representation of the FMA [25]. This work posits a development cycle for engineering large systems like the FMA. The authors apply consistency checks to part/whole role representations, formulating axioms which must be satisfied in the FMA. They found about 280 cyclic definitions, which were then correctable by the maintainers of the FMA. In Chapter 4 we will describe similar consistency axioms for the connectivity relations used in the descriptions of the lymphatic system, but will apply a procedural, path tracing methodology instead.

2.4.3 The Semantic Web

One thing that surely characterizes modern biology, medicine, and public health is the ubiquity of data in computerized form. So, too, a rapidly growing number of terminology systems and ontologies are appearing, many of which are published through the World Wide Web (for short, just "Web"). For example, the Gene Ontology (GO) is available for download, and accessible through a large variety of web-based and other browser tools. These resources are just a tiny part of the huge data and knowledge sources that are becoming available through the Web. The Semantic Web is an activity of the W3C consortium to somehow create standards, frameworks, and protocols that make all this information accessible, so that "intelligent" agents (computer programs) can find and use the information to answer queries, and do interesting and useful computations, that can't be done with just a single resource. Moreover, if this idea finds success, one knowledge source can uniquely and reliably reference others, further amplifying the power of these resources. In the words of the W3C Consortium,[30] "The Semantic Web provides a common framework that allows data to be shared and reused across application, enterprise, and community boundaries."

---

[30]W3C Semantic Web Activity, URL http://www.w3.org/2001/sw/.

The current Semantic Web activities focus on a few main areas: RDF, RDFS, and OWL. RDF (Resource Description Framework) is a language for describing resources. It is intentionally limited in syntax and semantics, and is a basis for extensions. RDFS (RDF Schema) is an extension that provides some vocabulary for describing or specifying the structure of RDF resources. OWL (Web Ontology Language) is built on and extends RDF and RDFS to provide enough semantics to write logical expressions and serve as a standard interchange language for ontologies as we have described them in this chapter. These ideas are all realized using the syntax of XML, because of its wide support and because although the ideas are not restricted to the Web, the Web is the main context for realizing actual instances of RDF resources and OWL ontologies.

RDF is implemented as a collection of XML tags with predefined semantics, to represent directed graphs, which in turn represent statements about resources, properties, and values. These are similar to EAV database entries (see Section 1.3.4 in Chapter 1), and can also be thought of as simple assertional sentences, with subject, predicate, and object. An example from the drug table would be.

```
Codeine has-drug-type narcotic
```

This example could be represented in RDF as

```
< rdf:Description about="Codeine">
  < has-drug-type > narcotic</has-drug-type>
</rdf:Description>
```

Strictly speaking, the "about" tag, which attaches an attribute to the "Description" element, should have a URI as its value. This URI would refer to some specific resource, though not necessarily web accessible. The "about" tag labels the graph node corresponding to the description. Thus, the resource is `Codeine`, the property attached to Codeine is the `has-drug-type` property, and its value is the simple string, `narcotic`.

It is also possible to have unnamed graph nodes, and a node can have multiple properties. The values can themselves be nodes with the same kind of structure, with arbitrary nesting allowed.

RDF Schema adds to the basic RDF tags some additional ones, enabling expressions that specify the class of a resource, and that a resource is a subclass of another. A good introduction to RDF may be found in [271, 134] and of course at the W3C web site mentioned earlier.

OWL adds to RDF and RDFS a more complete set of tags and semantics to be able to represent substantial biomedical terminology systems and ontologies. These together with their XML syntax implement a serial form of knowledge representation that can support knowledge interchange. A reasoning system would not operate on these verbose XML expressions directly, though. A reasoning system would have to read the textual representation and transform it into a suitable internal representation that can support the operations described earlier. Many logic systems do provide this ability to import and export ontologies in OWL XML format.

Here is an excerpt from GO that illustrates a little more how this plays out. This first entry, labeled as GO ID GO_0000018, is the biological process, "regulation of DNA recombination," as indicated by its `rdfs:label` property. It is an instance of an `owl:Class`, corresponding to the notion of "class" in our frame system, and "concept" in a DL.

```
<owl:Class xmlns:owl=" http://www.w3.org/2002/07/owl#"
            rdf:ID="GO_0000018">
    <rdfs:label>regulation of DNA recombination</rdfs:label>
<!-- regulation of DNA metabolism -->
    <rdfs:subClassOf rdf:resource="#GO_0051052"/>
    <rdfs:subClassOf>
      <owl:Restriction>
        <owl:onProperty>
          <owl:ObjectProperty rdf:about="#part_of"/>
        </owl:onProperty>
        <owl:someValuesFrom rdf:resource="#GO_0006310"/>
      </owl:Restriction>
<!-- DNA recombination -->
    </rdfs:subClassOf>
  </owl:Class>
```

The lines beginning with `<!--` are *comment* lines and are meant to be ignored by any computer program that processes this information (such as an XML parser or OWL parser). This resource has two more properties in addition to the `label` property, both `rdfs:subClassOf` properties. The first `subClassOf` property has as an attribute a reference to another GO term, as indicated by the `rdf:resource` tag, and the comment above it is correct, GO term 0051052 does indeed represent "regulation of DNA metabolism." The second `subClassOf` property may at first appear to be rather awkward. This property has as its value another description. This is the OWL version of normal DL use of descriptions as concepts in the same way as primitive (named) concepts. This description is of a class (unnamed) defined by a restriction on the "part of" property. It reads, "those things whose `part_of` property has a value which is another resource in GO, the one whose GO ID is GO_0006310." This (as the next comment line again indicates correctly) turns out to be "DNA recombination." So, in short, regulation of DNA recombination is a *part of* DNA recombination.

For comparison, here is how this same GO term might be represented as an expression in FaCT. We have left out the comment lines, and the reference to the OWL URI, since it is not relevant in this context.

```
(defconcept go-0000018
  (and (some label "regulation of DNA recombination")
       go-0051052
       (some part-of go-0006310)))
```

Here is another example, the very next entry in GO, the concept named "regulation of mitotic recombination." Again, we see exactly the same structure. In fact, there are no other properties in GO. The only relations that GO provides are a class hierarchy and a part-of hierarchy. In addition there are the labels, and in some versions, explanatory text which has no particular DL structure, but is just plain text.

```
<owl:Class xmlns:owl=" http://www.w3.org/2002/07/owl#"
            rdf:ID="GO_0000019">
    <rdfs:label>regulation of mitotic recombination</rdfs:label>
<!-- regulation of DNA recombination -->
    <rdfs:subClassOf rdf:resource="#GO_0000018"/>
    <rdfs:subClassOf>
        <owl:Restriction>
          <owl:onProperty>
             <owl:ObjectProperty rdf:about="#part_of"/>
          </owl:onProperty>
          <owl:someValuesFrom rdf:resource="#GO_0006312"/>
        </owl:Restriction>
<!-- mitotic recombination -->
    </rdfs:subClassOf>
    </owl:Class>
```

Despite the fact that GO does not use them, OWL provides a large collection of defined classes. It includes a collection of constructors similar to $\mathcal{SHIQ}$, and a significant number of built in axioms, forming a base ontology on which to build application ontologies. For more about OWL and related developments, there are several good sources [17,134]. For this also, the W3C web site provides a rich set of documents.

2.4.4 Example: Enhancing Queries With LOINC

Section 1.1.4 introduced LOINC, a terminology system for identifying laboratory tests. The information in LOINC can be represented as formulas in a description logic such as OWL. Through the UMLS and an OWL representation for both, it is possible to map LOINC entities to other medical terminology systems such as SNOMED. One important use of this mapping is the ability to audit these systems for internal and external consistency. Another is to provide a richer set of relationships and synonym identifications than is possible with either alone. Both of these are addressed in a recently reported work [4]. This section will briefly describe some aspects of this work.

It is possible to represent the content in Figure 1.15 in a description logic format. Each of the attribute names used to label fields in Figure 1.15 has an already assigned label to be used in an OWL representation. In Figure 2.11 this record is shown using the FaCT syntax we used earlier.

In OWL, one can include subclass relationships explicitly in a definitional expression, but FaCT does not support that. It is easy enough, however, to just add the assertion,

```
(implies loinc-14743-9 (some has_class chem))
```

where presumably we have defined chem with a specific code and definition. Once this representation exists and a mapping to SNOMED exists, the relations in SNOMED can be used to enhance the relations in LOINC.[31]

As can be seen in Figure 2.11, LOINC codes are compositional as well as hierarchical. Each code is a conjunction of specifications for its parts, labeled by the relationships illustrated in the figure. Adamusiak and Bodenreider created a separate DL hierarchy for the component parts of LOINC codes and assigned identifiers to them. The codes and parts are then mapped to SNOMED terms through the UMLS. This enables identification of implied relationships in LOINC that supplement the explicit ones.

For example, LOINC part LP30504-2, "Body fluid" is mapped to a corresponding term in SNOMED-CT, also named "Body fluid," SNOMED ID 32457005. Also, the part "Urine" in LOINC, ID LP7681-2, is mapped to the term of the same name in SNOMED-CT, ID 78014005. However, in SNOMED-CT, there is a subsumption relation stating that 78014005 is a subclass of 32457005, meaning that urine is a body fluid. However, this relationship is not found in LOINC. Having made the mapping, the relationship can be added to the DL representation of LOINC parts.

Another example is three parts in LOINC that turn out to be equivalent when mapped to the UMLS. These are "Erythrocyte," "Erythrocytes," and "RBC." Nothing in LOINC indicates a relationship between these terms.

Separating the parts from the codes and using the compositional aspect of the codes enable queries for codes to be expanded. A query for tests of carbohydrates in urine should return codes for many kinds of carbohydrates. However, there is only one such composition, LOINC code 16550-6. A search will not find, for example, code 2350-7, Glucose in urine, even though these codes are both classified as subclasses of sugars in urine. Once the component parts hierarchy

```
(defconcept loinc-14743-9
  (and
      (some has_component glucose)
      (some has_property scnc)
      (some has_time_aspect pt)
      (some has_system bldc)
      (some has_scale qn)
      ...))
```

Figure 2.11 An excerpt from a record in LOINC, rewritten as a Description Logic statement. Only a few fields are shown, for purpose of illustration only.

[31]The examples here are from Tomasz Adamusiak's presentation at the 2012 AMIA Annual Symposium and are included here with permission from Dr. Adamusiak and his co-author, Dr. Olivier Bodenreider.

is computed, the fact that the part "glucose" is a subclass of part "carbohydrates" can be used to include 2350-7 as a valid code in response to the original query. In addition many other codes will be returned. So the DL representation enables richer responses to queries that would otherwise return few results or fail altogether.

2.4.5 Back to the Future

We conclude this section by constructing a DL statement that somewhat captures the idea in the quote at the beginning of this section. This translation is a little more emphatic, leaving no room for doubt.

```
(implies (and (all :appearance duck-like)
              (all :walk waddle)
              (all :sound quack))
         duck)
```

which asserts that if it looks like a duck, walks like a duck, and quacks like a duck, it *is* a duck. Another way to write this is as a definition, where these properties are the necessary and sufficient conditions. In reality, of course, the subsumption relation is the reverse of the above. So we should write

```
(implies duck (and (all :appearance duck-like)
                   (all :walk waddle)
                   (all :sound quack)))
```

This is less emphatic, and allows that there might be other things that look, walk, and talk like a duck. Now, how could this actually be used to address a real life problem? If something fits the description, that does *not* mean it is a duck. However, we might posit that as a working hypothesis. So, we have come around once again to abductive, or explanatory, reasoning. How would a DL system help here? We are most inclined to give strength to the hypothesis if it is the only explanation or the closest explanation. So, in classifying the description, we would look to see what its direct superclasses are, and if all we find is duck, that's strong evidence. If on the other hand, there are lots of other possibilities, we need more evidence. The subsumption computation still is valuable since it tells us what alternative explanations are available, and what their consequences are as well (by classifying them, or looking for their direct subclasses).

All this computation touches on searching through graph structures, whether they are proof trees, frames, and links, or description (concept) hierarchies. Search methods and algorithms will be taken up in Chapter 5.

2.5 Summary

In exploring a variety of formal systems for representing biomedical knowledge as categorical assertions, we found a few general themes.

- Formal systems enable us to automate both reasoning and knowledge retrieval (answers to queries).
- Different formalisms have varying expressive power, and different computational demands or complexity.
- There are a range of choices regarding how much reasoning is put into procedures, and how much can be made implicit as part of a declarative formalism.
- No single formalism subsumes all biomedical knowledge representation and reasoning requirements.

The key point in this chapter is that once you formalize the knowledge in a particular domain, you can reason by symbol manipulation alone, without regard for the interpretation of the symbols. This is exactly what makes it possible for a computer system to perform reasoning.

In biology, medicine, and health, many people have attempted to formalize knowledge in their respective domains, with greater or less success. Similarly, the systems they built have found widely varying acceptance and use. For a long time some people have claimed that somehow biomedical knowledge is different in character than knowledge in the physical sciences, and cannot be represented in the same way. At the time of writing this book, however, we are seeing a rapid development in application of knowledge-based systems in many areas, within biology, medicine, and health, and in many areas of the humanities and social sciences.

The essential underlying theme of the molecular biology examples in this chapter is that cells and biomolecules are a kind of computing machinery that operates with symbols, or discrete units. Application of symbolic simulation to biochemical processes is not a new idea. Several examples [122,216,270,370] are reported in a much earlier book by Hunter [167]. The EcoCyc project [215] and its relatives [214] are widely used and actively developed and supported. Many other resources exist for larger scale experimentation.

It seems that the main problem was that we simply did not yet have the deep and precise knowledge that is required for these methods to be applicable. We still have a long way to go. But the main difference between the biomedical sciences and the physical sciences is that biological processes require different kinds of formalisms. Instead of differential equations (or perhaps in addition), we need to be able to build large-scale logic systems, large and rich representations of networks of entities and relationships, and be able to infer things from the connectivity of the networks.

A second important difference between physics and biology (which carries over into medicine) is that in physics there are only a few variables and parameters in a typical theory, and powerful relations between (or among) them, while in biology the number of independent variables, parameters, and entities to be named and reasoned about is huge. In biology, the *relations* are much more shallow and less uniform. The name used for a particular vessel in the lymphatic system of a human, representing an element of a theory of human anatomy, does not matter to the reasoning system, but that name must be unique in the context of the anatomic theory and any other knowledge structures used with it. When such an entity name is used in multiple places in the theory, it has to mean the same thing everywhere, otherwise the reasoner will infer false conclusions. This is why so much effort seems to be expended on terminology and on terminology management systems.

Another limitation of symbolic knowledge formalisms in biology, medicine, and public health is that the knowledge is not as categorical as the logic formulation would suggest. The notion of simple true or false values for assertions does not model well what seems to be going on in cells, organs, organisms, populations, diagnosis, disease, and treatment. Just as in physics some knowledge must be inherently expressed in the language of probability and uncertainty, so biology, medicine, and health need methods for representing processes and assertions that are uncertain, not for lack of detailed knowledge, but because of their inherent statistical nature. So, in the next chapter we discuss probability and uncertainty in modeling biomedical knowledge, and how to reason and answer questions with probabilistic knowledge models.

2.6 Exercises

1. Translating English Into Logic

   Write first order logic formulas that express the following. You can choose whatever symbols you want to correspond to each of the entities and relationships in the sentences. Some of the sentences below may each correspond to multiple FOL sentences.

   (a) The fluoroquinolones are potent bacterial agents against *E. coli* and various species of *Salmonella, Shigella, Enterobacter, Campylobacter,* and *Neisseria.*[32]

   (b) Levofloxacin is a fluoroquinolone.

   (c) Levaquin is a brand name drug corresponding to the generic, Levofloxacin.

   (d) Levofloxacin can cause tendonitis or ruptured tendons.

   (e) DNA sequences for genes in Eukaryotic cells include both introns and exons.

   (f) A Eukaryotic cell is one that has a nucleus (definition).

   (g) A Eukaryotic cell has mitochondria.

   (h) Kidney cancer is not treatable by radiation.

   (i) Jones went straight to the Emergency room because his mouth suddenly swelled up.

   (j) Smith said that Jones was probably bitten by a spider, because spider bites can cause swelling.

   (k) There was a huge outbreak of spider bites in Sequim in 1994.

   What sentences are entailed by the above but are not listed above? Use an appropriate inference method to prove your claims.

---

[32]Quoted from Goodman and Gilman's "The Pharmacological Basis of Therapeutics," Tenth Edition.

2. Reasoning about mountaineering

   I used to be a mountain climber, and learned the saying, "There are old climbers and there are bold climbers, but there are no old, bold climbers." Render this in first order logic. Then add "Ira is a climber" and from this KB, prove that Ira is not bold (this is true, actually). You will need one additional fact, which should be obvious from the picture on page xi, though I will be flattered if it isn't.

3. The murder mystery

   The following is adapted from Brachman and Levesque, "Knowledge Representation and Reasoning." [46] They in turn attribute it to an earlier work by Genesereth and Nilsson. V (the victim) has been murdered by a single individual. There are only three suspects, A, B, and C, and it is certain that one of them committed the crime. Here is what else is known:

   - A says that B was V's friend but that C hated V.
   - B says he was out of town the day of the murder, and anyway he didn't even know V.
   - C says that he saw A and B with V just before the murder.

   You can also assume that everyone, except possibly the murderer, is telling the truth.

   (a) Use logical inference to find the murderer, that is, write the facts as logic formulas, prove that there is a murderer, and identify him. Hint: after writing the formulas, you will need to convert them all to clause form using the rewrite rules we discussed, then use resolution to eliminate all but the murderer.

   (b) Suppose that there might have been a conspiracy, so that the assertion that there is a single murderer is not known. What does this do to your proof? In this case can the identity of the murderer(s) still be determined? If so, show how, but if not, explain why not.

   This can be done with just propositional logic. To get you started, you can use the symbols A, B, and C to stand each for "A is the murderer," "B is the murderer," etc. So, the assertion that only one is the murderer turns into three formulas, of which you should only need one, for example,

   $$A \Rightarrow \neg B \wedge \neg C \tag{2.37}$$

   You can use FB for "B was a friend of the victim," and negation to express that someone hated the victim, for example, $\neg FC$ would mean that C was not the victim's friend. Note of course that these formulas are true if A is not the murderer, so you can write a formula for that.

   In Brachman and Levesque, A, B, and C have names, but here I have abbreviated them in the style of high school "story problems." An essay by the late Canadian humorist, Stephen Leacock, "A, B, and C, the Human Element in Mathematics," deals with this at length, providing fascinating insights into these three characters, who have served mathematics teaching for so many years. It is reprinted in Clifton Fadiman's book, "The Mathematical Magpie," and can also be found on line—the entry is "A, B, and C" in en.wikisource.org.

4. Soundness of inference rules

   Establish that the following inference rules are sound, using truth tables.

   - Implication rewrite
     The rule is: $\alpha \rightarrow \beta$ is equivalent to $(\neg\alpha) \vee \beta$. Write out the truth tables for each to demonstrate that they are identical.
   - Modus ponens
     The rule is: from the two formulas, α and $\alpha \rightarrow \beta$, we can infer β. This is a little different, since it is *not* an equivalence. Instead, we combine them into the compound formula, $[\alpha \wedge (\alpha \rightarrow \beta)] \rightarrow \beta$. Show using truth tables that this compound formula is a tautology, meaning that it is true no matter what the truth values of α and β are.
   - Resolution
     Prove using truth tables that Resolution is a sound inference procedure. For simplicity you can just do the case where each clause has a single negative term and a single positive term, that is, from $\neg\alpha \vee \gamma$ and $\beta \vee \neg\gamma$ we can infer $\neg\alpha \vee \beta$.
     Hint: this set of formulas can also be written as a compound formula, which can be shown to be a tautology, using truth tables.

5. A taught-us race

   Read the Lewis Carroll dialog, "What the Tortoise Said to Achilles,"

       http://www.lewiscarroll.org/achilles.html

   and provide a resolution to Achilles' sad plight.

6. More on drug interactions

   In two articles by world reknowned drug interaction experts, Phil Hansten and John Horn [147,151], the following ideas are discussed:

   (a) including laboratory test results in the computation,
   (b) grouping drugs into a class hierarchy with higher level families and representing interactions among families and with lab tests,
   (c) handling generic and trade names for equivalent drugs,
   (d) handling patient conditions such as allergies.

   Expand the code from Section 2.2.5, or from Section 1.2.5 or make up a different representation, that accommodates several (at least two, if possible three) of the above ideas. Don't expect to create a complete functioning system. The goal is to have a consistent logical representation, with a test function like `interacts` that works on the data. The whole works should not be more than a page or two of code.

7. Lymphatic pathways via rules

In Chapter 4, section 4.3.5.2, the representation of the lymphatic system as a series of downstream relations is explained, and it is shown how pathway search can be used to determine if a lymphatic chain is downstream from another lymphatic chain, using the "efferent-to" relation. Now we will try to do this with rules and assertions.

(a) Create a small network of lymphatic chains from the FMA, as assertions in a rule language, either real Prolog, or the Graham mock Prolog. Your assertions will use predicates like `efferent-to` and you will have one or more rules for the `downstream` predicate.

(b) The search code in Chapter 5 applied to biochemical pathways traces and returns entire paths. Could this be done easily with the Graham code or simple Prolog? If so, write the required rules. If not, where are the complications or obstacles?

8. Exploring minimal biochemistry

In Section 2.2.1.5 we used propositional logic code and a representation of molecules and reactions to dynamically search for pathways. What you can synthesize depends on what you start with in the environment. Experiment with different starting environments, run the code, and see if you can find the minimum starting environment from which you can still metabolize glucose into pyruvate.

9. Exploring maximal biochemistry

Peter Karp's EcoCyc system (see URL http://biocyc.org/ for this and all the BioCyc resources and tools) has much richer representations for molecules and reactions than any here. Could you do the same computations in EcoCyc that are done in Section 2.2.1.5? If so, what parts would be the same, and what would be different? If not, what is missing or what could go wrong?

10. Description Logic—A clinical case

The following are descriptions of two hypothetical protocols for determining eligibility for cancer clinical trials. Using the basic Description Logic language defined above, translate each protocol into the DL format, normalize each description, and determine whether a patient eligible for one protocol is also eligible for the other (and if so, specify the order, i.e., which is subsumed by which).

Protocol A: Patients who have previously untreated squamous cell carcinoma of the oral cavity, histologically confirmed. The disease must be potentially curable by multimodality therapy. The disease must be at least stage 3. The patient's age must be at least 50.

Protocol B: Patients who have confirmed squamous cell carcinoma of the head and neck must be at least 18, and the disease must be treatable by some kind of conventional therapy.

You may need the additional facts that the oral cavity is a tumor site contained in the head and neck region, that multimodality therapy is one of many different conventional therapies. You can abbreviate terms, but include a table of definitions so these abbreviations can be understood.

Check your result for this exercise by putting the descriptions into a DL reasoner system of your choice and using the system's classifier to determine if one subsumes the other.

11. Using OWL

    Repeat the previous exercise using OWL as the DL representation language. Write the descriptions using the XML syntax associated with OWL. Optionally, you may use Protégé -OWL or any other implementation that supports OWL to check the correctness of your descriptions.

12. Frames and Drugs

    (a) Use the information in the article by Hanlon mentioned in an earlier exercise, and the frame system in this chapter, to create a small drug ontology, including in the frame for each drug class information about the diseases for which it is potentially dangerous, that is, the ontological class hierarchy comes from Table 1 and some additional information for each entry comes from Table 2.

    (b) Write a function that suggests alternate drugs or drug classes that could substitute for a specified drug if there is a problem.

13. Roll your own

    Pick a topic that you are familiar with, in biology, medicine, public health, or a related area, select or write a set of related sentences, translate them into FOL, and use the inference methods we have studied to prove the entailments of your sentences, i.e., what other sentences follow from the ones you selected. Similarly, translate the formulas into rules and assertions using the <- operator and use the Graham inference code, if possible, to prove your entailments. If the Graham code won't work, explain why.

CHAPTER 3

Probabilistic Biomedical Knowledge

Chapter Outline

Principles of Biomedical Informatics. http://dx.doi.org/10.1016/B978-0-12-416019-4.00003-2

Facts are stubborn, but statistics are more pliable

–*Mark Twain*
attributed

When representing knowledge in terms of symbols and formulas of symbols, uncertainty and probability come up in several different ways. One possibility is that our theory, formed by a choice of symbols and a symbol system language like first order logic, a set of axioms, an inference method, and an interpretation, might actually be incomplete. This sense of incompleteness is not the formal meaning relating to proof of entailment, discussed in Section 2.2.10. Here, incompleteness means that the theory is not able to infer all the important assertions we will observe in our biomedical or health environment to which the theory applies. This is a question of coverage, not of decidability. Sometimes this can be resolved by adding more axioms. However, sometimes the limitation of a logic representation for biomedical knowledge is that there is some uncertainty about the axioms and correspondingly about their entailments. This may be a reflection of lack of complete details or it may be that a biological process is inherently indeterminate, that is, only the probabilities of various outcomes are known. Thus methods are needed for representing knowledge that is probabilistic, uncertain, and subject to refinement as more information is obtained. Another aspect of application of biomedical knowledge to real-world decision making is that some variables or outcomes are more important than others. As well, then, it would be valuable to be able to model the significance of different outcomes that are consequences of our decisions. Thus the focus here is on *events*, or outcomes, and *decisions* that may lead to various outcomes, for events that are uncertain, either because they are of unknown status or they are actually nondeterministic.

In this chapter we introduce basic ideas of probability and statistics and then show how they can be applied to build probabilistic reasoning systems. This is followed by an introduction to utility theory and decision modeling, and a brief introduction to information theory.

Brief mention is made of some ideas from statistics, but the emphasis here is on the mathematics of graphical relationships among variables using conditional probabilities. Such graphs can compactly and meaningfully represent causal relationships in very complex systems with many variables. They are particularly suited to modeling pathology, prognosis, and treatment of disease.

3.1 Probability and Statistics

A statistician is a person who, with his head in the oven and his feet in the freezer, will say that on the average he feels comfortable.

—*anonymous*
well-known joke

This section introduces the basic ideas and formulas of probability, and some elementary concepts from statistics. More thorough developments can be found in standard probability and statistics textbooks [50, 430]. Another excellent treatment may be found in [298, Parts I and II].

In contrast to the concise but very far-ranging treatment in Wasserman [430] in particular, this chapter should probably be called "Some of Statistics."

3.1.1 Probability

Probability is about *events*. An event is something with a definite outcome or state that can be observed. Put another way, we are describing the state of a system by specifying the values of a set of variables. The possible values (or combinations of values) of the variables represent all the possible states. Often rather than predicting the specific outcome or state in each instance, we can only predict the *probability* of any particular outcome or observable state. The classic example is a coin which we toss to see if it will land head up or tail up.[1]

In any given situation, the set of all possible elementary outcomes, or results of measurements (for example in a laboratory, clinical setting, or public health surveillance), is known and specified. This set may be discrete or continuous, numeric or symbolic, finite or infinite. The set of all possible outcomes or observations is called the *sample space*. For a single coin toss, the sample space has two elements, `heads` and `tails`. For the observation of what base appears in some position in a DNA sequence, there are four elements in the sample space, A, G, C, and T, and many subsets. Each subset can be considered an event, for example, the occurrence of either G or C, or the occurrence of anything but T, and so on.

In some cases, the behavior of the biological or other system we are modeling is inherently nondeterministic. The things we observe are not uniquely predictable from other more basic characteristics. Taking cancer as an example, the process of spread of tumor cells from the primary tumor site to form metastatic lesions is inherently variable, or nondeterministic. For any particular cell, we cannot predict where it will go, but we can perhaps build a theory that predicts the frequencies with which cells migrate to one or another location. So we need probability-based modeling methods to represent such knowledge. This view that probability represents the reality of nondeterministic events is called the "frequentist" view.

Another important case is the situation where there may be an underlying deterministic behavior, but it is completely beyond our reach for now, and we must model our uncertain knowledge about the behavior of the system, as distinct from its actual behavior. Again referring to cancer as an example, we may have some diagnostic tests (for example, for prostate cancer) which can be used as evidence. The patient's status is definite, but unknown. He either has a prostate cancer or not, but we just don't know yet which is the case. In this situation we are modeling our *beliefs* about the situation, not the physical reality. As more evidence accumulates, our beliefs will be updated but the physical reality has not changed, only our knowledge about it has. This is called the "subjectivist" or "Bayesian" view.

---

[1] Actually, many coins do have a head, usually of a significant political figure, a president, monarch, or some such. However, I personally have never seen a coin with a tail depicted on it. The US "buffalo nickel" comes close, but the whole bison (not a buffalo) is depicted, not just the tail. By "tail" we generally mean the opposite side of the coin from the head.

These two views interact when we need to make decisions. We know that Interleukin-2 can be an effective agent for treating kidney cancer, but only 10–20% of the patients so treated respond to the treatment (with tumor growth arrest, regression, or complete ablation). We don't know which patient will respond (although one might imagine that at some future time we could discover which characteristics, perhaps genetic markers, would predict whether a patient with kidney cancer would respond to treatment or not). To decide whether to apply the treatment, we use what is known about the nondeterministic behavior of the treatment, and what we believe about a particular patient's likelihood of response.

3.1.2 Statistics

In many areas of biology and medicine, we describe observations in terms of *variables* that take on a range of values. We hypothesize that one variable has a functional dependency on another (or possibly several). This dependency typically reflects some kind of biological law. For variables with numerical values, for example, the relation could be linear, meaning that one variable is proportional to another, with possibly a constant offset. It could be more complex, like an S-shaped curve. The percentage of living cells that are killed by radiation as a function of absorbed radiation dose is an example of an S-shaped curve, described by a definite formula with parameters. Many statistical methods are designed to estimate the parameters of such a function, or even to estimate the shape itself, where there are few or no existing theoretical ideas to suggest a functional shape. However, despite the attractiveness of the idea of generating theories automatically through data analysis, there is no way to do this. Biological and medical theories, like their physical counterparts, are descriptions created principally by human imagination, though grounded in observation. Just as with theories based in logic formalism, we create models, compute their entailments, and check these predictions against the available data. In the realm of probabilistic theories or models, the process is the same, only the inference methods are somewhat different.

An important goal of statistics is to provide a mathematical structure for formulating such theories, computing their predictions, and deriving their parameters from experimental data. Since any particular observation has a particular value, how can we determine experimentally what the probabilities are for the various values? We can't measure probabilities directly. Two approaches are possible, one involving ensembles and the other involving time. In the ensemble approach, we have a large number of identical systems, such as a large number of identical coins. We toss them all, and count the number that came up heads and the number that came up tails. The fraction of the total that came up heads should approximate the probability of a "heads" result, and by definition (and guaranteed by the experimental procedure just described) the probability of a "tails" result should be 1.0 minus the heads probability. One might wonder how such experiments can really provide the information. The question is, "what sampling method provides an *estimate* of the probabilities in the probability distribution?" Alternatively, if the hypothesized probability distribution is specified by a parametric function, for example the Normal distribution, the question can be reformulated as how to estimate the parameters.

Sometimes the hypothesis is that a relation exists between two variables that can be observed together. If the functional form of this relationship is specified in terms of parameters, then these parameters can be estimated as well. One such commonly used method, in the case where the variables are assumed to be linearly related, is the Method of Least Squares.

In order for such procedures to yield an accurate approximation to the true probability, the number of systems tested or replicated event observations must be sufficiently large. Another important question that statistics addresses is how large a data set must be in order to claim that a result is significant.

3.1.3 The Laws of Probability

A *probability function* is a function that assigns a number between 0.0 and 1.0 to each subset of elements in the sample space. The term "event" is used to refer to any particular subset, including the subsets that each have only a single member, a single elementary outcome. Equation (3.1) expresses this requirement that the range of the probability function is between 0.0 and 1.0. This is the first *axiom* of probability. If x represents an event and $P(x)$ represents the probability that event will occur, then

$$\text{Axiom 1} \quad 0.0 <= P(x) <= 1.0. \tag{3.1}$$

Remember that (as noted earlier) the event variable x can have symbolic values instead of numeric values (as in the coin toss example). Many variables in biology and medicine have outcomes that are *named* rather than numeric.

The probability of an event tells us what is the likelihood that event will occur, out of all possible events. A probability of 0.0 means that the event will *never* occur, while a probability of 1.0 means that the event will *certainly* occur. For subsets with several outcomes as members, the probability function's value is the probability that *any* one of the outcomes occurs. Thus, events can be overlapping, or mutually exclusive (disjoint). For the coin toss, the head and tail outcomes are mutually exclusive. When we toss the coin it *always* comes up one way or the other, not both (we are not considering the possibility that it lands on edge—we will return to this later). With mutually exclusive events, there is no overlap. Since these events cover all possible outcomes the probabilities must add up to 1.0, because one of them certainly happens.

In the case of the coin toss event space, the requirement that the sum of all probabilities must be 1.0 is expressed in Equation (3.2). Here, x could take on the value we will denote by the symbol up, for the event where the coin landed head up, and then $\neg x$ will refer to the event where it landed head down (that is, *not* head up)

$$P(x) + P(\neg x) = 1.0. \tag{3.2}$$

The coin toss is an example of a Boolean sample space—it has only two elementary outcomes. Biomedical examples abound: 5-year survival in cancer patients, presence or absence of an epidemic outbreak, presence or absence of diseases in organisms, etc. When there are n possible (mutually exclusive) outcomes, x_i, Equation (3.2) generalizes to Equation (3.3). An example would be the selection of a colored ball from a bowl containing many balls, each of which is colored with one of n different colors. In this example, each of the x_i is one of the events, namely that a particular colored ball was drawn out. A biological example is the particular base that appears at a specified location in the genome of an organism. In that case, $n = 4$, since there are four possible bases that could appear at any point in a DNA sequence

$$\text{Axiom 2} \qquad \sum_{i=1}^{n} P(x_i) = 1.0. \tag{3.3}$$

Equation (3.3) is the *second* axiom of probability. It reflects the intuition that of all possible outcomes of an event, one of them certainly will occur.

Anything we measure or observe in biology, medicine, and public health can be formulated as a sample space, such as overall health state (Excellent, Good, Fair, Poor), identity of organism found in a blood culture or other bacterial culture (many possible states for this one), presence or absence of a point mutation in a particular place in a genome, immunization status of a child, etc. In addition to these *discrete* sample spaces, one can also have *continuous* observations, where the outcome is the measurement of a number (or numbers) that can vary continuously. Examples include medical laboratory test values such as the creatinine clearance level, blood glucose level, percent oxygen saturation in blood, rate of occurrence of new cases of a reportable disease (the number of cases is a discrete variable, but the rate may have a fractional component, and time is a continuous variable). Although we will briefly touch on continuous variables, most often the way a continuous variable is handled is to divide up its range into segments or bins, which can be considered to be discrete values. Typically, diagnostic tests are discretized in this way, so that a variable is assigned a "normal" range, and a test result is then classified as low, normal, or high.

In the case that we have multiple outcomes that are possible, one could have many subsets of outcomes, with overlapping elements. as well as individual outcomes. In the example of which base occurs at a location in a DNA sequence, there are four possible mutually exclusive outcomes, which we will call G, C, A, and T, as usual. Recall from Chapter 1, Section 1.1.2, that the different bases can be grouped by how many hydrogen bonds they form with each other. The A and T bases form two hydrogen bonds together, while G and C form three. So, we might ask what is the probability of occurrence of either G or C, rather than each alone. In general, to fully specify a probability function, we need to specify what are the probabilities of every possible subset of outcomes. Fortunately, there is a third axiom that constrains this function, namely that the probability of the union of disjoint subsets is the sum of the probabilities of the individual subsets. If we have a series of events (which could be subsets of more than one element, as in the DNA example), which we will call e_1, e_2, etc., and they are each pairwise disjoint (which is equivalent to saying no element is a member of more than one subset), then

$$\text{Axiom 3} \qquad P(\cup_{i=1}^{k} e_i) = \sum_{i=1}^{k} P(e_i). \tag{3.4}$$

Equation (3.4) is the third axiom of probability. It can also be expressed in a slightly different way, that the probability of an event consisting of a subset of *elementary* events, or single outcomes, is the sum of the probabilities of the individual outcomes.

When we consider the probability of two sets of outcomes occurring, it may be that the two sets overlap, or they may be disjoint. In the case that they are disjoint, for example, the probability of a base being either A or T, and the probability of it being G or C, it is easy to see that the probability of the union of the two sets is the sum of the probabilities. If the two overlap, for example, the set of G and C (the triple bonding pairs) and the purines (those are A and G), the sum rule does not apply. If we add the probabilities of the two subsets we will be counting the probability of occurrence of the overlapping part twice. So, denoting the two subsets as S_1 and S_2, the relation is instead,

$$P(S_1 \cup S_2) = P(S_1) + P(S_2) - P(S_1 \cap S_2), \tag{3.5}$$

where we have subtracted out the overlap. This can be easily proved from the three axioms (see [430, p. 6]).

In medical diagnosis it is common for a patient to have multiple symptoms, and possibly multiple diagnoses. These can occur separately or together. The sample space in cases like this is the *product* space of the individual outcomes. This means that each element of the sample space is a combination of outcomes for each of the separate measurements or outcomes. A medical example would be the problem of diagnosing pneumonia. One set of outcomes is the presence or absence of pneumonia. Another is whether the chest X-ray had positive findings or was negative. Yet another is whether the patient was experiencing cough or not. So, the sample space contains eight elementary events, as follows:

$$x_1 = \neg\text{pneum} \wedge \neg\text{xray} \wedge \neg\text{cough},$$
$$x_2 = \neg\text{pneum} \wedge \neg\text{xray} \wedge \text{cough},$$
$$x_3 = \neg\text{pneum} \wedge \text{xray} \wedge \neg\text{cough},$$
$$x_4 = \neg\text{pneum} \wedge \text{xray} \wedge \text{cough},$$
$$x_5 = \text{pneum} \wedge \neg\text{xray} \wedge \neg\text{cough},$$
$$x_6 = \text{pneum} \wedge \neg\text{xray} \wedge \text{cough},$$
$$x_7 = \text{pneum} \wedge \text{xray} \wedge \neg\text{cough},$$
$$x_8 = \text{pneum} \wedge \text{xray} \wedge \text{cough}.$$

Note that the event corresponding to the occurrence of one of the three items without considering the others is a subset of the above. So, the event, symbolized by "xray," is the subset consisting of x_3, x_4, x_7, x_8, and its probability would be the sum of the probabilities of those elementary events.

3.1.4 Conditional Probability

When multiple events occur, we can also express relations between them by measuring or computing the *conditional* probability that one event will occur given that another has already been determined to occur. The conditional probability that event x will occur, given that event y has already occurred, is denoted by $p(x|y)$.

The relations among the conditional probabilities, the individual probabilities, and the probability of both events are expressed by Equations (3.6) and (3.7). These equations provide a *definition* of conditional probability

$$P(x|y) = \frac{P(x \cap y)}{P(y)}, \tag{3.6}$$

$$P(y|x) = \frac{P(y \cap x)}{P(x)}. \tag{3.7}$$

Several things about Equations (3.6) and (3.7) should be noted. First, the definitions only apply when $P(y)$ and $P(x)$, respectively, are not zero. This is reasonable: if event y can't occur, it does not make sense to talk about the probability of another event, given that y has occurred, and so on. Second, if x and y are disjoint, then they don't occur together and $P(x|y)$ should be 0, which is consistent with the formula. Third, if y_i is the set of all possible (disjoint) outcomes, the conditional probabilities satisfy the following relationship:

$$P(x) = \sum_i P(x|y_i)P(y_i). \tag{3.8}$$

Finally, the two conditional probabilities are *not* equal in general, but since $P(x \cap y) = P(y \cap x)$, one can eliminate it from the two equations, giving a relation between the two conditional probabilities

$$P(x|y)P(y) = P(y|x)P(x) \tag{3.9}$$

and then we can solve the equation for one of the conditional probabilities in terms of the other

$$P(x|y) = \frac{P(y|x)P(x)}{P(y)}. \tag{3.10}$$

This formula is known as Bayes' Rule or Bayes' Theorem, named for the Rev. Thomas Bayes (1702–1761) who first derived it. We will use this formula extensively to support *probabilistic reasoning*.

One thing that can be done easily is to generalize this to the case of an arbitrary number of events, x_i. The joint probability for all the events, $P(x_1 \cap \ldots \cap x_{n-1} \cap x_n)$, is related by a kind of chain rule to the conditional probabilities, as follows:

$$P(x_1 \cap \ldots \cap x_{n-1} \cap x_n) = P(x_1|x_2 \cap \ldots \cap x_n)P(x_2|x_3 \cap \ldots \cap x_n) \ldots P(x_{n-1}|x_n)P(x_n). \tag{3.11}$$

This can be derived from the definition (generalized) of conditional probability. However, $2^n - 1$ numbers are still required (the one constraint remains that they all add up to 1). Later we will see that the requirements for determining the full joint probability can be drastically reduced in situations where the events and their relationships satisfy additional constraints.

Conditional probabilities are often used to express causal relations between events. What we mean by this is that when y occurs we expect most of the time that x will occur also, through some causal mechanism. However, we can define conditional probabilities whether or not we believe there is a causal relation between the two variables. In fact, the formulas ((3.6) and (3.7)) are symmetric with respect to x and y (although the two conditional probabilities are not equal) so there is no way to infer causality from the formulas or from raw data. Instead we construct a causal theory, test it, and if its predictions match the data, where no other theory does as well, we may accept the causal relation provisionally as true.[2]

When we hypothesize that y *causes* x, the conditional probability $P(x|y)$ is usually closer to 1.0 than to 0.0. The fact that it is not exactly 1.0 accounts for the nondeterministic nature of this causal relationship. For example, the presence of certain kinds of pollen usually causes allergy symptoms, but it is possible for the pollen to be present without the symptoms, even for the same individual.

3.1.5 Independence

Another idea concerning events is the idea of *independence*. If two events are independent, the probability that both occur is simply the product of the individual probabilities. The occurrence of one has no effect on the probability of occurrence of the other, in some sense. This is a *definition*, and it is stated mathematically in Equation (3.12)

$$P(x \cap y) = P(x)P(y). \tag{3.12}$$

In general, one can only determine if two events are independent by doing the calculations (if possible). Alternatively, one can posit the independence of two events as a statement about the biological or other process you are observing and modeling. In the pneumonia example above, we would expect that the three phenomena, presence of pneumonia, positive X-ray, and cough, are not independent. However, there are phenomena we believe really are independent, and we could build a statement to that effect (some form of Equation (3.12)) into our biomedical domain theory. Then we would derive consequences from that assumption.

The one condition under which you can determine if two events are independent or not, is the situation where the two events are disjoint. In that case, they cannot be independent, since the occurrence of one means that the other could not occur. Events being disjoint means that $P(x \cap y)$ is 0.0, but each of the individual probabilities is not 0, so their product is not 0, and therefore Equation (3.12) will not be satisfied.

One further thing to note is that independence does *not* mean that either of the conditional probabilities is 0. In fact just the opposite, if either of the conditional probabilities *is* 0, the events are disjoint, because then the probability of both occurring must be 0. Thus the two events are

---

[2]Often, there is only anecdotal data to support the theory, but for many people this is tantalizingly appealing, leading to the often used expression, "Coincidence? I think not."

not independent. If the conditional probabilities are *not* 0, the events might or might not be independent. If they are independent, combining Equations (3.12) and (3.6) (for example) gives

$$P(x|y) = P(x), \tag{3.13}$$

$$P(y|x) = P(y), \tag{3.14}$$

which states that the conditional probability of an event is independent of the other event.

It may happen that an event is not independent of another event in general, but given that a third event happens, the first two may become independent. Stated in terms of conditional probabilities, for events, x, y, and z, we say that x and y are *conditionally independent* given z if any one of the following conditions holds:

1. $P(x|z) = 0$,
2. $P(y|z) = 0$,
3. $P(x|y \cap z) = P(x|z)$ and $P(x|z) \neq 0$ and $P(y|z) \neq 0$.

This is a mathematical property of the probabilities, not something that has physical or biological significance. Its importance in the process of probabilistic reasoning is that a problem in which this condition is satisfied will be enormously simplified in terms of the number of data we need in order to calculate anything. In practice, one usually assumes conditional independence where it seems appropriate to the phenomena being modeled, rather than examining the data to determine it, because often there is not sufficient data, and the idea is to use what we know, to create a predictive model.

Thus, the biological or medical or health phenomenon that the probabilities represent may provide justification for claiming that events are conditionally independent. In the example above of pneumonia, we would expect that there is a causal relation between the presence of pneumonia and a positive finding on a chest X-ray, as well as one between the presence of pneumonia and the presence of a cough. Without knowing whether pneumonia is present or not, there may well be a correlation between cough and positive X-ray. The probability of a positive X-ray given cough likely will not be the same as the probability of a positive X-ray in general. However, once it is known that pneumonia is present, the probability of the positive X-ray likely *will* be the same, whether or not cough is present, so positive X-ray and cough would be conditionally independent, given the presence of pneumonia.

So, once again, we see that the mathematical and logical framework is the language in which we will express our theories of biology, medicine, and public health processes, but the particular shape these theories take must come from intuition, not just automatically from computations on data.

3.1.6 Random Variables and Estimation

Up to now we have been discussing the properties of probability values in relation to events, subsets of a sample space. When our sample space consists of a composite involving a number of (possibly related) observations, as in the pneumonia example, the sample space as a whole is difficult to describe. In addition, in many real-world applications, we don't have an easy way to define the sample space, but we can identify some property or properties of the sample space, perhaps subsets that can be well defined in terms of the values assigned to some variable. Most often in clinical trials or in biological experiments, we define variables, and we are interested in how one variable might depend on the other. For example, a biologist is interested in the possible effect of a gene function on metabolism of a certain cell type. She prepares cells with the gene present and cells without the gene, and observes whether there is any difference in the observable processes in the two cell groups. The presence or absence of the gene is a binary or Boolean variable, and there is some other variable measuring metabolic function of the cells, with perhaps many possible values. The events being observed are complex, but the subsets defined by the variable values are what is of interest, not the details behind them.

So, the idea of a *random variable* is that it provides an abstraction layer to some extent, over the sample space. Given a sample space and a probability function, the definition is as follows: a random variable is a function that assigns a real number to each element of the sample space. In mathematical notation, if ω is the sample space, R is the space of real numbers, and X is a random variable, then X is a function or mapping, as follows:

$$X : \omega \rightarrow \text{R}. \tag{3.15}$$

There could be many different mappings of the same sample space, corresponding to the fact that in an experiment, there are many ways to group the events, or assign a measurement to an event. An example is the staging of tumors. When a patient comes to the cancer clinic for radiation treatment, the radiation oncologist classifies the tumor according to its anatomic site, size, cell type, whether the cells look close to normal or have very deformed nuclei, etc. From this information they assign a stage, typically a number between 1 and 4. The data collected in a clinical trial are organized according to staging. The elementary events are the conditions of the individual patients, but in this case, stage would be a random variable on which the data analysis would be based. Questions to ask would be: how does the rate of tumor ablation (or even cure) depend on stage, for a given treatment regimen, or how does the recurrence rate depend on stage.

3.1.6.1 Probability Distribution Functions

Since the random variable assigns values to events in the sample space and we have assumed we have a probability function for the sample space, we can define *distribution functions* for the random variable.

First, we define the *cumulative distribution function*, or CDF. It is a function F whose domain is the real numbers and whose range is the interval $[0, 1]$. The value of $F(x)$ will be the probability of the event consisting of all the points in the sample space for which the random variable X is less than or equal to x,

$$F(x) = P(X \leq x), \tag{3.16}$$

where we have written $X \leq x$ to denote the subset of the sample space for which the value of X is less than or equal to x. This is well defined, since a random variable assigns real numbers to points in the sample space, so any real number will define a subset of the sample space. It could be the empty set if the random variable maps no outcomes in the sample space to a particular real number or range of real numbers. Every such subset has some probability assigned, so the function P has a value, and it will be between 0 and 1, by Axiom 1.

The cumulative distribution function has some useful properties that are easy to see. It is non-decreasing, since as x increases, more of the sample space is included and so the probability either stays the same or increases. As x tends to $-\infty$ the value of $F(x)$ tends to 0, since less and less of the sample space is included. As x tends to ∞ the value of $F(x)$ tends to 1, since eventually all of the sample space ω is included and $p(\omega)$ is defined to be 1.

We define two other functions for X, depending on whether X is *discrete* or *continuous*. We say X is discrete if the set of possible values for X is discrete, or put another way, countable.[3] For a discrete random variable we can define the *probability mass function* (sometimes called just the *probability function*), as the probability for the event consisting of all sample space points for which the value of X is one of these discrete values, x_i, labeled by an index, i

$$f(x_i) = P(X = x_i). \tag{3.17}$$

The CDF is related to the probability mass function in a simple way in this case. The CDF for a value x can be obtained from f by adding up all the values of $f(x_i)$ for x_i up to and including x

$$F(x) = \sum_{x_i \leq x} f(x_i). \tag{3.18}$$

In the case of a continuous random variable, the possible values will vary in some range, possibly the entire real line from $-\infty$ to ∞. In order to have a continuous random variable, there needs to exist a function, called the *probability density function*, or PDF, $f(x)$, satisfying three conditions, analogous to the probability mass function

1. for all x, $f(x) \geq 0$,
2. $\int_{-\infty}^{\infty} f(x)dx = 1$, and
3. for all real numbers a and b, where $a \leq b$, the probability that x is in the range from a to b is $P(X \in (a, b)) = \int_{a}^{b} f(x)dx$.

It follows from this definition that the CDF for a continuous random variable is just the probability of being in the range from $-\infty$ to x, or

$$F(x) = \int_{-\infty}^{x} f(t)dt. \tag{3.19}$$

---

[3]A set is countable if it is finite, or if it is infinite but able to be mapped one to one to the positive integers.

Note that the PDF, $f(x)$, is *not* the probability of the random variable X having the value x. In fact, that probability is 0. Unlike the probability mass function, the PDF can have values much greater than 1, and can even be infinite. What matters is its integral over some finite range.

Most of the time, we write statistical formulas in terms of the random variables, and we don't consider the underlying sample space. In general we are going to be writing formulas in terms of the probability mass function or density function.

Several probability distribution functions have become standard and widely used. Which one is applicable to describe a particular biomedical situation is a modeling choice that may be motivated by what you believe or hypothesize about the variables you are modeling. Here we just describe a few of the most common.

One of the simplest discrete distributions is the *Uniform* distribution. In the discrete case, the probability mass function describes a variable that takes on one of some fixed number, n, of values, all with equal probability. If x is one of the values, $1, \ldots, n$, then $f(x) = 1/n$, and it is 0 otherwise. The analog for the continuous case is similar, except that instead of a fixed *number* of values, the variable takes on a fixed *range* of values, from a to b inclusive. In this case, if $x \in [a, b]$ then the probability density function takes on the constant value, $f(x) = 1/(b - a)$, and it is 0 otherwise.

A popular contraption on display at the Pacific Science Center in Seattle, Washington, is a large box, tall, and wide but shallow in depth, with horizontal pegs set out in a vertical triangular array, equally spaced, with apex at the top, so that in each succeeding row, each peg in that row is just under the midpoint between the two upper pegs, or in the case of the end pegs arranged so that the peg above is above the midpoint of the two pegs just below. At the top, a conveyor drops hard wooden balls one at a time on the top peg. Each ball cascades down through the peg array to the bottom where there are (a discrete set of) bins to collect the balls. Figure 3.1 illustrates this schematically.

Roughly, the probability of going to the right or left when a ball hits a peg is 1/2, and the distribution at the bottom, the probability of a ball landing in a particular bin, is a special case

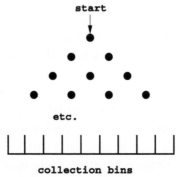

Figure 3.1 **A contraption that illustrates the Binomial distribution, for the special case that** $p = 0.5.$

of the *binomial distribution*. In general, the probability of one or the other choice at each point can vary from 0 to 1. For n levels and probability p (assumed the same at each choice point, or peg, in the illustration), the probability mass function is

$$f(x) = \binom{n}{x} p^x (1 - p)^{n-x}$$

for $x = 0, \ldots, n$ and 0 otherwise. The symbol $\binom{n}{x}$ is the *binomial coefficient*,

$$\binom{n}{x} = \frac{n!}{x!(n - x)!}.$$

In this formula, n and p are parameters of the distribution function. Most distribution functions are specified with parameters. One of the goals of statistics is to provide methods for estimating the parameters of a distribution from measured or observed data, where you have hypothesized that the thing being observed is modeled by the distribution you have chosen.

There is a continuous analog of the binomial distribution, called the *Normal*, or *Gaussian*, distribution. The probability density function for the Normal distribution is

$$f(x) = \frac{1}{\sigma\sqrt{2\pi}} \exp\left\{-\frac{(x - \mu)^2}{2\sigma^2}\right\}, \tag{3.20}$$

where μ can have any value, and σ has to be positive. This function has its maximum at $x = \mu$ and it is symmetric with respect to $x = \mu$ (meaning that $f(\mu + x) = f(\mu - x)$). It is the well-known "bell-shaped curve." When μ is 0 and σ is 1, it is called the *standard Normal distribution*. Applying Equation (3.19) would give the Cumulative Distribution Function corresponding to the Normal distribution, but this integral can't be done in closed form. Its shape, however, is well known, a sigmoid curve, starting from 0 at $-\infty$, rising quickly in the vicinity of μ, and then asymptotically approaching 1. The CDF has a value of 0.5 at $x = \mu$, and is also symmetric about $x = \mu$. The Normal distribution is important not only because it has some nice mathematical properties, but also because of the Central Limit Theorem, which says essentially that as we make larger and larger numbers of observations, the probability distribution of the mean of all the observations behaves more and more like a Normal distribution.

3.1.6.2 Expectation and Variance

The expectation of a random variable provides some information about the distribution function for that random variable. If we know or hypothesize what the form of the distribution function is, such as one of the standard well-known ones, there will be a relation between the parameter or parameters that specify the distribution function and the expectation. So, knowing the expectation we could recover the parameters and thus the full distribution function. Thus, estimating the expectation from actual observations is very important.

The expectation of a random variable is analogous to computing an average. However, in general, the different values that the random variable can take are not all equally probable. If we were to make a series of observations and they occurred with frequencies corresponding to the probabilities of each, when we average our observed values, we will expect to get a *weighted* average, not the midpoint of the range of possible values. Thus we define the *expectation* of a random variable, X, as the weighted average, where we multiply each value by the probability of it occurring. The following formula applies to a discrete random variable, X,

$$E(X) = \sum_{i=1}^{n} x_i f(x_i), \tag{3.21}$$

where $f(x_i)$ is the value of the probability mass function at $X = x_i$.

For a continuous random variable, we would use the probability density function, and do an integral instead of a sum

$$E(X) = \int_{-\infty}^{\infty} x f(x) dx. \tag{3.22}$$

For the example distributions described here, the expectations can be computed in terms of the parameters. For the Uniform distribution, in the discrete case, it is $n/2$ and in the continuous case for the interval $[a, b]$, the expectation is simply $(a + b)/2$, the midpoint of the interval. For the Binomial distribution, it is np, and for the Normal distribution, it is μ.

A reasonable estimate of the expectation of a discrete random variable is the *mean* of a collection of observational values. The observational values will (in a large enough sample) occur with repetitions, and a frequency histogram of these values should approximate the probability distribution

$$\mu = \frac{1}{n} \sum_{i=1}^{n} M_i. \tag{3.23}$$

Here we have used μ to denote the estimated value of the expectation. This (mean value) sometimes is written $<M>$. The Law of Large Numbers is an important theorem in statistics, which says that if we treat each observation as a sample of a random variable, and thus the mean is also a sample of a random variable, as the number of observations gets larger and larger, the probability distribution for the mean as a random variable gets more and more concentrated around the expectation of the distribution. This does not say anything about the shape of the distribution; the shape is addressed by the Central Limit Theorem, previously mentioned.

In addition to the mean, it will be useful to compute the expected value of functions of a random variable, for example, polynomials or other functions. They are computed in the same way, namely, if $Y = g(X)$, we compute the expectation of Y by

$$E(Y) = \int_{-\infty}^{\infty} g(x) f(x) dx. \tag{3.24}$$

One such function, the *variance*, is especially useful. The variance, σ^2, of a distribution is a measure of how wide or narrow it is. It is the expectation of the square of the deviation of the random variable value from the mean. For a discrete random variable, it is

$$\sigma^2 = \sum_{i=1}^{n} (x_i - \mu)^2 f(x_i) \tag{3.25}$$

and for a continuous variable, it is

$$\sigma^2 = \int_{-\infty}^{\infty} (x - \mu)^2 f(x)dx. \tag{3.26}$$

We can also get an *estimate* of the variance of a distribution, by averaging the squared deviation of the measurements or observations from the mean value. We compute the squared deviation because the deviation can be in either direction, positive or negative, and the average deviation will be 0.[4] This average squared deviation is also called the variance and its square root is called the *standard deviation*. We take the square root to get a number whose measurement units are the same as the measurements themselves, and the number then can be used as an indicator of how far from the mean the measurements typically will be found

$$\sigma^2_{\text{est}} = \frac{1}{n} \sum_{i=1}^{n} (M_i - \mu)^2 \tag{3.27}$$

and of course $\sigma = \sqrt{\sigma^2}$. We can code this pretty simply, for later use, for example in a syndromic surveillance system. For simplicity we represent the measurements as an array, m, and remember that in computer programs, arrays are zero-indexed while in mathematical expressions, sequences are usually one-indexed. In the variance function, mu is the computed mean, μ, and the variance function returns both the variance σ^2 and the standard deviation, σ as multiple values.

Knowing how to do iteration, one might be inclined to write the expressions for mean and variance as do constructs. The mean, for example, would look like this:

```
(defun mean (m)
  (let ((n (1- (array-dimension m 0))))
    (/ (do ((i 0 (1+ i))
            (stop n)
            (result 0))
           ((> i stop) result)
         (incf result (aref m i)))
       n)))
```

Expressions with sums and products occur often in probability and statistics. Rather than having to write out these do loops, it would be nice to have sum and product operators,

---

[4]It's simple to prove this and it is left as an exercise for the reader.

that can take expressions and transform them into iterative code as above. The Common Lisp standard does not include such operators, but we can write macros to do these jobs. Following an example in Graham,[5] we can define these operators, and then use them as if they were built in. They follow the form of the `for` macro in Graham, but each does the sum or product operation as part of the definition, so that makes their use simpler even than using the `for` macro. Once the macro is defined, we can use it as follows: to sum a series of values of an array, m, as in the mean function, we would write,

```
(sum i 0 n (aref m i))
```

In the above expression, the variable i does not need to be defined, just used. The code for sum will increment i from 0 to $n-1$ as in the original do expression. So, the value of i is *not* being passed to sum. The *symbol* i is used by the macro.

Here is an implementation of the sum macro, which evaluates the body of code that is input, for each value of the symbol passed in as var from the specified start value to the stop value, inclusive, summing up the results of evaluating body each time through as a progn.

```
(defmacro sum (var start stop &rest body)
  (let ((gstop (gensym))
        (gresult (gensym)))
    `(do ((,var ,start (1+ ,var))
          (,gstop ,stop)
          (,gresult 0))
         ((> ,var ,gstop) ,gresult)
       (incf ,gresult
             (progn ,@body)))))
```

The `let` form creates some unique local variables using gensym, and then uses them in the iteration. The symbols gstop and gresult are not literally used. Their *values* (the symbols created by gensym) are the symbols (variables) actually used in the do expression that the macro creates when it is called. Those are initialized to the values passed in with the macro call. This is to avoid the possibility that any particular name we might choose there would be also in the body code that the user of this macro provides. That is called variable capture and it is a source of subtle errors. On the other hand, the value of the first parameter, var, is exactly the name of a variable used for the iteration, and it is intended that it appear in the body.

When this new definition is used, it will be translated by macro expansion into an expression very similar to the first implementation above. In using this macro the symbol i above is the value of var in the macro definition, the number 0 becomes the value of start, and the array-dimension function is used to get the stop value (one less than the size of the array). The result of macro expansion of a sum expression, as it would be used in the mean function,

---

[5]We introduced the basic idea of how to write a macro with defmacro in Chapter 2, on p. 199. A brief introduction to macros can also be found in Graham [137, Chapter 10].

can be seen by using the Common Lisp function macroexpand-1. Its input is the expression as we would code it, and the transformed code is shown following, in all upper-case.

```
> (macroexpand-1
      '(sum i 0 n (aref m i)))

(DO ((I 0 (1+ I))
     (#:G16 N)
     (#:G17 0))
    ((> I #:G16) #:G17)
  (INCF #:G17
    (PROGN
      (AREF M I))))
```

The symbol G16 corresponds to stop in the original code, and G17 corresponds to result in the original code. These symbols are unique symbols that were created automatically by the macro expansion process. The result of the macro expansion is pretty much the same as the original do expression that we wrote. So, having this macro, we can just *use* it instead of writing out complicated do forms. The macro expander translates our code into the more complicated version for us. Although the body of this example is just the aref expression, the body of the macro call can be any sequence of expressions, as if the sum macro were just built into Common Lisp. So, the mean and variance functions can be coded very simply and readably.

```
(defun mean (m)
  (let ((n (1- (array-dimension m 0))))
    (/ (sum i 0 n (aref m i))
       n)))

(defun variance (m mu)
  (let* ((n (1- (array-dimension m 0)))
         (sigma-squared
           (/ (sum i 0 n (expt (- (aref m i) mu) 2))
              n)))
    (values sigma-squared (sqrt sigma-squared))))
```

A small amount of algebra can be used to derive an alternate form of the formula for variance. Expand the squared expression, then use Equation (3.23) to simplify the result. This gives

$$\sigma^2 = \frac{1}{n} \sum_{i=1}^{n} M_i^2 - \mu^2. \tag{3.28}$$

The first term is just the average of the squared measurements, $<M^2>$. The second term is the square of the mean, $<M>$, so the formula becomes the difference between the mean of the square and the square of the mean

$$\sigma^2 = <M^2> - <M>^2.$$

(3.29)

When handling large data sets, some attention must be given to efficiency of the arithmetic. In this case, we have two forms for the same computation. For a given n, one may be faster, or they may be about the same. This is not a question of whether the sum macro is faster than the hand-written do form, because they end up the same code. It is about which algebraic formula involves fewer operations. It is left as an exercise for the reader to determine which is computationally cheaper.

3.1.6.3 Regression: The Method of Least Squares

Now we can consider the case where two random variables may in fact be related somehow, perhaps because they represent physical phenomena, one of which may be caused by or strongly affected by the other. One of the variables is considered an independent variable and the other a dependent variable. In controlled experiments, values for the independent variable, X, are set by the experimenter and the values for the dependent variable, Y, are then observed in each case. So, we are considering a set of correlated pairs of observations, x_i, y_i. The question is, what is the relation between X and Y? Most often it is assumed that they are related by a straight line or linear formula, roughly speaking, of the form,

$$Y = mX + b.$$

(3.30)

Then our task is to estimate the parameters m and b from the set of observations. The most commonly used approach is to choose the values for m and b that minimize the average squared difference between what the formula would predict for Y, given that $X = x_i$ and the observed y_i. So the task is to minimize the sum of the squared differences between the computed Y values and the measured ones

$$R(m, b) = \sum_{i=1}^{n} [y_i - (mx_i + b)]^2.$$

(3.31)

The values for m and b that minimize R can be obtained by solving the equations that are obtained by setting the first partial derivatives with respect to each to 0

$$\partial R / \partial m = \partial R / \partial b = 0.$$

(3.32)

These are two equations in the two unknowns, m and b, solvable with a bit of algebra. The results are the formulas for the estimates of m and b

$$<m> = \frac{\sum_{i=1}^{n} (x_i - \overline{x})(y_i - \overline{y})}{\sum_{i=1}^{n} (x_i - \overline{x})^2}$$

(3.33)

and

$$<b> = \overline{y} - <m> \overline{x}.$$

(3.34)

There is nothing magical about the Method of Least Squares. Assuming a linear model is a statement of belief about the underlying biological or pathological process. The requirement of minimizing the sum of the squares of the residuals is a reasonable choice to give precision to the somewhat vague idea of "best fit" of the model to the data. When the relationship between X and Y is expected to be more complicated, similar methods can be used but the solutions for the estimates of the parameters may be difficult to obtain. This entire process is called Regression Modeling, and is treated more extensively in standard statistics texts such as [430].

3.1.6.4 Multiple Random Variables: The Joint Probability Distribution

In general, a domain has multiple random variables, not just one. The pneumonia example illustrates this situation. The sample space is a Cartesian product of the individual sample spaces. Instead of a single random variable we model the situation with three random variables, and the distribution functions are functions of multiple variables:

- An *atomic event* is an assignment of particular values to all the variables, $X_1, \ldots, X_n$.
- The table of probabilities for each combination of values is called the joint probability distribution, $P(X_1, \ldots, X_n)$.
- Often, each variable X_i is a Boolean, but this is not required.

An example of a simple joint probability distribution with just two variables, high PSA and prostate cancer, is shown in Table 3.1. A very common test that is used for screening for prostate cancer is the PSA test. PSA stands for "Prostate-Specific Antigen," a protein that may be present in blood in very small concentrations, but in the presence of prostate cancer the concentration is higher. A PSA below 4 ng/ml of blood serum is considered normal. A PSA of between 4 and 10 that is rising over time is a good indicator of the need for more extensive investigation. Above 10, an even higher level of concern is appropriate, but for our purposes we just model PSA as a Boolean event, with the two states, normal and high (above 4). The joint probability distribution table shows the probability for each combination of values of each of the variables. In general, for n Boolean variables, the joint probability distribution will have 2^n entries. For variables that can have more than two values, the table gets correspondingly larger. The size of the table is the product of the number of values each random variable can have.

There are some constraints on the values in the joint probability distribution table. First, since every entry in the table is the probability of one of the possible elementary outcomes, and the

Table 3.1 A Simple Joint Probability Distribution for Prostate Cancer and High PSA

	High PSA	¬High PSA
Prostate cancer	0.25	0.07
¬Prostate cancer	0.03	0.65

table covers all the outcomes, the sum of all the numbers in the table has to add up to 1.0. Table 3.1 satisfies this requirement.[6] Second, we expect to see some dependence between events in the table. If the PSA test has diagnostic utility, there should be some correlation between the presence of prostate cancer and a positive test, and vice versa, between the absence of prostate cancer and a negative test. This also appears to be the case in Table 3.1.

The PSA test is typical in medical practice in that it is not *pathognomonic*. A *pathognomonic* finding unequivocally signals the presence of a particular cause. One of the most common pathognomonic findings is the presence of two independent heart beat sounds in a woman. For sure, this indicates that the woman is pregnant.[7] Most tests have some level of accuracy that is well known. However, the notion of "accuracy" needs to have some precision as it can easily be misunderstood.

To describe the accuracy of a test, we need two numbers, not just one. First, the *sensitivity* of a test is defined to be the probability that a patient will test positive, given that the disease is present. This is the *true positive rate* relative to the disease positive population. It is the true positives divided by the total patients with the disease (the true positives plus the false negatives). In the prostate cancer example above it is $0.25/(0.25 + 0.07)$ or 0.78. The sensitivity tells you something about how well the test indicates the presence of the disease, but it does not tell you anything about the behavior of the test in the absence of the disease, namely the false positives. For that, the *specificity* is defined to be the *true negative rate*, the probability of a negative test when the patient does not have the disease. The specificity is computed by dividing the true negatives by the sum of the true negatives and false positives. For the prostate cancer example, it is 0.96 so it seems that the test is pretty specific.

Typically, we don't know the joint probability distribution function but we do know the sensitivity and specificity of a test. Since there is one numeric constraint on the table values (they have to add up to 1.0), if we had one more number, we could compute the table. That number is the overall rate at which the disease occurs in a particular population. This is called the *prevalence*, or prior probability. The sum of the numbers in the row representing presence of prostate cancer should be the probability overall of occurrence of prostate cancer in some specific population (of men), the population used to generate the numbers in the table. From Table 3.1, this is about 0.32, meaning that we expect that 32% of the men in that group will experience prostate cancer. In the next section, we will take another look at how to use Bayes' Rule in this and in more complex problems.

---

[6]The values in this table are *not* exactly the values found in the medical literature. In reality, there is wide variability in the definition and interpretation of PSA test results.

[7]This does not mean that the various chemical pregnancy tests are unnecessary. The pregnancy is well along by the time a fetal heart beat can be detected, and it may well be useful to know much sooner.

3.2 Bayes' Rule, Bayes Nets, and Probabilistic Inference

Bayes' Rule is the basis for modeling a large number of complex phenomena, from a simple causal relation between two variables, to a large and complex network of relationships among events. The events and the causal relationships among them form a graph, with the events as nodes in the graph and the relationships as directed arcs or edges connecting the nodes. The simplest possible graph is one with two nodes linked by a single arc, as shown in Figure 3.2.

The PSA test for prostate cancer is an example of this kind of relationship. It is a practical application rather than a scientific hypothesis. It is known that high PSA levels are associated with prostate tumors, so we can say that there is a causal link, with prostate cancer being the cause and the high PSA state being the effect. This is a case where we observe the positive test result and would like to draw some inference about the probability of the presence of prostate cancer.

In a rather different kind of application, we may wish to use the graph as a predictor of the effect, rather than a way to ascertain the cause. An example with many nodes and links is Figure 3.3, which is an example of a Bayes Net for modeling the prognosis of prostate cancer [281, 327, 389]. In this example, it is already known that the patient has prostate cancer and we wish to estimate (predict) the probability of effective treatment, or disease control.

The prognosis is dependent on findings about the disease, as well as treatment parameters (radiation dose to the tumor) and the lymph nodes at risk for metastasis. Control of prostate cancer is dependent on controlling the local disease—the tumor cells within the prostate itself— and any cells which have migrated to the nearby lymph nodes. The PSA level, Gleason score, and tumor stage are three clinical tests which describe the extent of the disease, and along with radiation dose, are predictive of cure rate. The nodes "Local Tumor Control" and "Lymph Node Control" each have states "yes" and "no." The node "Disease Control" reflects the fact that an individual whose lymph nodes are not cured has a 90% chance of further spread, regardless of the state of "Local Tumor Control," while if "Lymph Node Control" is "yes," only 1/3 of patients for whom "Local Tumor Control" is "no" will develop distant metastasis.

So the first consideration is how to use Bayes' Rule to compute one conditional probability when the other is known. The example is for a Boolean sample space, but the basic Bayes formula easily generalizes to the case of a sample space like the DNA bases, where there are

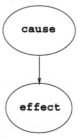

Figure 3.2 A simple causal relationship diagram.

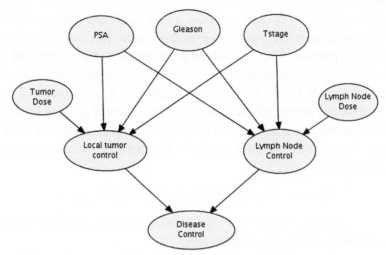

Figure 3.3 A Bayes Net for computing a prognosis for radiotherapy treatment of prostate cancer. *(Courtesy of Mark Phillips and Wade Smith.)*

many mutually exclusive possible events. Then we examine the properties of the larger networks or graphs, called Bayes Nets. The particular model shown in Figure 3.3 will be described in more detail in Section 3.2.4.

3.2.1 Simple Bayesian Inference

In most circumstances the joint probability distribution table is not available. However, the prostate cancer example illustrates a common situation in medicine. We have some test for the presence or absence of a disease. Usually what is known about the test is its sensitivity and specificity, collectively called the *accuracy* of the test. One might think if the test is 90% accurate, and you test positive, the probability that you have the disease is 90%. This is incorrect. The 90%, or probability of 0.9, is the probability of testing positive given that you have the disease, not the probability that you have the disease, given that you tested positive. The sensitivity and specificity are the conditional probabilities, respectively, of testing positive given that you have the disease, and of testing negative, given that you do not have the disease. These numbers are *not* the same as the probabilities we are really interested in, namely the probability that you have the disease given that you test positive, or conversely, the probability that you do not have the disease given that you test negative.

Fortunately, Bayes' Rule, Equation (3.10), provides a way to compute one conditional probability from the other. In the prostate cancer case let's suppose that we start with the following information:

- The PSA test for prostate cancer has a sensitivity of 0.78 and a specificity of 0.96, which would seem pretty useful.
- The overall rate of occurrence of prostate cancer in the male population being tested is 0.32 as mentioned earlier.

From this information, Bayes' Rule provides a way to compute the probability of having the disease given a positive test. However, we need one more piece of information, the overall rate of occurrence of positive PSA tests, which is the denominator in Equation (3.10). We do *not* need to go back to the joint probability distribution table. This quantity is related to the other quantities we already know. The event of positive PSA is the union of two things, the event of positive PSA *and* having prostate cancer, and the event of positive PSA *and not* having prostate cancer. Since those two are disjoint, the probability of positive PSA is the sum of these two probabilities. Using the definition of conditional probability, we can write them in terms of conditional probabilities, giving

$$P(\text{psa}) = P(\text{psa}|\text{pc})P(\text{pc}) + P(\text{psa}|\neg\text{pc})P(\neg\text{pc}). \qquad (3.35)$$

In Equation (3.35), psa is the event of having a positive PSA test and pc is the event of having prostate cancer. More generally, the formula is

$$P(y) = P(y|x)P(x) + P(y|\neg x)P(\neg x), \qquad (3.36)$$

where x and y are events (in the prostate cancer case, x is pc and y is psa). However, we *still* don't have all the numbers. In particular, we don't have $P(y|\neg x)$ or $P(\neg x)$. These are related to the numbers we do have, because Axiom 1 says that $P(\neg x)$ is just $1 - P(x)$, and similarly, $P(y|\neg x)$ is $1 - P(\neg y|\neg x)$. We know $P(\neg y|\neg x)$, which is the specificity. So the final version of Bayes' Rule becomes

$$P(x|y) = \frac{P(y|x)P(x)}{P(y|x)P(x) + [1 - P(\neg y|\neg x)][1 - P(x)]}. \qquad (3.37)$$

This equation now uses only the three numbers we typically would have available to us. So, doing the arithmetic for the prostate problem gives

$$\frac{0.78 \times 0.32}{(0.78 \times 0.32) + (0.04 \times 0.68)} = 0.90 \qquad (3.38)$$

and we can conclude that the probability of prostate cancer given a positive test is 0.90, not 0.78. So, it really pays to do the arithmetic, as the answer is significantly different from the naive (and incorrect) answer.

Now, let's suppose the test is much more accurate, but the disease we are testing for is very much more rare. As an example, use the number 0.99 for both sensitivity and specificity, but 0.0001 (1 in 100,000) for the disease prevalence. Then the calculation gives, for a positive test result,

$$\frac{0.99 \times 0.0001}{(0.99 \times 0.0001) + (0.01 \times 0.9999)} = 0.0098. \qquad (3.39)$$

This is dramatically different. The test is 99% accurate but the probability of having the disease if the test is positive is only about 1%. This result comes about because the disease is so rare that the occurrence of false positives overwhelmingly obscures the true positives.

Even more important, note that the sensitivity and specificity are properties of the relation between the test and the disease, and they do not change from one population to the other, but the prevalence may vary greatly from one population to another, and this will strongly affect the interpretation of test results.

Where do the numbers come from? The accuracy of diagnostic tests is the subject of medical (clinical) research. It is part of the work of public health practitioners to collect data to estimate the prevalence of various important diseases, so for many diseases also the overall prevalence or occurrence rate is known.

Implementing a formula like Equation (3.37) is straightforward in almost any computer programming language. Here is a version in Common Lisp.

```
(defun disease-given-test (sensitivity specificity prevalence)
  (/ (* sensitivity prevalence)
     (+ (* sensitivity prevalence)
        (* (- 1 specificity) (- 1 prevalence)))))
```

This is the simplest possible application. It is easy to generalize the formula to the case where there are many possible (mutually exclusive) cause states rather than just presence or absence of a disease.

3.2.2 Non-Boolean Variables

In some cases the causal hypothesis is not a Boolean variable, but can have any one of many mutually exclusive values, which we denote as c_i. We can generalize Bayes' Rule to handle this case. For each causal condition c_i, we have a probability, $P(c_i)$. The conditional probabilities for effect e, given each value of the cause variable c_i, similarly are represented by a conditional probability, $P(e|c_i)$. For each possible value of the causal variable, we would know the conditional probability of a positive effect. We would also need to know the prevalence rate or occurrence rate of each causal value. From this we can compute the probability of each of the causal values, given a positive effect.

This kind of situation occurs when a causal variable has continuous values but is typically divided up into discrete intervals for purposes of simplifying the calculations. An example of such a variable would be number of years of smoking, with the effect being any of the several diseases known to be linked with smoking (heart disease, COPD, emphysema, etc.). Rounding to whole numbers of years is a way to group the data in discrete bins, as is lumping still further into 5-year intervals.

In this case, the prior probability of the effect can be computed as a sum over all the cause values, of the products of the prior and conditional probabilities, just as for the Boolean case where we summed over the two cases, true and false

$$P(e) = \sum_i P(e|c_i) P(c_i). \tag{3.40}$$

Having a way to compute $P(e)$, we can now compute the conditional probability of each causal value given a single effect. This, together with Equation (3.40), is one form of a generalized Bayes' Theorem:

$$P(c_i|e) = \frac{P(e|c_i)P(c_i)}{P(e)}. \tag{3.41}$$

Now instead of simple true and false, we have a collection of values and probabilities, indexed by an integer labeling the values. A straightforward way to implement this computation is to represent the probabilities as arrays indexed by the value number, i. Our function to compute $P(c_i|e)$ will have as inputs, the index, i, the conditional probability array, which we will call sensitivity because its elements are the sensitivities of the effect to the occurrence of each causal value, and the array containing the prior probability of each causal value, which we will call prevalence. We will also need to implement formula (3.40) as a function summing over products of elements of the two arrays. Now we can see it was worth the trouble to define a sum operator, because we can use it in the implementation of formula (3.40). To be general, we would obtain the array dimension of the prevalence array to use as the iteration upper bound. Using the sum macro defined on p. 359, the effect-prob function, implementing the formula for $P(e)$ looks like this:

```
(defun effect-prob (sensitivity prevalence)
  (sum i 0 (1- (array-dimension prevalence 0))
      (* (aref sensitivity i) (aref prevalence i))))
```

Formula (3.41) can now be coded very simply by utilizing the function just defined for $P(e)$. Of course one could also write the body out in the following, but the layered design approach is likely to pay off as the formulas get further generalized.

```
(defun cond-cause (i sensitivity prevalence)
  (/ (* (aref sensitivity i) (aref prevalence i))
     (effect-prob sensitivity prevalence)))
```

You might wonder what happened to the problem of having to incorporate the specificity and the rule that probabilities add up to 1.0. They are present. The Boolean case is just the case of two causal factors, one the negative of the other. In the more general case, what was specificity, or the true negative rate, is now for each of the possible causal values, the true positive rate for that value. So one could think of specificity in the Boolean case as the sensitivity for testing negative.

3.2.3 Bayes Nets

A more complex graph such as Figure 3.3 represents a much larger collection of variables, each with a number of states. The joint probability distribution for such a graph is large, since every combination of values for each variable needs an entry. For the graph in Figure 3.3, the Tumor Dose node has five states, the PSA node has six (unlike our simple binary PSA test), the Gleason score has four states, and T-stage has three. The rest each are Boolean nodes, so altogether, the joint probability table for this set of variables would have $5 \times 6 \times 4 \times 3 \times 2 \times 2 \times 2 \times 2$, or 2880 states. To glean that much information from clinical trials data or retrospective data would be simply impractical. The point of the graph is that we can reasonably hypothesize causal dependencies among the variables, and thus the graph represents a kind of theory of prostate cancer, relating findings, and prognosis. The arcs connecting the nodes represent causal or predictive links. The PSA, Gleason score, and T-stage are all predictive of local tumor control and lymph node control. Tumor dose (from radiation treatment) also influences local tumor control, and whether we irradiate the lymph nodes will have an influence on lymph node control.

In the case of diagnosis, we still needed three numbers to determine everything we would want to compute, though they are more convenient than the joint probability table numbers. In this case, however, provided certain conditions are met, we can compute anything we want starting with far less than the full joint probability table.

First, consider a node with only incoming arcs, as in Figure 3.4. Each of the three variables is Boolean. We would like to compute the probability of the effect given the presence or absence of the causes. This is actually another version of the non-Boolean case just considered.

Although Cause-1 and Cause-2 are not mutually exclusive, the set of combinations of values for each does form a set of mutually exclusive events. In this case, the events are the four combinations of presence (T) or absence (F) for the two causes, namely, the set [TT, TF, FT, FF]. So, for this graph, we can just use formula (3.40) to compute $P(E)$, the probability of Effect, and formula (3.41) if we observe the presence or absence of Effect and want to know, for example, $P(c_1)$, the probability of one of the causes. Of course, $P(c_1)$ is the sum of two terms, $P(c_1 = T \cap c_2 = T)$ and $P(c_1 = T \cap c_2 = F)$, but each can be computed from formula (3.41).

In general, however, when we use a graphical representation to denote a causal model of a biological, clinical, or public health phenomenon, we will have a more complex graph. In

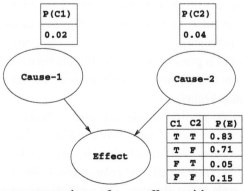

Figure 3.4 A graphical representation of an effect with two causal factors, showing the prior probabilities of the causes and the conditional probability for the effect given each combination of causes.

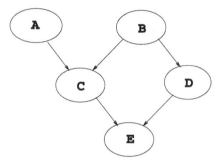

Figure 3.5   A graph with nodes that have both parents and children.

particular, we will have causal factors that lead to two or more effects, as well as multiple causes for a single effect, as illustrated in Figure 3.5.

In these cases the probabilities are much more difficult to compute. However, if each node is conditionally independent of all its non-parent nodes, given its parent nodes, we can write the joint probability in terms of relations of each local node to its parent nodes. For most reasonable graphs, this drastically reduces the number of probabilities that are needed.

To illustrate this idea we start with the simplest possible case, where a single cause has two effects. How do we combine them to estimate the conditional probability of a cause? We will use the example described earlier of testing for prostate cancer as an illustration. In addition to the high PSA test result, we consider the percentage of "free" or unbound PSA. Many observational variables have been considered as diagnostic markers for prostate cancer, but no single one seems to have both the sensitivity and specificity that is desired. Considering two (somewhat independent) variables should give some improvement, but we need a formula for updating our previous result of how to compute the probability of prostate cancer given a high PSA value.

Most of the PSA observed in serum is bound to a protein inhibitor, alpha-1-antichymotrypsin. The percentage of PSA that is unbound (non-complexed) is referred to as the percentage of "free" PSA, a fraction of the total. In diagnostic testing, two numbers are reported, the total PSA concentration and the percentage of it that is "free." The percentage of free PSA seems to be lower when prostate cancer is present. When the PSA is between 4 and 10, if the fraction of free PSA is less than 0.10, the probability of having prostate cancer is considerably higher than it is if the free PSA fraction is higher.

Here, rather than distinguishing all the levels and combinations, we just use simple categories. In the formulas, to be concise, we abbreviate "Prostate Cancer" as "PC," "High PSA," that is, greater than 4, as just "psa" and "abnormally low fraction of free PSA," less than 0.10, as "fpsa." Figure 3.6 shows diagrammatically the relationships. We have filled in the conditional probability tables for each node in the diagram. The prostate cancer node just has one entry, the prior, or overall, probability of occurrence of prostate cancer. The high PSA node conditional probability table is calculated from the joint probability distribution in Table 3.1. We have not

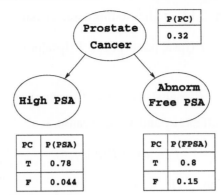

Figure 3.6 A Bayes Net for prostate cancer.

provided the corresponding joint distribution for fpsa, but only the conditional probabilities of observing fpsa when prostate cancer is present and when prostate cancer is not present.

Essentially the conditional probabilities for each of the tests give the true positive and false positive rates. From these and Bayes' Rule, we can compute what we are really interested in, that is, given a positive test, what is the probability of having prostate cancer

$$P(\text{pc}|\text{PSA}) = \frac{(0.78)(0.32)}{(0.78)(0.32) + (0.044)(0.68)} = 0.89. \tag{3.42}$$

This result is slightly different from the previous one because the false positive rate has an additional decimal place, giving a slightly lower result

$$P(\text{pc}|\text{fpsa}) = \frac{(0.8)(0.32)}{(0.8)(0.32) + (0.15)(0.68)} = 0.715. \tag{3.43}$$

Each test separately gives some significant indication, but we would expect that both tests together would give a stronger prediction. So, suppose we have high PSA *and* abnormal free PSA? How do we combine the information from both tests? We can use Bayes' Rule here too,

$$P(\text{pc}|\text{psa} \cap \text{fpsa}) = \frac{P(\text{psa} \cap \text{fpsa}|\text{pc})\, P(\text{pc})}{P(\text{psa} \cap \text{fpsa})} \tag{3.44}$$

but we don't have the conditional probabilities required here, the joint probability of both tests being positive or negative, etc., given the presence (or absence) of prostate cancer. In this case there are only a few combinations, but in general there are (as we saw with the prognostic graph) a large number of combinations, and direct observation of the joint causal probabilities simply will not scale to larger graphs.

Instead, following the strategy described for a dental reasoning case similar to this one, in one of the standard AI textbooks [360, p. 481], we observe that the two tests are actually independent, conditioned on the presence or absence of the disease. If you have the disease, the residual effects on the test outcomes would be expected to be independent of each other. Thus, we could claim that Equation (3.13) applies to this case, so that

$$P(\text{fpsa}| \text{psa} \cap \text{pc}) = P(\text{fpsa}|\text{pc}) \tag{3.45}$$

and

$$P(\text{psa}|\text{pc} \cap \text{fpsa}) = P(\text{psa}|\text{pc}). \tag{3.46}$$

More generally, for multiple effects e_j and a single cause, c,

$$P(e_i|c) = P(e_i|e_j, c). \tag{3.47}$$

Then, applying this to Equation (3.11), we get

$$P(e_1 \cap e_2 \cap \cdots \cap e_j|c) = P(e_1|c)P(e_2|c) \cdots P(e_j|c). \tag{3.48}$$

Now we can substitute for $p(\text{psa} \cap \text{fpsa}|\text{pc})$, giving

$$P(\text{pc}|\text{psa} \cap \text{fpsa}) = \frac{P(\text{psa}|\text{pc})\, P(\text{fpsa}|\text{pc})\, P(\text{pc})}{P(\text{psa} \cap \text{fpsa})}. \tag{3.49}$$

Now, all that remains is to simplify the denominator. This is a *normalizing* factor, just like the case with multiple causes and a single effect. It is the sum of the numerator in Equation (3.49) and the corresponding expression for when prostate cancer is not present, so,

$$P(\text{psa} \cap \text{fpsa}) = P(\text{psa}|\text{pc})P(\text{fpsa}|\text{pc})P(\text{pc}) + P(\text{psa}|\neg\text{pc})P(\text{fpsa}|\neg\text{pc})P(\neg\text{pc}).$$

Substitution in Equation (3.49) then gives the final formula for the prediction for the combined tests:

$$P(\text{pc}|\text{psa}\cap\text{fpsa}) = \frac{P(\text{psa}|\text{pc})P(\text{fpsa}|\text{pc})P(\text{pc})}{P(\text{psa}|\text{pc})P(\text{fpsa}|\text{pc})P(\text{pc}) + P(\text{psa}|\neg\text{pc})P(\text{fpsa}|\neg\text{pc})P(\neg\text{pc})}. \tag{3.50}$$

In the previous section, we saw how to generalize to a multi-valued hypothesis, or many possible causes, rather than just a Boolean. Now we see how to generalize to many pieces of evidence, when the items of evidence are conditionally independent, where we have a graph node that has no parents but several children.

$$P(c|e_i) = \frac{P(c)\prod_i P(e_i|c)}{P(c)\prod_i P(e_i|c) + P(\neg c)\prod_i P(e_i|\neg c)}. \tag{3.51}$$

In this formula, we have a *product* expression instead of a sum. We can support this by using a macro implementing a `product` operator, exactly analogous to the `sum` macro. Coding of this macro and implementation of formula (3.51) are left as an exercise for the reader.

Now, all that remains is to write an expression for the joint probability distribution for an arbitrary Bayesian graph, like the ones in Figures 3.5 and 3.3. This isn't simply a matter of combining the two formulas, but a theorem about such graphs [298, p. 39] provides a solution to the problem. If the graph has the property that each node (or variable) is conditionally independent of all its *non-descendents*, given its *parents*, then the joint probability distribution is the product of all the conditional distributions of each node given values for its parents. This means we can consider each node with its parents, independent of the rest of the graph. So for the graph in Figure 3.5 we could propose a joint probability distribution of the following form:

$$P(E, C, D, A, B) = P(E|C \cap D)P(C|A \cap B)P(D|B)P(A). \tag{3.52}$$

Such graphs are called Bayes Nets. They can be used to model both diagnostic and predictive problems. There are two big challenges in using these ideas. First, the topology of the network itself cannot be derived from the data, but generally comes from intuitions about relations among the variables. These are supported by the data, otherwise one would not have the conditional probabilities, but there are direct dependencies and indirect dependencies. For example, in Figure 3.3 disease control is indirectly related to T-stage, but one could also draw the graph with a direct connection there. Second, it is possible that for many nodes the local conditional probabilities are not known but other data are available and one can induce the missing information with suitable advanced methods [298].

3.2.4 Example: Radiation Therapy Planning

Even with heuristic methods for search, the trade-offs in radiation therapy between adequate dose to the tumor to sterilize it and overdose to surrounding structures make it difficult to decide which of a range of candidate plans is best. The problem is that the trade-offs are related to the individual's goals and preferences in terms of the treatment outcome and long-term (quality of life) outcomes.

Current strategies for plan optimization suffer from several shortcomings, many of which stem from incomplete knowledge. These approaches require quantitative models of effects, so that trade-offs between tumor coverage and normal tissue damage are accomplished by means of very accurate mathematical optimization of very imprecise surrogates (such as dose-volume constraints) or models. Examples of primitive models include Tumor Control Probability and Normal Tissue Complication Probability [244,245,263,365,366,433]. Inverse planning (optimization by starting from the desired dose distribution and back calculating the parameter settings that will achieve it) is currently plagued by the difficulties that arise from incomplete data (such as in complication probabilities), incomplete or vague prescriptions, and mutually contradictory constraints and objectives.

Bayesian belief networks are designed to handle incomplete information that is coupled with a causally based model, such as is often available in a clinical setting. At the University of Washington (UW) an influence diagram model was developed that addresses this problem [281,327]. This project illustrates yet another way in which a probabilistic reasoning system can be connected with a complex modeling system, the UW Prism radiation therapy planning system.

There are at least two applications of Bayesian networks in inverse planning. They can be used in place of a cost function used in the inverse planning process. They can also be coupled with utility functions to form decision networks to aid in the selection of solutions, particularly when no solution can achieve all of the objectives. Initial work utilized a Bayes Net implementation that functioned as a library callable from programs that create the specific network for the radiation therapy plan analysis. More recently, the project has been using a facility with a more elaborate user interface, as depicted in Figure 3.7.

As this and related work progresses, there will be software engineering challenges to incorporate these tools into the clinical environment. The National Cancer Institute recognized this problem in 1989, when it sponsored a project [182, 198] not only to develop software tools such as those described in this chapter, but also to address these software interoperability issues. Much work remains to be done.

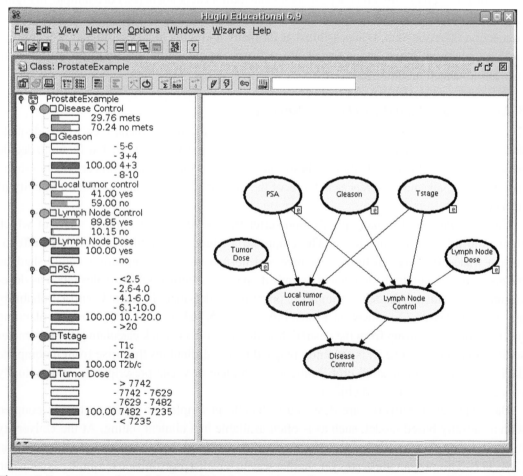

Figure 3.7   A screen shot of the use of Hugin to create and manage a Bayes Net model for decision making in prostate cancer treatment.
(Courtesy of Mark Phillips and Wade Smith.)

3.3 Utility and Decision Modeling

De gustibus non disputandum
—anonymous
Latin proverb

There's no accounting for taste.
—Mr. Spock
Star Trek episode, "The Trouble With Tribbles"

To make reasonable and practical decisions in clinical care and in public health, we will need to be able to incorporate into our models such factors as the relative importance of outcomes or effects, in cases where there are choices with multiple outcomes affected simultaneously by the choices. Again, referring to cancer, choices of treatments will often require deciding between a treatment that is more effective but more damaging to normal body functions, or a treatment that is less effective but less debilitating. How important is each of these outcomes to an individual, or to a population more broadly? We need ways to represent these importance factors, usually called *utilities*. A utility is a measure of how useful or important an outcome is. Once we have a framework for defining utilities, it is a considerable challenge to determine what are their values for any particular individual. *Utility theory* is a framework for creating models incorporating these values. Combining utility theory with probability models constitutes *decision theory*. We can use decision theory to represent trade-offs and compute the consequences of one choice as compared to another.

Utility is a single number that expresses the desirability of a state of the world. *Expected utility* combines the probability of any outcome with its utility. A *normative* theory of decision making asserts that a rational individual should make the choice that maximizes expected utility. This section presents two ways of representing and computing with utilities and probabilities, namely *decision trees* and *influence diagrams*.

A decision tree is a graph with a tree structure whose root or trunk is a decision to be made, where the branches represent the alternatives to be considered. Branching further from each decision option are different kinds of events that may happen when that decision choice is taken. These junctions are called chance nodes. There may be further chance nodes. Eventually the leaf ends of the tree represent outcomes with which some value can be associated. An example is shown in Figure 3.8. The structure of the tree comes from defining the problem to be solved. The values must be evoked from the person or population for whom the decision needs to be made. At each chance node, there are probabilities for each outcome or branch. The probabilities are estimates based on the conditions represented at that point, such as the likelihood of success of a surgery, the probability of recurrence of a tumor, etc.

Once the tree is created and values assigned, the *expected value*, or EV, at each node is computed. For an outcome, the value, or utility is multiplied by the probability of that outcome at that point in the tree (the same outcome could appear in several branches). At a chance node,

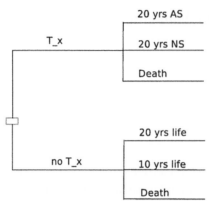

Figure 3.8   A decision tree representation of Mr. X's dilemma.

the expected values on each branch are summed to get the expected value for the node. At the chance nodes,

$$EV = \sum_i p_i \times EV_i.$$

This process is repeated until each decision option has an expected value assigned. Then we know which option has the most expected value, and can make a rational choice.

An influence diagram extends the idea of a Bayes Net. In addition to having probability nodes that represent random variables, it can have utility nodes, whose values can represent the utilities mentioned above, and decision nodes as in a decision tree. It is a very compact representation but complicated to work with.

Decision trees, as described here, are distinctly different from another tree structure that is sometimes also called a decision tree, but is more appropriately called a classification tree. Classification trees do not include probabilities but are an alternate representation of logical decisions based on categorical observations. They are briefly described later in this chapter, in Section 3.4.5.

The following section presents a detailed example of how a decision tree is constructed, as well as an alternate representation, an *influence diagram*. A book by Jensen and Nielsen [192] provides a more in-depth treatment of decision graphs, as well as an extensive treatment of Bayes Nets.

3.3.1 A Decision Analysis Vignette

These ideas can best be illustrated by a somewhat realistic medical example.[8]

---

[8]The material in this section, including the figures, is contributed by Richard Boyce, from his class notes in a class taught by Jason Doctor, on clinical decision making.

Mr. X has prostate cancer. His treatment options include chemotherapy, radiation, or surgery. He is against surgery and says that radiation therapy is not an option. He is well advanced in age and the cancer is a non-aggressive type so he is willing to consider not treating it at all. Chemotherapy comes with a chance of immediate death and may result in loss of his natural voice. Should he choose treatment or let nature run its course?

This is an artificial example contrived to represent a clinical decision involving a great deal of uncertainty. The course that Mr. X's illness will take is simply unknown as is the outcome of treatment. Uncertainty is a common feature of medical decisions at all levels; from individual care to public health to health policy. The study of decision making under uncertainty asks: "Which decision, from a set of decisions with uncertain outcomes, will result in the greatest *expected utility?*"

The normative theory of decision making can be applied to clinical decision making by decision analysis. We will walk through a simple decision analysis using the example above.

3.3.2 Graph Structure and Probability Assignment

First we need to structure the problem by deciding what the decision options and possible outcomes are. Possible outcomes can come from an expert, the literature, and/or clinical databases. Here we assume that the literature provides the following outcomes:

- *Decision options:* Treat or not treat.
- *Possible outcomes:*
    - Treat (T_x)—survive up to 20 years with loss of voice, survive up to 20 years with no loss of voice, immediate death.
    - Not treat ($\overline{T}_x$)—survive up to 20 years or immediate death.

We will use two *normative* techniques for representing this decision analysis problem. A decision tree representation is shown in Figure 3.8. An influence diagram representation is shown in Figure 3.9.

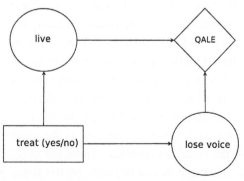

Figure 3.9   An influence diagram representation of Mr. X's dilemma.

Table 3.2   Probabilities for Possible Outcomes

	Outcome	Probability
$\bar{T}_x$	Live 10 years	0.8
	Live 20 years	0.1
	Immediate death	0.1
T_x	Live 20 years with artificial speech (AS)	0.6
	Live 20 years with normal speech (NS)	0.2
	Immediate death	0.2

These models were made up for this example. In practice the choice of model depends on factors such as the type of information available and the nature of the clinical decision. For example, different health states can be dynamic or static over time with implications on the decision model to employ. Dynamic states may be better modeled using Markov models or other time-based modeling strategies and methods.

Now, we need to assign probabilities to the outcomes. Like outcomes, probabilities can come from an expert, the literature, and/or clinical databases. The probabilities we assign need to fit within the constraints of probability theory. For example, the probabilities of all sub-branches of a particular outcome cannot add up to more than one. This may mean that we will need to perform some adjustment to the rough figures we get from our sources since the probabilities garnered from disparate sources may not neatly meet probability theory's constraints. One significant reason to use probability theory is that probabilistic beliefs that do not satisfy these axioms can lead to non-rational decisions at least some of the time. The interested reader is referred to [304] for more discussion. In this example we will use the probabilities shown in Table 3.2 for the outcomes in the models.

3.3.3 *Determining Utilities*

Now, we need to determine and assign the patient's utilities for the various outcomes. Simply put, a utility is an interval scale measure of the value an individual or group of individuals place on health states. It is a number that represents the subjective value of a given health state for one or more people affected by the decision. The derivation of utility is the focus of much discussion and research around *descriptive* and *normative* models for eliciting utility. *Descriptive* models attempt to acquire utilities based on how people really make valuations. They include prospect theory and rank-dependent theory [94]. Descriptive theories are designed to account for well-known heuristics and biases that individuals are subject to during decision making.

Normative models make the assumption that people follow expected utility in making decisions. That is, their value for each possible outcome is the product of their utility for each state and the probability of the state's occurrence. There are at least three different normative ways to elicit utilities. The best one to use depends on the focus population of the decision analysis (individual versus community), what methods make the most sense to the patient, and the time/resources that the decision analyst has to elicit utilities.

Time Tradeoff (TTO): The utility a patient has for a given health state is assigned the ratio of amount of time at which a patient is indifferent between X years in perfect health and Y years in that health state: $\frac{X}{Y}$. For example, if a patient is indifferent between 10 years of perfect health and 20 years of blindness then his/her utility for that health state is $\frac{X}{Y} = \frac{10}{20} = .5$. The *constant proportional tradeoff* assumption says that if a patient would give up x years of life to avoid y years in a given health state, (s)he would be also be willing to give up cx years of life to avoid cy years in that health state. TTO combined with the *constant proportional trade-off* assumption asserts that utility is linear with time. In reality a person's utility for a health state may not be linear over time, and in such cases TTO does not hold.

Standard Gamble (SG): It presents a patient with a choice—X years in a particular health state or a p chance of Y years in an improved health state versus a $1 - p$ chance of immediate death. Adjust the probability until the patient is indifferent between the certainty and the gamble. For example, the result of this process might show that the patient has no preference between 20 years in a poor health state and a .5 chance of 20 years full health versus a .5 chance of immediate death. The probability at which the patient is indifferent is his/her utility for the given health state. SG assumes independence between the utility of survival duration and the utility of health status. Since this may not always hold, multiple utilities can be derived for different durations of a health state approximating the decision maker's utility curve.

Utility-based questionnaire: There are standard, validated survey tools that provide utilities for specific health states over time for certain populations. For example, the "Health Utilities Index" or the "EQ-5D."

Suppose that we have two sets of utilities for Mr. X. One set, for years of life with natural speech (NS), was derived using standard gamble technique:

- $U_{ns}(0) = 0$—Mr. X's utility for death is 0.
- $U_{ns}(7) = 50$—Mr. X is indifferent to accepting 7 years of normal speech (NS) with 100% certainty and taking a gamble with $p = .5$ of obtaining 25 years of normal speech versus 0 years of normal speech.
- $U_{ns}(25) = 100$.

We will assume that we also have Mr. X's utility for artificial speech (AS), derived using TTO. Specifically, Mr. X's utility for 10 years of AS is .7 that is, he is indifferent to 10 years of AS and 7 years of NS $\left(\frac{X}{Y} = \frac{7}{10} = .7 \right)$. Also, his utility for 25 years of AS is 1 or, he is indifferent between 25 years of AS and 12.5 years of NS $\left(\frac{X}{Y} = \frac{12.5}{25} = .5 \right)$.

Using this relationship and the utilities derived from both methods we can establish the rough utility curves for both years of life with NS and years of life with AS. Without going into the details, the utility curves in Figures 3.10 and 3.11 were derived from the following relationships:

$$U_{ns} = \frac{100 - 50}{25 - 7}x + (50 - 2.8(7)) = 2.8x + 30.4, \tag{3.53}$$

$$U_{ns}(7) = U_{as}(10) = 50, \tag{3.54}$$

$$U_{ns}(12.5) = U_{as}(25) = 2.8(12.5) + 30.4 = 65.4, \tag{3.55}$$

$$U_{as} = \frac{65.4 - 50}{25 - 10}x + (50 - 1.03(10)) = 1.03x + 39.7. \tag{3.56}$$

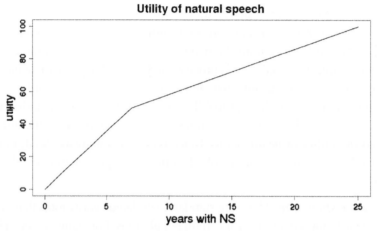

Figure 3.10 Patient's utility of natural speech.

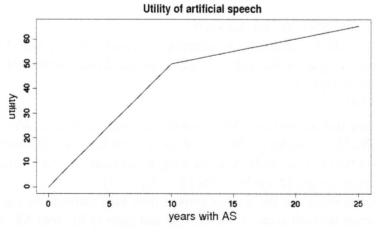

Figure 3.11 Patient's utility of artificial speech.

3.3.4 Computation of Expected Values

With these figures for probabilities of outcomes and utilities of states we can go back and add this information into our decision tree and influence diagram and compute the remaining values, which then identifies the decision that maximizes expected utility. Figures 3.12 and 3.13 show the two representations, this time with values assigned.

Both representations of Mr. X's decision problem can now be solved to determine the choice that will maximize expected utility. Solving the influence diagram is a bit involved and the reader is referred to [313] or [304] for details. The decision tree is easy to solve as it simply involves summing the products of the probabilities and utilities of each branch from leaf to root. This process, explained earlier, is called "folding back" the decision tree

$$T_x = .6(U_{as}(20)) + .2(U_{ns}(20)) + .2(0) = 53.5, \tag{3.57}$$
$$\overline{T}_x = .1(U_{ns}(20)) + .8(U_{ns}(10)) + 0 = 55.4. \tag{3.58}$$

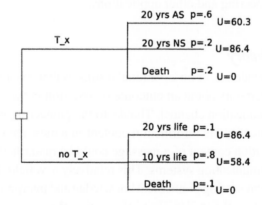

Figure 3.12 A decision tree representation of Mr. X's dilemma with utilities and probabilities.

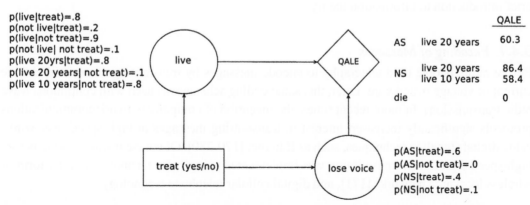

Figure 3.13 An influence diagram representation of Mr. X's dilemma with utilities and probabilities.

The process is shown in Equations (3.57) and (3.58). The result is that the decision not to treat gives slightly more expected utility than the decision to treat.

In the area of radiation therapy planning, mentioned earlier as an application of Bayes Net modeling, decision theory incorporating utilities as described here has been applied to the problem of ranking or evaluating treatment plans [188–190,445,446]. These ideas are important because as mentioned earlier, the goals or objectives of radiation therapy are multiple and contradictory or conflicting. No matter what is chosen for a cost function for optimization, the search for optimal plans does not yield a unique "best plan." More usually there are entire families of plans which represent optimality under choices of trade-offs between conflicting objectives. Therefore, the best strategy appears to be to generate a range of candidate plans, the so-called Pareto optimal set or Pareto front, and then use utility theory to evaluate which choice(s) represent the best trade-offs. A plan is said to be Pareto optimal if one can only improve on one objective by changes that worsen the results for some other objective. The idea was introduced and applied in economics by Vilfredo Pareto.[9] These ideas and methods are finding wide use in engineering and other applications.

3.4 Information Theory

Another way to think of evidence, or observational results, is that it is information, or a message, which decreases our uncertainty about an outcome or assertion in the same way as information coming through a communication channel. Thanks to the pioneering work of Claude Shannon [374], it is possible to quantify the information content in a message in terms of probabilities. The idea that the information content of a message could be quantified made a huge change in our ability to design communication systems. This relatively new field, *Information Theory*, has also found some use in representing biomedical knowledge and interpreting data. Its significance to biology and medicine is not yet clear, but information theory provides yet another way to look at representing uncertain or nondeterministic knowledge. In this section we give a very brief introduction to information theory.

3.4.1 Encoding of Messages

Before radio, people used telegraphy to encode messages by transmitting changes in electric current or voltage over a wire. Later, this same coding scheme was used with continuous wave radio transmissions. In more recent times, the invention of computer network communications protocols significantly increased interest in transmitting messages at high speed over wires, using digital signaling techniques, such as Ethernet [170,280]. Of course it was not long before high-speed digital data transmission was implemented using radio technology, in the form of wireless local area networks [171], and digital cellular telephone technology.

---

[9]See Wikipedia page, `http://en.wikipedia.org/wiki/Vilfredo_Pareto` for a biography of Pareto and discussion of his ideas.

- Morse code represents each letter of the alphabet by a group of dots and dashes, for example, for E, ...for S, − . − . for C, − − − for O, and so on. The symbols used for encoding are the dot and the dash. In radio transmissions, these were implemented as a short continuous wave transmitted signal and a longer one (ideally the long one, or dash, was three times the length of the short one). Skilled telegraphers used a telegraph key, a simple hand operated switch, to make the signals, and a receiver that generated a tone corresponding to the incoming signal.
- The ASCII code, introduced in Chapter 1, represents each letter by a pattern of seven binary digits, written as 0 or 1, such as 1000101 for E, 1010011 for S, 0110001 for the digit 1, and so on. In this case, the symbols are just 1 and 0. In electric circuits, the two symbols would be represented by two different voltages, sufficiently different enough to ensure that the value is unambiguous. On magnetic media (which also use ASCII), the magnetization direction and strength are used as a realization of the binary digits, 0 and 1.
- Other codes have been used to encode text on computer tapes and disks (for example, SIXBIT and EBCDIC, mentioned in Chapter 6, Section 6.1).

No matter what the encoding scheme is, we transmit messages by using *some* scheme. Even natural language can be considered an encoding scheme and human speech and writing are message sources. The basic symbols which are used to encode messages can be considered to be the common alphabetic characters. However, one can also consider words to be symbol units, in which case there will be many more individual symbols, but possibly higher efficiency of message transmission.

In order to quantify information content in messages, we need to characterize the source. What kinds of messages are possible? How often do different symbols occur? A message source may have completely unpredictable behavior, or it may be very regular. The idea of a "regular" message source can be given precise meaning.

- A *stationary source* is one in which the probability of a symbol occurring in a message (averaged over all possible messages) does not depend on the position in the message.
- An *ergodic* source is a stationary source in which the time-averaged occurrence rate of a symbol is the same as its ensemble average.

Information theory considers only ergodic message sources. In general, we are considering message sources that only send meaningful messages, not random nonsense. So, the symbols do not occur with equal probability, and they do not occur in random order. However, information theory is not so much concerned with the *grammar* of the message, as its encoding. Choosing an efficient encoding can help increase the rate of message transmission. When noise is present, some of the symbols received will not be the ones that were sent, so some of the message is lost. It is, as we will see, possible to make up for this by including some redundancy in the encoding, so that errors can be detected and corrected.

In English, the letter E has a frequency of about 0.13, while W has only 0.02. Similarly a table of word frequencies can be built up by counting words in large samples of text. Shannon

[374] experimented with random text generated by selecting words according to their frequency in English text. His first order approximation only used the base frequencies. His second order approximation used conditional probabilities, that is, given a particular word, what is the probability of a particular word following it? You can see that the second order process already starts to resemble real text. To really generate stochastic English text, one would want to incorporate a grammar-based generator, not only with word probabilities, but with structure. The problem in the second order text is that it tends to wander. It has no larger structure, as one might expect.

- First order approximation

    ```
    REPRESENTING AND SPEEDILY IS AN GOOD APT OR COME
    CAN DIFFERENT NATURAL HERE HE THE A IN CAME THE
    TO OF TO EXPERT GRAY COME TO FURNISHES THE LINE
    MESSAGE HAD BE THESE
    ```

- Second order approximation

    ```
    THE HEAD ON AND IN FRONTAL ATTACK ON AN ENGLISH
    WRITER THAT THE CHARACTER OF THIS POINT IS
    THEREFORE ANOTHER METHOD FOR THE LETTERS THAT THE
    TIME OF WHO EVER TOLD THE PROBLEM FOR AN
    UNEXPECTED.
    ```

Pierce [329] did a more extended experiment approximating fourth order correlation, by having friends write single words based on the last three. One of the more appealing sentences that resulted was "It happened one frosty look of trees waving gracefully against the wall."

It is relatively easy to create a profile of a body of text, by simply counting words. Using a large enough text, assuming it is representative, such a table would give an estimate of the first order word probabilities. When applied to a specific document rather than determining the frequency table for English text in general, this is called the Zipf distribution for that document. It is somewhat characteristic of the document, and useful for indexing. The `item-count` function described in Chapter 1 on p. 46 expresses the basic idea, that each word found in sequence in the text is added to the word table if not already there, and its count is incremented if it is already there. In Chapter 4, p. 408 there is a slightly improved version that uses a hash table instead of a list.

For the second order experiment, the word table should have an entry for each word in the text, but instead of a single number, the entry should be a list of words that directly follow that word in the text. Each of those would have with it the frequency with which it follows the indexed word. Here is code from Graham [137, p. 140] that will process a file of text, filtering for punctuation.[10]

```
(defparameter *words* (make-hash-table :size 10000))
(defconstant maxword 100)
```

---

[10]Note that Graham uses `car` and `cdr` instead of `first` and `rest`. Once again, "there's no accounting for taste."

```
(defun read-text (pathname)
  (with-open-file (s pathname :direction :input)
    (let ((buffer (make-string maxword))
          (pos 0))
      (do ((c (read-char s nil :eof)
              (read-char s nil :eof)))
          ((eql c :eof))
        (if (or (alpha-char-p c) (char= c #\'))
            (progn
              (setf (aref buffer pos) c)
              (incf pos))
            (progn
              (unless (zerop pos)
                (see (intern (string-downcase
                              (subseq buffer 0 pos))))
                (setf pos 0))
              (let ((p (punc c)))
                (if p (see p)))))))))
```

The idea here is to process a character at a time, determine if it is alphabetic or punctuation. The punc function does this by simple comparison.

```
(defun punc (c)
  (case c
    (#\. '|.|) (#\, '|,|) (#\; '|;|)
    (#\! '|!|) (#\? '|?|) ))
```

If it is alphabetic, put it in a buffer, which builds up the current word, character by character. If non-alphabetic, and some characters have accumulated, look up the previous word in the hash table. This is the job of the see function. If the previous word is in the hash table, check if the current word has an entry in the previous word's little table. If so, increment it, and if not, make an entry.

```
(let ((prev '|.|))
  (defun see (symb)
    (let ((pair (assoc symb (gethash prev *words*))))
      (if (null pair)
          (push (cons symb 1) (gethash prev *words*))
          (incf (cdr pair))))
    (setf prev symb)))
```

The see function captures the variable prev, which becomes a private but persistent variable, in which the previous word can be stored (and updated each time). The variable prev is only accessible from within see. This is a nice example of information hiding while still preserving global state.

Once the table is created and populated, new random text with second order word correlation can be generated by randomly picking a word, then randomly picking a successor from the list associated with it. That successor then becomes the current word, it is looked up and the process continues from word to word until the requested number of words are generated. Again from Graham, the code for generating text:

```
(defun generate-text (n &optional (prev '|.|))
  (if (zerop n) (terpri)
      (let ((next (random-next prev)))
        (format t "~A " next)
        (generate-text (1- n) next))))

(defun random-next (prev)
  (let* ((choices (gethash prev *words*))
         (i (random (reduce #'+ choices :key #'cdr))))
    (dolist (pair choices)
      (if (minusp (decf i (cdr pair)))
          (return (car pair))))))
```

Text produced this way has no meaning, but many people have a propensity to find meaning in everything, and so it can be the basis for much entertainment. Nevertheless, the point here is that the table represents all the possible symbols in a message, and something about the frequency of appearance. This will be the key to quantifying the information content of a message, for purposes of analyzing the message encoding, transmission, and decoding process, aside from whatever meaning it might have.

3.4.2 Entropy and Information

Borrowing from the field of physics, in particular statistical mechanics, there is a relation between the intrinsic order or organization of a message source and the amount of information it can provide. When there are lots of possible messages all with equal probability, a situation with almost total disorder or unpredictability, getting a message is very informative. On the other hand, if there is only one possible message (total order), when we receive it we have not gotten any new information. The concept of *entropy* is used to quantify the variety of states of a message source. The entropy of a message source is a measure of the information gained by receiving a message. For a message of one symbol, chosen from two possibilities, with probabilities p_1 and p_2 (note that $p_1 + p_2 = 1$), the entropy H is defined as

$$H = -(p_1 \log p_1 + p_2 \log p_2),$$

where usually base 2 is used for the logarithm function, giving the entropy in units of *bits*. When $p_1 = p_2 = 0.5$, we get $H = 1$, and we say the entropy is 1 bit. When the message is a choice from among n symbols, the entropy in bits per symbol is defined as:

$$H = -\sum_{i=1}^{n} p_i \log_2 p_i. \tag{3.59}$$

In the case that we have n symbols all of equal probability (recall that this is the discrete Uniform distribution), then $p_i = 1/n$. Every term in the sum is $\log_2 n/n$ and there are n terms, so the entropy is $\log_2 n$. So, to represent 32 different symbols, we would need 5 bits per symbol, and to represent 64 different symbols we need 6 bits per symbol. If the number of symbols is between those two the entropy includes a fraction between 5 and 6. It would seem useful to have a function to compute the entropy for a list of probabilities corresponding to symbols from some source.[11]

```
(defun entropy (probs)
  (- (apply #'+
       (mapcar #'(lambda (p)
                   (* p (log p 2)))
               probs))))
```

By aggregating the symbols into larger groups, we may be able to encode messages more efficiently, for example, by assigning codes to blocks of characters or to words. Pierce [329] provides some figures for English text, as follows:

- With about 50 typewriter keys, the entropy per symbol is around 5.62 bits. With 6 bits per character we waste some but not much information. Some early computer systems used a character set encoding with 6 bits per character, and supported only upper-case alphabetic characters along with the 10 digits, special symbols, and control characters (Chapter 6, Section 6.1).
- Encoding all possible blocks of three characters, with each block assigned a code, this is slightly reduced.
- At an average word length of 4.5 characters plus a space, with 5 bits per character, we have 27.5 bits per word.
- Encoding words instead of characters, using a vocabulary of about 16,000 words, allows this to be reduced to 14 bits per word.

Shannon found by experiment that the entropy of English text is about 9 bits/word, or around 1.7 bits per character (including the space). This may seem awfully small, until you recall the many humorous examples of text with letters left out which is still quite readable. For example, hr is a hlarius xmpl of txt wth ltrs lft ot bt wch is stil qt rdabl. The popularity of text messaging with cellular telephones also supports Shannon's findings.

---

[11]If `probs` were an array, it would be simpler to use the `sum` operator previously defined, but for small sets of numbers, lists, and the usual list operators seem straightforward. If you are by now an ambitious Lisper, try writing a macro, `mapsum`, that is like `sum` but operates efficiently on lists, instead of iterating on an index.

3.4.3 Efficient Encoding

Having determined the entropy of a message source, many questions are possible, such as:

- Is the entropy a limit on how efficient a coding scheme can be? Possibly with data compression, we can get arbitrarily efficient transmission if the message is large enough.
- What effect will noise or errors in transmission have on the rate of transmission of information?
- Is it possible to detect and correct transmission errors, thus guaranteeing reliable communication?
- If there is a limit on how efficient a coding scheme can be, can it actually be achieved? If so, how? What's the best way to do this?

The answer to the first question is "yes." Shannon's Noiseless Channel Theorem states that the transmission rate, the channel capacity, and the entropy of a source are all related. Channel capacity is the maximum rate at which a channel can transmit information, denoted by C. It is measured in bits per second. Every communication mechanism has a channel capacity. Shannon's Theorem then states that it is possible to transmit from a source of entropy H bits per symbol through such a channel, at the average rate of

$$\frac{C}{H} - \varepsilon$$

symbols per second. It is *not* possible to transmit at an average rate $> C/H$. Note that the rate at which information can be transmitted from a given source through a given channel depends on *both* the source entropy and the channel capacity.

So, what about data compression? A lossless data compression algorithm is one that is *reversible*. This means that the output of the compression can always be put through a decompression algorithm to recover a faithful copy of the original data. Such compression algorithms are most effective in reducing the size of files, when the files contain highly redundant data, so that their information content is low. Examples include medical images in which there are large areas of constant pixel values, such as large regions that are black because they are images of air. Files containing data that have higher entropy will not compress well.

The effect of noise is that some received symbols will not be the same as the corresponding transmitted symbols. When this happens, it reduces the rate at which information can be transmitted. Shannon derived a corresponding theorem for the noisy channel. In a noisy channel, the source has entropy,

$$H(x) = -\sum p(x) \log p(x)$$

but the output may have a different one,

$$H(y) = -\sum p(y) \log p(y).$$

$H(y)$ depends on both the input and the noise (transmission errors). If we knew both the transmitted and received messages, we could compute the entropy of the combination of transmitting *x* but receiving *y*.

$$H(x, y) = \sum_x \sum_y p(x, y) \log p(x, y). \tag{3.60}$$

We can write this in a clearer way by defining the notion of *conditional entropy*, somewhat analogous to conditional probability. Conditional entropy is defined from the probability of receiving any *y* given that *x* was transmitted.

$$H_x(y) = -\sum_x \sum_y p(x) p_x(y) \log p_x(y)$$

that is, given *x* what entropy is associated with *y*? Similarly, we define

$$H_y(x) = -\sum_x \sum_y p(y) p_y(x) \log p_y(x)$$

that is, given *y* what entropy is associated with *x*? Then we can write Equation (3.60) as

$$H(x, y) = H(x) + H_x(y) = H(y) + H_y(x).$$

The quantity $H_y(x)$ is called the "equivocation." The rate at which information is transmitted over the channel is then

$$R = H(x) - H_y(x).$$

Shannon proved that with a discrete channel of capacity C, and a discrete source of entropy per second H, if $H < C$, there exists a coding system to transmit information with arbitrarily small error rate. If $H > C$, it is possible to achieve equivocation that is less than $H - C + \varepsilon$ but there is no method of coding that will give an equivocation less than $H - C$. This is Shannon's *Noisy Channel Theorem*.

Now we turn to the problem of choosing an encoding for messages. It turns out that there is an optimal method of generating an encoding of arbitrary sets of symbols using binary digits, when the source has a discrete, finite set of symbols. It is called Huffman encoding, and we will illustrate how it works by following a very small artificial example (this is similar to an example in Pierce [329, pp. 94–97], but adapted to a public health reporting context).

Consider a hypothetical source of messages formed from a very small vocabulary. The vocabulary consists of the names of eight reportable diseases, and the messages consist of sequences of the disease names. Each day various clinics send messages to a central databank. The messages contain the names of each of the diseases that turned up in their clinic, in order, once for each patient, so there can be repeats. The public health agency will of course use these messages to create prevalence values, perhaps even with geographical maps. Table 3.3 lists a small set of disease names (the symbols or vocabulary of this message source).

The entropy per word can be computed from Equation (3.59). It turns out to be 2.073. Since we have eight choices, we could assign the eight possible combinations of three binary digits, one combination per word. This encoding scheme uses 3 bits per symbol. However, it can be

Table 3.3 An Example Set of Symbols and Probabilities for a Public Health Communication Problem

Disease	Probability
Influenza	0.55
Measles	0.13
Pneumonia	0.12
Lung-ca	0.10
Prostate-ca	0.04
Breast-ca	0.03
Anthrax	0.02
Kidney-ca	0.01

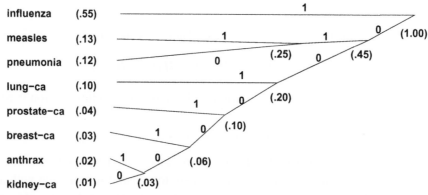

Figure 3.14   Algorithm for Huffman encoding.

done much more efficiently. The diagram in Figure 3.14 shows the derivation of a Huffman encoding, a method to generate a maximally efficient encoding for a discrete, finite source.

The procedure is this: connect together each of the two lowest probabilities, assigning the upper branch a 1 and the lower a 0. Do this successively until all are connected in a full tree. In the example in Figure 3.14, the branches go up in an orderly fashion, but with different probabilities, the connections may well cross each other. The procedure is to do a pair at a time. Once the graph is generated, the code for each symbol is the sequence of binary digits on its path going to the left from the apex of the graph. For the top symbol, `influenza`, its code is the single digit, 1. The code for `prostate-ca` is 0001. Table 3.4 shows the result of computing the Huffman encoding for all the symbols in this message source.

Adding the numbers in the last column of Table 3.4 gives the average number of digits per word, 2.09, which is much closer to the entropy of the source than 3. This might seem counterintuitive, since you would think that with only about 2 bits per symbol, you could encode only four symbols. This is true if the symbols all have equal probability. In the case of the message source we are considering here, some symbols occur only rarely, so it is efficient to use as many

Table 3.4   A Huffman Encoding for the Disease Transmission Problem

Word	Probability p	Code	Digits (N)	Np
Influenza	0.55	1	1	0.55
Measles	0.13	011	3	0.39
Pneumonia	0.12	010	3	0.36
Lung-ca	0.10	001	3	0.30
Prostate-ca	0.04	0001	4	0.16
Breast-ca	0.03	00001	5	0.15
Anthrax	0.02	000001	6	0.12
Kidney-ca	0.01	000000	6	0.06

as 6 bits per symbol. The savings of being able to encode the most frequently occurring symbol with only 1 bit make up for it, on the average. So, the efficiency of a variable length code is precisely related to the unequal probabilities of occurrence of different symbols. It is left as an exercise to implement the Huffman encoding procedure as a function on a list of symbols and their probabilities.

3.4.4 Error Detection and Correction

It appears that it is possible to transmit a reliable stream of information over a noisy channel. Indeed the functioning of all modern digital technology depends on this. The Noisy Channel Theorem says that the amount of redundancy required is related to the amount of noise, or alternately, the expected error rate. Here are brief descriptions of some methods for providing redundancy in binary encoded data.

- a "parity" bit can allow detection of single bit errors but not correction. On tapes and disks (and memory) this can trigger a "reread."
- repeating bits (sending three identical bits) can allow correction of single bit errors (1 in 3), but this drops the transmission rate to a third.
- "check digits" allow correction of single bit errors at lower cost, depending on the expected error rate. This technique is used in memory modules and in network transmission.

3.4.5 Information Theory in Biology and Medicine

This section describes a few examples of the application of information theory ideas in biology and medicine. These examples use information theory in methods to solve very specific problems, rather than as a representation of fundamental processes in cellular or system level biology. Maynard Olson, Professor of Emeritus of Medicine and Genome Sciences, Medicine, at the University of Washington, suggested at the University of Washington Annual Faculty Lecture in 2001 that the real revolution in modern biology is not the wonderful details we are discovering about the reactions and pathways of the cell and the specific properties of the genome, but rather the discovery that life is about the propagation of discrete messages in a noisy channel. However, the characterization of those messages and what they mean remain an enormous detailed problem.

3.4.5.1 Learning Classification Trees from Examples

The decision trees described in Section 3.3 are different from another kind of tree structure which is also sometimes called a decision tree, but really should be called a *classification* tree. A classification tree has no probabilities or expected values. It is a representation of a logic formula that incorporates many factors in a decision. In this case, the tree represents a series of conditional alternatives. At the base, it might be, "If PSA is higher than 4 and rising, get a biopsy. If not, then do nothing and retest in 6 months." Each of those branches then has a further conditional. For example, the biopsy branch might have, "If biopsy shows active cancer, get treatment." This branch in turn may have several treatments. At each point an unequivocal decision is made. Such tree structures can provide an efficient summary of what to do, when the number of variables is large. There are also learning algorithms for constructing trees from examples, but there is no sufficient space to discuss them here.

Classification trees can be induced, or automatically generated from examples. A well-known algorithm for generating an optimal classification tree is the ID3 algorithm [332,333]. In deciding at any point in the tree, which variable should be considered next, it chooses the one that has maximum entropy with respect to the data set, the variable that will by itself classify the most examples, or most reduce the uncertainty about the remaining examples in the training set. Examples of applications of classification tree learning to the solution of medical problems include neonatal intensive care [418] and recurrence of prostate cancer [451].

3.4.5.2 Medical Image Registration and Radiation Therapy Planning

Entropy plays a role in medical image processing as well, where it is used as a measure of correlation between image data sets. The problem is that different imaging studies (PET, CT, MRI) on the same patient need to be aligned with each other in order to correlate the physical locations of findings on the different kinds of images. This process is called *image registration*. Many algorithms for finding a suitable mapping or alignment have been developed [267,269, 425]. The most popular and successful ones use the entropy of the data sets as a cost function to search for a transformation between them that maximizes the correlation of pixel values.

Section 5.5.3 describes how a network representation of the lymphatic system (the UW Foundational Model of Anatomy) can be used to find pathways of dissemination of tumor cells from a primary tumor. The pathways are associated with Markov chains to obtain a probabilistic model to predict which nodal regions in a patient are at high risk for disease, and should be treated with radiation. However, that is not sufficient for radiation treatment planning. The next step is to determine where in any particular patient these regions actually are. Typically the task of drawing the target volume falls to the radiation oncologist. It is a very time-consuming one even with the best of computerized drawing software. So, another component of the CTV project is an investigation into how to automate the generation of an actual CTV once the required regions are known. The use of medical image registration methods is a key ingredient in solving the problem. This work was done by Chia-Chi Teng while a student at the University of Washington [410–413].

Image registration tools, that match different kinds of images on the same patient, such as CT to MR or PET, have been effective in assisting physicians to decide what regions to treat, but the actual contours still have to be drawn manually. Some things (soft tissue and bone) are better visualized by one kind of image, while others are more easily defined by others (the primary tumor). This use of image registration maps different kinds of images of the *same* patient to each other, to take advantage of the multiple visualization techniques.

The problem with the target volume is that lymphatics do not show up well on any kinds of images. However, it may be possible to build up a reference atlas, a set of contours drawn by experts on a "typical" set of images, and then use image registration to derive a mapping that can be used on the reference contours to map them from the reference atlas to the patient under consideration.

Image registration is a process of finding a geometric transformation g between two sets of images, which maps a point x in one image-based coordinate system to $g(x)$ in the other. By assuming that anatomy has similar characteristics between a specific patient and a reference person, we can transform a region from the reference image set to the patient image set. The algorithm and implementation employed by Teng was developed by Mattes and Haynor [269] for registering one patient's PET and CT image data. He adapted it for registering CT images of two different persons.

The registration algorithm is based on the idea of *mutual information*. Mutual information is an entropy-based measurement of image alignment derived from probabilistic measures of image intensity values [267,401,425]. It is calculated by estimating the marginal and joint probability distribution (histogram) of the intensity values of the test and reference images. The mapping of the reference image to the test image associates each pixel in one image with a pixel in the other image, but not necessarily at the same indices in the image array. The two arrays may be translated, rotated or even stretched, or distorted in other ways with respect to each other. If the mapping results in a high correlation of pixel values, the entropy will be very low. The mapping does *not* have to assign pixels in one image to pixels of the *same* value in the other. "Correspondence" means that pixels that all have the same value in the reference image get mapped to pixels that all have the same value in the test image, even though the two values may be very different. So, if all the dark areas in the reference got mapped to light areas in the test image, the correlation would be higher than if some dark pixels in the reference got mapped to dark areas in the test image and some got mapped to light areas.

The computation of probabilities is based on the histograms of pixels that are mapped to each other, and the entropy is computed similarly to Equation (3.60). For further details, and pictures of sample results of mapping on some test cases, the reader is referred to the publications cited earlier.

3.4.5.3 Coding and Noncoding DNA

Another interesting application of mutual information in biology is the process of distinguishing coding regions from noncoding regions in a DNA sequence. In a DNA sequence, each base might or might not have any relation to the next (or any other). The mutual information in a sequence, for a given separation k, can be defined as:

$$I(k) \equiv \sum_{i,j=1}^{4} P_{i,j}(k) \log_2 \left(\frac{P_{i,j}(k)}{p_i p_j} \right).$$

In this formula, p_i is prior probability of each of the four bases, and $P_{i,j}(k)$ is the probability of finding base i and base j separated by a distance k.

Suppose the sequence(s) are random, so that $P_{i,j}(k) = p_i p_j$ for all i, j, k. It is trivial to calculate $I(k)$ for this case. Since the fraction is always 1 and $\log_2(1) = 0$, the mutual information is also 0.

Suppose the opposite, that is, if you know the base at one position you can determine the one k spaces away. Assume all are equally probable, so $p_i = \frac{1}{4}$. What is $I(k)$ this time? The calculation is a little more complicated, as you know something about $P_{i,j}(k)$ but not the actual values. This is left as an exercise for the reader.

A study of what actually happens and whether there is any significant variation among different species was published in 2000 [139]. They found that there are some differences among categories of species in the patterns of codon usage, but no significant difference in average mutual information.

3.4.5.4 Redundancy and Error Correction in Biology and Medicine

The implementation of the genetic code in cells provides some conceptually simple mechanisms for error correction. The fact that the DNA in a cell is double stranded enables repair when a single strand is broken and some part of its sequence is lost. The complementary strand can assimilate nucleotides because of the matching process and can restore the missing section. Even double-strand breaks can sometimes be repaired.

In addition, the genetic code itself has some redundancy, as we saw in Chapter 1, in Table 1.5. Many amino acids are represented by several codons. There is some correlation between the frequency of appearance of amino acids in proteins and the number of codons that code for them. When a single point mutation occurs in the third base in a codon, it often does not affect which amino acid results in that position in the generated protein sequence. The biological significance of this is not well understood, but there is active research in this area.

Another application for error correcting codes is to ensure that a transmitted message has not been modified in transit. This is particularly important for software systems used for electronic medical records, and other systems that transmit, receive, and store sensitive or confidential data. Providing checksums, or some other kind of arithmetic function of all the bits in the message, can help detect alterations.

Information theory can play a role in determining the function(s) of a newly discovered protein by using sequence similarity [38]. Other interesting articles [2,3,364] provide for further exploration of the application of information theory in biology.

3.5 Summary

Modeling uncertainty and nondeterministic processes adds considerably to our ability to make biomedical knowledge computable. While the mathematics of probabilistic reasoning is quite different from that of logic, the themes are similar. We build predictive models, compute their entailments, and verify or reject hypotheses by comparing with experimental or observed data. Similarly, we build network representations that can explain findings, and perform classification tasks.

The MYCIN experiments [55,377] were noteworthy for the fact that MYCIN combined a probability-like representation of medical knowledge with a rule-based system. In MYCIN, as the rules were applied, the degree of certainty of a list of working hypotheses would increase or decrease. This is an example of a rule-based *production system*, in which the action of a rule can be any operation, not just the addition of an assertion to a set of propositions.

There are several good books on Bayesian inference, and Bayes Nets, including [192,298], and some chapters in [430]. Implementations of Bayes Net modeling, with very nice graphical user interfaces include a system from Microsoft Research, the Microsoft Bayesian Network Editor and ToolKit, MSBNx (http://research.microsoft.com/adapt/MSBNx/), and the Hugin system, used in the radiation therapy example in Section 3.2.4, at http://www.hugin.com/.

For more background and a deeper treatment of Decision Theory, see Hunink [166] and the classic work of Keeney and Raiffa [223]. Review articles by Nease [299] and Owens [313] may also prove useful.

The problem of combining logic-based models and numerical models in medicine has been addressed from the point of view of hybrid systems modeling [191,237]. An example of work combining probabilistic and logic-based systems is Daphne Koller's work combining frames and Bayes Nets [238]. More recently the ideas of Markov Random Fields [95] and Markov Logic Networks [340] are interesting possible modeling strategies for biomedical knowledge.

There is a lot to be done, but it points to the possible eventual unification of these seemingly different frameworks for biomedical theories.

3.6 Exercises

1. Prove (using the definitions) that the expectation of the deviation of a random variable from the mean is 0.
2. Derive formulas for the expectations of the random variable X and its square, as a function of the parameters, for each of the example distributions mentioned in Section 3.1.6.
3. Which of the two formulas, Equations (3.27) and (3.29) for computing the variance of a set of data, is computationally less expensive? Explain why.

4. Derive Equations (3.33) and (3.34) from the equations in (3.32) (do the derivatives and solve the resulting algebraic equations for m and b).
5. Write a macro, mapsum, that is like sum but operates efficiently on lists, instead of iterating on an index.
6. Write a function that generates a Huffman encoding for an arbitrary list of symbols and their probabilities of occurrence.
7. Compute the mutual information in a DNA sequence in which all four bases are equally probable and that has the property that given a base at any position, it is determined what the base is at a position k bases from it.

Biomedical Information Access

Principles of Biomedical Informatics. http://dx.doi.org/10.1016/B978-0-12-416019-4.00004-4

Biomedical information access is concerned with organizing text documents and other kinds of records and documents, to support searching such repositories to identify relevant and important documents as needed. Systems that address this task are called Information Retrieval (IR) systems, or more commonly, *Search Engines*. The key concepts we will address here are: how the structure and organization of information affect the efficiency of search, the construction and use of indexing systems, how a query is matched to items, how to rank order the results, and alternatives to rank ordering.

This chapter will introduce how such document repositories are constructed, and how search engines work. We will also examine how to access web-based information repositories from programs (as opposed to using web browser front ends). Some special considerations are needed for searching image collections and other non-text data for content. In describing the components of an information system we follow the outline of an excellent short article on the basics of search engines by Elizabeth Liddy [254]. Several good textbooks [22, 140, 155, 278] cover IR systems in more depth.

4.1 Information Retrieval Systems

There are some parallels in the design and construction of database systems and information retrieval (IR) systems. In both cases, the information to be searched must be structured into information units or *records* which are retrieved as whole units. In the case of a relational database, the units are rows in tables, or put more abstractly, tuples in relations. For bibliographic systems such as MEDLINE, the units are bibliographic records with tags labeling the various fields. Similar structures are used for biological information resources such as GENBANK and UniProt/Swiss-Prot. Library catalog systems also use a tagged record structure to store, retrieve, and display the information on file for the library's many holdings (books, journals, and other resources). So-called "full text" information systems may contain entire documents, such as articles, technical reports, etc. There is less structure in such collections, and a greater challenge in constructing efficient indexing and retrieval systems. Finally, the World Wide Web functions as a huge document repository, with highly variable structure, and presents its own unique challenges.

A commonly used standard for bibliographic record systems is the MARC (MAchine Readable Cataloging) record, a standard format for bibliographic information used by the Library of Congress and many other organizations [121]. Here is an example of a MARC record, an entry corresponding to one of the books on information retrieval mentioned above [278]. The text following the "!" characters are explanatory notes that I added to the original record.

```
LEADER 00000pam 2200385 a 4500
001     85828387 ! ID number
003     OCoLC ! ID type
005     20071010131615.0
008     061114s2007   enka  b 001 0 eng
015     GBA699284|2bnb
```

```
016 7   013608550|2Uk
020     9780123694126 ! ISBN 13
020     0123694124 ! ISBN 10
029 1   YDXCP|b2480931 ! some other ID number
035     (OCoLC)85828387
040     UKM|cUKM|dBAKER|dYDXCP
042     ukblsr
049     WAUW
082 04  025.04|222
245 00  Text information retrieval systems. ! title
250     3rd ed. /|bCharles T. Meadow ... [et al.]. ! edition
260     Amsterdam ;|aLondon :|bAcademic,|c2007. ! publisher info
300     xvii, 371 p. :|bill. ;|c24 cm. ! description
440  0  Library and information science ! series
500     Previous ed. : by Charles T. Meadow, Bert R. Boyce, Donald
        H. Kraft. 2000. ! the display says "note"
504     Includes bibliographical references and index. ! bibliography
650  0  Information storage and retrieval systems. ! LC subject keywords
650  0  Text processing (Computer science)
650  0  Information retrieval.
700  1  Meadow, Charles T. ! author
700  1  Boyce, Bert R.
700  1  Kraft, Donald H.
910     PromptCat
941     .b57945718
994     92|bWAU
```

Creating and managing a citation database for use in writing articles and books such as this one also involves use of structured records with tags labeling fields in each record. This book's citations were managed with BIBTEX, a bibliographic citation management program [317] that is part of the LATEX document preparation system [247]. The idea of BIBTEX is that one creates a structured entry for each item that one might cite, with tags labeling each of the fields of information. Thus a BIBTEX data file is yet another example of the use of metadata tags for structuring data. The representation of the data is chosen to keep maximum flexibility in relation to the possible uses of the data. Recently, some alternative formats for bibliographic data that use XML or other related syntax have been developed. The data can either be converted to BIBTEX format, or a program other than BIBTEX can be used to generate the bibliographic input for LATEX documents. This is possible because in the LATEX system the formatting of the information is done *after* retrieving the information from the bibliographic repository. A BIBTEX entry looks like this:

```
@article{kalet2003,
     author = "Ira J. Kalet and Robert S. Giansiracusa and Jonathan
```

```
                 P. Jacky and Drora Avitan",
    title = "A declarative implementation of the {DICOM}-3 network
             protocol",
    journal = "Journal of Biomedical Informatics",
    year = 2003, volume = 36, number = 3, pages = "159--176"}
```

The label `article` indicates the document *type*, and the tag that appears first after the left brace, `kalet2003`, is a label uniquely identifying that entry. Typically, such entries are simply stored sequentially in a text file, and the BIBTEX program fetches them by matching the label in a citation command with the label in the entry. Strictly speaking this is not an IR system, but the same organization of information can serve both purposes. There are many large repositories of citation metadata in BIBTEX format, each using its own conventions for the keywords identifying the entries. One difficulty in effectively using these resources is the potential for keyword conflicts, and the non-uniform vocabulary. Of course it is important to be able to search such repositories, to find the desired citation(s) rather than creating new entries for them.

For Internet web search, the units typically are pages, but the pages can be subdivided into parts as well. In both database systems and IR systems, to search efficiently, systems provide a way of *indexing* the information units. An extensive sorted index is sometimes called an *inverted file*, since it uses index terms from the information records to look up the record numbers or pointers. In the original files, the record numbers or pointers are used to retrieve the contents, while in an inverted file the contents are used to look up the record number. This is admittedly a bit simplified, but it is the essential idea. The structure and quality of the indexing system strongly affects the quality and efficiency of retrieval.

There are some difference between databases and IR systems as well. In a typical database application, queries are designed to produce some well-defined results, and are formulated with complete knowledge of the structure of the database and the nature of its contents. For example, a physician can query an electronic medical record to produce a complete list of exactly those patients that are on the hospital wards and under her care, so she can review all the cases. Another example would be a query for all the laboratory test results on a specific patient from a specified date up to the current date. The query must be sufficiently precise to specify exactly what is needed, and the system must return exactly what is asked for. To facilitate this, electronic medical record systems provide elaborate user interface facilities tailored to the kinds of information retrieval tasks that the designers think are most relevant to the medical practice setting.

On the other hand, the same physician may use PubMed to search for articles on a possible diagnosis pertaining to a patient. Many articles may be relevant, but there are likely also many that match the search query but are irrelevant. The physician may have a very precise question, but no retrievable record provides the precise answer, though there may be many records that provide relevant information from which the physician may determine an answer. More often, even the question is not precise, and the user's need is to find potentially relevant information that may help in deciding what to do next. In such systems, it can be helpful if the system can infer or compute a *relevance ranking* for the search results. However, relevance is not really computable from the query itself, but depends on the context, as well as the *meaning* of the query. Ultimately the person making the query will have to sift through the results to find what is most useful.

An IR system is thus designed to deal with a large collection of documents, with little global structure and highly variable content, much of which is simply unknown to the user of the system. In general, the user does not know exactly which documents he needs, and the query must somehow serve to identify which items are most likely to be useful. This is true even of systems such as PubMed, where the domain of discourse is more or less restricted to human biology and medicine. Since we are dealing with documents rather than highly regular tables, one might expect that natural language processing will be involved, and indeed there is a role for it.

All IR systems depend on building an index of the documents they will search for. The queries provided by the user are transformed into a form that can be used with the index to find matches for the query. The results are then typically ordered by some method that represents a best guess at relevance to the original query. From this, we can see that an information retrieval system is made up of four main processing components, whether you are building an Internet search engine or a specialized IR system with its own local document collection (such as a library catalog or bibliographic system). These components are:

- a document processor to generate and update the index,
- a query processor to compute something that can be used to match entries in the index,
- a matching process that actually searches the index, and
- a results ranking algorithm.

In addition most systems provide a nice user interface with many aids to the user to help formulate the most effective possible query, and to display the results in a form that is easy for the user to sort through. However, the user interface can be effective even if it is relatively simple. A stunning example of the ultimately simple user interface is the Google search engine. The user types a few words that make up the query, and the results are returned as a list (usually over many pages) of links (URLs) and short excerpts from the linked documents. A few buttons to select pages and go to the next or previous page suffice for navigation.

Before we look at how to build automated systems that perform these operations, we should consider how this was done before the advent of inexpensive and powerful computer systems. Indeed, the world's libraries had to solve this problem a long time ago. In modern times, two major systems of cataloging books and other items have been developed and are in use today, the Dewey Decimal System, and the Library of Congress Classification System.

Melvil Dewey invented the Dewey Decimal System while working at the library of Amherst College, where he was pursuing a Master's degree.[1] A facsimile of his treatise of 1876, "A Classification and Subject Index for Cataloguing and Arranging the Books and Pamphlets of a Library," is available through Project Gutenberg [91]. Dewey used three digit numbers to label items by subject category. For example, the range 500–599 was used for science and mathematics. The range 510–519 was specifically for mathematics, 530–539 for physics, and so on. The number system was extended by adding decimal parts, so that numbers could take forms such as 539.21 and so on, for finely divided subcategories.

The US Library of Congress Classification system was developed by Herbert Putnam in 1897. This system uses 21 of the 26 letters of the alphabet to designate major subject categories. These are subdivided further by using two letters together, and the letter or letter pair is then followed by a number representing even finer categorization. For example, the letter Q subsumes all works in the field of Science, but the works on physics specifically are labeled by the two letter combination QC. The classical mechanics books are in the QC120 and following numbers, while the books on quantum mechanics ("modern" physics) are in the QC770 range. Medicine in the Library of Congress system is categorized with the letter R.

The Library of Congress system does not use the range QS-QZ or the letter W. A third system in wide use was developed by the National Library of Medicine (NLM), of the US National Institutes of Health. The NLM's mission is focused specifically on medicine and biomedical basic sciences. The NLM classification system uses the letter codes that are not used by the Library of Congress system, so the two can coexist in the same library system. In the NLM system, for example, the code QS is used for Human Anatomy, QU for Biochemistry, WA for public health, WF for the respiratory system, WU for dentistry, WY for nursing, and so on. Books in biomedical informatics seem to be spread among the subject areas to which they most pertain, so bioinformatics books may be found in the QU section if they seem to have a biochemistry focus, but some books may also be found in the Library of Congress category QH where the subject seems to be more generally applicable to biology. Books with a more medical informatics orientation are generally found under W, Health professions.

In the following sections we will address each of the four components listed above. Finally, we will discuss to what extent these ideas are applicable in more exotic contexts, such as search and retrieval of image data (particularly medical images), and electronic medical records more generally.

4.1.1 IR System Design

In addition to the processing components mentioned above, an IR system will have a document repository, structured according to a model, and a logical view of the documents. The classic models on which many IR systems are designed are: Boolean (set theoretic), vector (algebraic), and probabilistic models. In each model, a document is represented by a logical view, or set of

---

[1]This information and some of the following comes from Wikipedia, the free online encyclopedia.

characteristics that can be matched with a user's query. In addition, in each model, there is a form for the allowable queries, a framework for executing the queries, and a ranking function that assigns a value, or real number, to each combination of a query and a document view. There are different kinds of logical views of documents, including ones based on index terms, full text, and a combination of full text and structure. A particular IR system will be based on one of the models, and one of the document view types, since these are independent dimensions.

The job of the document processor is to generate the logical view for each document. Essentially this means creating an index entry for it. The type of entry will depend on the model.

4.1.2 Indexing

The purpose of an index is to provide a way to find documents efficiently when you have a small amount of information indicating the content you are looking for. Before we had computerized information systems, indexing was done manually. Some of it still is. In addition to assigning the catalog numbers in the Dewey Decimal or Library of Congress system, librarians had the job of assigning subject terms to each item. For bibliographic reference systems such as MEDLINE this has also been done manually for many years. In this section we consider two different approaches to generating index terms. One is to use a predefined controlled vocabulary and assign index terms manually (or perhaps semi-automatically using natural language processing techniques), and another is to generate a list of index terms from the full text of the document.

4.1.2.1 Using Controlled Vocabularies for Index Terms

The LC and Dewey Decimal systems can be thought of as controlled vocabularies, predefined lists of terms whose meaning is generally understood. These terms are the call numbers assigned to the books or other items. However, when searching, we want terms that describe a subject area, not a particular item, so that we can find items that match the subject of interest. These subject terms are needed so that a search engine that uses a subject index can find the right books or articles easily. In such cases, we are using the "index terms" logical view of documents. To make the job of sensible retrieval easier, large controlled vocabularies have been created for use as index terms. Only those terms are used, and no others. This has the advantage that queries can be processed easily and if the human work of indexing was done well, the system is likely to provide high-quality results. It has the disadvantage of being fragile. If the query is not possible to express within the controlled vocabulary, there is no possibility of finding any documents.

The controlled vocabulary in use for MEDLINE is called "Medical Subject Headings," or "MeSH," for short [294]. The MeSH vocabulary has been under steady development since 1954. It currently contains over 22,000 terms. MeSH terms are organized into major subject hierarchies, such as anatomy, organisms, diseases, etc. Each of these has subcategories specializing their parent topic. Subheadings provide qualifiers for the heading terms. These are also organized into a hierarchy. Examples of subheadings are: diagnosis, therapy, metabolism, physiology, surgery. MeSH also includes a set of publication types that can be used to categorize entries, in addition to the subject matter terms.

The purpose of MeSH is to support document retrieval. It is a controlled vocabulary, not a formal knowledge representation of medicine in the sense of the ideas presented in Chapter 2. In MEDLINE each entry has a list of MeSH terms associated with it, which constitutes the document view for a "Subject keywords" search. In addition, MEDLINE entries can be searched for matches in the author or title fields, or other fields, including the abstract if available, so the full set of logical views of a MEDLINE document is extensive. However, documents are returned only if they match, so the IR model used here is Boolean.

Many other controlled vocabularies in medicine and biology have been created. The SNOMED (Systematized Nomenclature of Medicine) system, which originated in the field of pathology, is a large structured vocabulary used in electronic medical record systems to label information being stored about patients, and other data. Each SNOMED term is assigned a unique number, and has a defined relationship to other terms. It has been formalized using a Description Logic formalism, so that some inferencing or classification can be done with data labeled by SNOMED terms. Another controlled vocabulary that uses numeric codes and a classification hierarchy is the ICD (International Classification of Diseases) system [442]. ICD-10 is used in the US in medical records to unambiguously record diagnoses for individual patients. Its motivation and main use is to enable the generation of accurate and consistent billing information for medical services, in the face of ever more complex rules about billing and reimbursement. In addition to diagnosis, of course it is important to have a coding system for procedures, so that explicit bills can be generated for the work that was actually done for the patient. This is the goal and subject of the CPT codes [13]. The process of translating records of clinic and hospital visits, procedures, medications, etc., into billing codes is an enormously time-consuming and difficult task. It requires a large staff of highly trained coders.

Although the original wide use of computers in the medical setting was indeed to schedule clinic visits, hospital stays, and generate bills, the excitement in the medical informatics research community around electronic medical records was the possibility of providing decision aids, computer programs that would help improve the process of diagnosis and treatment decision making. One of the biggest challenges was interfacing the reasoning systems built with ideas from Chapters 2 and 3 to the sources of data, the emerging EMR systems. Another was transforming the knowledge from biomedical literature and textual electronic records into an appropriate symbolic form, motivating much research into controlled vocabularies and natural language processing (NLP). Several decades later, we are finally seeing some success, though no general solutions, to these two problems, interfacing to EMR data and applying NLP to automatically generate codes such as ICD-10 and CPT from textual transcribed dictation. It is amusing, and not really surprising that the first and principal use of such technology is to gain efficiency and accuracy in billing! Thus we have come full circle.

The many controlled vocabularies in use have large overlap in coverage, and it has been recognized that there needs to be some coordination or unification of these various systems. The Unified Medical Language System (UMLS) project [37,295] had as its goal to develop a framework for linking together all these different approaches to controlled vocabularies and terminology systems. It was not the intent of UMLS to provide yet another controlled vocabulary, nor was it the goal of UMLS to develop an ontology of medical knowledge. Nevertheless there have been many efforts to move in that direction, now that the importance of consistent knowledge models is better understood.

Many bioinformatics projects attempt to build controlled vocabularies for use in molecular biology, cell biology, and other basic science areas. The motivation here is similar to that in medicine, to provide a standard set of terminology for use in labeling (annotating) data extracted from the scientific literature and organized into data or document repositories that resemble the bibliographic record collections. An important example we already have introduced is the Gene Ontology [18, 19, 126].

Information such as author names, titles, MeSH keywords, etc., constitute *metadata* for a document. There has been considerable discussion of what the minimum or standard set of metadata should be for a document repository or a bibliographic information retrieval system. One such standard comes from the Dublin Core Metadata Initiative (http://dublincore.org/). This standard is especially motivated by the need to be able to identify World Wide Web resources with appropriate metadata, aside from indexing web pages in search engines such as Google^{TM}.

4.1.2.2 Query Models

We can characterize each document by which index terms it contains, whether we use terms from a controlled vocabulary, or we use terms generated by a full text analysis. Each document is represented by a feature vector, or list, in which each position in the list corresponds to a particular index term. Since the vocabulary is fixed, we can make this assignment once and for all, and compute a feature vector for each document. In the Boolean model, each entry in this feature vector is either a 0 or a 1, where 0 means the term is not present, and 1 means it is present. Then we similarly construct the feature vector for the query, and attempt to match the two vectors. A match means the document will be included, but if nothing matched the document will not be included in the results. There are at least two ways to perform the match operation. One is to match each individual component, and if *any* match, the result is a match. This is a *disjunctive* query, because it is essentially using a logical OR to combine the individual terms. This algorithm will return any document that has a match on *any* search terms. Another way is to require that *all* terms match, in which case far fewer documents will match, but the results may have a higher degree of relevance. This is a *conjunctive* query. In some IR systems, the query processor simply accepts a list of query terms as its query and uses one or the other method. This is typical of Internet search engines. However, more sophisticated or complex queries are possible in systems like PubMed,[2] where the user can specify how the matching of individual terms is to be combined to determine if the document overall should be included in the search result.

---

[2]URL: http://www.ncbi.nlm.nih.gov/sites/entrez/.

This idea can be expressed very simply in a function we will call `bool-or-query`, which takes as input a sequence of 0 and 1 values representing the feature vector of a document, and a sequence of the same sort, representing a query. It returns `t` if any pair of corresponding sequence elements both contain a 1.

```
(defun boole-or-query (query doc)
  (some #'(lambda (x y) (plusp (logand x y)))
        query doc))
```

The `some` function applies its first input (the anonymous function we have put in here) to the sequences that are the remaining inputs (here the two feature vectors, `query` and `doc`), taking an element from each sequence, and applying the operation. It continues until either the operation returns something not `nil`, meaning that the two elements are both 1, in which case the result is `t`, or it reaches the end of either or both sequences, in which case the result is `nil`. This function implements a disjunctive query, and is fast and simple. A conjunctive query is a little different because it has to test if all the features found in the query are found in the document. However, the document can have extra features that are not in the query, and they should be ignored. Here is a solution to the conjunctive query problem.

```
(defun boole-and-query (query doc)
  (every #'(lambda (x y) (if (plusp x) (plusp y) t))
         query doc))
```

This function examines every element of the query vector, but ignores the 0 entries, only checking a position in the document feature vector when the query feature vector has a 1 element in that particular position.

Another approach, the vector space model, assigns weights (usually between 0 and 1) for each of the index terms. The weights could represent how significant the term is for the document, and similarly how important each term is for the query. Weights can be based on term frequencies in the document, or on other factors. In this model, we can compute the degree of match of the query vector and the feature vector by computing an inner product and normalizing the result, giving a value between 0.0 and 1.0 where 1.0 is an exact match. This is the geometric equivalent of computing the cosine of the angle between two vectors in geometric space. Again, we assume we have two sequences, representing feature vectors, but this time they contain real values. We don't assume that the values all sum to 1.0 in each. The mathematical formula we are implementing is

$$s(q,d) = \frac{\sum_i q_i d_i}{\sqrt{\left(\sum_i q_i^2\right)}\sqrt{\left(\sum_i d_i^2\right)}}, \tag{4.1}$$

where we denote s as the *similarity* between the query and the document. It will be a number between 0 and 1, if all the vector components, or feature weights, q_i and d_i, are non-negative. This is easy to implement in code. If the feature vectors are implemented as *arrays*, you can use the sum operator defined on page 359. The expression, $\sqrt{\left(\sum_i q_i^2\right)}$, is also known as the *norm*, or length, of a vector, q. It would make sense to define a function, norm, that computes each of the two square root expressions that appears in the denominator above, since they have the same form. Having the norm function, one could then use it in the body of the code for s, where we have called this function similarity rather than just s.

```lisp
(defun norm (v)
  (sqrt (sum i 0 (array-dimension v 0)
              (expt (aref v i) 2))))

(defun similarity (v1 v2)
  (/ (sum i 0 (array-dimension v1 0)
          (* (aref v1 i) (aref v2 i)))
     (* (norm v1) (norm v2))))
```

Many other formulas and expressions have in them as components the norms of vectors. Once again, this is an example of *layered design*, where useful procedures are encapsulated into functional abstractions. One could go further, and encapsulate the numerator in Equation 4.1 as a function named dot-product, which also turns out to be a useful abstraction. The definition of similarity then takes on a very nice appearance. That is left as an exercise.

4.1.2.3 Computing Index Terms from Full Text

Instead of assigning index terms from a controlled vocabulary, we can take a different logical view of the documents, and use the full text of the document. In this case, we can still compute a feature vector for the document, and for the query. A simple kind of feature vector, often used, is a list of the words in the document, together with the frequency of occurrence of each word. This is called the Zipf distribution. In a practical system, the commonly occurring words such as "a," "an," "and," "the," and other so-called "stop" words, would be ignored, since their presence does not convey much about a document. The more valuable words are those that relate to the subject matter of the document, words like "cancer," "protein," and specific names. To compute the Zipf distribution of a text is easy, setting aside some issues like punctuation.

Essentially, we will read the text word by word, and let the Lisp reader turn each word into a symbol (thus ignoring distinctions between upper case and lower case). We will keep track of each symbol in a hash table, adding a new entry when we come across a new word, and incrementing the count associated with an entry when it is a word already in the table. A simple strategy for this was introduced in Chapter 1, the function item-count, defined on page 39. That function just uses a list, which works fine for small numbers of different words, and high numbers of occurrences of each. In a text document, however, we expect many more words and lower frequencies of occurrence. So, here is a version that uses a hash table instead of a list. This version reads words from an input stream rather than iterating over a list. The stream could be an open file or some other kind of stream as long as it can be read sequentially with the Common Lisp `read` function. The function takes a stream as input, and returns a frequency table of all the words found in the stream, a stream of text, using each word as a hash table key.

```
(defun item-count-hashed (stream)
  (let ((results (make-hash-table))
        (tmp (read stream nil :eof)))
    (until (eql tmp :eof)
      (if (gethash tmp results) (incf (gethash tmp results))
        (setf (gethash tmp results) 1))
      (setq tmp (read stream nil :eof)))
    results))
```

To use it, a program would open a file and pass the resulting open stream to the function `item-count-hashed`. Since it cannot be determined how long the text is, indefinite iteration is needed. The function will keep reading a word at a time until the `read` function returns the keyword `:eof`, as specified above, indicating the end of the file. There is only one problem. Common Lisp has no operator named `until`, to do what is intended above, that is, to keep repeating the body of the expression until the test `(eql tmp :eof)` is satisfied. However, it is easy to write a macro to do this. It is very similar to the `while` macro defined in Graham [137, page 164]. The only difference is that here we put the test directly into the do expression, instead of the negation. The `until` expression will just be translated into a do expression with no local variables, just a termination test and a body.

```
(defmacro until (test &rest body)
  `(do ()
       (,test)
     ,@body))
```

I suppose that by now the reader will feel very capable of writing out the do or perhaps a variant using `loop`, but macros such as these add some nice readability to code and are usually worth using. This macro can be used in many places as you write more code, and logically should become part of your collection of utility macros.[3]

Here is the result of using the `item-count-hashed` function on the first two paragraphs of this section, after removing all the punctuation and parentheses with a text editor. The hash table contents are displayed using the `printhash` function defined on page 204, but only the first few entries are shown here to save space.

```
> (setq zipf (with-open-file (s "zipf-test.txt")
                (item-count-hashed s)))
#<EQL hash-table with 167 entries @ #x716433ca>
```

---

[3] Actually, this macro and the `while` macro are included in the Allegro Common Lisp implementation from Franz, Inc. as an extension, one of many. Other Common Lisp implementations also have extensions, so one should check the documentation to see if you need to write your own.

```
> (printhash zipf)
Key: STREAM     Value: 6
Key: OF     Value: 20
Key: ASIDE     Value: 1
Key: OCCURRENCES     Value: 1
Key: BY     Value: 1
Key: OTHER     Value: 2
Key: VALUABLE     Value: 1
Key: COME     Value: 1
Key: 1     Value: 1
Key: VERSION     Value: 3
Key: READS     Value: 1
Key: CAN     Value: 3
Key: WHEN     Value: 2
Key: HIGH     Value: 1
Key: DIFFERENT     Value: 2
Key: MANY     Value: 1
Key: USES     Value: 3
Key: ITEM-COUNT     Value: 2
...
```

To be able to use this as an effective indexing tool, the entries would have to be sorted, stop words removed, and the resulting vector or list of words and frequencies would have to be standardized with the word lists of all other documents. This means effectively adding entries with 0 occurrences for words that did not appear in the document but which do appear in others. That way, every vector has the same list of components, so that two vectors can meaningfully be compared.

This simple code has many other limitations. It not only will not handle punctuation that makes text not simply a sequence of words that can become symbols. It also will not handle *stemming*, in which many different words can be identified as variant forms of the same root word. An example would be the word "gene." It may appear in plural form, as "genes," or the related adjective, "genetic." In building an index of feature vectors, one might also want to group together well-established synonyms, such as "heart" and "cardiac." Further, to get closest to the intent of a query, it is often desirable to extract multi-word phrases from a query, and match to multi-word phrases in a document. These are left as exercises for the reader.

4.1.2.4 Index Organization

Having chosen or generated the index terms for a set of documents, we have several options for how they are organized for efficient lookup. One obvious way to create a working index is to use a hash table, so that terms can be looked up in more or less constant time. Another is to build an inverted file, where the records contain the index terms and other information, so that the file can be easily searched. If this file is organized in alphabetical order by index terms, it is possible to use binary search on the file. This means that the time to find the record for a term scales with the logarithm of the number of terms, which is very reasonable for building large inverted files. Another way to organize the records is to create a tree structure, so that a search term can be efficiently found in the tree. Each node in the tree contains a link to a term that comes before it in the alphabetic order, and one that comes after it. One can search through the tree, comparing the search term with each term and either following the "before" link or the "after" link, depending on the result of the comparison. This structure also scales as $O(\log n)$, but is easier to maintain when adding terms (records). This tree structure and related functions are described in Section 1.3.2.

4.1.3 Processing Queries

A query can be just a list of terms, a logical expression, or even natural language, depending on what system is in use. The query will need some preprocessing before it can be directly used for searching for relevant records. Some of the same steps that are done on documents to generate index terms are also applied to queries. Any stop words in the query are dropped. The essential terms may need to be stemmed. In some systems (such as PUBMED) a controlled vocabulary can be used to match the query terms and determine if there are any synonyms, or if query terms are synonyms for standard index terms. Some systems will assign weights to query terms depending on their significance.

4.1.4 Searching and Matching

Search algorithms can vary from simple to extremely complicated. A Boolean model will involve simply looking for all the records that match, as described in the code earlier in this section. A system using the vector space model will collect results that give non-zero similarity, but may sort by similarity, and only include those that are above some threshold.

Searches may take place through a client-server network connection, for example, between a web browser and a web server, where the format of the query is a CGI type URL, and the query is processed by a script or program outside the web server context. A standard protocol, Z39.50, also has been developed by the Library of Congress to support bibliographic queries [15].

4.1.5 Ranking the Results

The final step in an information retrieval system, in response to a query, is to organize the results in a way that is most useful to the user of the system. Typically this means rank ordering the documents that successfully matched the query, according to some notion of relevance. Many factors can be used to do this ordering, including the frequency of appearance of the search terms in the document, the location of the matched terms, the date (perhaps more recent ones will be more relevant), or even the popularity of the document for large numbers of users. Each of these ranking methods has advantages and disadvantages.

An obvious connection exists between the information retrieval model and the possible ranking methods. A simple example is that in the vector space model, the similarity measure that identifies documents to include in the results, also provides a natural ranking measure. More complex models provide richer ranking methods.

One results ordering method has become very visible to the community due to the popularity of the Google^{TM} search engine and services provided by Google. The *page rank* algorithm [51] uses the idea that the relevance of a web page to a query is related to the number of web pages that link to it in some context. This is motivated by the idea in the academic research world that the importance of a journal article or other research publication can in some sense be measured by how many other publications cite it. In the academic community, there has been substantial discussion of using this as a better measure of qualification for promotion than the more usual number of publications. The rationale is that a few very important publications should be weighted as much as or more than a large number of obscure or unimportant ones. In the context of the Web, a page that is linked from many other pages is likely to be more useful than one to which no one else has links. Further, a page that has links to it from some really prominent or significant page, like the University of Washington Health Sciences Library main reference page, is likely to have more relevance than one linked from another page of the same author as the original page. This incorporation of where the links come from brings in the all important factor of human curation again, and one might wonder how such importance can be computed automatically. The basic formula for the page rank of a page, A, is

$$P(A) = (1 - d) + d \sum_{i=1}^{n} \frac{P(T_i)}{C(T_i)},$$
(4.2)

where T_1, T_2, ..., T_n are all the pages that point to page A, $P(X)$ is the page rank of page X, and $C(X)$ is the total number of links out of page X, to other pages. This clearly involves building a huge index of all the pages and links reachable from one to another. The engineering of this enterprise has been described in many places, and a good starting point is a page at the Web site of the American Mathematical Society.[4]

4.1.6 *Performance Evaluation*

Evaluation of performance of an IR system usually is done by computing two measures, recall and precision, that are similar to the sensitivity and specificity of diagnostic tests. Both measures depend on being able to define which documents are *relevant* to any particular query. The collection of relevant documents for a query is used as the "gold standard." This cannot be defined by any particular search algorithm, of course. A key part of creating the standard for relevance is human curation. When test document sets are created, a panel of experts has to consider the test queries and for each query, the panel must decide what are the relevant sets of documents. This reveals that the essence of information retrieval is an attempt to model the

---

[4] http://www.ams.org/featurecolumn/archive/pagerank.html.

intent of a person. It is in reality a kind of guessing game, not a theory of how things in the world actually work. Nevertheless, it is extremely important to understand how people express their information needs and how to guess well what they are asking for.

Once we have defined a set of relevant documents for a query, the recall and precision of a system for that query can be defined. The recall is the fraction of relevant documents that were actually retrieved in response to the query. So, if R represents the set of relevant documents, and S represents the set of documents retrieved, then the recall is the proportion of R that were returned as results. It is a measure of how many of the relevant documents were retrieved.

$$\text{Recall} = \frac{|R \cap S|}{|R|}, \tag{4.3}$$

where the vertical bars denote the number of elements in the set. This is analogous to the sensitivity for a diagnostic test, which is the true positive rate, that is, having applied the test (done the search) how many actual positives tested positive, or how many actually relevant documents were retrieved.

The precision is defined to give an indication of how much irrelevant stuff is also retrieved. The precision is the fraction of the *retrieved* documents that are relevant, given by the formula

$$\text{Precision} = \frac{|R \cap S|}{|S|}. \tag{4.4}$$

This is not exactly analogous to the specificity of a diagnostic test, but similar in intent. Rather than asking how many of the *non-relevant* documents were *not* retrieved, the precision is a measure of how few non-relevant results were also retrieved. Both of these numbers (recall and precision) will be between 0 and 1.

Having ranked the documents by some measure of relevance, one could imagine only considering the *most* relevant, or most highly ranked results. Depending on where you set this cutoff, this new set, the *selected* results, will have a somewhat different recall and precision than the entire set. It should be clear that there is a tradeoff between the two measures. The more of the results you include, the higher the recall will be. If the relevance measure is accurate, the fewer the number you include, the higher the precision, but the recall will go down, because you will lose some relevant documents that are ranked lower than others. This tradeoff is very similar to the tradeoff in diagnostic testing. In particular in using the PSA test for prostate cancer, one could choose a cutoff of higher than 4.0, which would increase the specificity of the test, but reduce its sensitivity. Similarly, lowering the cutoff will increase the sensitivity but decrease the specificity. Such trade-offs are part of many decisions in real life, which of course is the motivation for decision theory, briefly introduced in Chapter 3.

The definitions given here are the simplest possible ones. More sophisticated measures have been proposed and used. As you can see, the definitions of recall and precision interact with the algorithms for rank ordering search results. This remains a difficult problem for design of practical systems.

In addition to the quasi-objective measures of precision and recall, performance evaluation can also be based on measures of user behavior, including statistics on usage. An introduction to this can be found in Hersh [155, Chapter 7].

4.2 Beyond Text Documents

Although the focus so far has been on text data, it is also possible to organize non-text data using some of the same principles. This is especially important for handling biomedical image data, gene and protein sequence data, and electronic medical records. We've already examined some aspects of sequence data. Here we just make a few observations about medical images and EMR systems.

4.2.1 Biomedical Images

Biomedical images can be organized into a searchable document repository also, although searching the image pixel data directly for content is a very difficult problem. In general, images are stored with tagged or otherwise organized metadata, describing what is in each image. One example is the University of Washington Digital Anatomist Project Interactive Atlases [53,54]. These are a large collection of photos, medical images, hand contoured anatomic structures, and other kinds of images, with annotations. Representing the annotations by structures like MARC records and other bibliographic data makes them searchable, and thus since the images are referenced by the annotations, the images can readily be obtained. An example from the Digital Anatomist Brain Atlas is shown in Figure 4.1.

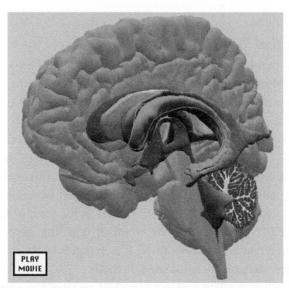

Figure 4.1   An image from the Digital Anatomist Brain Atlas, showing a cutaway view of the brain, with the caudate nucleus outlined.
(Courtesy of the Digital Anatomist Project at the University of Washington.)

These images are *annotated* in the same way that sequence data are annotated in the gene and protein sequence information repositories. All the search and retrieval methods that apply to bibliographic and other structured text records can be used with image repositories in which the images are annotated. What is especially useful about the Digital Anatomist images is that some of the annotations consist of outlines, pointers, and labels with coordinate locations keyed to the images. These are outlines and pointers to named structures, so one could query the system for images containing some structure, like the caudate nucleus. The images can then be displayed, with or without labeling of that object. The original application of the Digital Anatomist was to provide a supplement to textbooks and dissection in teaching anatomy to medical students. It has been considered for clinical applications such as radiation therapy planning as well [210].

Since the feature data of these images are textual, like the bibliographic data, they can be used to generate index information using the same methods mentioned previously. In addition, image segmentation, the detection, and delineation of objects in images, such as the caudate nucleus in Figure 4.1, results in sets of coordinates, which can be further processed to generate feature vectors for the images in which they appear. These feature vectors can be used to *classify* the image data. Potential clinical applications are numerous. One example is the possibility of using image data to identify craniosynostosis in children [255, 256]. This abnormal development of the skull is thought to be related to an increased risk of cognitive deficits. As medical applications of image analysis develop, and very large databases of image studies become available for research use, indexing and retrieval of images by content is going to be increasingly important. This project illustrates only one of many ways in which images might be indexed by content. A recent survey article provides a more systematic overview [9].

4.2.2 Electronic Medical Records

Although the primary purpose of EMR systems is to store data for use in clinical practice, they are clearly valuable repositories for research use. However, the indexing and retrieval capabilities of these systems are highly tailored to the workflow of clinical practice, on the hospital wards and in the outpatient clinics. There are no general search facilities. Nevertheless, EMR systems are being used to support research. The usual method for this is to create a separate data warehouse, in which data are extracted (copied) from the EMR and put into a suitable form, usually a separate relational database. Such data warehouses are then searchable in all the ways mentioned in Chapter 1, insofar as the data fit the relational model. However, a growing proportion of the content of EMR systems is in the form of unstructured text. Even more is in the form of PDF files that are the result of scanning, or output from some other system. The unstructured text is amenable to NLP techniques as described. The PDF files are at present essentially unreachable by automated methods.

In addition, since much of the data contained in EMR systems has a temporal aspect, another big challenge is to represent time-oriented data and query about time. Examples are series of vital signs measurements during an inpatient stay. Each should have a time stamp. Some way to describe trends, events, and other structures is needed. This is an active area of research and experimentation.

4.3 Networks and Biomedical Data Communication

Both biomedical research and health care involve many complex sources of computerized data, as well as many components that use the data. The data must be transferable between computer systems within a laboratory or hospital, as well as between different health care facilities. Experimental data generated by a laboratory is submitted to the National Center for Biotechnology Information (NCBI) for storage in data repositories that are used by other laboratories. NCBI is a branch of the National Library of Medicine (NLM), which is the institute of the National Institutes of Health (NIH) that is concerned with organizing, storing, and making available biomedical information as a national resource. In health care, patients routinely get medical diagnostic tests done at a family practice clinic, radiologic imaging studies at another facility, and then specialty care such as cancer treatment at yet a third facility. The patient's test results and image data need to be available to the other facilities. As the number, complexity, and magnitude of computerized data have grown, so has the problem of standardizing the computerized representation of the data. With the enormous growth of the Internet, rapid data communication has become within reach. Nevertheless, the complexity of network transmission of data has been a challenge, and many people still cling to the idea of writing files onto, carrying, and reading files from, removable computer media such as CDROM, "thumb drives," DVD-ROM, and other devices.

Since the mid-1980s, when networks grew and the Internet began to look like a realizable vision, there were many efforts to create network protocols for message and data interchange. Several prominent efforts focused specifically on medical data. These medical message and data interchange schemes are intended to be independent of how medical software might store the data in a file or database. This is a critical difference between the medical approach to data representation and the analogous problem in molecular biology. In bioinformatics much of the handling of data concerns storage in large files, so file formats become the framework for discussion, for example, the so-called FASTA format for sequence data. In Chapter 1 we showed how to deal with FASTA files, and in Chapter 5 we will see how to do some computations on the contents. Programs that read files must deal with parsing ideas and methods, but in a different context. The closest connection between the biological and medical data representations is access to data via web servers. The NCBI provides such access, for example, to GENBANK and related resources, through the Entrez Programming Utilities, described briefly later in this section.

Beyond the World Wide Web and XML, there are also protocols specifically for handling medical data. This section contains a brief introduction to how to design a network server or client program that uses TCP/IP sockets. Following that is an introduction to HL7 and some methods for processing HL7 messages. A detailed description of the DICOM protocol completes the discussion of medical data communication.

The two medical data communications protocols described in this section are intimately related to two major components of the environment of computing in health care, the "Electronic Medical Record" (EMR), and the medical image data repositories known as "Picture Archiving and Communications Systems" (PACS). Although these seem to be the focus of hospital computing, in reality there are hundreds of independent systems in a typical hospital, performing a wide range of functions, generating data, and using data from other systems. Examples include laboratory instruments, and laboratory data management systems that collect data from the instruments, accept orders for laboratory tests, and do data analysis and quality assurance computations; pathology systems that capture pathology reports, microscope images, and a variety of other data; anesthesia documentation systems that record information from procedures; ICU monitoring systems; specialized systems for radiation oncology that simulate radiation treatments; drug dispensing systems that are operated by the pharmacy department, and the myriad medical devices that are now computer controlled themselves, such as fetal monitors, infusion pumps (more about this in Chapter 6, Section 6.2), stress testing facilities, and so on.

The first widespread use of computers in medical centers was for scheduling, accounting, and billing. These were primarily to satisfy business needs, and were not "clinical computing" as we know it today. However, some specialized activities in medical centers were indeed handling much clinical data, notably the clinical laboratory, radiology, and radiation oncology. From the scheduling system, it was possible to generate admission, discharge, and transfer (ADT) data. This also required a registration system which stored basic information about patients, generally referred to as "demographic" information, name, address, telephone number, etc. This information had to be re-entered in the laboratory systems and others, until some means was invented to connect to a network and send the information via the network. The technical problems of getting the data encoded, transmitted, and decoded were compounded by the difficulties of duplicate names, misspellings, and a host of other problems. Regarding the names and information around laboratory tests, we already have touched on that with the discussion of LOINC, in Section 1.1.4.

Early research in medical informatics had a significant focus on an entirely different problem, that of getting computers to emulate medical diagnosis. The application of artificial intelligence ideas to diagnosis and treatment planning was then and is today a very challenging problem. It provided fertile ground for basic research, but found little clinical application. The use of computers in the hospitals in the 1980s was still mainly for patient appointment scheduling, billing, and other business needs. The idea of storing and retrieving clinical data for decision support was not widely embraced. There was considerable apprehension that computer phobia among medical care providers would stymie any efforts in this direction. In the 1990s this began to change. Today, EMR systems are widespread, and everyone seems to want more as soon as possible. Although these systems are mostly focused on storing and retrieving data in a user-friendly way, not on doing any interesting computations, one project caught my attention.

One of the most challenging problems in dealing with the medical environment is that a significant amount of critical information is contained in notes in a stylized free text form dictated by care providers. Much research and engineering has been devoted to methods for automatically extracting the information in these texts in a more systematic, symbolic form, which can be used (as you might guess) for all kinds of interesting reasoning applications. One might be to use this data as input to practice guidelines to generate suggestions for next steps in the patient's care. Another would be to use the information together with computerized order entry for drug prescriptions, to check for and alert for possible drug interactions.

But, it seems that these are not the targets of the project. One of the most urgent and most difficult problems facing medical practice today is to be able to choose the right diagnosis code(s) on which to base the billing codes. Hospitals typically hire large numbers of people to do this process manually. They are trained as "coders." The work is tedious, error prone, and very complex, because the billing rules are very complex. So, it seems that we have indeed found a need for advanced artificial intelligence methods in the context of the EMR, and it is for help with billing.

Beyond the problems of terminology and encoding, there is the problem of context. No laboratory test data or other clinical information can be interpreted by itself. The meaning depends on what else is known about the patient, as well as the reasons for ordering the test. This context problem makes it especially challenging to create and use what are generally called "guidelines." They are fragments of knowledge typically represented in the style of rules (as in Chapter 2) but are generally not designed to be combined in complex ways to do inference. It is beyond the scope of this book to address this contextual problem. Here the focus will be on some of the technical details of HL7, the protocol for transmitting, and receiving medical data more generally.

Nevertheless, work continues on encoding and using medical guidelines, and projects such as GLIF [321], Asbru [373], and GLEE [429] will become very important, when the environment is ready. A few comments about GLEE in particular may be helpful.[5] GLEE is a guideline execution engine that uses the GLIF3 guideline model. GLIF3 is implemented as a Protégé ontology, and actual guidelines are instances of the model classes. GLEE has an interface to the patient record; at the moment this is customized to the ECLIPSIS system at Columbia. The development of a more standardized, non-proprietary interface is an active research area. The "Virtual Medical Record," derived from (or based on) the HL7 Reference Information Model (RIM), is a possible basis for such an interface. It is intended to provide a way to translate between a standard model and the local EMR system. GLEE is about one hundred thousand lines of java, and is not available for general use, especially since it would involve a lot of custom programming to adapt it to any other EMR environment.

---

[5] Source: personal interview with the author of GLEE, Dongwen Wang.

In regard to medical images, the images that are stored in the PACS are generated by CT scanners, MRI scanners, digital radiography units, ultrasound, and many other computer controlled imaging devices. The images are used by the radiation treatment planning systems just mentioned, and are viewed in the emergency rooms, on the wards, and in the outpatient clinics. So, standard communications protocols are an absolute necessity in such a complex environment. No hospital gets all the various computer controlled equipment and computer applications from a single vendor. Even acquiring systems from a single vendor does not assure interoperability. The number and variety of applications of computers in the clinical practice environment are both rapidly increasing, so the problem is every day becoming more urgent.

The situation with PACS is somewhat more mature than with EMR systems. The DICOM standard evolved quickly in response to a much more focused need, though here too the history has some parallels. Historically, medical physicists and computer scientists in radiology have focused on complex algorithms for image processing. In the research community there has been a lot of interest in automated analysis of medical images to aid diagnosis. Of course there is a practical need for automated image segmentation (the automated delineation of organs and other objects in the images), particularly for radiation therapy planning. The push for DICOM, however, did *not* come from interest by radiologists in such advanced algorithms. It came about because radiology departments were running out of space for film storage, silver for film was becoming scarce and expensive, films could only be at one place at a time and films were getting lost. All these argued for network-based digital image storage and transmission systems. Since the early 1980s saw the rapid advent of digital imaging anyway, this was a pretty obvious need. After a few missteps, an effective protocol was developed, became a NEMA standard and was widely adopted.

4.3.1 Networks and Protocols

The purpose of the rest of this section is to provide a technical explanation of network protocols, and in particular, HL7 and DICOM. The level of detail will be at times substantially greater than in other sections. It is possible to get a very high level overview, for example, [30] for DICOM, but not really understand what is actually going on. My hope is that by providing sufficient detail you will appreciate how the high level ideas take shape and actually work. This subsection gives some basic background on networks, socket programming, servers, and clients. Next, you will see the anatomy of HL7 messages, and then the frightening complexity of DICOM.

4.3.1.1 Network Architecture

In network data packets, as in files, everything is really binary, and the distinction between text and binary is, as explained in Chapter 1, the interpretation of bytes as character codes in the case of protocols that are described in terms of text strings. DICOM is an example of a fundamentally binary protocol. The most common protocols that use plain text are SMTP (electronic mail),

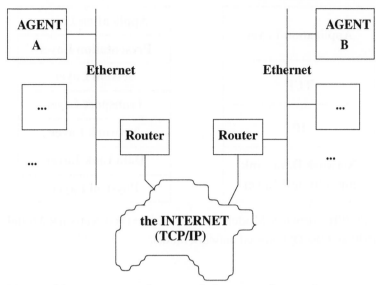

Figure 4.2   DICOM and HL7 are network protocols between client and server programs.

and HTTP (the World Wide Web protocol).[6] Both of these have been extended to support the transmission and reception of non-text data, in the form of attachments (for electronic mail), including data such as images, digital sound, and video. Web browsers can also receive and send files containing arbitrary data, which of course leads to security problems if the data consist of an executable program and the browser is configured to execute it.

Data interchange standards such as DICOM (for exchange of medical image and radiation therapy data) and HL7 (for exchange of clinical laboratory and other patient data) are about program to program communication between computers, not a standard file format, and not information retrieval for use by humans. Figure 4.2 illustrates this idea.

These standards each define an application layer protocol, in which the exchange of data follows a detailed plan, specifying the order and type of messages that can be exchanged.

The DICOM and HL7 standards are based on the ISO[7] Open Systems Interconnect (OSI) model, although actual implementations use TCP/IP protocols rather than OSI protocols. The OSI model is a layered design, in which the various concerns are separated, to enable inter-operability of different kinds of computers and different operating systems. The OSI model is described in many fine books on modern networking concepts, such as [75]. Figure 4.3

---

[6]These are both Internet standards maintained by the Internet Engineering Task Force (IETF). The mechanism for development of Internet standards is the RFC (Request For Comments). The full list is accessible at the official repository, the RFC Editor web site, `http://www.rfc-editor.org/rfc.html`. There are also "joke RFCs," most easily found via the Wikipedia page for "April Fool's Day RFC."

[7]ISO is a trademark of the International Organization for Standardization. The name "ISO" is not an acronym. ISO standards are used worldwide and are created in cooperation with various national standards bodies. In the US, the American National Standards Institute, ANSI, is the corresponding national standards body.

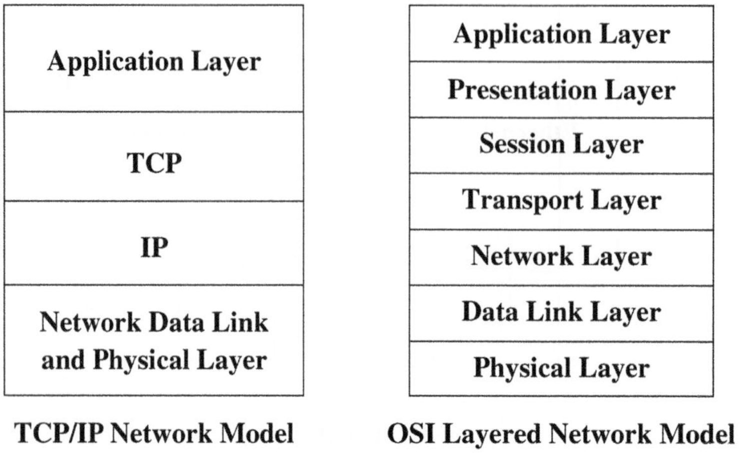

Application Layer	**Application Layer**
	Presentation Layer
TCP	**Session Layer**
	Transport Layer
IP	**Network Layer**
Network Data Link and Physical Layer	**Data Link Layer**
	Physical Layer

TCP/IP Network Model   **OSI Layered Network Model**

Figure 4.3   TCP/IP and the OSI layered network model.

illustrates how these concerns are organized into layers, with each layer only depending on the layer immediately below, and providing functions (services) only to the layer immediately above.

In the OSI model, the data link and physical layers are only concerned with getting packets of data from one computer to another, for example, the Ethernet standard with all its variants in speed and cable type, fiber optic cable, etc. The network, transport, and session layers provide logical pathways between clients and servers, find physical routes through the network, break up long messages into packets, reassembling them at the destination, and perform other functions that should not concern application programmers. The application and presentation layers contain the details of the biomedical application, which might have a fancy user interface, or might just be a server that sends or receives data on request from clients.

The OSI model does not specify particular protocols to be used at each layer, but ISO did define a series of protocols which are also referred to as OSI. These are not widely used. The TCP/IP protocols, developed earlier, only partly correspond to the OSI model. The IP protocol handles the problem of addressing, and getting information packets from one place to another (routing). It is a service used by the next layer up, which can be either TCP or UDP. The UDP (User Datagram Protocol) service is *stateless*, which means that the sender program sends a packet of information, the various routers attempt to get the packet to its destination, and no information is kept about the sender or receiver. This works for simple services, like time services for example. The time service just sends out time messages to the receiving clients. If a time message does not get delivered, it is not a problem. On the other hand, TCP (Transmission Control Protocol) is a *stateful* service. In a TCP application, the client program requests a connection to the server, and that connection information is maintained on both ends and through the system, so that the client and server can send messages back and forth, as if they had a private communication channel. When a program connects via TCP, the network software is set to insure that packets are guaranteed to be delivered from sender to receiver reliably and in the right order. Both HL7 and DICOM use TCP, not UDP.

In short, the problems of connection wires, signals, insuring error-free transmission, creating, and maintaining a logical path from a program on one computer to a program on another computer, are all relegated to a set of standards outside the scope of DICOM. The most common framework for this is TCP/IP (also described in [75] and illustrated in Figure 4.3 in relation to the OSI model). The idea of the layered approach is that the upper layers work with interchangeable lower layers and that the upper layer software can be designed without concern for solving the problems addressed by the lower layers.

4.3.1.2 Network Client and Server Design

For HL7, we will focus mainly on the structure and decoding of the information contained in HL7 messages, and not so much on the implementation of actual HL7 clients and servers. For DICOM, we will provide much more detail. This section is a general introduction to how client and server programs use socket libraries to actually implement message sending and receiving, as well as the basic logic of a client or server program.

What is needed in order to have a working network application? One must be able to open a connection from one program to another and do reading and writing with each end, in order to exchange messages. This is done in different ways in different programming languages but the abstract idea is that a socket is opened analogously to opening a file, with just a little extra detail. In gaining access to the NCBI databases via their Entrez API, we will use HTTP (Hyper Text Transfer Protocol) to write client programs that communicate with web servers. We will not have to be concerned with the details, because there are readily available functions that perform all the necessary operations to satisfy the HTTP protocol. We just call the library functions and we get the web pages (HTML or XML data) back. In Section 4.3.5 we will use a very simple protocol, after opening a socket connection, to get information from the UW Foundational Model of Anatomy through the FMS. The DICOM protocol involves a more extended and complicated dialog between client and server. For DICOM, we will show how the protocol works and can be implemented.

For a program running on one computer to be able to communicate with a program running on another computer, each must have some unique address or way to be identified, so that packets of data can be sent to the right destination. The IP protocol (version 4) uses addresses consisting of 4 bytes, represented as four numbers separated by dots, as for example, 128.95.181.24. An address usually corresponds to a particular interface device on a computer, such as an Ethernet wired connection, or a wireless network interface, so it is possible for a single computer to have multiple addresses. However, usually there are many programs running, providing many services, so one or a few IP addresses are not sufficient to distinguish one from another. A remote program needs a way to specify a connection to a particular service. This is done by specifying a *port number*, in the range of 1–65,535. The lower numbered ports (below 1024) are reserved for standard system services and require more than ordinary user privileges to be assigned to the server program. Commonly provided services are assigned to so-called *well-known ports*. For example, web servers usually use port 80, and mail handling services use a variety of ports depending on whether data will be encrypted and how (ports 993 and 587, for example). Servers that support the DICOM protocol for transferring medical image data and radiation therapy data usually use port 104.

When a server program is running, it waits for an incoming network connection request and provides services or functions requested by the remote program (the client). A client program is one that initiates a connection to a server, in order to provide the functions of the server to the user of the client program. These programs are generally written using the TCP/IP socket libraries provided for various operating systems. A *socket* is analogous to a file, in that a program opens a socket and performs read and write operations with it. However, the socket represents the input and output of a remote program that has also opened a socket for communication over the network. A server opens a "passive" socket, and several additional library procedures prepare it to accept incoming connections. The server "listens" on a particular *port*, an abstraction that allows multiple server and client programs to maintain logically independent virtual connections through the same physical interface. A client program opens an "active" socket and specifies the remote entity to connect to by providing an IP address and a remote port number. The TCP/IP socket library functions internally take care of all the details of device drivers, breaking data streams into packets, routing the packets, insuring that the transmitted data are received and reassembled in the correct order, etc. In short, TCP makes each program see the other program as a device with an input and output stream, and neither program has to have any concern about how the data are actually moved from one program to the other. The socket model is well described in standard networking texts such as [75].

In the University of Washington DICOM project [200], as for the FMS code in Section 4.3.5, we chose for simplicity to just use a Common Lisp vendor's implementation of a socket layer interface. This is analogous to using high level functions for network access in java and other languages. Most Common Lisp systems currently available have such interface functions. Here is the definition of a function provided in Allegro Common Lisp (ACL)[8] [117]:

`make-socket` &rest *args* &key *type connect address-family*   *[Function]*

creates and returns a socket object with characteristics specified by the arguments.

The `:type` argument can be either `:stream`, which makes a TCP socket, or `:datagram`, which makes a UDP socket. The `:connect` argument can be `:active` (for a client), or `:passive` (for a server). The `address-family` can be either `:internet` or `:file`. The `make-socket` function takes additional keyword arguments, `:local-port`, `:remote-port`, and `:remote-host` depending on whether the socket is active or passive.

---

[8]Allegro Common Lisp is a software product of Franz, Inc., Oakland, California.

The next step for a server is to wait for and accept an incoming connection request. The ACL function for this is:

accept-connection *sock &key (wait t)*   *[Function]*

If wait is t, waits for a remote connection request for this socket and returns a new stream object that can be used to do I/O to the connected socket. If wait is nil and there is no connection request pending it returns nil immediately, but if a connection request is pending, it completes the connection and returns a new stream object for input and output on the connected socket.

When the connection is accepted, our code will read and write data according to the specific application layer protocol, in our case, either HL7 or DICOM. Every protocol has some agreed on sequence of messages that server and client send in a specified order, with different options and alternatives depending on what operations are needed. In a generic server program, this would be implemented by a function, passed to the server when it starts. The same basic server code could then be used to implement many different network services.

Here is a simple implementation for a generic TCP server program. The function that implements the application layer protocol, DICOM or HTTP or HL7, is called here applic-fn. The port also has to be specified. This depends on the application. The registered port number for DICOM is port 104, but others could be used.

```
(defun tcp-server (applic-fn port)
   (let ((socket (make-socket :connect :passive
                              :address-family :internet
                              :type :stream
                              :local-port port))
         (stream nil))
      (unwind-protect
         (loop
            (setq stream (accept-connection socket :wait t))
            (funcall applic-fn stream)
            (close stream)))
      (close socket))) ;; only here if severe error occurs
```

This server will continue to wait for and accept connections until terminated by an external event such as an operating system intervention or a host computer shutdown. Another possibility is that an unrecoverable error internal to applic-fn occurs. In that case, the Common Lisp function unwind-protect catches the error condition, allowing the program to close the socket gracefully and exit.

There are other parameters to make-socket which would be used for a fully detailed implementation, but here we just focus on the essential elements. For a network client application we need a much simpler function than `tcp-server`, since the client does not wait for and repeatedly process connections.

```
(defun tcp-client (applic-fn port)
    (let ((stream (make-socket :connect :active
                                :address-family :internet
                                :type :stream
                                :remote-host host
                                :remote-port port)))
            (unwind-protect
                (funcall applic-fn stream))
        (close stream))))
```

In most applications, the server version of `applic-fn` is somewhat different from the client version, though the same function could be used for both, if suitably parameterized. The bulk of the work of writing network programs is in dealing with the application layer protocol, not in the details of TCP/IP networking. So the remainder of this section will describe some aspects of the two most important network protocols related to clinical medicine.

4.3.2 HL7: Clinical Data Encoding and Transmission

The HL7 standard for network interchange of clinical data introduces a framework or wrapper that enables the data to be encoded in a message. The effort to create a standard began in 1987, and has roots in a variety of more specific areas. There are three conceptual parts to the HL7 idea. First, systems exchange messages, so there is a *protocol* for how connections are made, which system sends first, and how the message dialog occurs, that is, what message(s) are sent and received and in what order. Second, the messages themselves have some general structure, which depends on the message types. This is the "grammar" of the HL7 language. Third, the message *contents* have some agreed on or predefined meaning (or not, as we will see). The possible contents should be specified by a data dictionary. This third area is the weakest part of the current (version 2.x) HL7 standard.

Since at the present time only Version 2.3.1 of the HL7 standard is available at no cost, downloadable from the HL7 organization (at `http://www.hl7.org/`), the details in this section and elsewhere in this book will follow Version 2.3.1.

4.3.2.1 The HL7 Message Exchange Protocol

The message exchange protocol specifies the order in which messages are sent and received under various circumstances. Examples of such circumstances are:

- a clinical laboratory system has a test result to send to an electronic medical record system that expects to receive it, and
- a physician's clinical workstation requests information (queries) about a prior clinic visit for a patient, or a status report on a previous medication order.

The rules for these situations are called "Application Processing Rules." The description here is based on Chapter 2, Section 12, of the HL7 Version 2.3.1 standard. The standard refers to the "initiating system" and the "responding system." It is relatively simple. In most cases, the protocol assumes that a channel is continuously open, the sender sends a message, the receiver sends an acknowledgment, the sender sends another message, and so on. The context for the messages is set up when the systems are configured and established when they start or when the TCP connection is opened.

In all cases, the initiating system sends first. The receiving system will examine the message to determine if it is consistent with the grammar (rules for message structure). If it is, the application system for which the message is intended generates a response message. If the message is not properly constructed, the receiving system sends a "reject" message. If the message is valid, the receiving system constructs an appropriate response (this could be just an acknowledgment or in the case of a query, a data message) and sends that. If there is some other detectable problem, an error response message is sent back.

The standard provides for a variety of more complex message sequences, including continuation messages, deferred responses, and other special protocols. They won't be described here. Instead we will focus on the problem of message decoding, focusing on the high level content of HL7 itself.

4.3.2.2 HL7 Message Structure

Many messages are sent as unsolicited messages, as a result of events. Examples of events are: the admission or discharge or transfer of a patient, the availability of a test result, the entry of a medication order into a system, or even the availability of billing data for the hospital's financial system. Each message has a *message type* associated with it. The message types are listed in Table 0076 in Chapter 2 of the HL7 standard document, as well as in Appendix A of that document. Each message type is described in detail elsewhere in the HL7 standard. Here, in Table 4.1, we list a few of them.

Figure 4.4 shows a sample HL7 message for a ficticious patient. The type of message illustrated here is specified by the message type ID, ORU, unsolicited observation result, which is in the first line of the figure.

Table 4.1 Some HL7 Message Types

ACK	General acknowledgment message
ADT	Admit, discharge, or transfer message
BAR	Add/change billing account
ORM	Pharmacy/treatment order message
ORU	Unsolicited transmission of an observation message
OSQ	Query response for order status
RDS	Pharmacy/treatment dispense message
REF	Patient referral

```
MSH|^~\&|lab_res_A||Mimi||20010201012546||ORU^R01||P|2.2|||<cr>
PID|||N123456||Smith^Jamie^||19701112|F||||||||||||||<cr>
PV1|||NAUB^||||||||||||||||||||||||||||||||||||||||||||||||||<cr>
ORC|RE||||||||||||||^|<cr>
OBR|1|2681988|W60872|CBDP^CBC, Diff/Smear Eval, Platelet|||
20010131131000|||||||20010201001800|^^&|267070^Tomgurg^Michael^|
||||||||H|F||^^^^^R|^^|||||||||<cr>
OBX|1|NM|WBC^WBC|1|5.86|THOU/uL|4.3-10.0|||||||||<cr>
OBX|2|NM|RBC^RBC|1|4.20|mil/uL|3.80-5.00|||||||||<cr>
OBX|3|NM|HB^Hemoglobin|1|13.3|g/dL|11.5-15.5|||||||||<cr>
OBX|4|NM|HCT^Hematocrit|1|39|%|36-45|||||||||<cr>
OBX|5|NM|MCV^MCV|1|92|fL|81-98|||||||||<cr>
OBX|6|NM|MCH^MCH|1|31.7|pg|27.3-33.6|||||||||<cr>
OBX|7|NM|MCHC^MCHC|1|34.5|g/dL|32.3-35.7|||||||||<cr>
OBX|8|NM|PLT^Platelet Count|1|210|THOU/uL|150-400|||||||||<cr>
OBX|9|FT|RBCMOR^RBC Morphology|1|\.nf\See RBC data|||||||||||<cr>
OBX|10|FT|PLTMOR^Platelet Morphology|1|\.nf\Normal|||||||||||<cr>
OBX|11|FT|WBCMOR^WBC Morphology|1|\.nf\See Diff|||||||||||<cr>
OBX|12|NM|TNEUT^Neutrophils|1|3.10|THOU/uL|1.80-7.00|||||||||<cr>
OBX|13|NM|LYMPH^Lymphocytes|1|2.17|THOU/uL|1.00-4.80|||||||||<cr>
OBX|14|NM|MONOC^Monocytes|1|0.47|THOU/uL|0-0.80|||||||||<cr>
OBX|15|NM|EOS^Eosinophils|1|0.06|THOU/uL|0-0.50|||||||||<cr>
OBX|16|NM|BASO^Basophils|1|0.06|THOU/uL|0-0.20|||||||||<cr>
```

Figure 4.4   A sample HL7 message, reporting the results from a CBC (complete blood count) test. *(Courtesy of Prof. David Chou, Department of Laboratory Medicine, University of Washington.)*

Every HL7 message is made up of one or more *segments*, each of which is terminated by an ASCII <CR> character. A segment is further divided up into *fields*. Each field may be blank or null or contain some datum. A field may have further structure, being made up of *components* and *subcomponents*. For each segment type, there is a three character code called the Segment ID. The segment types and IDs are also listed in Appendix A of the Standard, and each is described in detail elsewhere in the Standard. For example, the Message Header (MSH) segment is required as the first segment in all messages. The Patient identification (PID) segment contains the name of a patient, and other identifying information on the patient. The Observation request (OBR) segment has detailed specifications for a test or observation, such as a laboratory test or radiology procedure. The Observation/result (OBX) segment contains the data for one observation result. In the example message in Figure 4.4, there are many OBX segments corresponding to a single test request, because the test, a Complete Blood Count, has many different parts. They include counts of the white blood cells and the red blood cells, a measure of the hemoglobin content of the whole blood, and other information as well. Each item is specified in an OBX segment.

Compare this message with the data in Table 1.1 on page 12. In the message you will see that there are lines with labels corresponding to entries in Table 1.1, such as WBC, RBC, Hemoglobin, Hematocrit, and Platelet Count. However, in the OBX records there are some slight differences with the table. The units follow different conventions for use of upper case and lower case letters, and instead of the Greek letter μ for the "micro" prefix, the letter u is used instead. This is to simplify programming, since Greek letters are not part of the standard character set in use in most systems. Even more important, the standard ranges for the different tests do not match. So-called "standard ranges" are in fact not really standard, but reflect the practice adopted at a particular institution at a particular time.

Within a segment, the fields are separated by a *field separator* character, which is usually the vertical bar (|) character, but can be anything the sending application chooses. Within a field, if the field has components, the components are separated by a *component separator* character, usually the carat (^), but also can be anything the sending application chooses. The separators that are used in any particular message are provided at the beginning of the MSH segment of that message. So, the initial processing of a message can be uniformly handled by examining characters 4–8 of the first segment (which should begin with the characters MSH). The characters in positions 4 and 5 can then be used to divide the segments into fields and the fields into components. This nested structure can most easily be represented in a program as a list of segments, each of which is a list of field values, which are either a string or a list of strings (one string for each component). Then this nested list structure can be processed according to the specific message type and segment types.

4.3.2.3 Scanning HL7 Messages

In processing messages, as in natural language processing and in compiling computer programs from source code to machine code, there are two distinct steps. First, the input message, a sequence of bytes or characters, must be divided up into *tokens*. A token is a single logical unit, like a word in natural language, an identifier or operator in a programming language, or a message datum in a message. Then the sequence of tokens is analyzed to determine its structure (syntax) and meaning (semantics). The first step is called *scanning* or *lexical analysis*. The second step is called *parsing*. For HL7 messages, the scanning step is easy. Certain characters are used as field delimiters, and each token is a string that is between two successive delimiters. In a programming language the scanning step is more complex, because there are usually several kinds of delimiters or token termination characters.

Once the parsing is completed, we have a hierarchical structure containing the tokens and groups of tokens. The final step is to do something meaningful with the tokens, such as storing the data in appropriate fields and records in a patient database, or sending an alert message because certain data fields show that a patient's test results are abnormal.

To do this processing, we assume we have a long string containing a single HL7 message, including the <cr> segment terminators. Once we have the message separated into segments and fields as individual items, we can then parse the message according to the grammar specified in the standard.

For any given character used as a delimiter, it is easy to identify the first token, or substring, which is all the characters up to (but not including) the delimiter character. The `position` function tells us where the first delimiter is, and we just use that to extract the subsequence up to that point from a specified start. Specifying the start allows us to continue searching the string for successive tokens. Modular design suggests that this operation should be encapsulated in its own function, `next-token`. This function takes three inputs, a delimiter character, some text (a sequence of characters), and an index specifying where in the sequence to start. It scans the text sequence and returns two values: the first token, consisting of everything from the index, `start`, up to the first occurrence of the delimiter, and the position of the delimiter, if found, or `nil` if the delimiter was not found before the end of the sequence.

```
(defun next-token (delimiter text start)
  (let ((end (position delimiter text :start start)))
    (values
      (subseq text start end) end)))
```

In Chapter 1, Section 1.1.4, we took advantage of the fact that the Common Lisp function `read-from-string` returns multiple values. We used the Common Lisp function `multiple-value-bind` to capture and use them. Here we want to do this also, but with our own function, `next-token`. In order to have `next-token` return multiple values, we use the `values` operator, which returns all its inputs as multiple values.

Then we can use `next-token` in a recursive function that works its way down the string until the end is reached. To keep track of where the function is to start, it needs an input parameter, `start`, so we create an auxiliary function to do the work, `scan-aux`, using `labels`, and then define the `scan-h17` function to just use `scan-aux` with `start` equal to 0.

```
(defun scan-h17 (delimiter text)
   "returns a list of tokens from text that are demarcated
    by delimiter"
   (labels ((scan-aux (delim txt start)
              (multiple-value-bind (token end)
                (next-token delim txt start)
               (if end ;; non-null so more string is left
                   (cons token
                         (scan-aux delim txt (1+ end)))
                 ;; no divisions - just listify it
                 (list token)))))
     (scan-aux delimiter text 0)))
```

The `scan-aux` function uses `next-token` to find a token. After that token, either there is more string left to process or the token is the last one in the string. If there is more, the token is put on the front of the result of a recursive call, so `scan-aux` should always return a list. This means that if the token is the last one, it should be made into a list and returned.

The `scan-hl7` function can be used directly with the `<cr>` character as the delimiter, to split the message into segments, because that is fixed by the HL7 Standard. To then process the segments and split them into fields and components, we will need to examine the beginning of the message to find the delimiters the message actually uses. That's just a matter of retrieving the fourth through eighth characters in the message, using the `elt` function (recalling that in Lisp as in most programming languages, sequences are 0-indexed).

```
(defun delimiters (msg)
  "returns the 5 characters that are used as the field, component,
  subcomponent, etc. delimiters in the HL7 message msg as multiple
  values"
  (if (string= (subseq msg 0 3) "MSH")
      (values (elt msg 3) (elt msg 4) (elt msg 5)
              (elt msg 6) (elt msg 7))
    ;; this is probably a little severe ...
    (error "Wrong first segment type")))
```

Now to do the entire scan, we can use `mapcar` to apply the segment processing to each segment. Of course that segment processing will use `mapcar` to apply the field processing to each field, too. So, we call the `scan-hl7` function at three levels, on the outside with `<cr>` (remember, this is represented by `#\return`), and on the inner two levels with the separators retrieved from the message itself.

```
(defun hl7-message-scan (msg)
  "returns a list of segments, with each segment represented as a
  list of fields, and each field a list of components. Assumes that
  segments are separated by <cr> but uses the embedded delimiters in
  the message for the fields and components."
  (multiple-value-bind (fld cmp rep esc subcmp) (delimiters msg)
    (mapcar #'(lambda (seg)
                (mapcar #'(lambda (field)
                            ;; separate each field into components
                            (scan-hl7 cmp field))
                        ;; separate each segment into fields
                        (scan-hl7 fld seg)))
            ;; separate msg into segments
            (butlast (scan-hl7 #\return msg)))))
```

Because the `<cr>` character is used as a *terminator*, rather than a *separator*, at the end of the message the `scan` function will return a blank string. The Common Lisp function `butlast` removes this extraneous item from the end.

```
(("MSH" ("" "~\\&") "lab_res_A" "" "Mimi" "" "20010201012546" ""
  ("ORU" "R01") "" "P" "2.2" "" "" "")
 ("PID" "" "" "N123456" "" ("Smith" "Jamie" "") "" "19701112" "F" "" ""
  "" "" "" "" "" "" "" "" "" "" "")

...

 ("OBR" "1" "2681988" "W60872"
  ("CBDP" "CBC, Diff/Smear Eval, Platelet") "" "" "20010131131000" ""
  "" "" "" "" "" "20010201001800" ("" "" "" "&")
  ("267070" "Tomgurg" "Michael" "") "" "" "" "" "" "" "" "H" "F" ""
  ("" "" "" "" "" "R") ("" "" "") "" "" "" "" "" "" "" "" "")
 ("OBX" "1" "NM" ("WBC" "WBC") "1" "5.86" "THOU/uL" "4.3-10.0" "" "" ""
  "" "" "" "" "" "")
 ("OBX" "2" "NM" ("RBC" "RBC") "1" "4.20" "mil/uL" "3.80-5.00" "" "" ""
  "" "" "" "" "" "")
 ("OBX" "3" "NM" ("HB" "Hemoglobin") "1" "13.3" "g/dL" "11.5-15.5" ""
  "" "" "" "" "" "" "" "")
 ("OBX" "4" "NM" ("HCT" "Hematocrit") "1" "39" "%" "36-45" "" "" "" ""
  "" "" "" "" "")

...

 ("OBX" "9" "FT" ("RBCMOR" "RBC Morphology") "1" "\\.nf\\See RBC data"
  "" "" "" "" "" "" "" "" "" "")

...

 ("OBX" "11" "FT" ("WBCMOR" "WBC Morphology") "1" "\\.nf\\See Diff" ""
  "" "" "" "" "" "" "" "" "")
 ("OBX" "12" "NM" ("TNEUT" "Neutrophils") "1" "3.10" "THOU/uL"
  "1.80-7.00" "" "" "" "" "" "" "" "" "")
 ("OBX" "13" "NM" ("LYMPH" "Lymphocytes") "1" "2.17" "THOU/uL"
  "1.00-4.80" "" "" "" "" "" "" "" "" "")
 ...)
```

Figure 4.5 The result of applying h17-message-scan **to the sample HL7 message. The nested lists represent the segment, field, and component structure.**

The result of applying h17-message-scan to the message in Figure 4.4 is shown in Figure 4.5. Now that the segments, fields, and components have been separated into individual items (while preserving the grouping implied by the message structure), we can write code to examine the message type and take appropriate action with the contents.

4.3.2.4 HL7 Message Field Content
To do the parsing step, we need some representation of the message structure in terms of what segments appear in each kind of message. Assuming the segments are primitives, Figure 4.6 shows a set of BNF clauses[9] that define what segments are in a ORU message (and in what

---

[9]Backus-Naur Form [297,431] (BNF) notation is a formal syntax for describing a grammar for a language in terms of rules. It is commonly used to provide formal specifications for programming languages.

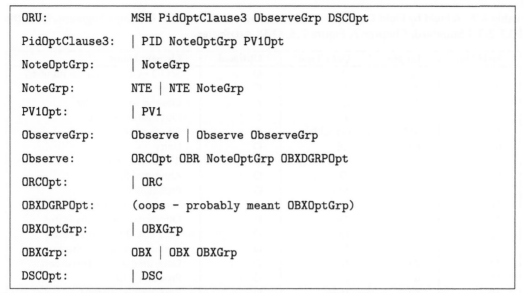

ORU:	MSH PidOptClause3 ObserveGrp DSCOpt
PidOptClause3:	| PID NoteOptGrp PV1Opt
NoteOptGrp:	| NoteGrp
NoteGrp:	NTE | NTE NoteGrp
PV1Opt:	| PV1
ObserveGrp:	Observe | Observe ObserveGrp
Observe:	ORCOpt OBR NoteOptGrp OBXDGRPOpt
ORCOpt:	| ORC
OBXDGRPOpt:	(oops - probably meant OBXOptGrp)
OBXOptGrp:	| OBXGrp
OBXGrp:	OBX | OBX OBXGrp
DSCOpt:	| DSC

Figure 4.6 A simplified BNF description of the structure of the ORU message in terms of message segments, taken from Appendix D of the HL7 2.3.1 standard. There is an error in the BNF specification where the undefined term, OBXDGRPOpt is mentioned.

order). This is taken from Appendix D of the HL7 2.3.1 standard. Note that there is an undefined term, OBXDGRPOpt. Following the mention of this term, a considerable number of lines down, the most likely term that was intended instead, OBXOptGrp, is defined. Another possibility is OBXDOptGrp which is defined *before* the mysterious term is mentioned, but this seems unlikely.

For each type of segment, the HL7 standard specifies what fields should appear in that segment, in what order, as well as the maximum length of the field, and the data type (one of the allowed data types specified in the standard). All fields are actually represented as character strings. However, the Standard makes no commitment to any particular encoding system. For some fields, the data that can appear are specified in HL7 Standard tables. For others, although the field has a description, it is not always sufficient for an application to usefully process it. The specific way in which data in fields are represented is in many cases left unspecified. This is unfortunate, as it means a lot of extra decisions have to be made by implementers, and a lot of discovery has to be done in order to make your software compatible with someone else's.

Table 4.2 shows the contents of each field in the OBX segment (from The HL7 2.3.1 Specification, Chapter 7, Figure 7.5, OBX Attributes). The two or three letter codes that appear in the Data Type column are defined in Chapter 2 of the HL7 standard, in Figure 2.2, "HL7 Data Types." Some of these are simple and straightforward, such as ST for String, which is a sequence of printable ASCII characters (except the delimiters mentioned earlier), generally less than 200 characters, or NM for Numeric, a number represented by printable ASCII characters. Others are more complex, such as time stamp, telephone number, or address, all of which have multiple

Table 4.2   A Field by Field Description of the Structure of the OBX Message Segment, from the HL7 2.3.1 Standard, Chapter 7, Figure 7.5, OBX Attributes.

Field No.	Length	Data Type	Optional	Element Name
1	10	SI	O	Set ID - OBX (sequence number)
2	3	ID	C	Value type
3	590	CE	R	Observation identifier
4	20	ST	C	Observation Sub-ID
5	65536	(variable)	C	Observation value
6	60	CE	O	Units
7	60	ST	O	References range
8	5	ID	O	Abnormal flags
9	5	NM	O	Probability
10	2	ID	O	Nature of abnormal test
11	1	ID	R	Observational Results Status
12	26	TS	O	Date last Obs normal values
13	20	ST	O	User-defined access checks
14	26	TS	O	Date/time of the observation
15	60	CE	O	Producer's ID
16	80	XCN	O	Responsible observer
17	60	CE	O	Observation method

components. Some, such as CE, for Coded Element, depend on a choice of coding system, which also has to be specified along with the value. Then, of course, one must have some agreement on the codes or names used to specify each of the coding systems.

What would we do with the data now that it is tokenized into lists? One possibility is that a laboratory order processing system would look for messages containing such orders, and extract the appropriate data from each message, in order for laboratory staff to execute the order. An HL7 message parser would look at the message header segment to see what the message type was. A computable version of the table of fields for each segment type would have tags identifying which items are which. The code would look up the field number according to the tag name. Then the parser would just use find to identify the needed segments, and process them similarly. An alerting system similarly could check prescription (pharmacy) orders against a drug interaction system, and generate an alert message when necessary. The challenge here is, as pointed out earlier, that the context is completely missing for doing such checking.

The fact that the HL7 standard leaves a lot unspecified is evident, if we examine the standard text itself. Here are some comments from the HL7 2.3.1 standard, Chapter 7, on Observations Reporting:

- "This standard does not require the use of a particular coding system…"
- "Local institutions tended to invent their own unique coding systems…"

- "Standard code systems such as LOINC[10] and SNOMED now exist...
  we strongly encourage their use..."

Actually Chapter 7 lists over 75 different coding systems for diagnosis, procedures, observations, drugs, and devices. The implementor is expected to encode in the messages themselves information about which coding system is being used, but that means that systems that *receive* HL7 messages have to be able to process any or all such encodings.

The ambiguities of HL7 version 2 were to be remedied or at least addressed by HL7 Version 3, and a component of Version 3 called the Reference Information Model (RIM). However, there are many logical problems with the RIM, too many to address here. For a provocative glimpse at some of these, the reader is invited to peruse the HL7 Watch web site, `http://hl7-watch.blogspot.com/`. Another useful resource is `http://users.tpg.com.au/kveil/FAQ.htm`, from the Australian HL7 community.

4.3.3 DICOM: Medical Image Data Encoding and Transmission

Of the many interesting aspects of medical imaging, the focus in this section is on the communication of medical image data between computer systems, including those that produce the images. This communication problem can be solved by thinking of the images and associated data as the subject of a message exchange. One way to implement the message exchange is to represent the grammar and syntax of the messages in a way that can utilize what is known about natural and formal language processing. Here we show how to do it for the medical image data exchange protocol called DICOM.

The development in this section is a much simplified summary of the actual DICOM implementation developed by Bob Giansiracusa in collaboration with the author [200], for use with the Prism Radiotherapy Planning System, described in Chapter 2, Section 2.2.8. Appendix A has information on how to obtain the complete original DICOM source code. Much of the remainder of this section is adapted with minor modifications and additions from [200].

4.3.3.1 A Short History of DICOM

Many file formats are used by image processing software. A typical approach is to put everything in one file, mixing text and binary representations for numbers, with all the descriptive information first, followed by the pixels in the same file. Each proprietary medical imaging system uses its own representation. These files are *not* in any standard image format used in general image processing software, such as TIFF, GIF, JPEG, etc. The files contain data about the image system settings (for example, CT scanner X-ray tube voltage), textual information such as the patient's name, hospital ID, date and time, etc.

In the early days of digital imaging, systems included a magnetic tape drive with 10 in. reels of magnetic tape, which would be used for backup and archival storage of medical images. Each manufacturer of digital imaging systems would have their own arcane format for how the data were organized on the tape. To use the data with some other computer program on

---

[10]See Chapter 1, Section 1.1.4.

another system required that a programmer write a special program that decoded the data that were on the tape, by reading the bytes sequentially and transforming them according to the manufacturer's specification of the format. A typical format would have the metadata (the data describing the image) in a header, the first 256 bytes (or some other proprietary number) of the file. The organization and size of the header would vary from one manufacturer to the next. To get the data you might need for your program, you would have to pick out items from the header, according to a description like the following:

- the number of images in the series is byte number 143 in the header, and is a one byte binary integer,
- the patient ID is a string occupying bytes 8–13, and consists of ASCII character bytes,
- the patient's name is in bytes 14–28, also an ASCII string,
- the date has the numeric month in byte 220, the day of the month in byte 221, and the (two digit!) year in byte 222 (this was prior to 2000), and all are binary integers,
- the image slice thickness is in byte 36, and is an integer, with value in millimeters, except if the slice thickness is 1.5 mm, this number will be 15, which is unambiguous, since there are no machines that do 15 mm slices,
- the image dimensions, horizontal and vertical, are in bytes 38 and 39, 40 and 41, respectively, except if the pixel size turns out to be 0.0, then the real pixel size and image dimensions are a standard value known only to those privileged to have the vendor's proprietary documentation,
- and on and on…

It's different for every machine type and model, even for the same manufacturer, and certainly different for different manufacturers. Then there was the problem of physically compatible magnetic tapes. For a few years, the industry standard half-inch 9-track magnetic tape[11] was widely used with imaging systems as an archival and backup storage medium. However, there were several different information densities at which tapes were written. If your tape drive could not read the tape density written by the imaging system, you were out of luck. Even with the same density, tape drives would write and read at different speeds, and had to be precisely adjusted. Occasionally, the imaging system drive and the viewing system drive would be just enough different in alignment so that a tape written on one was unreadable on the other. When vendors supported connecting the scanners to emerging local area networks (usually 10 Mbit Ethernet with TCP/IP), the tape compatibility problem was no longer important. Most of the imaging computer systems supported the popular Internet file transfer protocol, ftp, and the programs that implemented it, so we no longer had to write, carry, and read tapes (also then known as "sneakernet"). However there was still the chaos of all the file formats.

In the late 1980s as the Internet began to grow, a committee of radiologists, medical physicists, computer scientists, and engineers was formed to investigate how image data could be transferred from machine to machine electronically at high speed. It was known as the ACR/NEMA

---

[11]The big reels that you would see in the old time movies featuring mainframe computers.

committee, because it was sponsored by the American College of Radiology (ACR) and the National Electrical Manufacturers Association (NEMA). The latter is well known for developing and promoting standards for a very wide range of electrical and electronic applications.

By the early 1990s the work of the ACR/NEMA committee resulted in a standard for computer programs that transmit and receive medical image data between computer systems that have medical application software producing or using those images. These systems include diagnostic radiology systems such as CT and MRI scanners and radiation therapy planning systems as well as systems and software used in many other areas of medical practice. Chapter 1, Section 1.1.5 provided some background and history of digital medical imaging.

The first version of this standard for interconnection of digital imaging systems, known as ACR-NEMA version 1, proposed a new kind of point-to-point interconnection between computers, modeled on proprietary high-speed parallel port interfaces available from some computer vendors. This required invention of software protocols for handling the generic task of getting arbitrary data packets reliably from one computer to another, in addition to the digital imaging specific protocols that described the application data and functions. The idea was never implemented, mainly because of the cost of designing proprietary interface hardware and the fact that it was point to point. It was later abandoned in favor of using emerging world-wide standards for the interconnection hardware and general network software protocols for addressing, routing, and insuring reliable message delivery.

On their third try, the ACR-NEMA committee came up with a workable solution that solved both of the problems described above. First, they defined a network application layer protocol so that the data transfer could happen in one step rather than two (so that `ftp` was not needed), and second, the data were to be encoded in messages in a standard way, independent of vendor, and independent of how either sending or receiving system stored or used the data. This new approach also avoided the difficulty of having to reinvent all the lower level network technology, as it relied on existing network facilities such as TCP/IP.

The new standard, called Digital Imaging and Communications in Medicine (DICOM),[12] defines a network protocol for communication between computer systems that produce, store, display, and print medical images [12,30]. The DICOM standard provides a specification of what kinds of image data can be transmitted and received, how they may be encoded in digital form, and transmitted as messages over a computer network. DICOM was developed to relieve software authors of the burden of writing specialized programs to decode proprietary file formats used by manufacturers of imaging equipment. It was also intended to facilitate peer to peer communication between radiological imaging systems and application systems over a local or wide area network, without the need to handle intermediate media storage such as tapes or diskettes. Many implementations of DICOM exist, including demonstration and experimental software from university groups (the most notable being a software package developed at Mallincrodt

---

[12]DICOM is the registered trademark of the National Electrical Manufacturers Association (NEMA) for its standards publications relating to digital communications of medical information.

```
SEND-PDU: Sending A-Associate-AC PDU, 17-Jul-2002 18:13:02.
Dumping Outgoing PDU (all 183 bytes):

0200000000b100010000507269736d5f496d6167  ..........Prism_Imag
655f53727672504153535056f52545f5251202020  e_SrvrPASSPORT_RQ
2020000000000000000000000000000000000000      .................
0000000000000000000000000000010000016312e  ................1.
322e3834302e31303030382e332e312e312e3100  2.840.10008.3.1.1.
2100001a010000004000012312e322e3834302e  !.......@...1.2.840.
31303030382e312e3200050000031510000040001  10008.1.2.P..1Q.....
00005200001a312e322e3834302e313133393434  ..R...1.2.840.113944
2e3130302e31302e312e3100550000075044535f  .100.10.1.1.U...PDS_
312e30                                    1.0
```

Figure 4.7 An actual `A-Associate-AC` packet from an exchange between a DICOM client and the Prism DICOM Server.

Institute of Radiology known as CTN, for Central Test Node [288]), vendor implementations for their products (usually bundled with CT scanners and other imaging systems), and proprietary libraries for building DICOM applications, such as the Merge Technologies DICOM software libraries [279]. All the current DICOM standards documents are available via the World Wide Web [12].

Although the adoption of DICOM reduced considerably the programming work for achieving data interchange, the approach most programmers took was still to write traditional procedural code that processed the byte stream explicitly following the complex logic of the standard's description of the messages. An excerpt from one such implementation is shown in Figure 4.8. In this chapter you will see another approach, using the idea of expressing a network protocol as a language with a grammar and a vocabulary. The basic idea is that the knowledge about the structure of the messages and the structure of the content can be encoded in rules. Then the problem of parsing and message generation is one of writing a rule-based translator that could process many different kinds of message streams. There is a further simplification in that the message formats can more easily and directly be transformed into rules than into procedural code. The resulting program takes significantly less code than the conventional alternative and has less complexity.

4.3.3.2 About the DICOM Standard

For this section and the remainder of the chapter, you will find it very helpful to have a copy of certain critical sections of the DICOM standard documents available to look at as we go through the development. All parts can be downloaded as PDF files from the official NEMA DICOM web site, `http://medical.nema.org/dicom/`, via the link under "Products" called "The DICOM Standard." The most immediately useful will be Part 8, "Network Communication Support for Message Exchange." It is about 50–60 pages. The entire DICOM standard is over

```
CONDITION
parseAssociate(unsigned char *buf, unsigned long pduLength,
        PRV_ASSOCIATEPDU * assoc)
{   CONDITION cond;
    unsigned char type;
    unsigned long itemLength;
    PRV_PRESENTATIONCONTEXTITEM * context;
    ...
    (void) strncpy(assoc->calledAPTitle, (char *) buf, 16);
    ...
    (void) strncpy(assoc->callingAPTitle, (char *) buf, 16);
    ...
    cond = DUL_NORMAL;
    while ((cond == DUL_NORMAL) && (pduLength > 0)) {
        type = *buf;
    ...
        switch (type) {
        case DUL_TYPEAPPLICATIONCONTEXT:
            cond = parseSubItem(&assoc->applicationContext,
                                buf, &itemLength);
    ...
        case DUL_TYPEPRESENTATIONCONTEXTRQ:
        case DUL_TYPEPRESENTATIONCONTEXTAC:
            context = CTN_MALLOC(sizeof(*context));
    ...
            cond = parsePresentationContext(type, context,
                                    buf, &itemLength);
    ...
        case DUL_TYPEUSERINFO:
            cond = parseUserInfo(&assoc->userInfo,
                                buf, &itemLength);
    ...
```

Figure 4.8 An excerpt from the CTN DUL library, showing how an `A-Associate-RQ` **message is handled by procedural code.**
(Condensed and elided from the actual code, as indicated by ellipses.)

2000 pages, and is not easy to comprehend. The purpose of this section is to give the interested reader an orientation to some of the core details of the standard, and an acquaintance with where to find the rest of the detail if needed.

When a DICOM client program connects with a DICOM server program running on another computer, an exchange of messages takes place. This is the job of `applic-fn` in the TCP code. In order to exchange data, the client and server must agree on what operation will be performed and on the data format and syntax of the communication. A DICOM program may be able to send or receive data (or both), respond to queries about the contents of available data sets, or perform other services like printing. The data might be sent as binary numbers in big-endian or little-endian[13] byte order, and may have its data format explicitly specified in the data stream or implicitly inferred from its DICOM identification tag. So, before actual data are transferred, client and server exchange messages to settle which options will be used. Then the data are sent, and the exchange is concluded. A typical sequence of messages is as follows:

1. Client sends a request to server, asking if it (server) can store CT images, using little-endian byte order and implicit value representation.
2. Server responds that it can perform the requested operation.
3. Client sends the data in a sequence of messages.
4. Client requests that the communication be concluded.
5. Server responds that it is OK to close the connection.
6. Client is done (`applic-fn` returns).
7. Server is done (`applic-fn` returns), and the loop (in `tcp-server`) iterates with the server waiting for the next connection request.

DICOM includes other types of messages to handle the possibility that the client requests an operation that the server cannot perform or that something unexpected has happened in the network link.

The collection of message types and conditions under which each is sent are called the "DICOM Upper Layer Protocol," or DUL. The DUL messages and their sequences are described by a state table and by elaborate detailed byte by byte descriptions. These may be found in Part 8 of the DICOM standard. The states, actions, and state transition table are given in Part 8, Section 9.2, and the DUL message types and formats are defined in Part 8, Section 9.3.

The messages of DUL are incorporated into data packets called "Protocol Data Units," or PDUs. Some messages are single PDUs, but others, particularly the image data, may be composed of many separate PDUs sent in sequence. This causes implementation difficulties, as we will see later. It appears to have come from the fact that initially the ACR-NEMA committee did not decide to use existing network technology, and therefore had to invent a scheme for dividing long messages into packets, and reassembling them at the receiving end.

An example PDU will illustrate the idea. In the exchange above, the first message that would be sent from the client program is the `A-Associate-RQ` message, requesting the server program to establish an "association," an agreement about communication details and operations to perform. The byte by byte format of this data stream is diagrammed in DICOM Part 8, Figure 9.1.

The sequence of bytes contains a byte specifying the PDU type (a byte is sufficient because there are only 7 types so far), an unused byte, 4 bytes to be interpreted as a binary integer, the PDU length in bytes, 2 bytes for the protocol version (currently 1), two more unused bytes, 16 bytes

---

[13]These refer to the order of bytes for numbers that are larger than one byte. Computers are built to either one of two architectures, depending on whether such numbers are organized in memory so that the lower order byte comes first in the address sequence (little endian) or the higher order byte comes first (big endian). This is something like writing the date as day-month-year or year-month-day.

each for character strings (called Application Entity Titles) identifying the client software and the server software (each DICOM program is expected to have a name of some sort). There are 32 more unused bytes and then one or more other items, each with their own format in the byte stream, such as an Application Context Item, one or more Presentation Context Items, and other information. The Application Context Item contains the UID (Unique Identifier) for the DICOM standard, the string "1.2.840.10008.3.1.1.1". This identifies that the sender is using the DICOM protocol. It may seem unnecessary, but the designers anticipated that there might be multiple versions or variants, and this would identify which one is in use. Presentation Context Items specify what data and operations are being requested and/or agreed to, as well as what syntax to use in the data transfer. In this context, "syntax" refers to specification of things like byte order for binary data, whether data values will have their structure (or type) explicitly specified in the data stream, or implicitly specified by their ID tag and the data dictionary, whether image data will be compressed, and possibly other issues.

The server typically responds to the A-Associate-RQ message by sending an A-Associate-AC message, in almost identical format, acknowledging that it can perform the requested operation, and providing some identifying information about itself (the server). An actual byte stream, shown in hexadecimal notation (two hexadecimal digits per byte) on the left, with the corresponding characters on the right (according to the ASCII code, with dots showing non-printing byte codes), is shown in Figure 4.7. This byte stream printout was taken from the log file for an actual interchange between the Prism DICOM Server and a remote client application.

You can see in Figure 4.7 that this PDU is a response to a request addressed to the server program called Prism_Image_Srvr which came from a client program calling itself PASS-PORT_RQ. The server specifies that it can do the default transfer syntax, which is little endian byte order, and implicit value representation, by providing the UID "1.2.840.10008.1.2", the DICOM standard UID for the default transfer syntax. It refers to a Presentation Context ID of 1 that the client requested (you can find the 21 identifying the Presentation Context ID at the beginning of the sixth row of byte codes, then count inward one reserved byte, 2 bytes for the length of the item 1a, then the next byte is the 01 for the ID). The transfer syntax item actually begins where you find a 40 in the same row, then its length 0012 followed by the UID string. The rest of the PDU contains "user information" consisting of the UID of the server program, and its version name string PDS_1.0.

The seven types of PDUs that are specified by the DICOM standard are (DICOM Part 8, Section 9.3.1):

1. A-ASSOCIATE-RQ—request an association for the purpose of exchanging data.
2. A-ASSOCIATE-AC—accept a proposed association.
3. A-ASSOCIATE-RJ—reject a proposed association.
4. P-DATA-TF—send/receive commands and data.
5. A-RELEASE-RQ—request termination of an association.
6. A-RELEASE-RP—confirm termination of an association.
7. A-ABORT—abort an association because of some abnormal condition.

The standard specifies a variety of functions that a server or client program may provide, and a variety of image types and other kinds of data with which these functions operate. The functions are called "services" and are given names such as "C-STORE" (for the services that request or perform storage of data from a stream). The combination of a service and a particular kind of object is called a Service-Object-Pair (SOP) Class. Each such SOP class is assigned a UID, such as "1.2.840.10008.5.1.4.1.1.2", which is the UID for CT Image Storage. These are enumerated in DICOM Part 6. A requestor of a particular service is known as a Service Class User (SCU), while the application that performs the service is known as the Service Class Provider (SCP).

The kinds of data that DICOM supports, and their representation in message packets, are specified in the form of a data dictionary, based on a hierarchical model of the data elements. This model is not intended to dictate how an application stores data, but only to specify the relations between data items for the purpose of sending meaningful and consistent data. The DICOM Information Model, diagrammed in DICOM Part 3, Figure 7.2, illustrates that a DICOM object to be transmitted may consist of many other DICOM objects as component parts. DICOM Part 3 describes what data are included in each information object. For example, in Part 3, Annex C, Table C.7.2 lists the attributes of a study. A study in turn may include one or more series, described in Table C.7.4, and a series may contain some number of CT images, where each image has attributes described in Tables C.7.7 and C.8.3. The attributes are each assigned an identifier tag, which references a data dictionary (DICOM Part 6). The data dictionary specifies what type of datum each element is and how it is encoded.

The commands and data appear in the message stream as Protocol Data Values (PDVs) contained in one or more P-DATA-TF PDUs. A single PDU may contain multiple PDVs. The structure of a PDV in general is specified in DICOM Part 8, Section 9.3.5.1. The format of the value itself depends on what type of item it is, a patient name, numbers specifying the dimensions of the image, or the image pixels. That section includes a nice diagram which we don't reproduce here.

Each data element that can be part of a DICOM message has a unique tag, consisting of a group number and an element number, each a 16-bit unsigned integer. These data elements are also associated with a data type or value representation (VR), such as "Application Entity" (a 16 character string), "Date," "Decimal String," "Floating Point Single," "Floating Point Double," "Integer String," "Person Name," and "Sequence of Items." The details of each of these types are tabulated in DICOM Part 5. The data dictionary in DICOM Part 6 then lists the tag (in Hexadecimal numbers), data type, and other details for each of the known data elements. An example is "Image Position (Patient)," whose tag is (0020,0032), VR is "Decimal String" (DS), and there are three such values present (for x, y, and z).

Unlike HL7, the DICOM specification does not use BNF notation to describe the protocol. The specification is a narrative together with a series of diagrams describing byte by byte the syntax and contents of messages. The state machine and data dictionary are also described by text and tables.

The DICOM specification does include a kind of Entity-Relationship diagram model, which defines the relationships (containment and multiplicity) among the data, but the Entity-Relationship model provides no information to guide the parsing of messages, and only serves to help define what the text and tables refer to.

The standard describes many kinds of data and functions that a DICOM conformant program might support. Not all are required to be present. The standard also includes in Part 2 a specification for the format of a *Conformance Statement*. The purpose of the Conformance Statement is to provide a detailed description of exactly which elements of the standard an application claims to be able to perform. This includes which kinds of data objects will be handled, and what operations can be performed on them. By examining a properly written Conformance Statement for each of two DICOM implementations it should be possible to determine whether and under what conditions the two programs can communicate with each other. An example Conformance Statement is the Prism DICOM System Conformance Statement [199], available from the author's web site.

Despite the fact that the standard refers to a state table, includes a data dictionary, and talks about messages and sequences of messages, most DICOM implementations translate all this into procedural code, with the details of the message content and handling of different kinds of conditions represented by formatting operations and conditional expressions in the code. In the CTN implementation [288], for example, the structure of each kind of message is coded in a closed subroutine in the DUL library, along with `struct` declarations in the C programming language for parts of the message. Thus a message exchange consists of a series of explicit subroutine calls. Figure 4.8 illustrates this with an excerpt from the CTN code.

The `parseAssociate` subroutine decodes the contents of an A-ASSOCIATE message. It in turn calls a series of other subroutines (`parseSubItem`, `parsePresentationContext`, and `parseUserInfo`) which contain explicit code for each of these component parts (`parseSubItem` is used to parse application contexts). The sequence of message exchanges is similarly explicitly coded in other parts of the CTN DUL implementation. This procedural/imperative design style is commonly practiced and should not be surprising. Commercial software such as the Merge libraries [279] is similarly designed.

4.3.3.3 DICOM as a Language Specification

Faced with such a huge collection of possible operations and a large variety of data types and formats, it would seem to be a daunting task to write a working DICOM server or client for even a few of the possible functions. Indeed, the CTN core is over 30,000 lines of code. Experience with the design of the Prism Radiation Therapy Planning System [402] suggested that a more declarative approach could provide advantages, including ease of implementation, extensibility, and ease of maintenance. An idea that potentially could simplify the design is to create a new language in which to express these details. This section describes some of the ideas borrowed from language design and data representation. In Chapter 5 we describe the elements of this design in greater detail.

Table 4.3 The Correspondence Between DICOM and a Language

Language		DICOM Component
Sentence	⇔	DICOM message
Grammar	⇔	PDU structure
Vocabulary	⇔	DICOM objects and services

- The *DICOM Upper Layer protocol* (DUL) is the language used to make connections, compose, send, receive, and decode messages. This is represented in the DUL rules, rule interpreter, and state machine.
- *DICOM Services* are the operations DICOM programs can do, such as send, store, look up information. These roughly correspond to the verbs of the DICOM data language. They appear in PDUs as commands, together with the data on which they operate.
- *DICOM Objects* are the data that programs can send and receive, such as, patient data, CT, MR, and other images, radiation beams, anatomic structure contours. These are the nouns and adjectives, and they can be specified in a data dictionary, or lexicon.

Thus the task factors into writing an interpreter (a parser and generator) for the language, writing a grammar for the language in the form of rules, and creating a lexicon that describes all the data items. Table 4.3 expresses the idea of mapping DICOM concepts to a language implementation, showing the correspondence between the components of a language and the components of DICOM.

The DICOM protocol supports thousands of different kinds of data from medical images to radiation therapy treatment prescriptions. The messages that encode the data need to be constructed and on receipt decoded. As the transactions progress, information is generated or received, and will be needed in each succeeding state. How will it be maintained? One approach is to define a big and complex data structure, a class definition or a set of class definitions, with slots for all the different variables, etc. However, the DICOM standard is so huge, this would be unwieldy. Also, at any time, only a few of the items are needed. And, as the standard evolves, this data structure would have to be expanded and revised. If there were indeed a slot for each possible DICOM data element, there would be thousands of slots, an unwieldy object to keep around. Another approach is to assign global variables for each possible important datum. This similarly would be unwieldy and almost impossible to maintain, as new kinds of information are supported by the standard.

4.3.3.4 Organizing Data into an Environment

Fortunately, there is a better way, one that is easy to implement in Lisp. Labeling data with tags is a popular and more flexible approach than predefined structures, as we described in Chapter 1. An *association list* is a simple way of handling this. We can label items with symbols, like `command` for the ID of a command like `C-STORE`, `calling-AE-title` for the string containing the AE title of the client, and so on.[14] For individual DICOM data elements, each one has a two part tag, consisting of a group number and an element number, as in (0018 0050) for image slice thickness. We can represent these as two element lists, and these two element lists can also be used as lookup keys. The association list will grow as data come in, and will shrink as information is used and no longer needed. We call this the *environment*. It is passed as a parameter to all action functions, and they are all expected to return a new environment reflecting any modifications.

Therefore the environment is accessible to any operation that needs values of data fields in the decoded PDUs. Each component is a list headed by a keyword symbol naming the component, and the rest of the elements of the component are the values. Since a component contains a list of values, we can represent values of variable multiplicity simply by storing a variable number of elements in a component's value list. Also, this structure is recursive in that a value in a component can itself be another component (that is, another list headed by a symbol naming a field together with its list of values). Here is a sample, reconstructed from logging output of the Prism DICOM Server:

```
(:A-ASSOCIATE-RQ
  (PROTOCOL-VERSION 1)
  (CALLED-AE-TITLE "Prism_Image_Srvr")
  (CALLING-AE-TITLE "OP1DISK")
  (:APPLICATION-CONTEXT-ITEM
    (ACN-LEN 21)
    (ACN-STR "1.2.840.10008.3.1.1.1")) ;; DICOM
  (:PRESENTATION-CONTEXT-ITEM-RQ
    (PCI-LEN 100)
    (PC-ID 1)
    (:ABSTRACT-SYNTAX-ITEM-RQ
      (ASN-LEN 25)
      (ASN-STR "1.2.840.10008.5.1.4.1.1.2")) ;; CT
    (:TRANSFER-SYNTAX-ITEM
      (TSN-LEN 17)
      (TSN-STR "1.2.840.10008.1.2")) ;; default
    ...)
  ...
  (:USER-INFORMATION-ITEM
    (UII-LEN 47)
      (:MAX-DATAFIELD-LEN-ITEM
```

---

[14]"AE Title" is Application Entity Title. Each DICOM software product, or even individual copy of a DICOM program, is identified by a string, such as UW Prism DICOM Server. There is no requirement that these be unique, but some kind of name must be specified.

```
        (MAX-DATAFIELD-LEN 16384))
   (:IMPLEMENTATION-CLASS-UID-ITEM
    (IC-UID-LEN 20)
    (IC-UID-STR "1.2.840.113704.7.0.2"))
   (:IMPLEMENTATION-VERSION-NAME-ITEM
    (IV-NAME-LEN 11)
    (IV-NAME-STR "OmniPro-2.7")))))
```

The environment would contain these items after the state machine parsed an A-ASSOCIATE-RQ PDU requesting storage of CT images using the default syntax of "little endian" byte order. The request comes from a system named "OmniPro-2.7."

Functions which need to retrieve an atomic value (the ultimate number or string contained in a deeply-embedded environment structure like that above) can do so by supplying a list of field names (i.e., the keyword symbols naming each field) to a retrieval function that does a recursive search on each value found in a retrieved component, guided by the input list of component names. These lists of field names are like pathnames in a hierarchical file system, or a path specification in a tree-structured data structure, or a path accessible by XPATH in XML. A partial specification results in return of the entire subtree matching the specification provided. For example, a message parsed from a P-DATA-TF PDU usually consists of a nested sequence of many data fields. This entire structure, parsed into the syntax tree that represents the environment, can be returned as a single value in an environmental lookup. This nested structure can be passed to other routines (such as a message transmission routine which is formatting a reply) so that all the internal data fields are available using the same small set of environmental accessor functions.

Thus, we will need a function which is like find except that it searches the environment for a specific item accessible by a *path*. To retrieve such items, we will need to specify a path, or sequence of tags that lead to the desired item in its specific context (the same kind of item might appear in several places in the data). It will of course be a recursive search. If the path or environment is empty, the search of course fails. If there is only one item left in the path, we should be able to find it at the top level using assoc. Otherwise, we use the first tag in the path to begin the search, and if a structure with that tag is found, apply the same process with the rest of the path on the rest of the structure. This environment structure and path search is the same as the tagged document architecture and the search function path-query for finding tagged items in a parsed XML document, described in Section 4.3.4.

To add items to the environment, we can just use push. The structure of an item, with nested tags, etc., will come from the operation of parsing the input stream from which the item comes, or in some cases, just by construction explicitly in the code.

4.3.4 Network Access to Biomedical Documents and Data

Now we address the question of how to use the commonly available biomedical information repositories, not via their Web user interfaces, but for projects that need access to the actual records so that some further computing can be done with their contents. This section will focus on the Internet accessible resources maintained by the NCBI.

The data and documents available at NCBI are vast and are described in many other references, such as Baxevanis et al. [24]. The NCBI main web page is at `http://www.ncbi.nlm.nih.gov/` where you will find links to the major NCBI information resources (GENBANK, RefSeq, and many others) as well as bibliographic resources such as PubMed.

At first, when people wanted to use the results of queries to NCBI's web-based resources, programmers wrote code to construct an HTTP message that emulated a web browser. Their program would then read back the resulting HTML text. The HTML was of course full of tags and text that were irrelevant, including information to format the data on the web page display. The program would have to include some complex search and pattern matching to find the beginning of the data, and sift it out, ignoring all the other stuff. This was tedious and errorprone, requiring custom programming for almost every query.

A better way to explore the various databases is through the NCBI's Entrez search facility. Entrez includes a web front end through which you can search any of the dozens of resources. It also provides a network interface, the Entrez Programming Utilities (eUtils), that returns structured data using XML tags, in response to queries that can be submitted by programs that you can write, instead of HTML formatted data that is sent in response to a web browser. This program or script interface is very useful for retrieving and processing data automatically that would otherwise be tedious or impossible to do by manually clicking on web links, reading the results, and copying and pasting them into other web pages. The eUtils facility is not hard to use, and we will describe it here, with some examples. For more detailed documentation the reader is referred to the NCBI web page describing eUtils,

`http://www.ncbi.nlm.nih.gov/entrez/query/static/eutils_help.html`

Although XML syntax is verbose, having the information structured with tags and not having any extraneous information helps a lot. The big gain here is that instead of unnatural language text, the information coming back from a query is structured data with predefined structure and meaning.

We will present a very simple approach to sending queries to web sites and processing the resulting XML documents. More elaborate code for handling URIs and PubMed queries has been written by Kevin Rosenberg and is available at URL `http://pubmed.b9.com/`.

A program that sends a URL to a web server to get back information must be able to do several things, including:

1. open a network connection to the web server using a specified network address (a domain name or IP address) and port number (usually port 80 but not always),

2. construct and send a properly formatted HTTP request string, which will need to follow the HTTP standard,

3. include in the HTTP request string the right information for the particular web service, in this case, the data that constitute a query to one of the Entrez databases,

4. read back the response from the web server, which in our case will be an XML document, and

5. parse this response to find the desired information, which could in turn require sending more queries and processing more responses.

As noted earlier, most programming language implementations provide a straightforward way to open a network connection and then send and receive data using it. So, while it is not a bad idea to understand how network connections work, having a higher level facility already available is a great convenience. In addition, there are facilities for most programming languages that even take care of constructing the right HTTP message, so that you don't have to be concerned with learning the details of the HTTP protocol.

In the following, we will use a function from a web server and client library called Alle-groServe, written by John Foderaro.[15] The first and second requirements above are fulfilled by using the function called do-http-request, a function in the client code of AllegroServe. It is in the package net.aserve.client and we will include its fully qualified name, so that it is clear where it came from. This function has one required input, the URI, and many optional keyword arguments. In the usual case of an HTTP GET request, the default values will work and no additional information is necessary. Here is the function specification with the complete list of input parameters, as described in the AllegroServe documentation.

do-http-request *uri &key method protocol accept content content-type query format cookies redirect redirect-methods basic-authorization digest-authorization keep-alive headers proxy user-agent external-format ssl skip-body timeout [Function]*

Sends a request to *uri* and returns four values, the response body, response code, list of headers, and the URI object of the page.

The EUtils facilities of Entrez expect to see a URL that includes the search or other request parameters in the form of a CGI string appended to the base URL.[16] However, it is a little complicated to take a search string and convert it to the right format for a CGI addition to the web site URL, to make up the full URL to request. So, one of the keyword arguments to do-

---

[15] Available from http://opensource.franz.com/, the Franz, Inc. Allegro Common Lisp Open Source Center, or from Sourceforge, http://allegroserve.sourceforge.net/. Although this work was done originally for Allegro Common Lisp, there is an alternate version that runs in other Common Lisp implementations, and a link to that version is at both locations. We will refer to and use the Allegro Common Lisp version of AllegroServe here.

[16] CGI (Common Gateway Interface, http://www.w3.org/CGI/) is a standard syntax for constructing a URL that contains information to be passed to a "helper" program that runs outside the context of a web server, returning the results back to the web server to be further processed or passed on as a response to the original query.

`http-request` is available to make this easier. The `query` parameter is a list of dotted pairs, in which each dotted pair is a CGI data element. So, the list

```
(("db" . "pubmed") ("term" . "cancer AND radiation"))
```

will get translated into the CGI string

```
db=pubmed&term=cancer+AND+radiation
```

Why not just use `format` and do it directly? It's a good idea to build flexible tools when we write these functions, so that they can be used in many different circumstances, and can be used to build higher level tools as well. For example, the number of items that will go into a given query may vary greatly. A format control string that handles variable length lists of terms and other parameters would be an almost unreadable piece of code.[17]

Here is a function, `pubmed-query`, that will take a search string with terms separated by spaces, construct the query, send it, and return the result. This function call provides the URI parameter and the query parameter, and makes use of the default values for most other parameters. It includes the host and URL for the Entrez Esearch service as global parameters defined separately. In order to have all the symbols defined and needed packages loaded, we start by evaluating some `require` forms. First the AllegroServe code is loaded by evaluating a `require` form for the `aserve` package, followed by a `require` form for the Allegro XML parser package, and finally, the function definition for `pubmed-query`.

```
(require :aserve)
(require :pxml)

(defparameter +eutils-host+ "eutils.ncbi.nlm.nih.gov")
(defparameter +entrez-query-url+ "/entrez/eutils/esearch.fcgi")
(defun pubmed-query (searchstr &key start maximum)
  (let ((query-alist '(("db" . "m")
                       ("term" . ,searchstr)
                       ("mode" . "xml"))))
    (when maximum (push (cons "dispmax" maximum) query-alist))
    (when start (push (cons "dispstart" start) query-alist))
    (net.aserve.client:do-http-request
      (format nil "http://~a~a" +eutils-host+ +entrez-query-url+)
      :method :get
      :query query-alist)))
```

Four results are returned (as multiple values) by the `do-http-request` function. The first is a string containing the results of the search, the *body* of the response. The second is the response code, a number, such as 200 if the request succeeded, 404 if the requested "page" was

---

[17]For the reader who is a vintage programmer from the old minicomputer days, used to deciphering TECO macros, this might not be a problem, but for the rest of us, simpler is better.

not found, etc. The third is a list of headers in the form of dotted pairs. For each header (dotted pair), the `first` part is a keyword or string naming the header and the `rest` part is the value of that header. Keywords are used for names of "standard" headers, and strings are used for other headers. Finally the last item returned is the URI of the page. This is usually computed from the URI sent, but in the case that the query was redirected it will be the URI that is the (redirected) target.

The Entrez facilities return rich XML documents, whose structure and content depend on which facility you are using. The ESearch utility, in particular, will return a specified number of PubMed ID numbers for matches to the search parameters, the total number of matches (which could be much larger than the number actually returned, for practical reasons), and information about how the query was translated and processed. The exact set of results that are returned is specified in a Document Type Definition (DTD). For each of the Entrez facilities there is a DTD available at the Entrez eUtils web site. Figure 4.9 shows (with some blank lines and comments omitted) the DTD for eSearch.

The intended use of the DTD is for a computer program to be able to validate that the XML document returned is correctly formed. However, the DTD for a facility also can be used by you, the programmer, as a guide to how to extract the information you are looking for. The names of the data fields are meaningless to an XML parser, but you can read the NCBI documentation and determine, for example, that the tag `<IdList>` labels a list of PubMed ID numbers, and that each ID number will be labeled by a `<Id>` tag. This is the kind of information you will need in order for your program or script to decide what to do with the data returned from a query.

Now we show an actual example, searching for articles as shown above, on cancer and radiation. The code above should be loaded, then we use the `pubmed-query` function as shown. Although four values are returned, we only show the first, the search result string.

```
> (setq cancer (pubmed-query "cancer AND radiation"))
"<?xml version=\"1.0\"?>
<!DOCTYPE eSearchResult PUBLIC \"-//NLM//DTD eSearchResult,
   11 May 2002//EN\" \"http://www.ncbi.nlm.nih.gov/entrez/query/
DTD/eSearch_020511.dtd\">
<eSearchResult>
        <Count>160955</Count>
        <RetMax>20</RetMax>
        <RetStart>0</RetStart>
        <IdList>
                <Id>17962969</Id>
                <Id>17962911</Id>
                <Id>17962885</Id>
                ...
                <Id>17960564</Id>
        </IdList>
```

```
<!-- This is the Current DTD for Entrez eSearch
$Id: eSearch_020511.dtd 85163 2006-06-28 17:35:21Z olegh $ -->
<!ELEMENT        Count              (#PCDATA)> <!-- \d+ -->
<!ELEMENT        RetMax             (#PCDATA)> <!-- \d+ -->
<!ELEMENT        RetStart           (#PCDATA)> <!-- \d+ -->
<!ELEMENT        Id                 (#PCDATA)> <!-- \d+ -->
<!ELEMENT        From               (#PCDATA)> <!-- .+ -->
<!ELEMENT        To                 (#PCDATA)> <!-- .+ -->
<!ELEMENT        Term               (#PCDATA)> <!-- .+ -->
<!ELEMENT        Field              (#PCDATA)> <!-- .+ -->
<!ELEMENT        QueryKey           (#PCDATA)> <!-- \d+ -->
<!ELEMENT        WebEnv             (#PCDATA)> <!-- \S+ -->
<!ELEMENT        Explode            (#PCDATA)> <!-- (Y|N) -->
<!ELEMENT        OP                 (#PCDATA)> <!--(AND|OR|NOT|RANGE|GROUP)-->
<!ELEMENT        IdList             (Id*)>
<!ELEMENT        Translation        (From, To)>
<!ELEMENT        TranslationSet     (Translation*)>
<!ELEMENT        TermSet (Term, Field, Count, Explode)>
<!ELEMENT        TranslationStack        ((TermSet|OP)*)>
<!ELEMENT        ERROR              (#PCDATA)> <!-- .+ -->
<!ELEMENT        OutputMessage      (#PCDATA)> <!-- .+ -->
<!ELEMENT        QuotedPhraseNotFound    (#PCDATA)> <!-- .+ -->
<!ELEMENT        PhraseIgnored      (#PCDATA)> <!-- .+ -->
<!ELEMENT        FieldNotFound      (#PCDATA)> <!-- .+ -->
<!ELEMENT        PhraseNotFound     (#PCDATA)> <!-- .+ -->
<!ELEMENT        QueryTranslation   (#PCDATA)> <!-- .+ -->
<!ELEMENT        ErrorList          (PhraseNotFound*,FieldNotFound*)>
<!ELEMENT        WarningList        (PhraseIgnored*,
                                    QuotedPhraseNotFound*,
                                    OutputMessage*)>
<!ELEMENT        eSearchResult      (((Count,
                                        (RetMax, RetStart, QueryKey?,
                                        WebEnv?, IdList, TranslationSet,
                                        TranslationStack?, QueryTranslation
                                        )?
                                    ) | ERROR),
                                    ErrorList?, WarningList? )>
```

Figure 4.9 The DTD from NLM describing the structure of a query result when using eSearch. *(Slightly reformatted to fit here)*

```
<TranslationSet>
        <Translation>
                <From>cancer</From>
                <To>((&quot;neoplasms&quot;[TIAB] NOT
    Medline[SB])
        OR &quot;neoplasms&quot;[MeSH Terms]
        OR cancer[Text Word])</To>
        </Translation>
```

```
    . . .
    </eSearchResult>
    "
```

In the above, we have not included the full result, which would occupy several pages. Omitted parts are indicated by ellipses. What came back includes a count of how many items in the database were matches (160955), way too many to send back, so the default maximum number of results, 20, is used as a limit. Starting from the beginning, the first 20 matching PubMed IDs (PMIDs) are provided. In addition, the translation of the query into MeSH terms and other possible field values is returned. A look at the full query results shows that the search included the MeSH term "radiotherapy," as well as "neoplasms" (a term related to cancer, though not exactly a synonym).

Computer programs don't generally do computations directly on XML strings. The next task is to *parse* the string, to extract the tags and their associated values into some data structures that a program can manipulate. Since the basic structure of an XML document is a hierarchical or nested set of tags and values, the logical structure to use is a nested list, where the tags become symbols and their associated content can be a string or a number or another list containing tags and values.

We will use a format called LXML (Lisp XML), which is supported by an XML parser package available with Allegro Common Lisp. Several other XML parsers are available which run in any ANSI Common Lisp environment. They are listed in Appendix A, Section A.1.4. Section 1.2.6, in Chapter 1, described also the Document Object Model (DOM) parser, which is another option.

The `parse-xml` function takes as its one required input a string or an open stream (as for example a connection to an open file). It returns the equivalent LXML corresponding to the input string or stream. Here is an example. We retrieve the results of querying PubMed with the query "description logic cancer." This string does not generate a lot of hits.

```
> (setq result (pubmed-query "description logic cancer"))
gc: E=66% N=369200 O+=159624 pfu=0+17 pfg=0+39
"<?xml version=\"1.0\"?>
<!DOCTYPE QueryResult PUBLIC \"-//NLM//DTD QueryResult,
22 Jan 2002//EN\"
\" http://www.ncbi.nlm.nih.gov/entrez/query/DTD/pmqty_020122.dtd\" >
<QueryResult>
        <Count>15</Count>
        <DispMax>15</DispMax>
        <DispStart>0</DispStart>
        <Id>16952624 16398930 15797001 15360769 14728178 11879299
10513119 9267592 8686992 8648967 8121105 8337442 1574320 1015114
5559903</Id>
        <TranslationSet>
                <Translation>
```

```
                    <From>cancer</From>

              <To>((%22neoplasms%22%5BTIAB%5D+NOT+Medline%5BSB%5D)
    +OR+%22neoplasms%22%5BMeSH+Terms%5D+OR+cancer%5BText+Word%5D)</To>
                    </Translation>
                    <Translation>
                           <From>logic</From>
                           <To>(%22logic%22%5BMeSH+Terms%5D+
    OR+logic%5BText+Word%5D)</To>
                    </Translation>
              </TranslationSet>
    </QueryResult>
    "
    >
```

This result is a single long string with embedded new line characters, as well as embedded double quote characters.[18] Now we use the XML parser to convert it all to a list of items with tags and values.[19]

```
> (setq parsed-result (net.xml.parser:parse-xml result))
((:XML :|version| "1.0") (:DOCTYPE :|QueryResult|)
(|QueryResult| "
        " (|Count| "15") "
        " (|DispMax| "15") "
        " (|DispStart| "0") "
        "
(|Id|
  "16952624 16398930 15797001 15360769 14728178 11879299 10513119
9267592 8686992 8648967 8121105 8337442 1574320 1015114 5559903")
  "
      " ...))
>
```

It should not be surprising that the parsed result contains blank strings, or strings containing only whitespace characters (including new line characters). This is a consequence of the visual formatting contained in the return string, as explained in Section 1.2.6 of Chapter 1. So, the next step is to write a function we will call `strip-blanks` to remove all the blank strings. This is pretty easy to do. We need to specify which characters we consider "whitespace," or blank spaces. This should include space characters, line break characters, and tabs, at a minimum.

---

[18]In some of the XML returned in this example, instead of actual " and bracket [] characters a kind of escaped coding appears instead, the %22, %5B, and %5D. These can be detected and converted when writing code to deal with the lowest level of detail.

[19]Note that this will only work in the case-sensitive lower case version of Allegro CL^{TM}, the so-called "modern" lisp environment.

The predicate function, whitespace-char-p, takes a character as input and returns nil if it is *not* a whitespace character, otherwise it just returns the input.

```
(defun whitespace-char-p (ch)
  "Our own definition of which characters are whitespace"
  (find ch '(#\Space #\Newline #\Tab #\Linefeed #\Return
             #\Page #\Null)
        :test #'char=))
```

The strip-blanks function takes a list of strings, other objects, or possibly nested lists, tests each element to see if it is a blank string, removes it if it is, and recursively tests the element if it is a list. Otherwise, it just keeps the item and puts it on the front of the result of processing the remainder of the list. The end result is a nested list exactly like the original except with all the blank strings removed at all levels of nesting.

```
(defun strip-blanks (terms)
  "terms is a nested list possibly containing blank strings at
multiple levels. This function returns a new list with the same
items, each with blank strings recursively removed"
  (labels ((strip-aux (accum terms)
             (if (null terms) (reverse accum)
               (strip-aux
                 (let ((term (first terms)))
                   (typecase term
                     (string (if (every #'whitespace-char-p term)
                                 accum
                                 (cons term accum)))
                     (list (cons (strip-aux nil term)
                                 accum))
                     (t (cons term accum))))
                 (rest terms)))))
    (strip-aux nil terms)))
```

Here are the results of applying strip-blanks to the LXML that resulted from the XML parser above.

```
> (setq stripped-result (strip-blanks parsed-result))
((:XML :|version| "1.0")
(:DOCTYPE :|QueryResult|)
(|QueryResult|
  (|Count| "15") (|DispMax| "15") (|DispStart| "0")
  (|Id|
  "16952624 16398930 15797001 15360769 14728178 11879299 10513119
9267592 8686992 8648967 8121105 8337442 1574320 1015114 5559903")
  (|TranslationSet|
  (|Translation|
   (|From| "cancer")
```

```
(|To|
 "((%22neoplasms%22%5BTIAB%5D+NOT+Medline%5BSB%5D)+OR+
%22neoplasms%22%5BMeSH+Terms%5D+OR+cancer%5BText+Word%5D)"))
 (|Translation|
  (|From| "logic")
  (|To| "(%22logic%22%5BMeSH+Terms%5D+OR+logic%5BText+Word%5D)")))))
>
```

So, what did we get? The nested list structure above can be considered a query result object with the following properties:

Count: "15" (too bad this was not a number, but converting it is easy)

ID: a string containing 15 PubMed IDs

TranslationSet: a list consisting of the translation of each search term into a MeSH term or terms.

This nested list is easy to unpack, and is simpler than the DOM document we would have gotten by using the DOM parser for the same query result. Out of this structure, one can extract the list of PubMed IDs relatively easily, knowing that it is tagged by the |Id| symbol. The next step is to retrieve each of the complete records whose PubMed IDs were returned. Here we just show how to define a variant of the query function from earlier, to take an ID number and get the entry corresponding to it. We need an additional parameter, the Entrez eFetch URL.

```
(defparameter +entrez-fetch-url+ "/entrez/eutils/efetch.fcgi")

(defun pubmed-fetch (pmid)
  "Gets the XML for a single entry given the PubMed ID of the entry"
  (net.aserve.client:do-http-request
      (format nil "http://~a~a" +eutils-host+ +entrez-fetch-url+)
    :method :get
    :query
    '(("db" . "PubMed") ("report" . "xml") ("mode" . "text")
      ("id" . ,(format nil "~A" pmid)))))
```

The pubmed-fetch function will take a number (a PubMed ID), construct the right URL for Entrez, and send it, returning the full record for that ID, if it exists. Here is an example using the first ID in the list returned for the previous query. The nested function calls combine the three steps that were explained separately above.

```
> (setq doc (strip-blanks (net-xml.parser:parse-xml
                               (pubmed-fetch 16952624))))
((:XML :|version| "1.0") (:DOCTYPE :|PubmedArticleSet|)
 (|PubmedArticleSet|
  (|PubmedArticle|
   ((|MedlineCitation| |Owner| "NLM" |Status| "MEDLINE")
    (PMID "16952624") (|DateCreated| # # #) (|DateCompleted| # # #)
    (# # # # # # # # #) (|MedlineJournalInfo| # # #)
```

```
    (|CitationSubset| "AIM") (|CitationSubset| "IM")
    (|MeshHeadingList| # # # # # # # # # ))
   (|PubmedData| (|History| # # #) (|PublicationStatus| "ppublish")
    (|ArticleIdList| # # #)))))
```

The resulting list here is printed in a truncated form, with # marks and ellipses indicating that more deeply nested list contents are not displayed. The pprint (pretty print) function will give a full display. The complete list looks like this:

```
> (pprint doc)
((:XML :|version| "1.0") (:DOCTYPE :|PubmedArticleSet|)
(|PubmedArticleSet|
 (|PubmedArticle|
  (((|MedlineCitation| |Owner| "NLM" |Status| "MEDLINE")
   (PMID "16952624")
   (|DateCreated| (|Year| "2006") (|Month| "09") (|Day| "05"))
   (|DateCompleted| (|Year| "2006") (|Month| "10") (|Day| "20"))
   ((|Article| |PubModel| "Print")
   (|Journal| ((ISSN |IssnType| "Print") "0022-5347")
    (((|JournalIssue| |CitedMedium| "Print") (|Volume| "176")
      (|Issue| "4 Pt 1") (|PubDate| (|Year| "2006") (|Month| "Oct")))
     (|Title| "The Journal of urology")
     (|ISOAbbreviation| "J. Urol."))
   (|ArticleTitle|
    "Significance of macroscopic tumor necrosis as a prognostic
indicator for renal cell carcinoma.")
   (|Pagination| (|MedlinePgn| "1332-7; discussion 1337-8"))
   (|Abstract|
    (|AbstractText|
    "PURPOSE: We investigated the prognostic significance of
macroscopic tumor necrosis in renal cell carcinoma. MATERIALS AND
METHODS: We retrospectively analyzed the records of 485 patients who
underwent surgical treatment for organ confined or metastatic renal
cell carcinoma. The presence or absence of tumor necrosis was
evaluated based on macroscopic description of the tumor,
and tumors were considered necrotic only if they exhibited more than 10%
macroscopic necrosis. RESULTS: Macroscopic tumor necrosis was
identified in 27% of total patients. Patients with macroscopic
necrotic renal cell carcinoma were more likely to have larger tumor,
metastatic disease, higher local stage and higher tumor grade (all p
< 0.001). Pathological features of microvascular invasion (p = 0.026)
and sarcomatoid differentiation (p = 0.002) along with several
laboratory findings were also observed to be associated with
macroscopic tumor necrosis. Among the total subjects those without
macroscopic tumor necrosis had significantly higher progression-free
(p < 0.0001) and disease specific survival (p < 0.0001). When
survival analysis was limited to nonmetastatic tumors only, the same
```

logic applied, which was not the case for the patients with
metastatic disease (p > 0.05). Among the different histological
subtypes of renal cell carcinoma, macroscopic tumor necrosis was
observed to have a significant impact only for the clear cell
subtype. In patients with nonmetastatic RCC multivariate analysis
revealed that macroscopic tumor necrosis (p = 0.004) was an
independent prognostic predictor of disease specific survival along
with pathological T stage, tumor grade and tumor size. CONCLUSIONS:
Our results suggest that macroscopic tumor necrosis may be a reliable
prognostic indicator for nonmetastatic clear cell renal cell
carcinoma which should routinely be examined for during pathological
analysis."))
  (|Affiliation|
"Department of Urology, Seoul National University Bundang
Hospital, Seongnam, Korea.")
  (((|AuthorList| |CompleteYN| "Y")
  (((|Author| |ValidYN| "Y") (|LastName| "Lee")
   (|ForeName| "Sang Eun") (|Initials| "SE"))
  (((|Author| |ValidYN| "Y") (|LastName| "Byun")
   (|ForeName| "Seok-Soo") (|Initials| "SS"))
  (((|Author| |ValidYN| "Y") (|LastName| "Oh")
   (|ForeName| "Jin Kyu") (|Initials| "JK"))
  (((|Author| |ValidYN| "Y") (|LastName| "Lee")
   (|ForeName| "Sang Chul") (|Initials| "SC"))
  (((|Author| |ValidYN| "Y") (|LastName| "Chang")
   (|ForeName| "In Ho") (|Initials| "IH"))
  (((|Author| |ValidYN| "Y") (|LastName| "Choe")
   (|ForeName| "Gheeyoung") (|Initials| "G"))
  (((|Author| |ValidYN| "Y") (|LastName| "Hong")
   (|ForeName| "Sung Kyu") (|Initials| "SK")))
  (|Language| "eng")
  (|PublicationTypeList| (|PublicationType| "Journal Article")))
 (|MedlineJournalInfo| (|Country| "United States")
  (|MedlineTA| "J Urol") (|NlmUniqueID| "0376374"))
 (|CitationSubset| "AIM") (|CitationSubset| "IM")
 (|MeshHeadingList|
  (|MeshHeading| ((|DescriptorName| |MajorTopicYN| "N") "Adult"))
  (|MeshHeading| ((|DescriptorName| |MajorTopicYN| "N") "Aged"))
  (|MeshHeading|
   ((|DescriptorName| |MajorTopicYN| "N") "Aged, 80 and over"))
  (|MeshHeading|
   ((|DescriptorName| |MajorTopicYN| "N") "Carcinoma, Renal Cell")
   ((|QualifierName| |MajorTopicYN| "Y") "mortality")
   ((|QualifierName| |MajorTopicYN| "Y") "pathology")
   ((|QualifierName| |MajorTopicYN| "N") "surgery"))
  (|MeshHeading| ((|DescriptorName| |MajorTopicYN| "N") "Female"))
  (|MeshHeading| ((|DescriptorName| |MajorTopicYN| "N") "Humans"))

```
  (|MeshHeading|
   ((|DescriptorName| |MajorTopicYN| "N") "Kidney Neoplasms")
   ((|QualifierName| |MajorTopicYN| "Y") "mortality")
   ((|QualifierName| |MajorTopicYN| "Y") "pathology")
   ((|QualifierName| |MajorTopicYN| "N") "surgery"))
  (|MeshHeading| ((|DescriptorName| |MajorTopicYN| "N") "Male"))
  (|MeshHeading|
   ((|DescriptorName| |MajorTopicYN| "N") "Middle Aged"))
  (|MeshHeading| ((|DescriptorName| |MajorTopicYN| "N") "Necrosis"))
  (|MeshHeading|
   ((|DescriptorName| |MajorTopicYN| "N") "Neoplasm Staging"))
  (|MeshHeading|
   ((|DescriptorName| |MajorTopicYN| "N") "Nephrectomy"))
  (|MeshHeading|
   ((|DescriptorName| |MajorTopicYN| "N") "Prognosis"))
  (|MeshHeading|
   ((|DescriptorName| |MajorTopicYN| "N") "Retrospective Studies"))
  (|MeshHeading|
   ((|DescriptorName| |MajorTopicYN| "N") "Survival Rate"))))
 (|PubmedData|
  (|History|
  ((|PubMedPubDate| |PubStatus| "received") (|Year| "2005")
   (|Month| "10") (|Day| "3"))
  ((|PubMedPubDate| |PubStatus| "pubmed") (|Year| "2006")
   (|Month| "9") (|Day| "6") (|Hour| "9") (|Minute| "0"))
  ((|PubMedPubDate| |PubStatus| "medline") (|Year| "2006")
   (|Month| "10") (|Day| "21") (|Hour| "9") (|Minute| "0")))
  (|PublicationStatus| "ppublish")
  (|ArticleIdList|
  ((|ArticleId| |IdType| "pii") "S0022-5347(06)01387-5")
  ((|ArticleId| |IdType| "doi") "10.1016/j.juro.2006.06.021")
  ((|ArticleId| |IdType| "pubmed") "16952624"))))))
```

This looks complicated and it is, but it is really just nested lists, where each one begins with a tag identifying it. Recall that to deal with the Document Object Model, we had to write code that looked things up in slots, and recursively go through the structure. Here, we just need to use the assoc function to find a list beginning with a tag. Again, using recursion, we can specify a *path* that identifies the item we are looking for, and a traversal with the path elements as a guide will return the result. Here is a very simple first pass at such a search function.

```
(defun path-query (path tree)
  "returns the first match to path in tree, where path is a list of
  tags referring to nested lists"
  (cond ((null tree) nil)
        ((null path) nil)
        ((null (rest path)) (assoc (first path) tree))
```

```
              (t (path-query (rest path)
                             (rest (assoc (first path) tree))))))))
```

The idea here is that the `path` parameter is a sequence of tags, that identify the labels along some series of branches in the tree structure. The function looks for the first tag, and if found, it recursively searches the rest of the list associated with that tag, using the rest of the path. If there is no more path to search, whatever is left in the tree is the result we are looking for. Here it is applied to an artificial example:

```
> (setq test '((foo a b c) (bar (tag1 g f d) (tag2 q r s)) (baz 5)))
((FOO A B C) (BAR (TAG1 G F D) (TAG2 Q R S)) (BAZ 5))
> (path-query test '(bar tag2))
(Q R S)
```

Now, before jumping to apply it to the PubMed query results, it should be noted that there are a few complications. First, some tags are not simply symbols, but are themselves lists. This corresponds to the tags that have *attributes* associated with them. This makes search a little more complex. Instead of just a call to `assoc`, some checking to see if the tag appears as a list rather than by itself is needed. The structure is something like this:

```
(result (tag-1 (tag-1a stuff ...)
               (tag-1b more-stuff ...))
        ((tag-2 (attrib-1 foo)
          (atrib-2 bar))
         (tag-2a stuff-2 ...)
         (tag-2b more-stuff-2 ...)
         ...)
        ...)
```

So, using the simple function above, a search for the path

```
(result tag-2 tag-2b)
```

would not be successful, because `tag-2` would not be found. It should not be too hard to enhance the simple code above to check for tags that appear with attributes.

The second problem is that the version above only finds the first path that matches, and returns all the associated information. This is more straightforward to fix, and is also left as an exercise. It should be possible to see how this idea can be elaborated to be able to search these tree structures for arbitrary path expressions, including "wild card" matches. Of course there is support for this kind of pattern matching in XML as well, with the XPath facility.

Now it should be possible to envision how to build elaborate programs that can correlate data from different records, or perform specialized queries that the normal search facilities don't provide. Having the data in such a structured form, readable by a computer program, provides the basis for such customized operations. You can imagine correlating the results of queries across multiple data sources, even with very different structures. For example, you might need to get the results of a search of the protein sequence database in GENBANK, cross reference it to the Gene Ontology, to find out what related gene products are there, look those up in PubMed for articles about functional annotation, etc.

Data that have hierarchical structure are easily labeled by unique paths or tag sequences. An example of a very practical application of this idea is the internal design of a system that implements the DICOM medical image data transmission protocol described earlier in Section 4.3.3. The detailed implementation of DICOM is developed in Chapter 5, Section 5.4.4.

Documents or data encoded using XML tags with nested elements also have this structure, in that an element can be referenced or indexed in terms of the path, or sequence, of tags that lead to it. XPath is a syntax and semantics for specifying such paths for XML documents, though it is not itself part of XML. This and other facilities are described in [149].

Another approach to representing data in terms of path labels is the use of B-trees and tries. A B-tree generalizes the notion of a binary tree to have more than two branches. This trades between the cost of computation at each node and the number of levels needed to traverse. It also provides a way to do multilevel indexing. Another kind of tree structure that provides this kind of efficiency is the "trie" (see [305, pages 343 and 472]). Tries are also called "discrimination nets" [65]. They reappear in yet another form as *path* specifications, in trees that represent highly nested XML documents. Tries can be used to index anything that can be represented as (or indexed by) a sequence of components. It is possible, for example, to index protein sequence records using tries. A protein is essentially a word (perhaps a long one) made up of the letters corresponding to the 20 amino acids. So Uniprot records can be reorganized into a trie and thus possibly reduce the search time to look up a particular sequence.

Is this really useful? Trie lookup is $O(m)$ where m is the length of the protein sequence that is the subject of the lookup, rather than $O(\log(n))$ where n is the number of entries. For Uniprot, typically $\log(n)$ is less than m so that may not be a speedup. However, the SwissProt accession number is a more commonly used lookup key, and is much shorter, so reorganizing SwissProt as a trie could well improve performance.

4.3.5 *Network Access to an Ontology: the FMA*

In Chapter 5 we will present an illustration of how combining the ideas of Markov chains and representing lymphatic paths as sequences of lymphatic nodes or vessels could produce a predictive model of the regional dissemination of tumor cells. Such predictions on a patient-specific basis then could be used to improve the accuracy of radiation treatment. This section explains how the requisite connectivity information from the Foundational Model of Anatomy can be obtained through a network service.

For some time, the FMA project has supported a network query server for the information contained in the frame system. It is known as the Foundational Model Server (FMS). When connected it accepts a query string as input, and returns result text in a specified format. This cycle repeats until the client program sends a "quit" request. So, it is really easy to write a client program to query the FMS, and in this section we will provide the entire code to do it.

The FMS accepts queries in a Lisp-like syntax, using a functional style something like OKBC [66], but much simpler. An FMS query consists of a string that looks like the printed representation of a Lisp list, starting with a command symbol (analogous to a function name), then zero or more parameters, which, depending on the command, may be strings, numbers, `t`, or `nil`. Here is an example query, asking for the constitutional parts of the heart (constitutional parts are those that make up an anatomical entity, while regional parts are *spatial* subdivisions),

```
(fms-get-children "Heart" "constitutional part")
```

and here is the response from the FMS:

```
("Coronary sinus" "Interatrial septum" "Interventricular septum"
"Atrioventricular septum" "Tricuspid valve" "Mitral valve"
"Aortic valve" "Pulmonary valve" "Wall of heart"
"Cavity of right ventricle" "Cavity of left atrium"
"Cavity of left ventricle" "Fibrous skeleton of heart"
"Cavity of right atrium" "Cardiac vein" "Right coronary artery"
"Left coronary artery" "Systemic capillary bed of heart"
"Lymphatic capillary bed of heart" "Neural network of heart")
```

The query above is not a function call in a Lisp program. It is a string of characters sent to the FMS through a network connection. Similarly, the return value is a long string (without newline characters) that is read back from the socket. However, it is easy to construct strings like this in a Lisp program, and even easier to parse the result string (the Lisp built-in function, `read`, does this). In order to perform this operation we need a way to make a connection to a remote server like the FMS. Earlier we showed how to use proprietary extensions to Common Lisp to send queries to a Web server and get back text in the form of HTML or XML. In this case, we are dealing with a lower level operation, rather than a layered implementation of the HTTP web communications protocol. This is done by using the *socket* model, also described earlier, and its implementation in socket libraries.

Our plan will be to create a programmatic interface that operates as a client. The client process will make a connection to the FMS, handle arbitrary queries, and return results. With this abstraction layer, we can write complex and powerful programs that use the FMA as a knowledge resource, through the FMS. This interface layer will consist of three basic functions. The `fms-connect` function will connect to the FMS and return an open stream that can be used to send and receive queries. The `fms-query` function will take a list specifying a query according to the FMS query language, convert it to a string, send the string, read back the result, and return it as a list of strings (terms from the FMA). Finally, the `fms-disconnect` function will close the stream and socket connection when the application no longer needs to refer to the FMA.

Since we are using strings, or terms, to label the entities that are represented in the FMA, we can write some functions that handle the details of making queries from a reasoning program.

In particular, we will need to open a socket to the host system of the FMS. First we define some useful constants.

```
(defconstant *data-flag* 4)
(defconstant *end-flag* 3)
(defconstant *fms-host* "fma.biostr.washington.edu")
(defconstant *fms-port* 8098)
```

As there is no ANSI standard function for opening a socket, this will depend on the Lisp implementation. Most Lisp implementations provide extensions to support socket communication. In Allegro Common Lisp, there is a package named `socket` and a function, `make-socket`, that returns an open stream if the remote connection is successful. In Lispworks, the corresponding function is in the `comm` package and is called `open-tcp-stream`, while in CMUCL, it is a two-step process, in which a call to `connect-to-inet-socket` in the `extensions` package returns an input for the function `make-fd-stream` in the `system` package, which returns the open stream to the specified remote host and port. Once we have an open stream connected to the FMS server program, we can use the `read` and `write` (and `format`) functions of standard Common Lisp.

The following is a function to connect to the FMS with particular variants for some Lisp implementations. Note that for Lispworks^{TM}, a non-standard package must be explicitly loaded. For the others, the package will be "auto-loaded" or already in the running environment by default. We don't want the Lisp reader to attempt to read an expression not meant for the particular implementation in use, so in this code, the #+ *reader macro* is used. When #+ appears in program code, the symbol following it is looked up in a list that is the value of the standard environment variable `*features*`. If it is found, the next expression is read and evaluated. If it is not found, the next expression is ignored. Each implementation will have some identifying symbol(s) in the `*features*` list, so we can conditionalize the code to automatically include the right expressions for whichever Lisp implementation is in use.

```
#+lispworks (require "comm")

(defvar *fms-socket* nil "open socket to the fms")

(defun fms-connect (&optional (host *fms-host*) (port *fms-port*))
  (let ((stream #+allegro (socket:make-socket :remote-host host
                                              :remote-port port)
                #+lispworks (comm:open-tcp-stream host port)
                #+cmu (system:make-fd-stream
                        (extensions:connect-to-inet-socket host port)
                        :input t :output t :element-type 'character)
                ))
    (format t "~A~%" (fms-readback stream))
    (setq *fms-socket* stream)))
```

We also defined a global variable, *fms-socket*, to have the open stream as its value. The reading and writing functions can refer to it, rather than having to pass the open stream as a parameter in all our functions.

The fms-connect function opens a text stream connection to the FMS server program running on the computer named host, at the port numbered port. It writes the server's greeting to *standard-output*, and returns the open stream, as well as setting the global variable, *fms-socket*. It uses another function, fms-readback, which is called when expecting output from the FMS. We'll show how that works shortly. The fms-readback function is called here because the FMS protocol specifies that when the connection is made, the FMS sends a greeting message, then waits for a query. After that, the sequence is a simple two-step cycle: the client sends a query, and the FMS sends a response.

The protocol for the FMS says that whenever a query is made (or when the connection is first made), the server's response will be one of the following:

    `<message> <data_char> <sexpression> <end_char>`

or

    `<message> <end_char>`

and in either case, our fms-readback function should return the last non-flag item sent. This will be a list if the result is a valid response, or a symbol corresponding to a message, possibly an error message, if there was an error or end of file. The data flag and end flag characters are defined above. Strictly speaking, this is actually not the protocol—the documentation promises that the Control-C and Control-D characters serve these functions, and the printable 3 and 4 are just there to provide a human readable indicator. However, since both appear, we must process both. We use the printable characters as the flags, and in this code, the read-line calls are then needed to read and discard the real flag characters.

```
(defun fms-readback (stream)
  (let (item data)
    (loop
      (setq item (read stream nil :eof))
      (cond ((eql item :eof) (return data))
            ((eql item *data-flag*) (read-line stream)
             (setq data (read stream)))
            ((eql item *end-flag*)
             (read-line stream) (return data))
            (t (setq data item))))))
```

The input variable, stream, is an open stream, connected to the FMS. This is the value returned by a call to fms-connect. Two local variables, item and data, are used in the code body. The protocol is to read from the stream repeatedly until either an "end" flag is received or there is an end of file indication from the read operation. First, the function checks for an end of file condition, in which case it just returns any data previously read. If there is no end of file indicated, then the function checks if what was read is the printable data flag, indicating that some data object will be sent next. In that case, the call to `read-line` is there to read past the control character that is the real data flag, then the call to `read` returns the actual data. This works because the string that is sent by the server is in the form of a Lisp list so it is seen by the Lisp reader as a single object. If the printable "end flag" is seen, that means the server is done sending and the receiver (our function) should simply return the data previously read. The call to `read-line` again is there to read past the control character that is the real end flag. Finally, if none of those conditions hold, the item read must be the data sent back by the server, so the function just saves it for now and attempts another read operation.

So this function, `fms-readback`, serves to return the server's response to a connection or query. With this, we can then define a function, `fms-query`, that takes an expression in the right format, either a list, or a string that looks like a list, sends it to the FMS, gets the result with a call to `fms-readback`, and returns the result. The input, `expr`, is either a list in the form

```
(fms-get-children "Brain" "subdivision")
```

or the equivalent string (with embedded " characters). If the input is a list, it needs to be converted to a string for sending to the FMS server. If already a string (or after conversion using `format`), our function can just use `write-line` to send the string. To insure that the function does not wait for buffers to fill, we call the Common Lisp function, `finish-output`, which forces any buffers to empty and the pending output to be completed. Then the function just reads back the result, using `fms-readback`. Here is the function, `fms-query`.

```
(defun fms-query (expr)
  (if (listp expr)
      (setq expr ;; convert to string with embedded " around terms
        (format nil "(~A~{ \"~A\"~})" (first expr) (rest expr))))
  (write-line expr *fms-socket*)
  (finish-output *fms-socket*)
  (fms-readback *fms-socket*))
```

This function, `fms-query`, uses the open stream stored in the global variable, `*fms-socket*`, so the next layer does not need to be concerned with passing this value in.

Finally, it will be nice to have another wrapper function, `fms-disconnect`, that sends the right command to the FMS to terminate the interaction, and that also closes the open socket stream. The FMS closes the connection on its end in response to the string, `(quit)`, so the `fms-disconnect` function just uses `fms-query` to send this string and read back any termination message. Then a call to `close` closes the stream on the client end. The call to `format` just writes a short message to the terminal screen for the user to see that the action completed.

```
(defun fms-disconnect (&optional (stream *fms-socket*))
  (fms-query stream "(quit)")
```

```
(close stream)
(format t "~%FMS Connection closed~%"))
```

The FMS protocol supports many kinds of queries, but we will just describe a few here. The idea is to write wrapper functions that look like the FMS query strings, but are actual functions in Lisp. Then a program that uses the FMS just needs to do these function calls. For the `fms-get-children` query, we define a wrapper function as follows:

```
(defun get-children (term relation)
  (fms-query (list 'fms-get-children term relation)))
```

This function returns a list corresponding to the "fms-get-children" query for the anatomical entity named `term` using the hierarchy named `relation`. We can then use this to define yet higher level functions that can be used to retrieve specific structural information. For example, a function to fetch information about the *constitutional parts* of an anatomic entity would take the form

```
(defun const-parts (term)
  (get-children term "constitutional part"))
```

Here is how it works. First a connection to the FMS is established, using `fms-connect`. Then the `parts` function is used to get the parts of the anatomic entity called "Heart." The whole list returned by the FMS has in all 37 items, each a string naming some entity in the FMA. In the code below, it is assumed that all the functions are defined in a file and loaded before typing the expression you see at the > prompt.[20]

```
> (fms-connect)
; Autoloading for ACL-SOCKET:CONFIGURE-DNS:
; Fast loading from bundle code/acldns.fasl.
CONNECTED
#<MULTIVALENT stream socket connected from 192.168.0.2/1407 to
   xiphoid.biostr.washington.edu/8098 @ #x7168776a>
> (const-parts "Heart")
("Coronary sinus" "Interatrial septum" "Interventricular septum"
  "Atrioventricular septum" "Tricuspid valve" "Mitral valve"
  "Aortic valve" "Pulmonary valve" "Wall of heart"
  "Cavity of right ventricle" ...)
```

Some other queries supported by the FMS include queries to find out what attributes are specified for a given term, what attributes or relations are supported by the FMA in general,

---

[20]Note that we specified the domain name `fma.biostr.washington.edu` but actually got `xiphoid.biostr.washington.edu`. This happened because `fma.biostr.washington.edu` is an *alias* for the real system containing the FMS, and it gets translated when the connection request is run. This allows the maintainers of the FMS to move it to a different system if necessary and just update the alias name. Client programs won't need to change anything and will continue to reach the right server.

and what are the parent terms in a given hierarchy (relation tree) for a particular term. Here are wrapper functions to construct these queries. The `get-attributes` function will return all the attributes or relations that are defined for a given term (the names of the attributes, not their values).

```
(defun get-attributes (term)
   (fms-query (list 'fms-get-attributes term)))
```

The `get-parents` function will return all the parents of a given term that are related by a specified relation in the FMA.

```
(defun get-parents (term relation)
   (fms-query (list 'fms-get-parents term relation)))
```

The `get-hierarchies` function will return the hierarchies (relationships) that include the specified anatomic entity, or if no term is supplied, it returns a list of all hierarchies or relationships in the FMA.

```
(defun get-hierarchies (&optional term)
   (fms-query (if term (list 'fms-get-hierarchies term)
                 (list 'fms-get-hierarchies))))
```

In the FMA there are currently 138 different hierarchies, including the built-in relations for classes and instances in the Protégé ontology development system. Not every relation corresponds to a hierarchy. Here is the output of `get-hierarchies`.

```
> (setq rels (get-hierarchies))
(":TO" ":DIRECT-INSTANCES" ":ANNOTATED-INSTANCE" ":DIRECT-DOMAIN"
 ":FROM" ":DIRECT-TYPE" ":DIRECT-SUPERCLASSES" ":DIRECT-SUBCLASSES"
 ":SLOT-INVERSE" ":DIRECT-SUBSLOTS" ...)
```

As you can see, at first, the Lisp printer only shows the first few entries, with an ellipsis to show there is more. Using the pretty-printer (with the function `pprint`) will show the entire list.

```
> (pprint rels)
(":TO" ":DIRECT-INSTANCES" ":ANNOTATED-INSTANCE" ":DIRECT-DOMAIN"
 ":FROM" ":DIRECT-TYPE" ":DIRECT-SUPERCLASSES" ":DIRECT-SUBCLASSES"
 ":SLOT-INVERSE" ":DIRECT-SUBSLOTS" ":ASSOCIATED-FACET"
 ":DIRECT-TEMPLATE-SLOTS" ":ASSOCIATED-SLOT" ":DIRECT-SUPERSLOTS"
 ":SLOT-CONSTRAINTS" "branch" "part" "tributary" "Preferred name"
 "Synonym" "inherent 3-D shape" "related part" "has derivation"
 "has fate" "has shape" "contained in" "orientation" "has adjacency"
 "anatomical plane" "related to" "has location" "related concept"
 "related object" "part of" "general location" "bounded by"
 "is boundary of" "location" "attributed part" "contains"
 "attributed continuous with" "surrounded by" "surrounds"
 "arterial supply" "venous drainage" "lymphatic drainage" "innervation"
 "sensory nerve supply" "member of" "has wall" "input from" "output to"
```

```
"muscle attachment" "fascicular architecture" "muscle origin"
"muscle insertion" "branch of" "tributary of" "member" "bounds"
"arterial supply of" "segmental supply" "nerve supply"
"nerve supply of" "attaches to" "receives attachment" "union"
"unites with" "form" "segmental contribution from"
"anatomical morphology" "venous drainage of"
"segmental contribution to" "attributed branch" "variation"
"attributed tributary" "germ origin" "cell shape"
"attributed cell type" "cell surface specialization"
"NN abbreviations" "chromosome pair number" "cell connectivity"
"connection type" "connecting part" "connected to" "regional part"
"constitutional part" "systemic part" "regional part of"
"constitutional part of" "systemic part of" "clinical part"
"clinical part of" "NeuroNames Synonym" "epithelial morphology"
"2D part" "attributed regional part" "attributed constitutional part"
"Non-English equivalent" "lymphatic drainage of" "2D part of"
"segmental composition" "segmental composition of"
"segmental supply of" "primary segmental supply"
"secondary segmental supply" "primary segmental supply of"
"secondary segmental supply of" "origin of" "muscle insertion of"
"volume" "mass" "viscosity" "temperature" "ion content"
"continuous with" "branch (continuity)" "homonym for"
"continuous with distally" "continuous with proximally"
"tributary (continuity)" "custom partonomy" "custom partonomy of"
"primary site of" "primary site" "Metastatic site of" "efferent to"
"afferent to" "adjacent to" "attached to" "nucleus of origin of"
"receives input from" "nucleus of termination of" "sends output to"
"has related dimensional entity" "gives rise to" "develops from")
```

The relations in upper case preceded by colons, that look like keywords in Common Lisp, are actually the Protégé built-in relations. These can be used as relation parameters in the above functions, just as one would use the FMA-specific relations. So, with this interface one can write a program that obtains implementation details of the Protégé classes themselves, and perform computations on them. So, for example, one can get a list of all the FMA entities whose type is "Lymphatic chain", by querying as follows:

```
> (get-children "Lymphatic chain" ":DIRECT-INSTANCES")
("Pulmonary lymphatic chain"
 "Subdivision of pulmonary lymphatic chain"
 "Axillary lymphatic chain" "Subdivision of axillary lymphatic tree"
 "Posterior mediastinal lymphatic chain"
 "Tracheobronchial lymphatic chain"
 "Tributary of tracheobronchial lymphatic chain"
 "Left cardiac tributary of tracheobronchial lymphatic chain"
 "Brachiocephalic lymphatic chain"
 "Right cardiac tributary of brachiocephalic lymphatic chain" ...)
```

Altogether there are currently 101 entries. However, this is not the entire collection of everything that is a lymphatic chain, since there are more that are classified under "Subdivision of pulmonary lymphatic chain." This was done to help authors who are adding entries to the FMA to keep track of their progress by regions. In this case they wanted to keep the subchains (subdivisions, or regional parts) of bigger chains in a class or category separate from the long list of individual chains, some of whose subchains are not yet identified and entered.

To get all the instances of a class, it seems that we can take advantage of the fact that every FMA class is both an instance of its parent, and a subclass of it also. So, by recursively querying for all the subclasses, we should also end up with all the instances. Here is such a recursive query.

```
(defun all-subclasses (term)
  (let ((subs (get-children term ":DIRECT-SUBCLASSES")))
    (cond ((null subs) nil)
          (t (append subs
                     (mappend #'all-subclasses
                              subs))))))
```

The recursive query terminates when there are no more subclasses. It uses append to combine the direct subclasses with the list of subclasses resulting from the recursive call. The recursive call is done with mappend to apply it to every subclass, and combine those results. Applying this to the class "Lymphatic chain" shows that there are actually 353 classes altogether.

```
> (setq alltest (all-subclasses "Lymphatic chain"))
("Pulmonary lymphatic chain"
 "Subdivision of pulmonary lymphatic chain"
 "Axillary lymphatic chain" "Subdivision of axillary lymphatic tree"
 "Posterior mediastinal lymphatic chain"
 "Tracheobronchial lymphatic chain"
 "Tributary of tracheobronchial lymphatic chain"
 "Left cardiac tributary of tracheobronchial lymphatic chain"
 "Brachiocephalic lymphatic chain"
 "Right cardiac tributary of brachiocephalic lymphatic chain" ...)
> (length alltest)
353
```

4.3.5.1 *Semantic Consistency Checking and the FMA*

Because of the sheer size of biomedical ontologies, there is considerable motivation to find ways to do automated auditing. Several kinds of auditing are possible. First and foremost, for an ontology built on a formal logic framework, such as a description logic, it may be possible to test for formal contradictions [25]. Insofar as a given ontology is intended to satisfy some set of axioms, it may be possible to check for consistency with those axioms. The axioms may be relatively generic, or they may be very domain specific. An example of a more generic axiom is the requirement that part-whole relationships are transitive but not circular. A more specific one might be the restriction of slot values to certain types. This latter requirement can often be enforced by the frame system itself, if the model supports it. Finally, some constraints are really content-specific and cannot be checked or enforced by the frame system, but can possibly be automated by writing suitable procedures. This last kind of constraint is one that we will develop here, specifically for the lymphatic trees.

The human body has two lymphatic trees, which together provide a drainage system to bring interstitial body fluids into the blood stream. Material contained in the lymph fluid can then be filtered by the kidneys, processed by enzymes, or broken down by the action of the immune system. The lymphatic system consists of two tree structures, whose branches are vessels, with terminal ends that are permeable to interstitial fluids and are embedded in or in the vicinity of the different tissues throughout the body. These collect fluid from the surrounding regions, then join together to make larger vessels, until finally the fluid reaches one of two trunks, the Thoracic duct or the Right lymphatic duct. These in turn connect respectively to the left and right brachiocephalic veins. Figure 4.10 shows a part of one such tree, as specified by the "regional part" relation.

For many anatomical entities it makes sense to define their lymphatic drainage, the leaf ends of the lymphatic system that are embedded in these entities and thus collect fluid from that area. Not all anatomical structures have a lymphatic drainage, so this slot is not introduced in the class, "Anatomical structure." Instead, the "lymphatic drainage" slot is introduced in each of several subclasses, including "Organ," "Organ system subdivision," "Cardinal organ part,"

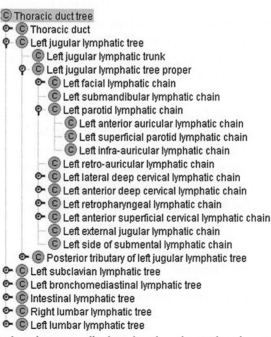

Figure 4.10 A Protégé class browser display showing the regional parts of a lymphatic tree, a subdivision of the lymphatic system.

and some others. Anatomical structures such as cells and biological macromolecules are also anatomical structures but clearly they do not have lymphatic drainage properties.

Two relationships will be particularly central to our consistency checking in the FMA. They are the *efferent to* and *afferent to* relationships. These relations only apply to components of the lymphatic system. If an entity, A, is efferent to entity B, that means that B is (in the lymphatic system) *downstream* from A. In the FMA, where applicable, for each lymphatic chain or lymphatic vessel, the ones that are next to it going downstream, are listed in the *efferent to* slot, so B will be in the *efferent to* slot of A. Thus, a particular chain has in its *efferent to* slot all those chains which it is efferent to. The *afferent to* relationship is just the opposite, so that if A is afferent to B, that means B is upstream from A, or closer to the organ from which the lymphatics are draining. This means that B will appear in the *afferent to* slot of A.

For those entities that have a lymphatic drainage, it is a starting point for path tracing. Starting with an organ or organ part that has some element of the lymphatic system as its drainage, asking what each element is efferent to will produce the next downstream link in a pathway through the lymphatic system. If the representation of the lymphatic system is complete and consistent, following the "efferent to" relation from any instance of a lymphatic drainage should eventually end with either the thoracic duct or the right lymphatic duct. An automatic path generating function should be able to identify lymphatic paths that are consistent with this requirement, and ones that are not. Several kinds of problems could occur. One is that a path reaches a dead end, because a lymphatic chain has not had an "efferent to" relation entered for it. Another is that a loop is found, that is, an "efferent to" relation leads to a lymphatic chain that is upstream in the same path. A third possibility is that the type of entity that appears in the "efferent to" slot is not a lymphatic chain or lymphatic vessel. This third constraint should be enforceable by the frame system at the time of data entry, but in fact in the FMA there are several different partitionings of the lymphatic system, with some overlap.

4.3.5.2 The Lymphatic System: an Example

To actually perform consistency checking using the FMS interface described in Section 4.3.5, we will need some more abstraction layers so that we have some insulation from the lower level details. Ultimately the right abstraction layers would enable replacing the lowest layers with an interface to OKBC or to a Protégé plug-in facility. The following functions provide a thin layer to provide specific queries for anatomical entities and relationships pertaining to the lymphatic system.

```
(defun lymphatic-drainage (term)
  (get-children term "lymphatic drainage"))
```

Applying this function to an anatomic structure should produce the names of the lymphatic chains that drain it.

```
> (lymphatic-drainage "Soft palate")
("Superior deep lateral cervical lymphatic chain"
```

```
"Right retropharyngeal lymphatic chain"
"Left retropharyngeal lymphatic chain")
```

We may need to look up the parts of an entity because the lymphatic drainage information may not be contained in the object itself, but in some part or subdivision. Earlier we defined a function to fetch constitutional parts. Here, more likely, we will need regional parts.

```
(defun regional-parts (term)
  (get-children term "regional part"))
```

The `efferent-to` function returns the nodes or chains that are downstream from the entity specified by the lymphatic chain named `term`.

```
(defun efferent-to (term)
  (get-children term "efferent to"))
```

Similarly, `afferent-to` is an *upstream* relation, and can be used to generate trees rather than tracing paths back to the trunk. The `afferent-to` function returns the nodes or chains that are upstream from the entity specified by its input.

```
(defun afferent-to (term)
  (get-children term "afferent to"))
```

Now that we have convenient functions for access, for the downstream relation, and for obtaining starting points for a specific anatomical entity, its lymphatic drainage, we need a way to recursively traverse the lymphatic network and generate paths. This should be a straightforward application of the `path-search` function developed in Chapter 5. In this case, the `goal?` function, operating on a path, will need to check if there are any more next nodes, or if we have reached the end of the path (no more next nodes). It will have to use the `successors` function, so it will be a closure in a call to `path-search`. Here is `find-all-paths`, which takes a starting *path* and generates all the paths from it to their ends.

```
(defun find-all-paths (start successors extender merge)
  (path-search start
               #'(lambda (current) ;; goal - stop when no more
  nexts
                   (null (funcall successors current)))
               nil ;; get all the paths
               successors extender merge))
```

In order to apply this function to the lymphatic paths tracing problem, we need to provide starting paths, which we obtain for a named anatomic entity by calling the `lymphatic-drainage` function. We need a `successors` function. This will be a closure that uses `efferent-to` on the most recently generated node in the current path. We use `cons` as the extender function, which means that a path is a list of names of the lymphatic chains, and we put the most recent or new chain on the front of the list to get a new path. Finally, we can do depth first search here, since all the paths should terminate, either at the Thoracic duct or the Right lymphatic duct, so the merge function is just append. The find-all-paths function will be applied to each lymphatic drainage entry and will produce a list of paths. Since we want to apply it to all the lymphatic drainage entries and append the lists together, we use mappend, defined on page 124.

```
(defun lymphatic-paths (site)
  (mappend #'(lambda (start)
               (find-all-paths start
                               #'(lambda (path)
                                   (efferent-to (first path)))
                               #'cons #'append))
           (lymphatic-drainage site)))
```

Here is an example of how to use the lymphatic-paths function.

```
> (fms-connect))
> (setq results (lymphatic-paths "Soft palate"))
```

and here are the results. Each list represents a path, and its components are named in reverse order from the end back to the beginning.

```
(("Jugular lymphatic trunk"
  "Inferior deep lateral cervical lymphatic chain"
  "Superior deep lateral cervical lymphatic chain")
 ("Jugular lymphatic trunk" "Jugular lymphatic chain"
  "Superior deep lateral cervical lymphatic chain")
 ("Right lymphatic duct" "Right jugular lymphatic trunk"
  "Right inferior deep lateral cervical lymphatic chain"
  "Right superior deep lateral cervical lymphatic chain"
  "Right retropharyngeal lymphatic chain")
 ("Right lymphatic duct" "Right jugular lymphatic trunk"
  "Right superior deep lateral cervical lymphatic chain"
  "Right retropharyngeal lymphatic chain")
 ("Thoracic duct" "Left jugular lymphatic trunk"
  "Left inferior deep lateral cervical lymphatic chain"
  "Left superior deep lateral cervical lymphatic chain"
  "Left retropharyngeal lymphatic chain")
 ("Thoracic duct" "Left jugular lymphatic trunk"
  "Left superior deep lateral cervical lymphatic chain"
  "Left retropharyngeal lymphatic chain"))
```

There are two paths from each of the three lymphatic chains that drains the soft palate. So the lymphatic system is multiply connected, but at least so far, not cyclic.

Provided that the representation of the lymphatic system is consistent, and complete enough in coverage, this is a means by which the models described in Section 5.5.3 can be generated as needed. Altogether, it is a powerful clinical application of the FMA, the prediction of spread of tumor cells to regional metastases. This is a potential first step in automating the definition of the target for radiation treatment of cancer. In the following sections this and some other code to be developed shortly, will show how to do some consistency checking in the representation of the lymphatic system.

4.3.5.3 The State of the Lymphatics in the FMA

The following code and results are an illustration of how one might do semantic consistency checking in an ontology like the FMA. We specifically focus on the lymphatic system because it will be revisited in Section 5.5.3. Here are two specific requirements for consistency and completeness in the lymphatic system:

1. Every lymphatic chain should have `efferent-to` relations only to other lymphatic chains or to lymphatic vessels such as the Thoracic duct or the Right lymphatic duct.
2. Starting with any lymphatic drainage from an organ or organ part, all paths using the `efferent-to` relation should end up at either the Thoracic duct or the Right lymphatic duct.

The first requirement is a slightly more stringent requirement than the current FMA specifies in the "efferent to" relation, but it is necessary in order to get meaningful pathways. In the FMA specification of the "efferent to" relation, the following are allowed classes:

- Lymphatic chain
- Lymphatic vessel
- Anodal lymphatic tree
- Lymphatic plexus
- Lymph node

The Thoracic duct and Right lymphatic duct are lymphatic trunks, which are subclasses of Lymphatic vessel, so they should be allowed, along with other trunks. Anodal lymphatic trees are small networks of lymphatic vessels that directly drain structures.

The tests will be performed in a sequence of steps. Of course for fully automated checking these steps could easily be combined. First, for the lymphatic chains and vessels, we get all the instances, using the `all-subclasses` function previously defined.

```
> (setq allchains (all-subclasses "Lymphatic chain"))
("Pulmonary lymphatic chain"
 "Subdivision of pulmonary lymphatic chain"
 "Axillary lymphatic chain" "Subdivision of axillary lymphatic tree"
 "Posterior mediastinal lymphatic chain"
 "Tracheobronchial lymphatic chain"
 "Tributary of tracheobronchial lymphatic chain"
```

```
"Left cardiac tributary of tracheobronchial lymphatic chain"
"Brachiocephalic lymphatic chain"
"Right cardiac tributary of brachiocephalic lymphatic chain" ...)

> (setq allvessels (all-subclasses "Lymphatic vessel"))
("Variant lymphatic vessel" "Lymphatic capillary"
 "Tributary of lymphatic trunk" "Tributary of lymph node"
 "Superficial lymphatic vessel" "Deep lymphatic vessel"
 "Lymphatic trunk" "Incomplete right lymphatic duct"
 "Absent thoracic duct" "Absent cisterna chyli" ...)
```

Checking the lengths of the two lists that result, it appears there are 353 chains and 670 vessels. For each of the items in the two lists, we retrieve the values in the "efferent to" slot. This is easily done by writing a function to go through each list, get the "efferent to" contents, and pair it with its chain or vessel.

```
(defun get-efferents (terms)
  (mapcar #'(lambda (x)
              (list x (efferent-to x)))
          terms))
```

We apply it to both lists, giving the following results.

```
> (setq chain-slots (get-efferents allchains))
(("Pulmonary lymphatic chain" ("Bronchopulmonary lymphatic chain"))
 ("Subdivision of pulmonary lymphatic chain" NIL)
 ("Axillary lymphatic chain"
  "Subclavian lymphatic trunk" "Subclavian lymphatic tree"))
 ("Subdivision of axillary lymphatic tree" NIL)
 ("Posterior mediastinal lymphatic chain"
  ("Thoracic duct" "Tracheobronchial lymphatic chain"))
 ("Tracheobronchial lymphatic chain"
  ("Bronchomediastinal lymphatic trunk"
   "Bronchomediastinal lymphatic tree"))
 ("Tributary of tracheobronchial lymphatic chain" NIL)
 ("Left cardiac tributary of tracheobronchial lymphatic chain" NIL)
 ("Brachiocephalic lymphatic chain"
  ("Bronchomediastinal lymphatic trunk"
   "Bronchomediastinal lymphatic tree"))
 ("Right cardiac tributary of brachiocephalic lymphatic chain" NIL)
 ...)

> (setq vessel-slots (get-efferents allvessels))
(("Variant lymphatic vessel" NIL) ("Lymphatic capillary" NIL)
 ("Tributary of lymphatic trunk" NIL)
 ("Tributary of lymph node" NIL)
 ("Superficial lymphatic vessel" NIL) ("Deep lymphatic vessel" NIL)
 ("Lymphatic trunk" NIL) ("Incomplete right lymphatic duct" NIL)
 ("Absent thoracic duct" NIL) ("Absent cisterna chyli" NIL) ...)
```

Now we can just iterate through these lists checking if the entries match our criteria above. In every entry, the second item must be a list of only lymphatic chains or vessels, in which case we put it on a "good" list, or it is nil, meaning no information is entered, and we put it on a "not done" list, or it is some other things, in which case we put it on a "bad" list.

```
(defun check-efferents (termslots allowed)
  "termslots is a list of chains or vessels and their efferent to
   slot values. allowed is the complete list of allowed values"
  (let (good bad not-done)
    (dolist (chain termslots)
      (let ((efferents (second chain)))
        (cond ((null efferents) (push chain not-done))
              ((every #'(lambda (x)
                          (find x allowed :test #'string-equal))
                      efferents)
               (push chain good))
              (t (push chain bad)))))
    (list good bad not-done)))
```

Strictly speaking there is nothing in this function about the "efferent to" relation. It can be used to check any list of (term slot-value) pairs against a list of allowed values. Here are the results for the list of chains:

```
> (setq chain-checks (check-efferents chain-slots
                        (append allchains allvessels)))
((("Left submental lymphatic chain"
   ("Left submandibular lymphatic chain"
    "Left jugulo-omohyoid lymphatic chain"))
  ("Right submental lymphatic chain"
   ("Right submandibular lymphatic chain"
    "Right jugulo-omohyoid lymphatic chain")) ...)
 (("Left parasternal lymphatic chain"
   ("Left bronchomediastinal lymphatic tree"))
  ("Right parasternal lymphatic chain"
   ("Right bronchomediastinal lymphatic tree")) ...)
 (("Left level VI lymphatic chain" NIL)
  ("Right level VI lymphatic chain" NIL)
  ("Left level V lymphatic chain" NIL) ...))
```

We consider a slot good if its values come from either the chain list or the vessel list. Now, how many are there of each?

```
> (length (first chain-checks))
140
> (length (second chain-checks))
11
> (length (third chain-checks))
202
```

So, 140 of the chains are correct, 11 have some problem, and 202 are still not completed. If we look at the 11 that have a problem, it is apparent that these entries are mostly ones where a lymphatic tree has been entered. It seems problematic that a chain could be efferent to a subtree of the lymphatic system, although this is anatomically correct. A chain that is *part of* a tree connects to the tree and the lymphatic fluid flows from the chain into that branch of the tree. However, for path tracing, it is not useful, because the flow from that chain does not go through the whole tree, but only through a subset of branches. So this needs to be fixed in order to do sound path tracing. There are so few that it is relatively easy to fix this modeling problem.

```
> (pprint (second chain-checks))
(("Left parasternal lymphatic chain"
  ("Left bronchomediastinal lymphatic tree"))
 ("Right parasternal lymphatic chain"
  ("Right bronchomediastinal lymphatic tree"))
 ("Left tracheobronchial lymphatic chain"
  ("Left bronchomediastinal lymphatic trunk"
   "Left bronchomediastinal lymphatic tree"))
 ("Right tracheobronchial lymphatic chain"
  ("Right bronchomediastinal lymphatic trunk"
   "Right bronchomediastinal lymphatic tree"))
 ("Lymphatic chain of lower lobe of left lung"
  ("Left bronchopulmonary lymph node"))
 ("Lymphatic chain of lower lobe of right lung"
  ("Right bronchopulmonary lymph node"))
 ("Parasternal lymphatic chain"
  ("Bronchomediastinal lymphatic trunk"
   "Bronchomediastinal lymphatic tree"))
 ("Infraclavicular lymphatic chain"
  ("Subclavian lymphatic tree" "Apical axillary lymphatic chain"
   "Subclavian lymphatic chain"))
 ("Brachiocephalic lymphatic chain"
  ("Bronchomediastinal lymphatic trunk"
   "Bronchomediastinal lymphatic tree"))
 ("Tracheobronchial lymphatic chain"
  ("Bronchomediastinal lymphatic trunk"
   "Bronchomediastinal lymphatic tree"))
 ("Axillary lymphatic chain"
  ("Subclavian lymphatic trunk" "Subclavian lymphatic tree")))
```

Now the same checks can be performed on the vessels list, and the counts for the three categories come out as follows:

```
> (length (first vessel-checks))
13
> (length (second vessel-checks))
2
> (length (third vessel-checks))
655
```

On inspection of the two that are flagged as bad, the same problem appears. A tree has been entered where we would expect a trunk, vessel, or chain.

The second check, to determine whether all paths terminate at the Thoracic duct or Right lymphatic duct, is more challenging. The fact that many "efferent to" relationships are still not done makes it highly likely that there are many incomplete paths, so this check is probably premature. Nevertheless, it should be clear how to perform it. From the list of chains and vessels, one traces the paths, and examines their end points. Looking back to the example for the Soft palate, you can see that starting there, a path ends at the Jugular lymphatic trunk. However, there are two jugular lymphatic trunks, a right and a left. One goes to the thoracic duct and one goes to the right lymphatic duct. So it is correct to have no entry at the higher level of generality. This is yet another complication in checking consistency and completeness, as well as in applying the knowledge to clinical problem solving. There are many such places where anatomical structures are described as general classes, and then have more specific right and left instances (subclasses). This is very important, since geography is important and (especially with radiation therapy) one must specify on which side the entities of interest are. The presence of the general and right/left instances makes reasoning difficult to automate. There is room for some further innovation here.

So, none of the results above should be considered as deficiencies in the FMA. Rather, our checks have identified some further complexities in the model itself. Another possibility is that the path check may reveal circularities. The path tracing code described previously would not terminate in this case, and a different kind of check is needed, where one keeps track of nodes already visited and flags any returns to such nodes.

The query interface described here, while still operational, is peculiar to the FMA, and not well matched to standards and methods used widely in Semantic Web research and implementation. The FMA can also be represented as a graph structure using RDF [271], and a generalization of the SPARQL query language [426] has been developed to support the complex queries that are becoming important in medicine [90,375].

Anatomy, important as it is, is sometimes viewed as a "dead" subject, with little or nothing new to be discovered. However, despite several thousand years of dissection, study, and development of terminology, it seems that much remains to be done. The FMA project, with its goal to develop a consistent computational theory of anatomy, is one of the most ambitious ontology building projects in biomedical informatics. We have learned from it that the expressivity of the metaclass idea, and the representation of relations themselves as entities, are both key elements of complex biomedical theories. Another very important realization comes out of this work as well. Biomedical informatics is not just straightforward application of well-established computer science methods. As Mark Musen has said, "ours is the discipline that cares about the content" [291]. Biomedical informatics gives formal shape to biomedical content, and in the struggle to get it right, feeds back to computer science and information science new ideas and challenges. This is exactly parallel to the relation between theoretical physics and mathematics.

4.4 Summary

Some key general ideas are evident from the development in this chapter, as well as some points specific to biological and medical information storage and retrieval systems.

- Biomedical Information Access is more about understanding how people think, express their needs, and what is useful, than it is about models of the actual information content, though content models help.
- Terminology systems provide assistance with searching and organizing information, even though they do not meet the requirements for sound inference systems.

For a deeper and more detailed treatment of the concepts and issues around the design and use of information retrieval systems there are several excellent books:

- Baeza-Yates [22] is a very readable textbook covering many more topics than we touch on here, including a mathematical treatment of information retrieval models and query processing.
- Grossman [140] includes mathematical details for algorithms and a substantial treatment of integration of IR and database systems.
- Hersh [155] develops further biological and medical examples, as well as application of natural language processing in IR, including his early work on mapping concepts in text to terminology systems such as MeSH [156].
- Meadow et.al. [278], is a more recent addition, complementing the above references.

4.5 Exercises

1. Writing useful utility functions

   The dot product of two vectors is defined as the sum of the pairwise products of their components, so for example, two three-dimensional vectors in space having components A_x, A_y, A_z and B_x, B_y, B_z would have a dot product, $A \cdot B = A_x B_x + A_y B_y + A_z B_z$ and in general if the components of a vector are indexed numerically, the dot product can be written as $A \cdot B = \sum_i A_i B_i$. Implement dot-product and rewrite the code for similarity in terms of it.

2. Adding attributes to tags

   Rewrite the function `path-query` in Section 4.3.4 so that it can find entries in the nested list structure described there, even if the tag for an entry is a complex tag, with attributes.

3. Finding all the matches

   Augment the function `path-query` in Section 4.3.4 so that it can find all matching entries for a path specification, not just the first match.

4. Wilder and wilder

   Augment the function `path-query` in Section 4.3.4 so that it can handle "wild-card" path specifications that use the * symbol to mean "anything matches here."

5. The Drug Interaction Protocol: a project

   The drug interaction protocol described here is one possible scheme by which a client program can query a drug interaction checker running as a network server.

   The server waits for an incoming connection on a configurable port number. As each message comes in, it processes the message, sends a reply, and waits for the next message. When the QUIT message is received, the server closes the connection after responding, and waits for the next incoming connection.

   The client sends requests, reads the returned results, and continues in this loop until the user decides to terminate the program. At that point the client sends a QUIT message, reads the response, and then closes the connection and exits.

   The server recognizes the following messages:

   - HELLO <client-id>
     <client-id> is a quoted string identifying the client, followed by the client's IP address in octet notation (also a quoted string), for example,

     ```
     HELLO "Drugscape 1.3" "206.122.23.134"
     ```

   The server responds with a numeric message ID, a quoted string including its identification, the server's IP address as above, and a greeting as in the following example:

     ```
     204 "DIS 1.1" "128.95.181.5" "Pleased to meet you"
     ```

   If the message from the client is incorrectly formed, the server responds:

     ```
     401 "Error - malformed message"
     ```

   If the client is unknown, the server has the option to reject the message and close the connection.

     ```
     402 "Error - client not authorized"
     ```

   - FIND <drug-name>

In this message, <drug-name> is an unquoted string. If the drug of that name has an entry in the database, the server responds with the drug interaction record on this drug,

```
206 <drug-name> <effect-list> <interactions-list>
```

where the two lists are Lisp lists in the form in the original drug-db-data, enclosed in parentheses, containing symbols. If the drug is not found, the server sends,

```
404 "Error - <drug-name> not found."
```

- INTERACTS <drug1> <drug2>
  In this message, <drug1> and <drug2> are symbolic names of drugs, presumably in the database. If the lookup is successful, the server responds with either of the following two messages:

  ```
  207 <interactions-1> <interactions-2>
  208 "no interactions"
  ```

  where <interactions-1> is a list of symbolic names of effects of drug 1 that are on the interactions list of drug 2, and <interactions-2> is the corresponding list for drug 2. If either drug has no entry in the database, the server responds with the drug not found message above (404).

- QUIT
  The server responds with 240 Goodbye and closes the connection.

The client will have to generate these messages when appropriate, and be able to recover if an ill-formed or inappropriate response is returned by the server. Similarly, the server will have to be able to recover if an ill-formed or out of order message is received.

As with all specifications, formal or otherwise, this one may need additional clarifications. Modify or augment the specification so that it is sufficiently detailed and unbiguous that a programmer can reliably implement a client or server program supporting the protocol.

6. Now write the code

Use the socket library code (the file sockets.cl available at the book web page URL http://faculty.washington.edu/ikalet/pbi/) to write a network client or server program. Make the implementation conform to the DIP specification above. The additional code may be a few pages, but should not be more than ten. Client programs should use a text only interface, that is, prompting in a terminal window for keyboard input. The server code should use the drug interaction database code also available at the book web page.

CHAPTER 5

Computational Models and Methods

Chapter Outline

Principles of Biomedical Informatics. http://dx.doi.org/10.1016/B978-0-12-416019-4.00005-6

Earlier chapters show that much can be done with suitable representations of data and knowledge, to support inference, probabilistic modeling and prediction, and information organization. This chapter introduces additional methods for deriving useful results from data and for creating complex models of biological processes. These methods include: search through data suitably organized, as sequences, or as networks, natural language processing, and modeling with state machines.

5.1 Computing with Genes, Proteins, and Cells

Molecular biology has become a rich area for applications of computing ideas. Much of the focus has been on matching and searching representations of gene and protein sequences, to identify new genes and proteins, to relate functions of different genes and proteins, and to augment the annotations (the properties of each gene or protein). Some of this work involves organizing gene and protein knowledge into computable structures. The biology is itself fundamentally a computational process, that can be modeled by computer programs and data. In this section we look at some of these applications, to illustrate how symbols and structures can capture biological knowledge in a useful way. The main point here is that much useful information is in the annotations, so we make them into computable data and knowledge, in addition to the sequences themselves.

A wide range of software tools are available to do these computations. A good survey of such tools is a book edited by Baxevanis [24]. Another useful reference [428] focuses specifically on proteomics. We present only a small sampling of these operations.

5.1.1 Mapping Genes to Proteins

In Chapter 1 we developed a basic representation or encoding of sequence data. Using this representation, we can emulate the basic machinery of the gene to protein process. If that can be automated, we can then make point alterations in genes and compute what effects such alterations might have on the proteins corresponding to those genes. In order to do this we need to be able to transcribe a DNA sequence into an RNA sequence (easy), splice out the parts of the genomic DNA that do not code for the protein, producing mRNA (very hard), translate the mRNA into a protein (easy), and finally look up domain information and other properties to see if the alteration caused a significant change in the protein structure or function (somewhat hard). Here we will develop some simple code to do some of the more straightforward mapping steps from DNA to RNA and from mRNA to proteins.

The transcription of a DNA sequence into an RNA sequence requires that we be able to generate a new sequence from an old sequence, using a transformation function that changes a DNA base to the complementary RNA base. The `mapcar` function transforms a sequence into a new sequence and all we need to do is provide the DNA to RNA transformer. There are two ways to do this. One is to model the actual biological process by computing the complementary DNA from the coding DNA and then computing the complementary RNA corresponding to that DNA. The other is to just transcribe directly, making a copy of the coding DNA except changing each T to a U. They are both simple; we will do the direct method here.

```
(defun dna-to-rna-base (base)
  "for any dna base, returns the corresponding rna base"
  (if (eql base 't) 'u base))
```

```
(defun dna-to-rna (seq)
  "produces an rna sequence corresponding to the dna sequence seq"
  (mapcar #'dna-to-rna-base seq))
```

Then, if dna-seq is a sequence of symbols corresponding to one strand of DNA, for example, the sequence shown earlier, we can do the transcription as follows (first we set the value of dna-seq to a list of the first few bases in the BRCA1 excerpt from Chapter 1):

```
> (setq dna-seq '(C A C T G G C A T G A T C A G G A C T C A))
(C A C T G G C A ...)
> (setq mrna-seq (dna-to-rna dna-seq))
(C A C U G G C A U G A U C A G G A C U C A)
```

However, the transcription from DNA to RNA so far does not account for the fact that some parts of the RNA (or corresponding DNA) are non-coding regions, or introns. This phenomenon is found in eukaryotic cells, but not in general in prokaryotic cells. So, in eukaryotic cells, before translation takes place, the introns must be spliced out or removed from the RNA produced from a gene. Then the resulting mRNA can be used by ribosomes to translate the sequence into a protein. How does the cell's transcription machinery know where to splice out the introns? Some features appear to be common to introns that can help recognize them. However, the algorithmic recognition of introns remains a largely unsolved problem. Given a gene's DNA sequence and the amino acid sequence of a protein that gene is known to code for, it is more tractable to search for a match between the two via an mRNA assignment. Several projects have developed algorithms and programs to do that. Two examples are the SIM4 project,[1] which does alignments between mRNA and corresponding genomic sequences, taking into account signals that are usually associated with the beginning and end of an intron, and Spidey,[2] a similar project hosted at NCBI.

From this point we assume that we have a sequence of symbols representing a (single) strand of mRNA. This sequence will contain the symbol U, where there was a T in the DNA (as described previously, we are assuming that the DNA is complemented and the RNA is generated from that). The next step would be to translate the codons into amino acids, giving a sequence representation of a protein coded by the gene. The protein sequence can be searched to see what domains it contains, or if it is a known protein, it can be examined to see if an important domain is missing or mutated.

---

[1] http://pbil.univ-lyon1.fr/members/duret/cours/inserm210604/exercise4/ sim4.html.
[2] http://www.ncbi.nlm.nih.gov/spidey/.

To translate the sequence we apply another transformation function that takes a codon, a three base sequence, and returns the corresponding amino acid symbol. This is simply a lookup in a table, arranged as an inverse of Table 1.5 on page 39. Instead of a linear table, we arrange it as a three-dimensional array, where each dimensional index can have values from 0 to 3, depending on which base is in which position. This is therefore a $4 \times 4 \times 4$ table, with a total of 64 entries, one for each possible three base sequence.

```
(defconstant +rna-to-amino-table+
    (make-array '(4 4 4) :initial-contents
                    '(((gly gly gly gly)
                       (glu glu asp asp)
                       (ala ala ala ala)
                       (val val val val))
                      ((arg arg ser ser)
                       (lys lys asn asn)
                       (thr thr thr thr)
                       (met ile ile ile))
                      ((arg arg arg arg)
                       (glu glu his his)
                       (pro pro pro pro)
                       (leu leu leu leu))
                      ((trp STOP cys cys)
                       (STOP STOP tyr tyr)
                       (ser ser ser ser)
                       (leu leu phe phe)))))
    "The genetic code table")
```

In this arrangement, the bases correspond to array indices: G is 0, A is 1, C is 2, and U is 3. So the codon GCC would be array element (0,2,2) and the codon AGU would be the array element (1,0,3). This table, +rna-to-amino-table+, has its contents organized inversely to Table 1.5 in Chapter 1. The two tables are not strictly inverses of each other, since the mapping of a codon to an amino acid is unique, but the mapping of an amino acid to a codon is in many cases multi-valued, that is, for a given amino acid, there may be several codons that correspond to it.

```
> (aref +rna-to-amino-table+ 0 2 2)
ALA
> (aref +rna-to-amino-table+ 1 0 3)
SER
```

We can use a lookup table to convert base symbols to indices, so that the higher level code need not depend on how we organized the mapping array. The rna-index function takes a base symbol and returns the index, which will be used as an array index in the `rna-to-amino` function. The indices are not needed for anything else, so we use a `let` expression to create a local variable for the `rna-index` function to use. The `rna-indices` list will be retained through all invocations of `rna-index` but cannot be accessed or changed by any other code, so this method is a very safe way to create and protect local information.

```
(let ((rna-indices '((g 0) (a 1) (c 2) (u 3))))
   (defun rna-index (base)
      (second (assoc base rna-indices)))))
```

The `rna-to-amino` function then just does a lookup in the table, using the indices corresponding to the three base symbols input to it. The same information hiding idea using a `let` expression can be used here to make the `rna-to-amino-table` private. After evaluating the `let` expression, the previously defined global constant can be set to `nil`, or the `make-array` expression can just be included here with no global variables at all, even temporary.

```
(let ((rna-to-amino-table +rna-to-amino-table+))
   (defun rna-to-amino (b1 b2 b3)
      "takes three bases and returns the corresponding amino acid"
      (aref rna-to-amino-table
            (rna-index b1)
            (rna-index b2)
            (rna-index b3))))
```

The codons illustrated above would then be used with `rna-to-amino` as follows:

```
> (rna-to-amino 'g 'c 'c)
ALA
> (rna-to-amino 'a 'g 'u)
SER
```

The symbols are quoted because we need the symbols themselves as values; we are not using the symbols as variables with values assigned to them. Normally, `rna-to-amino` would be used within a larger function, so we would not be typing symbols with quotes, but would be passing values from some sequence of symbols.

Finally, to translate an entire sequence of mRNA starting from some particular place in the sequence, we have to process it three bases at a time and use `rna-to-amino` on each set of three. Rather than deal with tracking an index into the sequence, we just take the first three bases, make them into a triple, translate that, and put it on the front of the result of recursively processing the remainder of the sequence.

```
(defun rna-translate (mrna-seq)
   "takes an mrna sequence and converts it to a polypeptide,
    a sequence of amino acids"
```

```
(cond ((null mrna-seq) nil)
      ((< (length mrna-seq) 3) nil)
      (t (cons (rna-to-amino (first mrna-seq)
                             (second mrna-seq)
                             (third mrna-seq))
               (rna-translate (nthcdr 3 mrna-seq))))))))
```

Here is an illustration of the application of `rna-translate` to the example mRNA sequence (mrna-seq) generated previously.

```
> (rna-translate mrna-seq)
(HIS TRP HIS ASP GLU ASP SER)
```

Recall that the sugar-phosphate backbone is not symmetric. The end where the first phosphate connects to the $5'$ carbon in the sugar (see Figure 1.9 on page 37) is called the $5'$ end of the mRNA sequence, and the other end is the $3'$ end. Translation always proceeds from the $5'$ to the $3'$ end. For this direction, there are three choices, since each codon is made up of three bases. The translation can start at the end, one base in and two bases in from the end. Each of these choices determines a *reading frame*. The translation will terminate when a stop codon is reached. A given piece of mRNA could have several stop codons. The subsequence found between two successive stop codons is an *open* reading frame. In any given mRNA sequence, there may be many open reading frames, so before doing the translation, it is necessary to scan the sequence to find the stop codons, and construct open reading frames from that information. The search for stop codons should be done in each of the three reading frames. For any three bases we can do a simple lookup to determine if these three bases form a stop codon.

```
(defun is-stop (b1 b2 b3)
  (eql (rna-to-amino b1 b2 b3) 'stop))
```

Using the `is-stop` function, we can check the first three bases in a sequence. If they form a stop codon, add the location and a number designating which ORF it is in. Similarly, check if they form a stop codon when read in reverse, and add that to the list. Then move one base down the sequence, incrementing the position counter, and repeat, until there are less than three bases left in the remainder of the sequence. This is the basis for defining a function `find-stops` whose input is a mRNA sequence and whose output is a list of the stop codon locations. Each location is a sequence position and a number, 0, 1, or 2, to indicate which reading frame it belongs to. In the usual style, we use `labels` to define a local tail-recursive function to do the work. It takes an accumulator as input, which will start out empty, and a position, pos, to keep track of where in the sequence the function is looking (the current position in the sequence, where we are checking for stop codons). The `find-stops` function just calls this local function starting at the beginning, at position 0.

```
(defun find-stops (mrna-seq)
  (labels ((fs-local (seq pos accum)
```

```
                   (if (< (length seq) 3)
                       (reverse accum) ;; no complete codon left so done
                   (cond ((is-stop (first seq) (second seq) (third seq))
                          (fs-local (rest seq) (1+ pos)
                                    (cons (list pos (mod pos 3))
                                          accum)))
                         (t (fs-local (rest seq)  (1+ pos) accum))))))))
           (fs-local mrna-seq 0 nil)))
```

The result is a list of stop codon locations. Each location is a pair of numbers. The first number is the sequence index or base at which the stop codon is located, and the second number is either 0, 1 or 2, to indicate in which of the three reading frames the stop codon appears.

Here is the result of applying the `find-stops` function to the mRNA resulting from transcription of the DNA fragment from BRCA1 in Chapter 1.

```
> (setq stops (find-stops (dna-to-rna brca1)))
((8 2) (30 0) (46 1) (73 1) (109 1) (124 1) (130 1) (161 2))
```

Now that we have a list of where the stop codons occur, the next step would be to organize them into descriptions of the open reading frames. This enables us finally to select subsequences of the mRNA that will translate into polypeptides. Each reading frame will give a different set of polypeptides. The complexities of how this happens, the role of promoter regions, and more details of the translation process can be found in molecular biology texts [10].

5.1.2 Computing with Protein Sequences

Some simple properties of proteins can be predicted, or computed, from the primary sequence of the protein and known properties of the various amino acids. Many software tools exist to aid biological researchers in performing these calculations. The purpose of this section is to illustrate how the data can be represented and how the calculations could be performed. A big value of the tools (see for example the ExPASy web site, `http://www.expasy.ch/ tools/`, of the Swiss Institute of Bioinformatics) is the fact that they look up data in existing bioinformatics databases such as Swiss-Prot, and they provide a nice World Wide Web interface for specifying or uploading the protein sequence to analyze. In addition to the web site documentation, some of these tools are described in a chapter [124] in the proteomics book mentioned earlier [428]. Here we just focus on the computations and we leave the database access and user interface issues for later.

5.1.2.1 Representing Amino Acids and Proteins

Certainly, the representation of a protein sequence as a list of amino acid symbols is sufficient to start with, since the calculations we will do only depend on the primary structure. However, we will need some information about the amino acids, so we need a way to associate amino acid specific data with each amino acid symbol. Thinking in terms of object-oriented modeling, it is reasonable to define a class, `amino-acid`, of which each of the 20 known amino acids will be instances. The class definition will include a slot for each important property of the amino acid. These include: the molecular weight, volume, a pK value, and the hydrophobicity. We also include the one and three letter abbreviations here for convenience.

```
(defclass amino-acid ()
  ((name :accessor name
         :initarg :name
         :documentation "Symbol, the full name")
   (abr1 :accessor abr1
         :initarg :abr1
         :documentation "Symbol, single letter abbreviation")
   (abr3 :accessor abr3
         :initarg :abr3
         :documentation "Symbol, three letter abbreviation")
   (mass :accessor mass
         :initarg :mass
         :documentation "Average molecular mass in Daltons")
   (volume :accessor volume
           :initarg :volume)
   (surface-area :accessor surface-area
                 :initarg :surface-area)
   (part-spec-vol :accessor part-spec-vol
                  :initarg :part-spec-vol)
   (pk :accessor pk
       :initarg :pk
       :documentation "Ionization constant")
   (hydrophobicity :accessor hydrophobicity
                   :initarg :hydrophobicity)))
```

The mass, also called molecular weight, is that of the amino acid unit as it appears in a polypeptide, not as a free molecule, so it is roughly the sum of the atomic weights of all the atoms in the chemical formula, less one water molecule (a Hydrogen from the NH_2 group and a Hydroxyl (OH) from the COOH group). The average mass, or molecular weight, as listed at the ExPASy tools web site[3] will be used when we make each amino acid instance. The non-ionizing amino acids will have their pK set to 0.

Here is the definition of +amino-acids+, in which all are created and put on the list.

```
(defconstant +amino-acids+
    (list
     (make-instance 'amino-acid
```

---

[3] http://www.expasy.ch/tools/.

```
    :name 'Alanine :abr1 'A :abr3 'ala
    :mass 71 :volume 88.6 :surface-area 115
    :part-spec-vol .748 :hydrophobicity .5)
(make-instance 'amino-acid
    :name 'Arginine :abr1 'R :abr3 'arg
    :mass 156 :volume 173.4 :surface-area 225
    :part-spec-vol .666 :hydrophobicity -11.2)
(make-instance 'amino-acid
    :name 'Asparagine :abr1 'n :abr3 'Asn
    :mass 114 :volume 117.7 :surface-area 160
    :part-spec-vol .619 :hydrophobicity -.2)
(make-instance 'amino-acid
    :name 'Aspartate :abr1 'd :abr3 'asp
    :mass 115 :volume 111.1 :surface-area 150
    :part-spec-vol .579 :hydrophobicity -7.4)
(make-instance 'amino-acid
    :name 'Cysteine :abr1 'c :abr3 'cys
    :mass 103 :volume 108.5 :surface-area 135
    :part-spec-vol .631 :hydrophobicity -2.8)
(make-instance 'amino-acid
    :name 'Glutamine :abr1 'Q :abr3 'Gln
    :mass 128 :volume 143.9 :surface-area 180
    :part-spec-vol .674 :hydrophobicity -.3)
(make-instance 'amino-acid
    :name 'Glutamate:abr1 'E :abr3 'glu
    :mass 129 :volume 138.4 :surface-area 190
    :part-spec-vol .643 :hydrophobicity -9.9)
(make-instance 'amino-acid
    :name 'Glycine :abr1 'G:abr3 'gly
    :mass 57 :volume 60.1 :surface-area 75
    :part-spec-vol .632 :hydrophobicity 0)
(make-instance 'amino-acid
    :name 'Histidine :abr1 'H :abr3 'his
    :mass 137 :volume 153.2 :surface-area 195
    :part-spec-vol .67 :hydrophobicity .5)
(make-instance 'amino-acid
    :name 'Isoleucine :abr1 'I :abr3 'ile
    :mass 113 :volume 166.7 :surface-area 175
    :part-spec-vol .884 :hydrophobicity 2.5)
(make-instance 'amino-acid
```

```
     :name 'Leucine:abr1 'L :abr3 'leu
     :mass 113 :volume 166.7 :surface-area 170
     :part-spec-vol .884 :hydrophobicity 1.8)
  (make-instance 'amino-acid
     :name 'Lysine :abr1 'K :abr3 'lys
     :mass 128 :volume 168.6 :surface-area 200
     :part-spec-vol .789 :hydrophobicity -4.2)
  (make-instance 'amino-acid
     :name 'Methionine :abr1 'M :abr3 'met
     :mass 131 :volume 162.9 :surface-area 185
     :part-spec-vol .745 :hydrophobicity 1.3)
  (make-instance 'amino-acid
     :name 'Phenylalanine :abr1 'F :abr3 'phe
     :mass 147 :volume 189.9 :surface-area 210
     :part-spec-vol .774 :hydrophobicity 2.5)
  (make-instance 'amino-acid
     :name 'Proline :abr1 'P :abr3 'pro
     :mass 97 :volume 122.7 :surface-area 145
     :part-spec-vol .758 :hydrophobicity -3.3)
  (make-instance 'amino-acid
     :name 'Serine :abr1 'S:abr3 'ser
     :mass 87 :volume 89.0 :surface-area 115
     :part-spec-vol .613 :hydrophobicity -.3)
  (make-instance 'amino-acid
     :name 'Threonine :abr1 'T :abr3 'thr
     :mass 101 :volume 116.1 :surface-area 140
     :part-spec-vol .689 :hydrophobicity .4)
  (make-instance 'amino-acid
     :name 'Trypophan :abr1 'W :abr3 'trp
     :mass 186.21 :volume 227.8 :surface-area 255
     :part-spec-vol .734 :hydrophobicity 3.4)
  (make-instance 'amino-acid
     :name 'Tyrosine :abr1 'Y:abr3 'tyr
     :mass 163 :volume 193.6 :surface-area 230
     :part-spec-vol .712 :hydrophobicity 2.3)
  (make-instance 'amino-acid
     :name 'Valine :abr1 'V :abr3 'val
     :mass 99 :volume 140.0 :surface-area 155
     :part-spec-vol .847 :hydrophobicity 1.5)
  ))
```

To do computations of protein properties that depend on the properties of individual amino acids, we will need to be able to look up these amino acid instances. One way to do this is to make a hash table that is indexed by all the ways of expressing amino acid names, including the full name, the one letter code, and the three letter code. The following code creates the hash table and defines a function, lookup-amino-acid, that carries a reference to the table, even though it is not a global variable or accessible outside this function.[4]

```
(let ((amino-acid-table (make-hash-table)))
  (dolist (amino-acid +amino-acids+ 'done)
    (setf (gethash (name amino-acid) amino-acid-table)
      amino-acid)
    (setf (gethash (abr1 amino-acid) amino-acid-table)
      amino-acid)
    (setf (gethash (abr3 amino-acid) amino-acid-table)
      amino-acid))
  (defun lookup-amino-acid (aa-name)
    (gethash (if (stringp aa-name)
                 (read-from-string aa-name nil nil)
               aa-name)
      amino-acid-table)))
```

Having this lookup function, we can write sequences in terms of name symbols, including all three types of amino acid names. Then, when we need properties of particular amino acids, we just look up the amino acid instance by name in the table, and get the slot data with the usual accessor functions.

5.1.2.2 Simple Properties of Proteins from Sequence Data

Having data for amino acids, some properties of proteins (polypeptides in general) can be calculated from the primary sequence in a straightforward way. A variety of software tools have been developed and deployed to do these calculations, using various protein sequence knowledge bases. One such is the ExPASy web site, mentioned in Chapter 1. Examples of such calculations include the molecular mass of a protein, the isoelectric point, the optical extinction coefficient, and the amino acid composition. These are not only useful by themselves but also as components of a profile for a protein. It is possible to identify close matches for unknown amino acid (polypeptide) sequences with known proteins by comparing their amino acid profiles rather than the more expensive comparison through sequence alignment and matching. Such tools are also provided at ExPASy. In this section we show how to implement these calculations by reading sequence data from FASTA files downloadable from ExPASy and elsewhere.

The molecular mass (in Daltons) of a polypeptide (protein) can easily be computed, by summing the mass of each amino acid in the sequence and adding one water molecule for the

---

[4]Larry Hunter kindly contributed this code.

hydrogen and hydroxyl group at the ends. The `mass` function has already been defined as a generic function by declaring it as an accessor for the `mass` slot in the `amino-acid` class. We can define additional methods for other types of inputs. For a polypeptide, we add the mass of its sequence to the mass of a water molecule. To compute the mass of a sequence, we add up the masses of each of the elements of the sequence. The method here actually applies to any list, but we are assuming it will only be used on lists of amino acid symbols. Each element is a symbol referring to an amino acid, so we look up the actual amino acid instance in the hash table created earlier, using `lookup-amino-acid`, and get its mass (with the accessor declared in the class definition).

```
(defconstant +mass-water+ 18.0)

(defmethod mass ((pp polypeptide))
  (+ (mass (seq pp)) +mass-water+))

(defmethod mass ((seq list))
  (apply #'+
    (mapcar #'(lambda (x) (mass (lookup-amino-acid x)))
            seq)))
```

Here is the computation of the extinction coefficient of a polypeptide and its absorbance (optical density) following the formula described in the documentation for the ProtParam tool [428].

```
(defconstant ext-tyr 1490)
(defconstant ext-trp 5500)
(defconstant ext-cys 125) ;; this is actually cystine, not cysteine

(defun extinction-coeff (seq)
  (+ (* (count 'tyr seq) ext-tyr)
     (* (count 'trp seq) ext-trp)
     (* (count 'cys seq) ext-cys)))

(defun absorbance (seq)
  (/ (extinction-coeff seq) (mass seq)))
```

5.1.2.3 Amino Acid Profiles

Proteins can sometimes be uniquely identified by the profile of percentages of occurrence of the various amino acids. This is also easy to compute from the polypeptide sequence. We use the general `item-count` function presented in Chapter 1 on page 46. The `profile` function takes a sequence that is a list of name symbols of the amino acids in a polypeptide or protein, and returns a table with an entry for each symbol that occurs in the sequence. Each entry has three elements. The first element is the name, the symbol that is used in the sequence representation. The second element is the number of occurrences of that symbol in the sequence, and the third is the corresponding frequency. The call to `item-count` generates the basic profile and the anonymous function used by `mapcar` just adds the percentage after each name and count.

```
(defun profile (seq)
  "returns a profile table for any sequence of items - each table
  entry has the item, the occurrence count and the corresponding
  percentage or frequency."
  (let ((size (length seq)))
    (mapcar #'(lambda (x)
                (list (first x) ;; the name
                      (second x) ;; the count
                      (/ (second x) size))) ;; the percentage
            (item-count seq))))
```

The profiles produced in this way are hard to use for direct comparison, since the order of amino acids may be different for different sequences. Also in some profiles, amino acids may be absent, when what is needed for comparison purposes is an entry with a 0 count. We can remedy this most simply by providing functions to do additional processing.

To sort the profile list, just use the built-in Common Lisp `sort` function, with appropriate comparison and key parameter. Here are two examples. To sort by percentage, use the numerical comparison `<`, and the lookup key function `third`, to reference the percentages. To alphabetize by the single letter code or name, use the comparison function `string-lessp`, and a lookup key function that combines `symbol-name` with `first`, to reference the names as strings. This use of `sort` works for all three kinds of names, the one letter, three letter, and full name versions.

```
(defun sort-profile-by-percent (profile)
  (sort (copy-list profile) #'< :key #'third))
```

```
(defun sort-profile-by-name (profile)
  (sort profile #'string-lessp
        :key #'(lambda (x) (symbol-name (first x)))))
```

Two software tools available at ExPASy use amino acid composition to identify proteins. One, `AACompIdent`, takes as input one of a variety of amino acid profiles (percent composition of some subset of the 20 amino acids), and returns a list of sequences in Swiss-Prot, ranked according to the closeness of match. Closeness of match is computed between the unknown and each database entry as the sum of the squared difference for each amino acid in the profile. One of the profiles is the entire list of amino acids (constellation 0). The others omit some amino acids and in some cases combine pairwise Asparagine with Aspartic Acid as a single number, and Glutamine with Glutamic Acid also as a single number. The code above can generate the complete profile for an unknown sequence, and then you can just omit some entries (and combine others) to get the various constellations. Computing the mean square sum is easy.

```
(defun sum-square-diff (profile-1 profile-2)
  "Each profile is an alist sorted the same way. The sum of the
  squared differences of the values is returned."
```

```
(apply #'+ (mapcar #'(lambda (x y)
                       (expt (- (third x) (third y)) 2))
                    profile-1 profile-2)))
```

As noted earlier, before using this function, of course each profile needs to be sorted by the amino acid code, and any that were not present need to have a 0 entry added. This is easy too. The function `make-full-profile` goes through a list of symbols that are the amino acid codes, here defaulting to `+aa-symbols+`, adding a 0 entry for any not already present, and then returns the augmented list, sorted using `sort-profile-by-name`.

```
(defconstant +aa-symbols+
    (list 'a 'b 'c 'd 'e 'f 'g 'h 'i 'k 'l 'm
          'n 'p 'q 'r 's 't 'u 'v 'w 'x 'y 'z))

(defun make-full-profile (profile &optional (names +aa-symbols+))
    "adds any not found in sequence and sorts by name"
    (let ((full-profile (copy-list profile)))
        (dolist (name names)
          (unless (find name profile :key #'first)
            (push (list name 0 0) full-profile)))
        (sort-profile-by-name full-profile)))
```

To use `sum-square-diff`, apply `make-full-profile` to each of the two inputs, `profile-1` and `profile-2`. The sorting operation can easily be incorporated into a wrapper function for convenience. Alternately the `make-full-profile` function could be used when generating the profiles initially, so that the profiles in the knowledge base would already be complete and have their values sorted.

The `AACompIdent` tool can also select entries by keywords and incorporate other numerical parameters (like molecular mass and isoelectric point as computed above). Implementation of these ideas is left as an exercise for the reader.

How could the Swiss-Prot knowledge base handle this efficiently? Certainly it would not be efficient to go through the records each time and do the profile computation for each sequence in the knowledge base. It would be a good idea to precompute the full amino acid profiles once for every entry in the knowledge base, then just use this profile database instead of the original sequence data.

To create the profile reference file, we get a FASTA format version of Swiss-Prot, and process it with the following code. It uses the `read-fasta` function defined on page 52. We have parametrized the allowable codes and the function for obtaining the accession number, so this function can also be used with DNA sequences.

```
(defun fasta-profiler (filename &optional (codes +aa-codes+)
                                          (ac-parser #'sp-parser))
    (let ((profiles '())
          (seq nil)
          (n 0))
```

```
(with-open-file (strm filename)
  (loop
    (setq seq (read-fasta strm codes nil ac-parser))
    (if seq
        (progn
          (format t " ~&Record ~A accession number ~A~%"
                    (incf n) (first seq))
          (push (list (first seq)
                      (profile (second seq)))
                profiles))
      (return profiles))))))
```

Unlike the original sequence data, the profile data have a completely regular structure and would logically go into a table in a relational database. While this is a natural inclination for the typical programmer with database background, it may not really be the most efficient for use in something like the AACompIdent tool. The computation to be performed will use *every* record in the profile database, and the order does not matter, as the *results* have to be sorted, not the inputs. In this case, it is likely more efficient to simply read records from a sequential file or a list in memory and process them one by one, accumulating a list of results to be sorted by a single call to the sort function.

The Uniprot/Swiss-Prot FASTA file contains about 276,000 entries. All the entries can be read and profiles generated without exceeding the memory limits on my aging Linux laptop.[5] Here is the timing result for running the fasta-profiler function on this file.

```
; cpu time (non-gc)  439,090 msec (00:07:19.090) user, 6,320 msec system
; cpu time (gc)    1,420,290 msec (00:23:40.290) user, 5,900 msec system
; cpu time (total) 1,859,380 msec (00:30:59.380) user, 12,220 msec system
```

However, so much memory is required that it is difficult to do any other computations. To use this data, the profiles should be stored (saved) in a file, and each profile record retrieved as necessary. The profile calculation only needs to be repeated when new data become available, such as new or revised protein or other polypeptide sequences, each time a new version of Swiss-Prot becomes available. In fact, a version that reads a record, processes it and writes the results, would use very little memory, and likely run faster overall because the garbage collection time (gc) will be far less. Here is a version that does this.

```
(defun fasta-profiler-to-file (&optional (infile *standard-input*)
                                         (outfile *standard-output*)
                                         (codes +aa-codes+)
                                         (ac-parser #'sp-parser))
  (let ((seq nil)
        (n 0))
    (with-open-file (in-strm infile :direction :input)
      (with-open-file (out-strm outfile :direction :output)
        (loop
```

---

[5] A Sony VAIO PCG-V505BX, with 1 GB of memory, running Allegro Common Lisp™ Version 8.1 (as of March, 2008)

```
                (setq seq (read-fasta in-strm codes nil ac-parser))
                (if seq
                    (progn
                      (format t "~&Record ~A accession number ~A~%"
                                 (incf n) (first seq))
                      (pprint
                        (list (first seq)
                               (profile (second seq)))
                        out-strm))
              (return n)))))))
```

Here is the timing result for this version. As expected, far less time is spent in garbage collection, and a little more time is spent to write the results out.

```
; cpu time (non-gc) 567,180 msec (00:09:27.180) user, 9,510 msec system
; cpu time (gc)      39,290 msec user, 430 msec system
; cpu time (total)  606,470 msec (00:10:06.470) user, 9,940 msec system
; real time   762,828 msec (00:12:42.828)
```

Now that the profiles are stored and can be read in as needed, we can take a single new sequence, compute its profile, and compare with all the profiles in the knowledge base, returning a list sorted from smallest difference to largest. The compare-profiles-from-file function takes a single profile, p, and compares it with all the ones stored in the file specified by filename. Again, to keep memory usage (and consequently, garbage collection and paging) to a minimum, we read each profile record in from the profiles file, do the comparison, and just keep the result. To use the sum-square-diff function, we need to make sure each profile lists every amino acid symbol in the same order, including entries with zero values for amino acids not in the sequence corresponding to the profile. The function make-full-profile, described earlier, takes care of this. It is done "on the fly," as we process each record, so that we do not need a huge amount of temporary storage.

```
(defun compare-profiles-from-file (p filename)
  (let ((p2 (make-full-profile (second p))))
    (pprint p)
    (pprint p2)
    (with-open-file (strm filename)
      (do ((profile (read strm nil :eof) (read strm nil :eof))
           (results '())
           (n 1 (1+ n)))
          ((eql profile :eof) (sort results #'< :key #'second))
        (format t "~&Processing profile ~A~%" n)
        (push (list (first profile)
                    (sum-square-diff p2 (make-full-profile
                                          (second profile))))
              results)))))
```

The file named `filename` contains a series of profiles, each written out separately, so that each profile can be read in by a single call to the `read` function. The `compare-profiles-from-file` function would be used as follows, assuming `test1` was the profile (with name) of an unknown or test protein sequence, and the file containing the profiles was `profiles.txt`.

```
> (setq results1 (time (compare-profiles-from-file
                        test1 "profiles.txt")))
```

The version of `sum-square-diff` from above uses the ratios directly and does ratio arithmetic, producing an exact result for each comparison. This gives pretty big numbers in the numerators and denominators, and takes some time that could be saved by converting the ratios to floats first and then doing the arithmetic. The lost accuracy is insignificant and the time speedup is about 30%. So, the fast version of `sum-square-diff` looks like this:

```
(defun sum-square-diff (profile-1 profile-2)
  "Each profile is an alist sorted the same way. The sum of the
  squared differences of the values is returned."
  (apply #'+ (mapcar #'(lambda (x y)
                         (expt (- (coerce (third x) 'single-float)
                                  (coerce (third y) 'single-float))
                               2))
                     profile-1 profile-2)))
```

The only differences are the added calls to `coerce`. Timing tests show that the cost of type conversion on the fly is more than made up by the faster arithmetic. Here is the timing of a sample run comparing the first protein in the Uniprot knowledge base with all the others (including itself, which of course returns a squared difference of 0).

```
; cpu time (non-gc) 167,630 msec (00:02:47.630) user, 4,210 msec system
; cpu time (gc)       6,230 msec user, 120 msec system
; cpu time (total)  173,860 msec (00:02:53.860) user, 4,330 msec system
; real time 245,037 msec (00:04:05.037)
```

Here is a function that displays the results of `compare-all-profiles` in a little nicer format than just printing out the list. The optional parameter n specifies how many comparisons to list. If n is omitted, the entire data set is displayed.

```
(defun display-rank (data &optional n)
  (do* ((temp data (rest temp))
        (d (first temp) (first temp))
        (i 0 (1+ i)))
       ((if n (= i n) (endp temp)))
    (format t "~&~A ~10,7F~%" (first d) (second d))))
```

Finally, here are the results displayed using `display-rank`. They are in order from closest (most similar, or smaller difference) to most different. Only the closest 20 are shown here.

```
> (display-rank result 20)
Q4U9M9    0.0000000
P15711    0.0018922
Q9UKF7    0.0022550
Q8K4R4    0.0026742
Q8K3E5    0.0026889
Q9LDA9    0.0029408
P36534    0.0030399
Q6DTM3    0.0032246
P90971    0.0032871
Q6FP52    0.0034160
O13475    0.0034380
P47045    0.0034395
O93803    0.0035768
P05710    0.0035998
Q8R4E0    0.0036943
Q95Q98    0.0037262
Q96330    0.0037376
Q6BM93    0.0037383
Q7YQL5    0.0037762
Q7YQL6    0.0037762
```

You may notice that the comparison for the test sequence and Q7YQL5 came out exactly the same (to 7 decimal places) as for Q7YQL6. Checking in the Swiss-Prot knowledge base, you will find that they are in fact identical sequences. This should give some confidence that the idea of matching proteins by computing profiles gives usefully accurate results.

Further performance enhancements are possible. One option would be to store structured profiles in a relational database, with possibly faster retrieval time than reading from a file. Storing the numbers as floats would avoid calls to coerce. Storing the completed profiles would eliminate the need for a call to make-full-profile each time the knowledge base is used. The additional storage space required is insignificant relative to the time saved.

5.2 Search

> *When you come to a fork in the road, take it.*
> *–Yogi Berra*
> *as quoted in [29]*

Some of the data and knowledge we have described so far can be thought of as nodes and connections in graphs. For example, the frames that make up a class hierarchy are each elements of a graph, where the links between the elements (or nodes) are the subclass and superclass relations. Answering questions about indirect class-subclass relationships involves a kind of search of this graph.

Similarly the structure of some subsystems of human anatomy are clearly networks connected by links. An example is the cardiovascular system, consisting of a network of veins, arteries, capillaries, and the heart and lungs. Each vessel is connected to others by relations such as `tributary-of` or `branch-of`, expressing upstream and downstream relationships. Such a network can be searched for connectivity, much as the frame traversing functions do that were developed in Chapter 2, for example, using the `part-of` relationship.

Biochemical reactions at the cellular and subcellular levels also can be thought of as forming *pathways*, representing metabolic or other cellular processes. A biochemical reaction can be represented as a link between substrates (the substances that go into the reaction) and products (the substances that result). The molecules and their reactions then form a searchable network. To determine whether a given substance can be produced from a set of starting molecules and a set of reactions involves searching through the reaction network to see if there is a path from the start to the desired product.

5.2.1 A General Framework for Search

Many algorithms for searching graphs have been discovered and studied. Some involve systematic exploration of the graph without any overall knowledge about the graph. Others depend on having *heuristics* or knowledge about the graph, either locally or globally. These two types of processes are called *uninformed* and *informed* search, respectively. However, most search methods have some general features in common, and it is possible to parametrize aspects of the search. This section describes how to do search on graphs in a general framework that allows specialized search methods to use the same general code, but just vary the order in which the graph is explored, and how the progress is evaluated.

What is needed in order to do search? We need a representation of state. This will be a representation of the starting position or node in the graph. To get from one node to the next, a function, `successors`, needs to be provided that, given a node, generates a list of next nodes. The `successors` function and the data structure(s) it uses define the search space or graph. A very simple way of explicitly representing a graph is to label each of the nodes, and then define the connectivity of the graph by listing each node with the names of its immediately adjacent nodes, as in Graham [137, Chapter 3, Section 15]. This is very compact, since it relies only on local information. From the aggregate local information, global properties of the graph can then be computed.

Here is a simple example of a graph and a representation in the form we described. Figure 5.1 is an abstracted map of the area around Seattle, showing the towns, the ice rinks where the author plays ice hockey, and the location of the author's house. The numbers are mileage between junction points. The ice rinks are marked with an R and my home is at the location marked H, with my name also next to it.

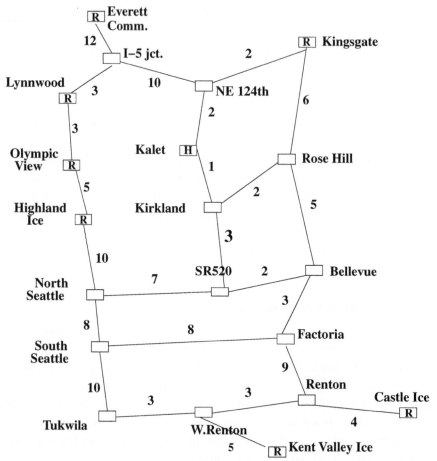

Figure 5.1 A diagrammatic representation of the major roads, towns, and ice rinks in the Seattle area, showing also the author's home.

The graph nodes and connections will be a big list that is the value of a parameter called `*graph*`. Each node (town, junction, rink, etc.) will be represented by a list in which the first element is a symbol naming the node and the rest of the list consists of the names of the next neighboring nodes. The `successors` function in this case just looks up the node and returns the list of neighbors. Here, our `successors` function refers to the global variable containing the graph, but later we can parametrize this too.

An important use of this graph is to search for the shortest path to the rink at which a particular game will be played. The distance numbers are not represented in this simple rendition, but could be added so that each neighbor is a two element list, the name and the distance. This would make the `successors` function a little more complicated but will allow for more sophisticated search methods.

```
(defparameter *graph*
    '((everett-comm i5-junct)
```

```
(i5-junct everett-comm lynnwood ne124th)
(ne124th i5-junct kalet kingsgate)
(kalet ne124th kirkland)
(kingsgate ne124th rose-hill)
...))

(defun successors (node)
  (rest (assoc node *graph*)))
```

We will need to maintain a queue of nodes from which to expand the search. Each element in the queue represents a node in the graph or a path from the start to a node. Each step in the search will examine the element at the front of the queue at that step. The order in which nodes are kept in the queue may vary. The order will be different for different search strategies, for example, depth-first, breadth-first, or various orders using heuristic evaluation functions. In a depth first search, as the graph is explored and new nodes are generated, the new ones are examined and expanded first, until dead ends are reached, and then the search backs up and visits the other nodes at the next higher level, and so on. This is simple to implement, though it has some undesirable properties, one of which is that if the graph is infinite, it is not guaranteed to terminate even if the goal node is reachable.

If the graph or search space has *loops* in it, that is, some successor nodes are actually nodes previously visited, the graph is effectively infinite. Such graphs are called *cyclic*. The Seattle area graph above definitely is cyclic, as the links connecting the nodes are bidirectional. A graph without loops is called *acyclic*. Even an acyclic graph may be infinite, in that the successors function can always generate more new nodes regardless of how many have already been examined. A simple example is the search for prime numbers by examining the integers one by one. There are an infinite number of integers. An odd but conceivable biologic example would be to search for interesting protein structures by putting together amino acids. Each node is a polypeptide chain. Since there is no limit on their length, we can always generate new chains to explore by adding one more amino acid to any existing one.

The basic algorithm will be to check the beginning of the queue to determine if a goal has been reached. If a goal has not been reached, continue the search by generating new nodes with the successors function, merging the new nodes with the remaining ones in the queue. Then continue with the node that is now at the top of the queue. The initial version will be really simple. To determine if the goal is reached we compare the node at the top of the queue with a goal node, provided as an input. The successors function is used to generate the next nodes to visit from the node at the top of the queue. The new nodes generated by the successors function are put on the front of the queue (using append) so they will be examined next (this is depth-first search).

We will implement the search using an auxiliary function that takes as input the search queue, so that we can pass the search queue through recursive calls as the search continues. Rather than defining a separate function, we define it within the body of the main function by using labels. The auxiliary function, search-inner, does the real work, starting with the initial state node, and just returning the symbol `success` if the goal node is found and the symbol `fail` if the search terminates without finding it.

```
(defun simple-search (initial-state goal)
  (labels
      ((search-inner (queue)
         (if (null queue) 'fail
           (let ((current (first queue)))
             (if (eql current goal)
                 'success
                 (search-inner (append (successors current)
                                       (rest queue)))))))))
    (search-inner (list initial-state))))
```

Of course we have to define the `successors` function. For the simple representation of a graph as a list of nodes and their neighbors, the `successors` function just looks up a node and returns the remainder of the list containing it, as described above.

As we saw with traversing collections of frames, specific traversal functions like `subsumed-by` and `part-of` could be parametrized (see page 299) by making the link traversal function an input to a generalized version, `connected-upward`, rather than an explicit call to a particular function. In the following code we generalize even further, by putting the search control into functions that are passed in as parameters as well. The representation of nodes and links, as well as the state of the search, all then become abstractions, whose particulars are in the functions that are inputs. The search function itself needs no details about how the graph or the state are represented, but uses the input functions to manipulate the data.

We need to specify the goal of the search, but this is not necessarily simply the node which we want to reach. So, a good generalization would be to represent the goal by a predicate function that returns `t` if a goal is reached. This also allows the possibility that there could be more than one node in the graph that, when reached, is a success. Alternately we may be looking for multiple paths to a node, or have some other success criterion. So `goal?` is a functional representation of the goal, allowing for multiple goals and repeated arrival at the goal. Thus, instead of having a goal node as an input parameter, we pass a test function, `goal?`, which will be applied to the first node in the queue, using `funcall`.

Similarly, we add the `successors` function to the formal parameters list of our search, instead of defining a named function. We use `funcall` to call this function, too, which is now the value of a variable named `successors`.

```
(defun better-search (initial-state goal? successors)
  (labels
      ((search-inner (queue)
         (if (null queue) 'fail
           (let ((current (first queue)))
```

```
                        (if (funcall goal? current) 'success
                         (search-inner (append (funcall successors
                                                          current)
                                          (rest queue)))))))))
            (search-inner (list initial-state))))
```

Now, it seems that depending on what we are searching for, we might not want to just stop searching on the first success, but be able to keep a list of successes. If the goal *is* reached, the program needs to determine whether to search for more successes, or stop and return the results. This is the role of the enough? function, which we now add as a parameter. The enough? function is a predicate on the list of states that have reached the goal (the successes so far). It determines whether search should stop: nil means don't stop but generate all possible answers, while t means stop on the first answer. An enough? function might just stop when a specified number of successes is found, or it could have a very complex criterion for when to stop searching.

```
    (defun multi-search (initial-state goal? enough? successors)
      (labels
          ((search-inner (queue wins)
             (if (null queue) wins
               (let ((current (first queue))
                     (remains (rest queue)))
                 (cond ((funcall goal? current)
                        (setq wins (cons current wins))
                        (if (or (eq enough? t)
                                (and (null enough?)
                                     (null remains))
                                (and enough?
                                     (funcall enough? wins)))
                            wins
                          (search-inner remains wins)))
                       (t (search-inner (append (funcall successors
                                                          current)
                                          remains))))))))
        (search-inner (list initial-state) '())))
```

Finally, we should consider alternative search strategies for where to look next. We are adding the new nodes to examine on the front of the queue, but we have already pointed out that this strategy, depth-first search, while simple, also can be problematic. In addition to possibly not terminating, even when it does terminate it is unlikely to be the most efficient way to conduct the search. If the graph is finite and we search exhaustively, we can then examine all the results to see which is best, but there are many other search strategies that can do better. We will see that it is possible to guarantee termination if a goal node exists, and to find it efficiently.

In order to implement different search strategies, we now need to parametrize the way the new nodes are put in the search queue. So, we add one more parameter, merge, which will take as inputs the new nodes generated by successors, and the remaining states in the queue. The merge function adds the new nodes to the queue in an order reflecting the search strategy. For depth-first, we already know it is just append; we put the list of *new nodes* in front of the remaining ones. Another common strategy is breadth-first search, in which we examine all the nodes already on the queue before exploring the new ones. This is easy too. Instead of using append directly, we reverse the order of the two arguments, and use append to put the *remainder* in front of the list of new nodes. Similarly, we can do more sophisticated searches by using heuristic knowledge about the nodes and the graph to do more complex merging of new nodes to explore with those already on the list.

We will shortly explain each of the various search strategies in terms of how the merge function works. The merge function combines new states with the remainder of the queue. This is where most of the action is, since it decides in what order the search proceeds.

```lisp
(defun gsearch (initial-state goal? enough? successors merge)
  (labels
      ((search-inner (queue wins)
         (if (null queue) wins
             (let ((current (first queue))
                   (remains (rest queue)))
               (cond ((funcall goal? current)
                      (setq wins (funcall merge (list current)
                                                  wins))
                      (if (or (eq enough? t)
                              (and (null enough?)
                                   (null remains))
                              (and enough?
                                   (funcall enough? wins)))
                          (values wins remains)
                          (search-inner remains wins)))
                     (t
                      (search-inner (funcall merge
                                             (funcall successors
                                                      current)
                                             remains)
                                    wins)))))))
    (search-inner (list initial-state) '())))
```

This general search function, `gsearch`, is then used with particular kinds of merge methods to get the different kinds of search strategies.

5.2.2 Uninformed Search Strategies

An uninformed search strategy is one which does not use any information about the graph other than connectivity. Uninformed searches can be systematic, but they do not use knowledge such as the cost of a path from one node to the next, or the number of nodes between the current node and the goal, or the cost of reaching the goal (these costs are rarely known with precision, but are likely only estimated in any case). Two basic kinds of uninformed searches are: depth-first, already mentioned, and breadth-first, which examines all the nodes already in the queue before examining the new ones that are generated. Depth-first is the search strategy used in the Prolog run-time environment, to do backward chaining.

To do depth-first search, we examine the newest nodes first, so as each set of new nodes is generated we put it at the top of the queue. Since the new nodes are a list, and the queue is a list, the new queue can be produced by using append to put the new nodes on the front. So, the `merge` function is just `append`.

```
(defun depth-first-search (initial-state goal? enough? successors)
  (gsearch initial-state goal? enough? successors #'append))
```

Figure 5.2 shows how a depth-first search might proceed. In this figure, nodes 5, 7, 14, 15, 21, 22, and 24 have no successors, so the search backtracks to a node with successors that have not yet been examined. A significant advantage of depth-first search is that it does not require a lot of memory or temporary working storage. Only the nodes on the current path from the beginning are needed so that the backtracking can be done. So, when the search tree has finite depth, the search queue does not grow unreasonably large. In general, it is proportional to the average depth of the search tree. On the other hand, it is possible that one or more branches of the search tree is actually infinitely deep. If there were an infinite number of nodes to examine, or there was a loop in the graph, the search would not terminate and might never reach a goal node.

In contrast, an advantage of breadth-first search is that it will not get stuck in a branch that is infinitely deep. Breadth-first search is guaranteed to find goal nodes if they exist. To do breadth-first search we put the new nodes on the end of the queue, so they will only get examined after the existing nodes are checked. This can also be done with append but the order of the arguments needs to be reversed. So, for the `merge` function we provide an anonymous wrapper function to pass the arguments to `append` in reverse order, with the remainder of the queue first, then the list of new nodes.

```
(defun breadth-first-search (initial-state goal? enough? successors)
  (gsearch initial-state goal? enough? successors
    #'(lambda (new-states queue) (append queue new-states))))
```

Figure 5.3 shows how a breadth-first search might proceed. All the nodes at a given level are examined before examining any nodes at the next level of branching. If a goal node exists,

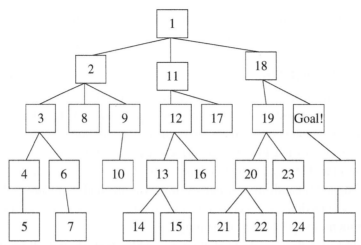

Figure 5.2 A network of nodes being explored depth-first. The numbers on the boxes represent the order in which the nodes are visited by the search.

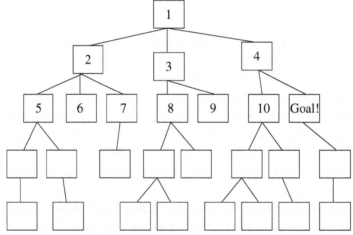

Figure 5.3 A network of nodes being explored breadth first. The numbers on the boxes represent the order in which the nodes are visited by the search.

that level at which the goal node is located will be reached without going down any infinite paths. Loops will not be a problem either, since the path through a loop involves a series of levels, and the loop is not traversed more times than the current level allows. The main disadvantage of breadth-first search is that the search queue can get huge, because all the nodes up to the current search depth must be kept on the queue. So the queue grows exponentially with depth.

5.2.3 Informed Search Strategies

I know just where you are, Joey. You're not too far. But don't go the other way, come this way.
—Yogi Berra, giving directions on the telephone to his friend, Joe Garagiola
as quoted in [29].

In many cases, the problem at hand includes some information about the network other than the connectivity. One kind of information is the length or cost of each path or step from one node to the next in the network. Another is an estimate of the distance or cost remaining to reach a goal node from the current node being examined. Such estimates can be conservative, always larger than the actual cost, or the opposite, always smaller than the actual, or the estimates can be variable, sometimes lower and sometimes higher. There are many different ways to use this information, and we will just examine a few of them here.

The simplest method is to examine each of the successor nodes from the current node, and always take the path that has the least estimated cost to reach a goal. This is called *hill-climbing*, or *greedy* search. One can think of it as ascending by the (likely) steepest path, or in terms of distance, the apparently shortest path. The idea is to get to the goal as quickly as possible. The problem is that the estimates can be very misleading. A nearby node can appear to be closer, but may be a dead end (a local maximum, or in climbing terms, a false summit), or actually lead to a goal node by a more round-about path. Nevertheless, if the structure of the problem is reasonably regular, greedy search may work very well. Like depth-first search, hill-climbing is not guaranteed to terminate, because it puts new nodes ahead of existing ones, although it tries to be "smart" about which to pursue first.

Hill climbing requires a function that estimates the distance or cost remaining to reach a goal. This is called a *heuristic* function. The only absolute requirements on this function are that its value should be 0 for a goal node and that it should never have a negative value. Thinking of the heuristic function as a cost function, we are giving preference to paths that appear to have the lowest cost. Sometimes the main difficulty in solving a search problem is to find or invent a good heuristic function. We will see shortly that "good" can be given some useful meaning in this context. In the following, `estim-fn` is the heuristic function, and it will of course depend on what the goal is in any particular application. It still puts the new states on the front of the queue, so it is similar to depth-first, but it sorts the new states from most promising to least promising.

```
(defun hill-climb-search (initial-state goal? enough?
                          successors estim-fn)
  (gsearch initial-state goal? enough? successors
           #'(lambda (new-states queue)
               (append (sort new-states
                             #'(lambda (s1 s2)
                                 (< (funcall estim-fn s1)
                                    (funcall estim-fn s2))))
                       queue))))
```

A problem with hill-climb is that it does not take into account whether any of the new states might actually be less promising than some of the remaining nodes to explore. A better approach would be to merge the pending states and the new states, in some "best first" order. For this more sophisticated informed search we thus need some additional help, a priority merge function that merges two priority queues. The priority merge is going to be smart about getting to a node by multiple paths. It merges the queues (lists in ascending order of val-fn) using a collate function, called same?. The purpose of the function, same?, if given, is to determine if a new node is the same node as one already on the queue. If it is, check if the cost is less by the new path to it than the previous path to it. If the new way of getting to it is better, use the new one, otherwise use the old one.

```
(defun priority-merge (a b val-fn same?)
  (cond ((null a) b)
        ((null b) a)
        ((and same? (funcall same? (first a) (first b)))
         (cons (if (<(funcall val-fn (first a))
                      (funcall val-fn (first b)))
                   (first a)
                 (first b))
               (priority-merge (rest a) (rest b) val-fn same?)))
        ((< (funcall val-fn (first a))
            (funcall val-fn (first b)))
         (cons (first a)
               (priority-merge (rest a) b val-fn same?)))
        (t (cons (first b)
                 (priority-merge a (rest b) val-fn same?)))))
```

Then we can do what is sometimes called *best-first* search. Best-first is much like hill-climb, except that we do not insist on pushing forward from the most recent node. Estim-fn is the estimated quality function; as before, smaller is better, so this might be considered shortest path or least cost. This may be either an estimate of the distance to the goal, as for hill-climb, or the work already done on a path; in the latter case, this works more like breadth-first.

```
(defun best-first-search (initial-state goal? enough?
                          successors estim-fn same?)
  (gsearch initial-state goal? enough? successors
           #'(lambda (new-states queue)
               (priority-merge (sort new-states
```

```
                    #'(lambda (s1 s2)
                       (< (funcall estim-fn s1)
                          (funcall estim-fn s2))))
                queue estim-fn same?))))
```

The above definition of best-first search in fact implements branch-and-bound search because the priority queue mechanism assures that any path whose cost estimate exceeds the cost to reach the goal along some other path will sort behind that other path in the queue. Nevertheless, the first goal path will not be found in the queue until all the lower-cost paths have been expanded. This is the essence of branch and bound.

So far, we have either used the cost so far for a path, or the estimated cost to the goal. To really get the best path, it makes sense to take both of these into account. An algorithm that does this is A* search. For this we need two heuristic functions, one which computes the cost already incurred for each path, and the other estimates the remainder.

g is the cost-so-far function, which must be computable from the state, so, we will need to keep track of accumulated cost,

h* is an underestimate of the remaining cost, also computable from current state.

If indeed h* never overestimates the cost remaining, this algorithm is guaranteed to find the lowest cost path to the goal. A proof of this may be found in standard AI texts. The A* algorithm implements a kind of dynamic programming by finding newly generated states that are same? as ones already on the queue, and retaining only the one with lowest cost estimate. Note the similarity to "best-first" search. In fact best-first is A* with g set uniformly to 0.

```
(defun a*-search (initial-state goal? enough? successors g h* same?)
   (labels ((estim-fn (state)
              (+ (funcall g state) (funcall h* state))))
     (gsearch initial-state goal? enough? successors
              #'(lambda (new-states queue)
                  (priority-merge (sort new-states
                                        #'(lambda (s1 s2)
                                           (< (estim-fn s1)
                                              (estim-fn s2))))
                                  queue #'estim-fn same?)))))
```

A* search has been used very effectively in the PathMiner project [275] to generate biochemical pathways from basic reactions. To set the stage for a simple version of the idea of applying search to find biochemical pathways, we consider next what it would take to represent pathways rather than just states, successors and goals.

5.2.4 Path Search

Using any of the above search strategies, we can create a path search, which returns not just success or fail, but the path(s) that got to the goal. This is what is needed to find biochemical pathways for metabolic processes. It is also exactly what we will need for the Clinical Target Volume project described later in this chapter, in which we use anatomic knowledge to predict the regional spread of tumor cells. We need to generate successor paths instead of just successor nodes. We do this by maintaining each item in the queue as a path, a list of nodes beginning with the most recent. For each path, we generate new paths by adding each new node to the current path, so when there are several successor nodes, we will get several paths from one current path. The function extender does this operation of constructing a new path from each successor node and the current path. We still also require that the caller provide a function to generate the successor nodes from the current node, because that describes the connectivity of the graph. The extend-path function, which we define here, will use these functions, successors, and extender to generate a list of new paths from an existing path.

```
(defun extend-path (path successors extender)
  (mapcar #'(lambda (new) (funcall extender new path))
          (funcall successors path)))
```

This function, extend-path, will then be used in a closure (a lambda expression) as the actual successors function in the call to gsearch. All the search control strategies previously described can be used with this function, path-search, by providing the appropriate merge function as above. We also retain the parametrization of how to determine if a goal is reached, and how many paths to find (the goal? and enough? functions).

```
(defun path-search (start goal? enough? successors extender merge)
  (gsearch (list start) ;; makes start into a path of length 1
          goal? ;; keep goal parametrization
          enough? ;; and when to stop
          #'(lambda (current)
                (extend-path current successors extender))
          merge)) ;; keeps parametrization of control strategy
```

To find all paths, the goal function should just check if there are any more successors for the current path, and if there are none, call it a success. This traces each path to its end or leaf node.

```
(defun all-paths-goal (current successors)
  (null (funcall successors current)))
```

Since it takes both the current path and the `successors` function as inputs, it needs to be called inside a closure, so the `successors` parameter can refer to the right thing in the surrounding environment. Also worth noting is that for path search, the `successors` and `extender` functions need to take a *path* as an input, not just a single node.

For finding metabolic pathways in the biochemistry of cells, we will design a representation of biochemical reactions and molecules as forming a graph that can be searched. We will define a `successors` function for the biochemical pathways graph and provide appropriate functions for the other search parameters. Then we will apply the whole works to a simple example, to generate pathways representing the process of glycolysis, or metabolism of sugar molecules. Similarly we show how to use `path-search` to find lymphatic pathways for spread of tumors.

All the code in this section except the path search variant was provided by Dr. Peter Szolovits, Professor at the MIT Computer Science and Artificial Intelligence Laboratory. For a more extensive discussion of search strategies and algorithms, the reader should refer to one of the many excellent artificial intelligence textbooks, for example Norvig [305], Russell and Norvig [360], Nilsson [303], and Tanimoto [408].

5.2.5 Example: DNA Sequence Alignment

One of the most commonly needed operations on DNA and Protein data is comparison of two sequences, to determine how good a match they are to each other. The process of finding a good match, allowing for possible insertions, deletions, or modifications at particular points in one or the other sequence is called *sequence alignment*. In this section we show how pairwise sequence alignment can be done using the search methods in this section.[6]

The basic idea is to formulate sequence matching as search. Each state in the graph to be searched is an instance of a four element structure. This structure contains the current remainder of sequence A to be matched, the remainder of sequence B, a score representing the cost (we are going to use informed search strategies, based on "goodness of match"), and the history so far. The history part consists of a record showing one of six possibilities: either the two sequences matched (the common element will appear there), or sequence B is modified (a mutation), or sequence A has an insertion or deletion, or sequence B has an insertion or deletion. Each possibility generates a new path in a search tree, and from each of those, the search continues. So, we will define a structure for the states, a successors function to generate each of the six possibilities, an estimation function to assign a score to each state, and a function to determine if two states represent the same point in the graph (to use in merging new states with the ones already on the search queue.

We parametrize the list of symbols that can be in a sequence, so that the same code can be used for both proteins and genes. For example, here is the alphabet for DNA (genes).

```
(defparameter alphabet '(a c t g))
```

---

[6]The code in this section was also kindly provided by Peter Szolovits.

The cost function will use a lookup table. For insertions and deletions there will be some cost, and for mutations also. The cost will be different depending on which possible element is substituted for which. So, we need a table of the relative costs of single point mutations. The dimensions of this table will depend on the size of the alphabet. For DNA, the table will have 16 entries (4×4). For proteins we would need a 20×20 table, since there are 20 amino acids. This cost table is an essential element of the knowledge that goes into typical alignment algorithms. It is implemented here as a double association list. This means that we can look up for a given base, what the cost is for that base to change to any other base. In this implementation we'll use an ordinary list for the first lookup and dotted pairs for the second lookup.

```
(defparameter mutation-penalties
  '((a          (c. 0.3)  (g. 0.4) (t. 0.3))
    (c (a. 0.4)           (g. 0.2) (t. 0.3))
    (g (a. 0.1) (c. 0.3)           (t. 0.2))
    (t (a. 0.3) (c. 0.4)  (g. 0.1)          )))
```

We also need a cost for insertion and deletion. We provide a maximum penalty or cost parameter, `infinity`, for the case that there is no entry in the table corresponding to one or the other sequence element.

```
(defconstant infinity 10000000.0)
(defconstant omit-penalty 0.5)
(defconstant insert-penalty 0.7)
```

Then, the `mutation-penalty` function returns 0.0 for a match, and if the two elements (one from sequence A and one from sequence B) don't match, it just does a table lookup using `assoc`.

```
(defun mutation-penalty (from to)
  (if (eql from to) 0.0
    (let ((from-entry (assoc from mutation-penalties)))
      (if from-entry
        (let ((to-entry (assoc to (rest from-entry))))
          (if to-entry (rest to-entry)
            infinity))
        infinity)))))
```

We use a structure named `ms` to hold states in the search tree. Its name is an abbreviation for "match state." Each state holds the remainder of each of the two sequences, a score and the search history. Keep in mind that the slot accessors will have the prefix `ms-` because we are using a structure rather than a class.

```
(defstruct ms
    seq1 seq2 score history)
```

The search needs a goal function. The goal is reached when we are at the end of both sequences.

```
(defun ms-goal? (ms)
  (and (null (ms-seq1 ms))
       (null (ms-seq2 ms))))
```

Two states are considered the same for the purpose of the merge function when the corresponding remaining sequence fragments are the same identical sequences.

```
(defun ms-same? (ms1 ms2)
  (and (eql (ms-seq1 ms1) (ms-seq1 ms2))
       (eql (ms-seq2 ms1) (ms-seq2 ms2))))
```

The successor nodes in the search tree are generated by looking at what is left in each sequence, and comparing the first element of each with the other. There are six possibilities: a match, a mutation, an insertion into the first sequence, a deletion from the first, an insertion into the second, and a deletion from the second.

```
(defun generate-successors (ms)
  (let ((x (ms-seq1 ms))
        (y (ms-seq2 ms))
        (sc (ms-score ms))
        (hx (ms-history ms)))
    (if (not (null x))
        (if (not (null y))
            (if (eql (first x) (first y))
                (list (make-ms :seq1 (rest x) :seq2 (rest y)
                               :score sc
                               :history (cons (list 'match
                                                    (first x))
                                              hx)))
                (list (make-ms :seq1 (rest x) :seq2 (rest y)
                               :score (+ sc (mutation-penalty
                                             (first x)
                                             (first y)))
                               :history (cons (list 'mutate
                                                    (first x)
                                                    (first y))
                                              hx))
                (make-ms :seq1 x :seq2 (rest y)
                         :score (+ sc omit-penalty)
                         :history (cons (list 'omit (first y))
                                        hx))
                (make-ms :seq1 (rest x) :seq2 y
```

```
                     :score (+ sc insert-penalty)
                     :history (cons (list 'insert (first x))
                                    hx))))
         (list (make-ms :seq1 (rest x) :seq2 y
                        :score (+ sc insert-penalty)
                        :history (cons (list 'insert (first x))
                                       hx))))
         (list (make-ms :seq1 x :seq2 (rest y)
                        :score (+ sc omit-penalty)
                        :history (cons (list 'omit (first y))
                                       hx))))))
```

Finally, we need a function that implements a search termination condition. The search termination condition is simple—one success is enough.

```
(defun found-one? (wins)
  (not (null wins)))
```

Now we can create some example functions that use the above together with any of the informed search functions. For example, here is how to use hill-climbing search with the above definition of the search space.

```
(defun match-hc (one two)
  (hill-climb-search
    (make-ms :seq1 one :seq2 two :score 0.0 :history nil)
    #'ms-goal?
    #'found-one?
    #'generate-successors
    #'ms-score))
```

The search starts with the two sequences, and an accumulated score of 0. The parameters one and two are the sequences to align, implemented as lists of symbols.

To do best-first search, instead of hill-climbing, we need to provide a same? function, that will identify two nodes that are the same but were reached with different costs. The merge function can then put the lower cost one ahead in the queue.

```
(defun match-bf (one two)
  (best-first-search
    (make-ms :seq1 one :seq2 two :score 0.0 :history nil)
    #'ms-goal?
    #'found-one?
    #'generate-successors
    #'ms-score
    #'ms-same?))
```

Finally, to do A* search, we need to have two cost functions, one that evaluates the cost so far on each path, and another which estimates the remaining cost from any node. In this case, we have no idea how to estimate, but A* works with any function that underestimates the cost remaining, so let's just say it is always 0.

```lisp
(defun match-a* (one two)
  (a*-search
    (make-ms :seq1 one :seq2 two :score 0.0 :history nil)
    #'ms-goal?
    #'found-one?
    #'generate-successors
    #'ms-score ;; f
    #'(lambda (s) 0.0) ;; g*
    #'ms-same?))
```

Now we can consider some short test sequences, and see how these various algorithms do. We would expect somewhat different results for each.

```lisp
(defparameter s1 '(a a t c t g c c t a t t g t c g a c g c))
(defparameter s2 '(a a t c a g c a g c t c a t c g a c g g))
(defparameter s3 '(a g a t c a g c a c t c a t c g a c g g))
```

Here are the results of some test runs for hill climbing search. All three combinations of the test sequences were run, and it seems that the best match is between s1 and s2.

```lisp
> (setq h12 (match-hc s1 s2))
(#S(MS :SEQ1 NIL
       :SEQ2 NIL
       :SCORE 1.8000002
       :HISTORY ((MUTATE C G) (MATCH G) (MATCH C) (MATCH A)
                 (MATCH G) (MATCH C) (MATCH T) (MUTATE G A)
                 (MUTATE T C) (MATCH T) ...)))
> (setq h13 (match-hc s1 s3))
(#S(MS :SEQ1 NIL
       :SEQ2 NIL
       :SCORE 3.0
       :HISTORY ((MUTATE C G) (MATCH G) (MATCH C) (MATCH A)
                 (MATCH G) (MATCH C) (MATCH T) (MUTATE G A)
                 (MUTATE T C) (MATCH T) ...)))
> (setq h23 (match-hc s2 s3))
(#S(MS :SEQ1 NIL
       :SEQ2 NIL
       :SCORE 2.0
```

```
                 :HISTORY ((MATCH G) (MATCH G) (MATCH C) (MATCH A)
                          (MATCH G) (MATCH C) (MATCH T) (MATCH A)
                          (MATCH C) (MATCH T)
                          ...)))
```

The A* search algorithm takes longer running time, but each run took only a few seconds at most, even with uncompiled code. The result for matching s1 and s2 was the same, but A* found a little better match between s1 and s3, and a lot better match for s2 and s3.

```
> (setq a12 (match-a* s1 s2))
(#S(MS :SEQ1 NIL
       :SEQ2 NIL
       :SCORE 1.8000002
       :HISTORY ((MUTATE C G) (MATCH G) (MATCH C) (MATCH A)
                 (MATCH G) (MATCH C) (MATCH T) (MUTATE G A)
                 (MUTATE T C) (MATCH T) ...)))
> (setq a13 (match-a* s1 s3))
(#S(MS :SEQ1 NIL
       :SEQ2 NIL
       :SCORE 2.8
       :HISTORY ((MUTATE C G) (MATCH G) (MATCH C) (MATCH A)
                 (MATCH G) (MATCH C) (MATCH T) (MUTATE G A)
                 (MUTATE T C) (MATCH T) ...)))
> (setq a23 (match-a* s2 s3))
(#S(MS :SEQ1 NIL
       :SEQ2 NIL
       :SCORE 1.2
       :HISTORY ((MATCH G) (MATCH G) (MATCH C) (MATCH A)
                 (MATCH G) (MATCH C) (MATCH T) (MATCH A)
                 (MATCH C) (MATCH T)
                 ...)))
```

Of course it would be nice to have some code to print the resulting sequence alignments in an easy-to-understand format, with color and labels. This kind of user interface code often is the bulk of the work in building effective bioinformatics software tools, but it is outside the scope of this book.

Although the complexity of the above application of search might seem inefficient, it will terminate in a reasonable amount of time for sequences up to some 1,100 bases each.[7] For amino acid sequences it is not practical to run this search algorithm on sequences of more than 20–30

---

[7]This was determined by Nancy Tang and conveyed to me in her (unpublished) project report for the "Life and Death Computing" course at the University of Washington, Autumn 2012.

peptides. The analysis of how the algorithm scales with the sequence length in each case (DNA and proteins) is left as an exercise for the reader. The result of that analysis should explain the difference between the two cases.

5.2.6 Example: Biochemical Pathways

We have shown how biological sequence alignment can be formulated as a search problem. It is one of many examples of biological and medical problems that can be described in terms of graphs or networks. Other examples include biochemical pathways, signaling and regulatory networks, anatomic structure (such as arteries, veins, and the lymphatic system), and applications of anatomic structure such as spread of tumors through the lymphatic system. Work done on modeling biological and medical phenomena with networks is a very large field. Here we will develop just a few examples to illustrate the principles and methods, rather than try to create a comprehensive survey of this subfield of biomedical informatics.

In Chapter 2, Section 2.2.1.5, we showed how to solve the problem of determining if a pathway exists from a set of starting molecules to a final molecule to be synthesized by using propositional logic and theorem proving. Here we will deal with this by formulating the pathway problem as a search problem, with nodes and arcs, so that we can just use the search code developed in this section. A biochemical reaction can be represented as a link between substrates (the substances that go into the reaction) and products (the substances that result). The molecules and their reactions then form a searchable network. To determine whether a given substance can be produced from a set of starting molecules and a set of reactions involves searching through the reaction network to see if there is a path from the start to the desired product.

We'll represent a biochemical reaction as a structured type, so we can refer to the reactants and products with slot references. Molecules will just be represented by symbols.[8] The structure for a reaction will include a slot for the reactants, the significant products (here just called `products`), and the other products. The significant products are the ones that become the reactants for the next step in the pathway. The other products are also used but it is assumed that in general there is plenty of each of those anyway.

```
(defstruct reaction
        reactants ;; Lefthand side
        products ;; Righthand side, significant products
        other-products) ;; non-significant products
```

We'll define a global variable, `*reactions*` to hold all the reactions for later reference. The `add-reaction` macro just provides a convenient way to specify the reactions we will use in running the code. The reactions following are the ones that are relevant to glucose metabolism. To examine more complex processes, this list would be expanded to include more reactions.

---

[8]The code in this section is based on Jeff Shrager's BioBike [380] tutorial, "Metabolic Simulation I: Introduction to dynamic analysis of metabolic pathways," but differs in that we construct functions to be used by our search code instead of writing custom search code.

```
(defvar *reactions* nil)

(defmacro add-reaction (reactants products &optional other-products)
  '(push (make-reaction :reactants ',reactants
                        :products ',products
                        :other-products ',other-products)
         *reactions*))
```

For convenience, we write all the calls to `add-reation` into a `make-reactions` function that can be called to initialize the knowledge base of reactions. In a practical system, there would be more elaborate support for adding reactions to the reaction collection, as part of the process of "knowledge management."

```
(defun make-reactions ()
  (add-reaction (fru) (f1p))
  (add-reaction (f1p) (glh dap))
  (add-reaction (glh) (g3p))
  (add-reaction (glu atp) (g6p) (adp))
  (add-reaction (g6p) (f6p))
  (add-reaction (f6p atp) (fbp) (adp))
  (add-reaction (fbp) (dap g3p))
  (add-reaction (dap) (g3p))
  (add-reaction (P NAD+ g3p) (bpg) (NADH H+))
  (add-reaction (bpg adp) (3pg) (atp))
  (add-reaction (3pg) (2pg))
  (add-reaction (2pg) (pep) (H2O))
  (add-reaction (pep atp) (pyr) (adp))
  (add-reaction (pyr NAD+ coa) (aca) (NADH H+ CO2))
  (add-reaction (cit) (ict))
  (add-reaction (ict NAD+) (akg) (NADH H+ CO2))
  (add-reaction (akg NAD+ coa) (sca) (NADH H+ CO2))
  (add-reaction (sca gdp P) (suc coa) (gtp))
  (add-reaction (suc FAD) (fum) (FADH2))
  (add-reaction (fum H2O) (mal))
  (add-reaction (mal NAD+) (oxa) (NADH H+)))
```

The idea of pathway search will be to find applicable reactions, given available reactants or substrate molecules. Thus, we would like to look up reactions by specifying one or more substrates. We can go through all the reactions and create an index (using a hash table), so that for any given molecule, the reactions using it as a reactant are listed in its hash table entry. This provides an easy way to find applicable reactions for any given molecule.

```
(defvar *reactant-table* (make-hash-table))

(defun lookup-by-reactants (prod)
  (gethash prod *reactant-table*))

(defun init-reactant-table (rxns)
  (dolist (rxn rxns)
    (let ((reactants (reaction-reactants rxn)))
      (dolist (mol reactants)
        (pushnew rxn (gethash mol *reactant-table*))))))
```

The applicable reactions for a given set of molecules will depend on what else is available in the biochemical environment of the cell, so we will need to keep a list of the molecules that are present, called the *environment*, here a formal parameter, env. For each molecule in the set, we look up the reactions that include that molecule in its reactants. Then we check the environment to see if all the other reactants are available. If so, that reaction is an applicable reaction. This process gives a list of reactants for each molecule. If we use mapcar to do this operation on all the molecules we are checking, we will get a list of lists of reactions. The result we want is a flattened list, so we need to append the lists together. This is easy to do by just using apply with append to combine the lists.

```
(defun applicable-rxns (mols env)
  (apply #'append
         (mapcar #'(lambda (mol)
                     (remove-if-not
                      #'(lambda (x) (every #'(lambda (y)
                                               (find y env))
                                           (reaction-reactants x)))
                      (lookup-by-reactants mol)))
                 mols)))
```

This combination of append and mapcar should look familiar. It is implemented by the mappend function defined on page 124. Using mappend makes applicable-rxns a little simpler. This function returns a list of reactions that are possible for the given molecules in the specified environment.

```
(defun applicable-rxns (mols env)
  (mappend #'(lambda (mol)
               (remove-if-not #'(lambda (x)
                                  (every #'(lambda (y)
                                             (find y env))
                                         (reaction-reactants x)))
                              (lookup-by-reactants mol)))
           mols))
```

The nodes in the search space should have information about what molecules are available at the time they are reached, in addition to the reaction that got the search to that node. So we will need to be able to apply a reaction to make a new environment that includes the products of the applicable reaction. This is the job of the `apply-rxn` function.

```
(defun apply-rxn (rxn env)
  "returns a new environment with the products of rxn added to
   the current environment env. This function uses but does not
   modify env."
  (labels ((add-prods (prods accum)
              (if (null prods) accum
                (if (find (first prods) accum)
                    (add-prods (rest prods) accum)
                  (add-prods (rest prods)
                             (cons (first prods) accum))))))
    ;; note we want to add the small molecules too if new, like ADP
    (add-prods (reaction-other-products rxn)
               (add-prods (reaction-products rxn) env))))
```

The structure of a node in our reaction or pathway graph will be a simple two slot structure. One slot holds the reaction and the other the environment that is obtained *after* applying the reaction to the current environment. In order to use our search function we will need a *successors* function that generates the new nodes reachable from each of the current nodes in the search queue. The `next-nodes` function does this. It finds the applicable next reactions based on the products of the reaction in the current node. For each of these next reactions, `apply-rxn` generates the new environment, and a new node is made for the reaction and the environment.

```
(defstruct node rxn env)

(defun next-nodes (path)
  "returns a list of next nodes in the pathway graph from a given
   path, where path is a list of nodes, most recently added first."
  (let* ((current (first path))
         (last-rxn (node-rxn current))
         (env (node-env current))
         (rxns (applicable-rxns (reaction-products last-rxn)
                                env)))
    (mapcar #'(lambda (rxn)
                (make-node :rxn rxn :env (apply-rxn rxn env)))
            rxns)))
```

Finally, all this can be put together in a call to `path-search`. We assume we have a single starting node, which is an initial environment and a reaction. The goal function, a closure here, will determine if the end product we are looking for is in the environment of the most recent node of the current path. The enough? function here is also a closure, that sets an upper limit on the number of paths to be found. The next-nodes function serves as successors. For an extender, we just use cons, building up the path by adding new nodes on the front of the list. Finally, for a search strategy in this version, we use breadth first search, by appending new paths to the queue at the end rather than the beginning.

```lisp
(defun metabolic-paths (end-product start-env max-paths)
  (path-search start-env
               ;; check if end-product is in the current
environment
               #'(lambda (current) ;; current is a path
                   (find end-product (node-env (first current))))
               #'(lambda (wins) ;; no more than max-paths
                   (>= (length wins) max-paths))
               #'next-nodes ;; see above
               #'cons ;; just put new nodes on front
               #'(lambda (x y) (append y x)))) ;; breadth-first
```

As a convenience for testing, we provide an initialization function that takes a list of starting molecules and produces a list of initial nodes for the search.

```lisp
(defun initial-nodes (mols env)
  (mapcar #'(lambda (rxn)
              (make-node :rxn rxn :env (apply-rxn rxn env)))
          (applicable-rxns mols env)))
```

To display the results, it is useful to have a function that will print a reaction in a somewhat human readable format, and a function that will print out an entire pathway, one reaction per line of text.

```lisp
(defun pprint-reaction (rxn)
  (format nil "~A --> ~A + ~A"
          (reaction-reactants rxn)
          (reaction-products rxn)
          (reaction-other-products rxn)))
```

Here are some examples of reaction structures when printed using this function:

```lisp
> (pprint-reaction (first *reactions*))
"(MAL NAD+) --> (OXA) + (NADH H+)"
> (pprint-reaction (third *reactions*))
"(SUC FAD) --> (FUM) + (FADH2)"
```

Having a way to print each reaction, it would also be nice to have a way to print out pathways. For this, we create the printed version of each reaction in a path (recall that a path is just a list of reactions, so we can use `mapcar` to apply a function to each element and generate a list of results in the same order as the original list).

```lisp
(defun print-path (path)
  (let ((nice-rxns (mapcar #'(lambda (node)
                               (pprint-reaction (node-rxn node)))
                           path)))
    (dolist (rxn nice-rxns)
      (format t "~A~%" rxn))))
```

Now we just need some code to initialize the environment, create all the reactions, and then we can test our system. In the following test code, we reverse the pathways that are generated, since the `metabolic-pathways` function generates the paths with the reactions in reverse order. This is a consequence of using `cons` to extend each path by adding a reaction. We could generate them in forward order by using a function that appends the new path at the end, for example the anonymous function `(lambda (new path) (append path (list new)))` or something similar.

```lisp
(defvar *init-env* '(atp adp gdp gtp fru glu NAD+ FAD P))

(defvar *init-mols* '(glu fru))

(defun init ()
  (make-reactions)
  (init-reactant-table *reactions*)
  (initial-nodes *init-mols* *init-env*))

(defun test (init-node n)
  (mapcar #'reverse ;; or make metabolic-paths reverse stuff
          (metabolic-paths 'pyr init-node n)))

(defun test2 (init-node n)
  (mapcar #'reverse
          (metabolic-paths 'mal init-node n)))
```

The simplest possible tests start with the above initial environment and the glucose and fructose molecules. The test functions search respectively for all the paths leading to the production of pyruvate (`pyr`) or malate (`mal`).

```lisp
> (setq starts (init))
(#S (NODE :RXN #S (REACTION :REACTANTS (GLU ATP)
                            :PRODUCTS (G6P)
                            :OTHER-PRODUCTS (ADP))
```

```
                    :ENV (G6P ATP ADP GDP GTP FRU GLU NAD+ FAD P))
 #S (NODE :RXN #S (REACTION :REACTANTS (FRU)
                            :PRODUCTS (F1P)
                            :OTHER-PRODUCTS NIL)
       :ENV (F1P ATP ADP GDP GTP FRU GLU NAD+ FAD P)))
```

The `starts` variable contains a list of two pathway steps that could start off the chain. They are the only reactions possible initially with the environment and starting molecules specified. We can then give each to the `test` function to see if there is a pathway that can synthesize pyruvate. We can also run the `test2` function to see if we can synthesize malate. The pyruvate run succeeds in two ways, just slightly different. The parameter n just limits the number of pathways found. In our case, only two pathways were found.

```
> (setq result (test (first starts) 10))
((#S(NODE :RXN #S(REACTION :REACTANTS (GLU ATP)
                           :PRODUCTS (G6P)
                           :OTHER-PRODUCTS (ADP))
         :ENV (G6P ATP ADP GDP GTP FRU GLU NAD+ FAD P))
   #S(NODE :RXN #S(REACTION :REACTANTS (G6P)
                            :PRODUCTS (F6P)
                            :OTHER-PRODUCTS NIL)
         :ENV (F6P G6P ATP ADP GDP GTP FRU GLU NAD+ FAD ...))
   ...)
 (#S(NODE :RXN #S(REACTION :REACTANTS (GLU ATP)
                           :PRODUCTS (G6P)
                           :OTHER-PRODUCTS (ADP))
         :ENV (G6P ATP ADP GDP GTP FRU GLU NAD+ FAD P))
   #S(NODE :RXN #S(REACTION :REACTANTS (G6P)
                            :PRODUCTS (F6P)
                            :OTHER-PRODUCTS NIL)
         :ENV (F6P G6P ATP ADP GDP GTP FRU GLU NAD+ FAD ...))
   ...))
```

A close examination shows that one of the pathways has one more step that is not strictly necessary, but is possible so it was included. Here is the first pathway result:

```
> (print-path (first result))
(GLU ATP) --> (G6P) + (ADP)
(G6P) --> (F6P) + NIL
(F6P ATP) --> (FBP) + (ADP)
(FBP) --> (DAP G3P) + NIL
(P NAD+ G3P) --> (BPG) + (NADH H+)
```

```
(BPG ADP) --> (3PG) + (ATP)
(3PG) --> (2PG) + NIL
(2PG) --> (PEP) + (H2O)
(PEP ATP) --> (PYR) + (ADP)
```

and here is the second path, where you can see that the extra step of producing g3p from dap was not necessary, but certainly possible.

```
> (print-path (second result))
(GLU ATP) --> (G6P) + (ADP)
(G6P) --> (F6P) + NIL
(F6P ATP) --> (FBP) + (ADP)
(FBP) --> (DAP G3P) + NIL
(DAP) --> (G3P) + NIL
(P NAD+ G3P) --> (BPG) + (NADH H+)
(BPG ADP) --> (3PG) + (ATP)
(3PG) --> (2PG) + NIL
(2PG) --> (PEP) + (H2O)
(PEP ATP) --> (PYR) + (ADP)
```

The test for malate, however, fails, showing that it is not possible to synthesize malate from the given environment and starting points. We know, though, that this ought to succeed. It is part of the Krebs citric acid cycle, the well-known process of sugar metabolism. Looking through the reactions, it appears that we need citrate and Coenzyme A. Adding these two to the starting environment, does *not*, however, solve our problem. The reason is that the pathway search serially chains together reactions from a starting molecule to the final product. To synthesize malate, however, we need to have two pathways that converge, since there is a reaction that needs two inputs to be synthesized. So, this kind of pathway search has some basic limitations. In fact the notion of a linear biochemical pathway is too simple to model even as central and relatively simple a process as glucose metabolism.

One way to fix this problem with malate is to modify next-nodes so that it considers the reaction products of the most recent reaction, and in addition any other possible reactions in the current environment. The breadth-first control strategy will prevent reaction loops from being a problem, though inefficiency could be significant when the number of molecules and reactions get large. This approach is similar to the forward chaining idea (as implemented starting on page 177 in Chapter 2). Other alternatives exist too, as we saw in the example of using backward chaining in Chapter 2.

Scaling up to really interesting biochemical environments takes some more sophistication, but is possible. One approach is to incorporate some knowledge of the distance between molecules, and to use an estimating function to guide informed search. McShan and colleagues did this for a large collection of biomolecules, using a distance function based on the differences of the chemical formulas of the current state molecule and the goal molecule [275]. This cost function was used with A* search. They used similar techniques to discover new metabolic pathways in mammalian systems [276].

Although this all makes biochemical sense, it seems that the idea of a pathway is really about *explanation*. The existence of a pathway means that certain things are possible for the cell to accomplish. In reality, all these reactions are going on in parallel, and not sequentially in some kind of assembly line.

Finally, having previously shown how to use propositional logic (predicate calculus) and proof methods to solve the same problem should suggest that proof methods are in fact forms of search over a graph whose nodes are logic formulas and whose links or arcs are the inference steps (applications of individual inference rules).

5.3 Natural Language Processing

Earlier we mentioned that natural language processing can have a role in building index terms for documents. It has many other applications in biology, medicine and public health. In searching full text documents, or even the text abstracts from bibliographic records, the simple presence of significant terms is not enough to draw conclusions. They occur in a context. Similarly, the processing of full text medical dictation (history and physical exam notes, other clinic notes) can help in automating the derivation of billing codes, as mentioned previously. This section gives just a very basic introduction to the ideas of natural language processing. More thorough treatments can be found in some of the standard Artificial Intelligence texts [305,360].

Text (articles, abstracts, reports, etc.) in ordinary human languages is the recording medium for much biomedical data and knowledge. Thus, searching computerized text is important for identifying documents or data records containing relevant information for example about disease and treatment, about protein interactions, or biological pathways, or about patterns of disease outbreaks. The text could be abstracts of publications, full text of articles, text-based knowledge resources like GENBANK, or other kinds of data and knowledge resources. As part of the process of building index information about such records, computer programs have to be able to decode the text stream into "words" (also called lexical analysis) and then organize the words in some useful way. After decoding, it is usually necessary to *parse* the sequences of words, to identify sentence structure and ultimately meaning.

Just looking at the parsing process for now, assuming that we have sentences as sequences of words, we can represent each word as a symbol, and each sentence as a list. If we are looking for sentence structure in terms of grammar, an effective way to represent knowledge about that is to create a *lexicon*, or dictionary of words in the language, along with their structural usage, recording in the dictionary whether the word is a verb, noun, adjective, etc. Grammatical structures can be represented by *rules*, that make a correspondence between a grammatical structure and its parts. Here is a small example, where in each expression the name of a structure or word type is the first symbol in the list, the = is a separator for human readability, and the remainder consists of all the possible words or structures that would match the structure or type.

```
(S = (NP VP))
(NP = (Art Noun PPS)
      (Name) (Pronoun))
(VP = (Verb NP PPS))
(PPS = () (PP PPS))
(PP = (Prep NP))
(Prep = to in by with on of)
(Adj = big little long short)
(Art = the a)
(Name = Ira Terry Sam Protein-25)
(Noun = molecule metabolism protein patient surgery cell)
(Verb = regulates survived metabolises)
(Pronoun = he she it this these those that)))
```

In the above, the symbol S refers to "sentence," NP is "noun phrase," VP is "verb phrase," PP is "prepositional phrase," PPS is a sequence of prepositional phrases, and ADS is a sequence of adjectives. The lists following the individual parts of speech constitute the dictionary. The idea is that you take the words in the sentence in sequence and find them in the right hand side, one or more at a time, try to construct structures from them (using substitution by the symbols on the left hand side), and then continue with the next words, combining structures (which you can also look for on the right hand side).

Here are some sentence parsing examples using a small subset of the above rules. We imagine a function called `parse` that takes as input a list of words and returns the sentence structure with the words organized according to the rules and the dictionary.[9]

```
> (parse '(the molecule))
((NP (ART THE) (NOUN MOLECULE)))

> (parse '(the patient survived the surgery))
((S (NP (ART THE) (NOUN PATIENT))
    (VP (VERB SURVIVED)
        (NP (ART THE) (NOUN SURGERY)))))
```

Many sentences can be read several ways, all of which are grammatically correct. Sometimes the intended meaning is obvious to the reader, but for a computer program this presents a hard problem. This kind of ambiguity leads to more than one parse for a given sentence. The simplest approach is to initially generate all the possibly grammatically correct structures.

```
> (parse '(this protein regulates metabolism in the cell))
((S (NP (ART THIS) (NOUN PROTEIN))
```

---

[9]The symbols that are mixed case in the rules and dictionary have been converted to all upper case by the Lisp reader when they were read in. This is one of many details that are not critical to the idea but can be dealt with in a fully engineered natural language system.

```
        (VP (VERB REGULATES)
            (NP (NP (NOUN METABOLISM))
                (PP (PREP IN) (NP (ART THE) (NOUN CELL))))))))
    (S (NP (ART THIS) (NOUN PROTEIN))
        (VP (VERB REGULATES)
            (NP (NOUN METABOLISM))
            (PP (PREP IN) (NP (ART THE) (NOUN CELL))))))))
```

The ambiguity in this sentence is that the phrase "in the cell" could be where the protein does its regulation, or it could mean that the metabolism being regulated is in the cell.

The work of the parser code is to examine each word in the text sequentially, determine its type (or possible types) from the lexicon, then look at the current structure to see if it matches the right hand side of a grammar rule. It may extend a tentative structure, such as a noun phrase, or it may be the first word of a new substructure. The parser builds a parse tree, using the rules to determine what to expect next. This structure is a tree because it reflects the grammatical structure, and because there may be alternate parse paths through the rules, as in the second example above.

In addition to rule-based grammars, other kinds of analyse can be applied, such as probabilistic or statistical methods. Grammar-based methods also can be enhanced to include *semantic* parsing, which associates actions with sentence parts, so that text can be interpreted as implying action. For example, the text, "Schedule a chest CT for Mr. Jones no later than three weeks from today," can be determined to be a request for an action (create a schedule entry), with associated constraints, and the NLP program becomes a front end to an appointment scheduling assistant.

It is beyond the scope of this book to develop these techniques, but fortunately they are already well described in Norvig [305]. The interested reader is referred there for working code as well as more discussion in depth. An overview of the ideas of NLP as applied to medical text, written by Carol Friedman and Johnson, can be found in Chapter 8 of the classic text edited by Shortliffe [378].

5.4 State Machines and Dynamic Models

The knowledge representation ideas discussed in Chapters 2 and 3 provide a static view of knowledge in biomedicine, that is, what entities exist and what are their relationships. Even with deduction systems based on first order logic or some other inference system framework, they do not represent change with time, though they *are* able to represent *spatial* relationships (as exemplified by work on the Foundational Model of Anatomy from the University of Washington [353] and recent work on a logic of spatial anatomy relations [96]). Much work has been done on temporal abstraction in analysis of clinical data from electronic medical record systems [74,196,393], modeling of physiologic processes and population dynamics using differential equations, modeling of the behavior of genes and gene products using state machines, and on modeling dynamic systems using Petri nets [23,323,322]. However, the temporal abstraction ideas described in Chapter 1, Section 1.3.3 posit an essentially static view of data, only adding time as an attribute, not as an independent variable in a dynamic system model.

As we have seen in the application of first order logic, we express general knowledge in terms of relations among *variables*, whose values represent the particular state of an individual, but whose relations, true for all individuals within some range of applicability, represent a biomedical *theory*. The relations can take the form of differential equations, transition formulas, or rules for change over time. Each of these modeling methods has applicability in biology, medicine, and public health. When the process we are modeling has continuous numerical values for its variables, and we want to consider time as continuous as well, differential equations are the usual framework and methodology. Where time can be treated as a series of steps, a state machine style of model works well. In either case, *time* is considered to be an *independent* variable, while the other state variables are considered to be *dependent* variables. They are functions of time, but not vice versa, and they are not functions of each other (though the equations or relations usually represent some constraints among the variables).

Although differential equations represent continuous processes, many situations are not solvable analytically, that is, one cannot by paper and pencil methods derive solutions of the equations that can then be applied to generate answers for any value of the time variable. So, there are methods for using digital computers to numerically approximate the solutions of such equations.

In many biological and medical process modeling problems, we do not have enough detailed knowledge to construct differential equation models. Even more importantly, the kinds of processes that we wish to model may be better understood using *qualitative simulation* [241]. In a qualitative simulation, the variables take on discrete values, sometimes just binary values such as on and off, and time is modeled as a series of discrete steps. There are many kinds of systems for discrete modeling, including for example Petri nets, Markov models, and Augmented Transition Networks, but we will not address these here.

5.4.1 State Machines

A state machine describes a process in terms of state variables. The idea is that time is viewed as a series of discrete steps. The variables start out with some initial values. In each time step, the variables are changed. The new value of a variable may depend on the current values of other variables. It may also depend on the history of its values and on the histories of the other variables. The framework we will present for modeling a state machine will be very general, and we will abstract out the state transition process, later putting in the details for particular simulation systems.

When there are a finite number of state variables, and each can take on only a finite number of values (not necessarily numeric), the state machine is called a finite state machine. Otherwise it is an infinite state machine. Examples of infinite state machines are those in which some of the variables can take on continuous values, or real numbers. A radiation treatment machine has continuously varying parameters such as the angle of tilt (the gantry angle) of the machine with respect to the patient, the rate at which the machine produces radiation, etc. An example of a finite state machine is a genetic signaling network, in which genes are either active or inactive (in reality, there can be continuous activation levels, but this is a good approximation in many cases).

We will now show how a state machine can be simulated in a modest amount of code. We will write this code without committing to *any* details about the state variables or the way in which the next state is computed from the current state or history. We will only assume that variables can be labeled and represented by symbols, and that we can keep their values in lists or hash tables associated with the symbols. Once we have the general simulation code, we will develop a few examples, showing how to put in the details for a given system. At this point we are not committing to whether the machine is finite or infinite.

To start with, we will use a hash table to store the state variables and values. The variables will be the keys to look up entries in the state table, and the associated data in the table will be their values. So, for now, we just make a shadow function, to hide the hash table.

```
(defun make-state-table (&optional (size 100))
  (make-hash-table :size size))
```

We can also write a simple function for adding a variable to the state table. It requires that we specify an initial value for the variable.

```
(defun add-variable (table state-var initial-val)
  (setf (gethash state-var table) initial-val))
```

However, so far we have made no commitments about the kinds of values, since a variable (a symbol) in Lisp can have any value, and similarly, any value can be stored in an entry in a hash table. This value can be a representation of the current state, or a history list, or something more complex.

Next we need a transition function that computes the next state from the current state (and possibly from the history as well). Some of this machinery will be independent of the particular process we are modeling. However, a way to specify the details for a particular process need to be provided here. So, we will write a `next-state` function, that will take as input the current state (the hash table) and a single variable transition function, and from them generate the next state. Now, there are two possibilities here. One is to just compute the next state and update the hash table entries by replacing the old values with the new values. However, we may need the previous values in future calculations. So another way is to maintain the state as a collection of history lists. For each variable, we can either store a single value, or a list of values that the variable has taken for each time step since beginning the simulation run.

```
(defun next-state (table trans-fn history)
  "updates the state table in parallel, by computing the new state
   values, then adding them after doing all of them."
  (let (new-states)
    (maphash #'(lambda (var val)
                 (push (list var
```

```
                        (funcall trans-fn var table))
                    new-states))
            table)
    (mapc #'(lambda (new)
            (if history
                (push (second new)
                        (gethash (first new) table))
                (setf (gethash (first new) table)
                    (second new))))
        new-states)))
```

The next-state function takes the state table and the single variable transition function and updates the state table by one time step. The transition function needs to be written in the appropriate way depending on whether it is using just current values or history. The third parameter for next-state, history, is a boolean, that specifies whether to keep a history or not (in other words, just keep the current state variable values). This function has two parts. The first part, the call to maphash, computes the new values for all the variables in the hash table, using the variable transition function, trans-fn. These values are accumulated on a list, as variable-value pairs. In the second part of the function, the mapc operator goes through the list of variables and new values and updates the hash table, either by adding each new value to a history list for its variable, or just setting the new value in place.

This strategy updates the variables in parallel. Each new value is computed from the current values and the history, in an order-independent way. If we immediately stored the new value of each back into the hash table, some variables would be computed from new values of others, and some would be computed from current values. This could lead to inconsistent results. So the parallel update makes for greater stability and is a better match to what happens in a real system.

The transition function, trans-fn, takes a single variable name and computes its next state value from the current contents of the state table, which could include looking at the history, not just the current state. When the transition function depends only on the current state and not the history, it is called a Markov process, which will be discussed in some more detail later in this chapter.

Finally, to actually run the simulation, we need to provide an initialized table, a transition function, and we need to specify whether to keep a history, and how many time cycles to run. The run-state-machine function takes care of this iteration. It might be nicer to have a more flexible termination facility, but for now, a simple iteration count will be good enough.

```
(defun run-state-machine (table trans-fn
                            &key history (cycles 1))
    (dotimes (i cycles table)
        (next-state table trans-fn history)))
```

Figure 5.4 A self-regulating box, that turns itself off when you try to turn it on.
(Photo courtesy of Hanns-Martin Wagner, `http://www.kugelbahn.ch/`*)*

A state can be as simple as a symbolic or numeric value, but more often a state machine has many state variables. Each state is represented by an assignment of one possible value to each variable. To start with, we will illustrate the idea with a simple but amusing (and surprisingly relevant) example simulation model.

5.4.2 A Simple State Machine

An amazing novelty gadget from my youth was a plain wooden box with a lid and a toggle switch on the front. When you flipped the switch to the "ON" position, the lid would open, and a plastic or rubber model of a hand would come out, over the front and down to the switch, flipping the switch to the "OFF" position, then the hand would retract into the box and the lid would close. This is not only entertaining but it also has a kind of feedback loop that makes the model somewhat interesting. Such self-regulation is also found in biological systems. Figure 5.4 illustrates one example of this box, a larger "trunk" style version.[10] The photographer and others attribute the idea for the box to Claude Shannon, the father of *Information Theory*, whose pioneering work [374] is the foundation of much of modern computing and communications technology. A brief introduction to Information Theory was presented in Chapter 3.

---

[10]More photographs and a small movie clip of the box in action can be seen at Hanns-Martin Wagner's web site, URL `http://www.kugelbahn.ch/sesam_e.htm`.

We can model this strangely biological system with three state variables, representing the states of the switch, the hand, and the lid. Each can be in two states. For the switch, the two states are on and off. For the hand, they are in and out, and for the lid, they are open and shut. We make a state table with make-state-table.

```
> (defvar *box* (make-state-table 3))
*BOX*
```

The initial states can be set up by calling add-variable.

```
> (add-variable *box* 'switch 'off)
OFF
> (add-variable *box* 'hand 'in)
IN
> (add-variable *box* 'lid 'closed)
CLOSED
```

Now we have to define a transition function. For each variable we need to decide what is its next state, depending on the current state of everything (and possibly the history). Since we are considering one state variable at a time, the shape of our transition function will be a case form with a clause for each variable.

```
(defun box-transition (var box)
  (case var
  (switch (switch-transition))
  (lid (lid-transition))
  (hand (hand-transition)))) 
```

Now, we have to replace each of the xxx-transition forms with what actually has to happen in each case. We will soon find out that what happens next does *not* depend on the history, but the transition rules take care of the proper sequencing. So, we will just keep values for current state, not history, in this state table.

First we deal with the switch. If the hand is out and the switch is on, the next state of the switch is off; the hand turns off the switch. If the switch is already off, its state does not change. Also, if the hand is in, the state of the switch does not change. You might think we have to consider the lid, but the lid does not affect the switch. It may be that the lid has to be open for the hand to transition to out but that is a different matter. One of the most important things about describing state transitions is to carefully describe only the direct relationships, and let the transition network as a whole handle the indirect relationships.

```
(defun box-transition (var box)
  (let ((switch-state (gethash 'switch box))
        (hand-state (gethash 'hand box))
        (lid-state (gethash 'lid box)))
```

```
(case var
  (switch (if (and (eql switch-state 'on)
                   (eql hand-state 'out))
              'off
              switch-state))
  (lid (lid-transition))
  (hand (hand-transition)))))
```

Next we consider the lid. If the switch is on and the lid is `closed`, the lid should go to the open state. If the switch is `off` and the hand is `in` and the lid is open, the lid should go to the `closed` state. In all other cases, the lid should not change state. For example, if the lid is open and the switch is `off` but the hand is `out`, the lid should not close, as it will break the hand. In any case, the box is not designed that way. You can check all the other combinations to be sure this description is right.

Finally, we consider the hand. If the hand is `in` and the switch is on, the hand should come out, but only if the lid is already open. If the switch is `off` and the hand is `out`, it should go in. In all other cases the state of the hand should not change. So, here is the complete transition function, with the code for the lid and the hand added in.

```
(defun box-transition (var box)
  (let ((switch-state (gethash 'switch box))
        (hand-state (gethash 'hand box))
        (lid-state (gethash 'lid box)))
    (case var
      (switch (if (and (eql switch-state 'on)
                       (eql hand-state 'out))
                  'off
                  switch-state))
      (lid (if (and (eql switch-state 'on)
                    (eql lid-state 'closed))
               'open
               (if (and (eql switch-state 'off)
                        (eql lid-state 'open)
                        (eql hand-state 'in))
                   'closed
                   lid-state)))
      (hand (if (and (eql switch-state 'on)
                     (eql hand-state 'in)
                     (eql lid-state 'open))
                'out
                (if (and (eql switch-state 'off)
```

```
                    (eql hand-state 'out))
          'in
      hand-state))))))
```

To run the simulation, we use the `run-state-machine` function, with `history` set to `nil`, the default, since we are just recording symbols representing the state of each variable, and updating them, rather than keeping a history. Adding some output operations using `format` enables us to see how the system goes through various states. We start by turning on the switch, and then run through some number of cycles. Here is the result of a sample run.

```
> (setf (gethash 'switch *box*) 'on)
ON
> (run-state-machine *box* #'box-transition:cycles 10)
Computing transition for LID
Lid opening
Computing transition for HAND
Computing transition for SWITCH
Computing transition for LID
Computing transition for HAND
Hand coming out
Computing transition for SWITCH
Computing transition for LID
Computing transition for HAND
Computing transition for SWITCH
Switch turning off
Computing transition for LID
Computing transition for HAND
Hand going back in
Computing transition for SWITCH
Computing transition for LID
Lid closing
Computing transition for HAND
Computing transition for SWITCH
Computing transition for LID
Computing transition for HAND
Computing transition for SWITCH
  . . .
```

The behavior is as intended: the lid opens, the mysterious hand comes out, turns the switch off, goes back in, and the lid closes. After that no further state changes happen. To activate the box, we have to turn on the switch externally again with our `setf` expression.

This approach encodes all the state transition knowledge in the logic of a function, or, hard coding. This does not scale well as the number of variables increases. Other approaches to writing the transition function are possible. One is to encode the transition logic in rules in a format like predicate calculus or FOL, and make the transition function use a rule interpreter (inference engine). Another is to create a state transition table, indexed by the state variables, and write the transition function as a table lookup function. Section 5.4.4 describes a state table approach in the design of a state machine implementing the DICOM medical image transmission protocol. Reimplementing the box with a state table instead of state variables is left as an exercise.

In the case of the box, the model allows any combination of values for the state variables as a state for purposes of description, even though certain states are not physically achievable or desirable. This is especially important when discussing computer controlled medical devices. The manufacturer may have conceived of the device as having a few states that are the normal operating states, but neglect the fact that many other states may in fact be achievable, though they are *unsafe*. The objective of safety analysis and state machine modeling is precisely to identify the undesirable states and to determine by analysis whether a given design can ever get into an unsafe state, rather than pretending these states can't happen. This idea is illustrated more in detail in Section 6.2.6.

Many embellishments of this basic framework are also possible. One is to provide a way to print out progress information as illustrated above, or even a display of the entire state description at each time step. Another is to provide some other kinds of debugging hooks. These things could be done by adding lines of code in `run-state-machine` or `next-state` to print out information. They could also be done in the transition function. It's difficult to do this in a general way because we have very little restriction on the actual simulation state details. A better way would be to include an optional input parameter, yet another function to be provided, that deals with the details of the data. We will leave this also as an exercise for the reader.

5.4.3 Example: Simulating the Cell Cycle in Yeast

Now we apply the state machine code to a real biological problem, that of simulating gene regulation. This section is based on a tutorial created by Jeff Shrager, as part of his BioBike [380] biocomputing project. The tutorial in turn is based on a report by Li et al. [251], who used state-based simulation to investigate the robustness of the cell cycle regulation network in yeast.

The cell cycle is a growth and division process that cells go through as a part of their normal activities. For single celled organisms, the division process, mitosis, is the means of reproducing, giving rise to two new organisms from each parent organism. The new cells then grow and eventually divide, and thus the process continues. In multi-celled organisms, cell growth and division is part of normal tissue growth and regeneration. The rate at which this happens varies according to the tissues. Skin cells reproduce rapidly and are lost to the surrounding environment.[11] Nerve cells reproduce slowly or not at all.

In all eukaryotic cells, the cell cycle process consists of four phases, labeled G1 (the phase in which the cell is growing), S (the phase of DNA synthesis and chromosome replication), G2 (a resting phase before the actual cell division phase), and M (the phase in which mitosis takes

---

[11] Actually a significant component of common house dust consists of dead skin cells and debris.

place and the cell divides into two "daughter" cells). Each daughter cell, following mitosis, then enters the G1 phase and the process begins anew.

In yeast, *Saccharomyces cerevisiae*, the genes involved in cell cycle regulation have been thoroughly studied [392]. Li's paper identifies 11 genes that are the most important regulatory elements, out of about 800 genes that have been reported to be involved. The genes are related pairwise, with each gene having either an activation effect or an inhibition effect on each of a small number of other genes. Some genes have no negative or inhibitory inputs. These genes are hypothesized to "self degrade" if they do not receive positive input for a specified time interval. This means that if the gene does not receive some activation input after a while, it turns off by itself.

Li and colleagues asked the question, can the cell cycle regulation process be demonstrated to be robust? If there are random perturbations or some variables start out with the wrong values, will the cycle still progress to an appropriate end point? By creating a state machine and running it with all possible combinations of initial states, they were able to show that this particular regulatory network is extremely stable.

Here we will develop code and a state representation that can reproduce the results of Li and can be used to do other kinds of theoretical investigations. For each gene there is a corresponding protein, the gene product. In the following we refer to genes and proteins interchangeably. We will use the same state machine code as for the box with the hand that comes out, but a slightly more complicated representation of the state variables and state transition function.

The model will include a state variable for each gene. A gene can be either active or inactive, represented by a numeric value of 1 or 0 for its variable. A gene's effect on another can be either positive, negative, or zero, depending on its position and on the arcs in the model, as well as its state. If a gene is inactive, it will not affect other genes, even if it is linked by an arc in the model graph.

This time we need to have the state incorporate history so that the state transition function can determine if a protein or gene needs to "self-degrade." So, the state representation for each protein will be a list, with an element for each time step. Each new state will go on the front of the list, so that the oldest state value is at the end of the list and the most recent at the front. Each state value will be a two element list. The first element of this list will be a 1 or 0 representing the current value of the state for that protein. The second element will be the total influence applied to that protein during the previous cycle. Thus, a protein that alternates on and off each time step, and has only one steady positive influence and no others would have a history something like this:

```
(1 1) (0 1) (1 1) (0 1) (1 1) (0 1)
```

Now we need a way to represent the network, Figure 5.1B in the Li article. Each gene when active produces a protein that is the agent actually participating in the process, so we will refer to the proteins in the following description. Each protein will be represented by a symbol, whose print-name is the usual abbreviation for the protein. The parameter *model-graph* will hold a list of all the arrows with their endpoint proteins. Each of the arrows or connections between proteins in the network will be represented as suggested by Shrager, by a triple (a three element list). The first element is the protein at the tail of the arrow, the protein that will *affect* another protein. The third element is the protein at the head of the arrow, the protein that is *affected*. The second element is a + if the tail protein positively affects the head protein, and – if the tail protein negatively affects the head protein. As in the graph, one protein can affect several others, and similarly, one protein can be affected simultaneously by several others.

```
(defparameter *model-graph* ;; (fig 1b)
   '((size + cln3)
     (cln3 + mbf)
     (cln3 + sbf)
     (sbf + cln12)
     (cln12 - sic1)
     (cln12 - cdh1)
     (sic1 - clb56)
     (sic1 - clb12)
     (mbf + clb56)
     (clb56 + mcm1/sff)
     (clb56 - sic1)
     (clb56 + clb12)
     (clb56 - cdh1)
     (cdh1 - clb12)
     (clb12 - cdh1)
     (clb12 + mcm1/sff)
     (clb12 + cdc20&14)
     (clb12 - sic1)
     (clb12 - swi5)
     (clb12 - mbf)
     (clb12 - sbf)
     (mcm1/sff + cdc20&14)
     (mcm1/sff + swi5)
     (mcm1/sff + clb12)
     (cdc20&14 + swi5)
     (cdc20&14 + cdh1)
     (cdc20&14 - clb12)
     (cdc20&14 + sic1)
     (cdc20&14 - clb56)
     (swi5 + sic1)
     ))
```

Now, we need to create a way to compute what the new value or state should be for each of the proteins. This will be the function that will serve in the state machine as `trans-fn`.

We will put all the arrow information in a hash table, not the state table, but one that can be used by the transition function. This hash table should be indexed by the target proteins, since we will want to look up how each target protein is influenced by the other proteins. Li represents this step as a mathematical sum, but that means doing matrix multiplication with a sparse matrix. It is not so much that it is inefficient, as it is a less direct way to represent the graph than just writing down a list of all the arcs.

```
(defun make-graph-hash (graph)
  "returns a new hash table indexed by the target proteins,
  containing the influencing protein and its sign."
  (let ((graph-table (make-hash-table)))
    (dolist (arc graph graph-table)
      (push (butlast arc)
            (gethash (third arc) graph-table)))))
```

The `butlast` function gives a list of the first two elements of each arc in the graph. This becomes the content of the transition table entry. The `third` function gives the target protein at the head end of each arc, and it is the index for looking up or setting the contents.

To create the initial state of the simulation, we need to be able to turn on (or off) a list of proteins all at once, so we write a function to do that. The `set-proteins` function takes a list of proteins (genes), a state to set, and the state table, and for each protein, it adds a new current state to that protein's history, consisting of the new state and a total influence of 0.

```
(defun set-proteins (proteins state table)
  (dolist (p proteins)
    (push (list state 0) (gethash p table))))
```

We need to be able to initialize the network so that any arbitrary subset of proteins is initially on. In order to do this, we need to also initialize the rest to be initially off. This is easy if we generate a list of the nodes (the proteins) from the graph. The `graph-node-list` function will extract all the protein names from the graph, giving a list of all the state variables. The use of `pushnew` insures that we don't have duplicates.

```
(defun graph-node-list (graph)
  (let (nodes)
    (dolist (arc graph nodes)
      (pushnew (first arc) nodes)
      (pushnew (third arc) nodes))))
```

From this list we can initialize the states of all the proteins to be initially off. Then we turn on the ones that are to be initially on. These operations are performed by the `init-cell-cycle` function.

```
(defun init-cell-cycle (graph initially-on-list)
  (let* ((nodes (graph-node-list graph))
         (state-table (make-state-table (length nodes))))
    (set-proteins nodes 0 state-table)
    (set-proteins initially-on-list 1 state-table)
    state-table))
```

The next problem to solve is how to deal with the self-degrading proteins. We just make a list of them for now, so that we can include that in the transition function as one possible way in which a transition might take place. We also parametrize the degradation time for the self-degrading proteins, to make it easy to experiment with varying the degradation time.

```
(defparameter *self-degrading-proteins*
  '(cln3 cln12 mcm1/sff swi5 cdc20&14))

(defparameter *degrade-time* 5)
```

Keeping in mind layered design and providing a little abstraction layer, here is a predicate function that checks if a protein is a self-degrading protein.

```
(defun self-degrading (p)
  (find p *self-degrading-proteins*))
```

Finally, here is how the self-degradation is computed. We retrieve the history list for the protein from the state table, and determine if all three conditions are met for the protein to degrade: it must be one of the self-degrading proteins, the history must be at least as long as the degrade time, and the most recent states as counted by the degrade time have to be the combination of "on" with no influences, namely, the value (1 0).

```
(defun self-degradation (p table)
  (let ((history (gethash p table)))
    (and (self-degrading p)
         (>= (length history) *degrade-time*)
         (every #'(lambda (state)
                    (equal state '(1 0)))
                (subseq history 0 *degrade-time*)))))
```

Now we can compute the new state for any protein, given the state table, the hashed form of the graph and the protein's current state (or history). First compute the sum of all the influences on the protein, from the graph hash table. If it is positive, the new state is a 1, if negative the new state is a 0, and if the influence is 0, check if the protein should self-degrade, otherwise it just stays in its previous state. Of course what is actually returned is a list of the new state value and the influence sum.

```
(defun new-protein-state (p state-table graph-hash)
```

```
    "returns a list containing a 1 or 0 for the next state and the
    influence sum, for a particular protein, p, based on the
current
    state. If p is a self-degrading protein, its history is checked
    and if it is time to degrade, its state becomes 0."
    (let ((influence-sum 0))
      (dolist (p2-arc (gethash p graph-hash))
        (let* ((p2 (first p2-arc))
               (p2-state (first (gethash p2 state-table)))))
          (case (second p2-arc)
            (+ (incf influence-sum p2-state))
            (- (decf influence-sum p2-state)))))
      (list
        (cond ((> influence-sum 0) 1)
              ((< influence-sum 0) 0)
              (t (if (self-degradation p state-table) 0
                     (first (gethash p state-table)))))))
        influence-sum)))
```

Finally, we can now write the code for the simulator. The code consists of starting with the graph, a list of what proteins should be initially on, and a number of cycles to run. First we create the state table from the graph, create the influence hash table, run one cycle with the initially on proteins, then turn them off, then run the specified number of cycles.

```
(defun run-cell-cycle (graph initially-on n)
  (let ((state-table (init-cell-cycle graph initially-on))
        (graph-hash (make-graph-hash graph)))
    (run-state-machine state-table
                       #'(lambda (var table)
                           (new-protein-state var table
                                              graph-hash))
                       :history t :cycles 1)
    (set-proteins initially-on 0 state-table)
    (run-state-machine state-table
                       #'(lambda (var table)
                           (new-protein-state var table
                                              graph-hash))
                       :history t :cycles n)))
```

This function will run a specified number of iterations (n) of the state machine described by the gene regulation graph provided as the graph parameter, with the specified initially on genes. When the function terminates it returns the state table as its result. A test run for the 11 gene network described by Li is shown here.

```
> (setq cell (run-cell-cycle *model-graph* '(size) 25))
...
#<EQL hash-table with 12 entries @ #x714e1f9a>
> (print-cell cell)
SBF ((0 0) (0 0) (1 1) (1 1) (1 1) (1 0) (1 0) (1 0) (0 -1) (0 -1)
     (0 -1) (0 -1) (0 -1) (0 -1) (0 -1) (0 -1) (0 -1) (0 0) (0 0)
     (0 0) (0 0) (0 0) (0 0) (0 0) (0 0) (0 0))
MCM1/SFF ((0 0) (0 0) (0 0) (0 0) (1 1) (1 2) (1 2) (1 2) (1 2) (1 2)
          (1 1) (1 1) (1 1) (1 1) (1 1) (1 1) (1 1) (1 0) (1 0) (1 0)
          (1 0) (1 0) (0 0) (0 0) (0 0) (0 0) (0 0))
CLN12 ((0 0) (0 0) (0 0) (1 1) (1 1) (1 1) (1 1) (1 1) (1 1) (1 0)
       (1 0) (1 0) (1 0) (1 0) (0 0) (0 0) (0 0) (0 0) (0 0) (0 0)
       (0 0) (0 0) (0 0) (0 0) (0 0) (0 0) (0 0))
SIZE ((0 0) (1 0) (1 0) (0 0) (0 0) (0 0) (0 0) (0 0) (0 0) (0 0)
      (0 0) (0 0) (0 0) (0 0) (0 0) (0 0) (0 0) (0 0) (0 0) (0 0)
      (0 0) (0 0) (0 0) (0 0) (0 0) (0 0) (0 0))
CDH1 ((0 0) (0 0) (0 0) (0 0) (0 -2) (0 -3) (0 -2) (0 -2) (0 -2)
      (0 -2) (0 -1) (0 -1) (0 -1) (0 -1) (0 -1) (0 0) (0 0) (1 1)
      (1 1) (1 1) (1 1) (1 1) (1 1) (1 1) (1 1) (1 1) (1 1))
CLB12 ((0 0) (0 0) (0 0) (0 0) (1 1) (1 2) (1 1) (1 1) (1 1) (1 1)
       (1 0) (1 0) (1 0) (1 0) (1 0) (1 0) (0 -1) (0 -1) (0 -2) (0 -2)
       (0 -2) (0 -2) (0 -2) (0 -3) (0 -3) (0 -3) (0 -3))
SIC1 ((0 0) (0 0) (0 0) (0 0) (0 -2) (0 -3) (0 -2) (0 -1) (0 -1)
      (0 -1) (0 0) (0 0) (0 0) (0 0) (0 0) (1 1) (1 1) (1 2) (1 2)
      (1 2) (1 2) (1 2) (1 2) (1 2) (1 2) (1 2) (1 2))
CDC20&14 ((0 0) (0 0) (0 0) (0 0) (0 0) (1 2) (1 2) (1 2) (1 2) (1 2)
          (1 2) (1 2) (1 2) (1 2) (1 2) (1 2) (1 2) (1 1) (1 1) (1 1)
          (1 1) (1 1) (1 1) (1 0) (1 0) (1 0) (1 0))
MBF ((0 0) (0 0) (1 1) (1 1) (1 1) (1 0) (1 0) (1 0) (0 -1) (0 -1)
     (0 -1) (0 -1) (0 -1) (0 -1) (0 -1) (0 -1) (0 -1) (0 0) (0 0)
     (0 0) (0 0) (0 0) (0 0) (0 0) (0 0) (0 0))
SWI5 ((0 0) (0 0) (0 0) (0 0) (0 0) (0 0) (1 1) (1 1) (1 1) (1 1)
      (1 1) (1 1) (1 1) (1 1) (1 1) (1 1) (1 1) (1 2) (1 2) (1 2)
      (1 2) (1 2) (1 2) (1 1) (1 1) (1 1) (1 1))
CLB56 ((0 0) (0 0) (0 0) (1 1) (1 1) (1 1) (1 0) (1 0) (1 0) (0 -1)
       (0 -1) (0 -1) (0 -1) (0 -1) (0 -1) (0 -1) (0 -2) (0 -2) (0 -2)
       (0 -2) (0 -2) (0 -2) (0 -2) (0 -2) (0 -2) (0 -2) (0 -2))
CLN3 ((0 0) (1 1) (1 0) (1 0) (1 0) (1 0) (1 0) (0 0) (0 0) (0 0)
      (0 0) (0 0) (0 0) (0 0) (0 0) (0 0) (0 0) (0 0) (0 0) (0 0)
      (0 0) (0 0) (0 0) (0 0) (0 0) (0 0) (0 0))
```

We ran the process for 25 iterations. For each gene, the succession of states is shown. The first number in each pair is the state, where a 0 means "off" and a 1 means "on." The second number is the influence sum at that time step. So, for example, CLB12 starts in the off state, but after four time steps it turns on, stays on for 12 more time steps, and then turns off. After about 20 time steps, it seems that the process has settled down, and all the genes have reached a quiescent or final state, beyond which no further change will happen until the size signal is turned on.

In this case, there is just one "initially on" protein, the size signal. If there is more than one initially on protein, and we don't want to turn all of them off, just the external signal, the initialization steps will have to be more complex. This is left as an enhancement to be added by the reader.

The project reported by Li went on to examine the paths followed by every possible starting state (2048 in all). They found that a large majority of them eventually converged to the steady state, and at some point joined up with a path that they identify as the canonical cell cycle regulation process. Their conclusion was that the cell cycle regulation system is very robust. They did similar experiments with perturbations of the network, for example, leaving out one or more interaction arrows, and found similar behavior.

Building a larger scale system of this sort would enable a large variety of such "in silico" experiments, not only with cell cycle regulation but with other gene network processes as well. For more information the reader is referred to [39] which describes differential equation modeling systems as well as logical and stochastic modeling.

As with other examples, this code can be improved by introducing some abstraction layers for getting and setting state, as well as other enhancements.

5.4.4 Example: Implementing DICOM with a State Machine

Section 4.3.3 provided an introduction to the basic ideas of the DICOM medical image data transfer protocol. DICOM is a network communication protocol, as well as a standard for medical imaging data representation in network messages. DICOM is implemented by writing network client and server programs, as described in Section 4.3.1. In particular, the DICOM standard describes the DICOM Upper Layer (DUL) Protocol as a state machine, so this section describes how to implement it as such.

Because DICOM has two layers of encoding, one for the structure of a PDU, and the other for the data contained in a P-DATA-TF PDU, the design we will develop similarly will have two layers, each with a grammar, a dictionary, and an interpreter.

The core code will implement a *finite state machine*, in which the system can be in any of a number of predefined states. As conditions change (messages arrive, results are stored, messages are sent), *events* are generated, which lead to other states via specified *actions*. In this implementation we will need to:

- create a representation of the state transition table and write code for a state machine, and write the action functions that perform all the different kinds of actions needed depending on the current state and event,
- create a representation of the syntax and semantics of the different PDU types, and implement an interpreter (parser and generator) for PDUs,

- create a representation of the syntax and semantics of the different commands and data contained in a P-DATA-TF PDU, and implement an interpreter (parser and generator) for that.

We will now develop each of these ideas in detail.

5.4.4.1 The DICOM State Machine

In the basic design for a server or client, we need to write a function, applic-fn, that will implement the application protocol, in this case, DICOM. The first problem we face is that the simple server and client code of Section 4.3.1 will not work for DICOM. In that code, the opening and closing of connections is handled outside the implementation of an application protocol. This is an example of what is meant by "layered design." The protocol simply assumes it has an open channel and deals only with the data that flow through it. Unfortunately, the DICOM protocol does not conform to this design, but mixes up the process of opening and closing connections with the processing of messages. So the DICOM server's version of applic-fn includes the call to accept-connection and close for the stream. Similarly, the client code opens the connection as part of the protocol. We can make the same code do for both, by modifying the server and client code. Our applic-fn will be called dicom-state-machine. One additional input it will have is a mode parameter, :server or :client depending on which is needed. Also, we will keep a lot of working information for our state machine in an environment variable, which, for the server starts out empty (nil), but for the client, it contains the remote connection information. Here is the server:

```
(defun dicom-server (port)
   (let ((socket (make-socket :connect :passive
                              :address-family :internet
                              :type:stream
                              :local-port port)))
      (unwind-protect
          (dicom-state-machine socket :server nil))
      (close socket)))  ;; only here if severe error occurs
```

The client code is very simple, since the make-socket call will be done inside the dicom-state-machine function.

```
(defun dicom-client (host port)
   (let ((env (list (list 'remote-host-name host)
                    (list 'remote-port port))))
      (unwind-protect
          (dicom-state-machine nil :client env))))
```

The standard defines the *control structure* or sequencing of actions via a *state table* which relates States, Events, and Actions. *States* name situations the system can be in (waiting for a certain PDU, sending data, etc.), *events* name conditions triggering certain actions, and *actions* name the operation to be performed in the appropriate situation. A *state table* (or *state transition table*) is simply a table containing an entry for each legal state/event combination. Each entry indicates what action to perform under that condition and what state to enter for the next cycle.

Representing States and Transitions

In addition to the environment association list, which is used for incoming and outgoing data of DICOM messages, some global variables are needed for the state machine. In particular, some of the action functions will signal an event. The event is represented by a symbol, and we will use a global variable, *event*, to hold this value. The state variable does not need to be global in the same sense, as it is only set in the body of the state machine main loop. A first sketch of dicom-state-machine might then look like this:

```
(defvar *event* nil)

(defun dicom-state-machine (socket mode env)
   (let ((tcp-stream nil)
         (state 'state-01)
         (trans-data nil)
         (action-fn nil))
     (loop
       (setq trans-data (get-transition-data state *event*)
             action-fn (first trans-data)
             env (funcall action-fn env tcp-stream)
             state (second trans-data)))))
```

The system starts in state 1, events are signaled (presumably by the action functions), and the system moves to the next state, processing the next event, and so on. The function get-transition-data is not yet defined. It should return a two element list consisting of the name of the action function to call for the state transition, and the name (a symbol) of the next state. The form of this function will depend on how we store information about the states and what to do for each event.

We have left out some important details. How do events actually happen? When (where in the program) are PDUs read from the client, and when are they written? Some of this is easy. Certain action functions will include the writing of a PDU; the action function does not return until the PDU is read by the remote system. An action may also involve storing or retrieving data, or even opening or closing the connection. Action functions in some cases will have to provide an event value, to be used in the next loop iteration. So the operation of the server is expected to be synchronous. We will return to the question of what the actions do and how to code them, after creating the representation of the state table.

Unfortunately, this first attempt at the `dicom-state-machine` function will not work. The action functions write PDUs but do not read them. A look at the definitions of the states suggests that in some states, a PDU read is posted, and the result of reading and processing the incoming PDU also can signal the next event. So, we add code to handle this. In state 1, the system has to behave differently depending on whether it is operating as a client or server. In server mode it is waiting for a connection, so that is where we will have the call to `accept-connection`. In client mode, it simply signals `event-01`. In states 2, 5, 7, 10, 11, and 13, the system is waiting for either an incoming PDU or for the connection to be closed by the remote system. In either case, we post a call to `read-pdu` and then handle the resulting event. Here is the augmented code:

```
(defvar *event* nil)

(defun dicom-state-machine (socket mode env)
  (let ((tcp-stream nil)
        (state 'state-01)
        (trans-data nil)
        (action-fn nil)
        (iteration 0))
    (loop
      (case state
        (state-01 (if (eq mode :server)
                      (setq tcp-stream
                            (accept-connection socket :wait t)
                            *event* 'event-05)
                    (if (= iteration 0)
                        (setq *event* 'event-01)
                      (return))))
        ((state-02 state-05 state-07 state-10 state-11 state-13)
         (setq env (read-pdu tcp-stream env))))
      (setq trans-data (get-transition-data state *event*)
            action-fn (first trans-data)
            env (funcall action-fn env tcp-stream)
            state (second trans-data)
            iteration (1+ iteration)))))
```

In client mode, the state machine should only go through one complete sequence of operation, and when it returns to `state-01` after the first time, it should exit. The function defined above will work well as both a server and a client. The `read-pdu` function will use our new (yet to be defined) language for DICOM messages.

An important goal in any large software project is to design a modular system, one in which component parts can be understood on their own, without too much (hopefully without any) involvement with the context in which they are used in the system. In our case, we can set as goals the following:

- state transitions (what will be the next state) should only depend on the current state and the current (most recent) event, and not on details of action functions that signal events,
- action functions should perform well-defined operations independent of the state or event that triggered them,
- the reading of a PDU should not depend on the current state, and should only have two effects, to add to the environment and to signal an event.

These goals are consistent with the design of the DICOM Upper Layer Protocol, as specified in Part 8, with a few exceptions that we will address later.

Now we consider how to represent the state transition table (Table 9–10 in DICOM Part 8). There are 13 states, and we label them with symbols of the form, `state-nn` where the `nn` is a number, like 06 or 11. We similarly label all the possible events with symbols like `event-05`, and for each transition there will be an action function as well. The action functions will get a variety of different labels, depending on the kind of actions they represent, but the idea is the same, each action function is identified by a symbol. For ease of reference to the DICOM standard, we use symbols that match the names in the standard (except that "state" is spelled out instead of being abbreviated "sta").

A state table then consists of a list of entries, one for each state. Each entry needs to contain the state name, and for each applicable event, the event name and the name of the appropriate action function and next state. For human readability we can add that each state entry will have a string containing lines of text to explain which state it is. We use a string instead of comment lines, because the string will be machine readable, and perhaps can be used later for an online documentation system.

We can now make a direct translation of the standard's state table into a Lisp data structure consisting of a list of entries as described above and illustrated in Figure 5.5. This state table representation serves both to document the control structure of the system to the programmer and as machine readable data to be used directly by the state machine code.

The table entry example shown in Figure 5.5 shows the information associated with the state labeled `state-06`. The table entry includes a descriptive comment, the string "Association established, ready for data transfer." For each event that may occur in this state, there is a list including the event label, the action function to invoke, and the next state.

In some cases, the new state is the same as the original, and in others the new state is different. For example, in state `state-06`, the response to the signaling of `event-09`, which is a request for data, is to invoke action function `dt-01`, which constructs and sends the appropriate data packet, and to transfer next to (actually to remain in) `state-06`, ready to send more data. On the other hand, the response to the signaling of `event-11`, which is a request to release the

```
(state-06
 "Association established, ready for data transfer"
 ((event-09) dt-01 state-06)
 ((event-10) dt-02 state-06)
 ((event-20) dt-03 state-06)
 ((event-21) dt-04 state-06)
 ((event-11) ar-01 state-07)
 ((event-12A event-12B) ar-02 state-08)
 ((event-16) aa-03 nil)
 ((event-17) aa-04 nil)
 ((event-15) aa-01 state-13)
 ((event-03 event-04 event-06 event-13 event-19)
         aa-08 state-13))
```

Figure 5.5   State table entry for data transfer state.

communication arrangement, is to invoke action function `ar-01`, which is the action to release the association, followed by a change to `state-07`, "Awaiting Associate-Release response."

Where several events are grouped in parentheses (for example, the row containing `event-12a` and `event-12b`), the meaning is that any of the listed events signals the invocation of the indicated action function and state transition.

The next state is listed last in each row. A next state of `nil` means that the server exits the DUL interpreter loop, closes the TCP connection to that client, and listens for another connection request. For a client, it means that control will return to the application part of the client code, which can then take appropriate action on behalf of the user.

We create a global variable for the state table, a list of all the entries.

```
(defparameter *state-table*
   '((state-01 "Awaiting establishment of connection"
        ((event-01) ae-01 state-04)
        ((event-05) ae-05 state-02))
     (state-02 "Awaiting A-Associate-RQ"
        ((event-06) ae-06 state-03)
        . . .
        ...)))
```

Now we can implement the accessor function, `get-transition-data`.

```
(defun get-transition-data (state event)
   (let* ((state-entry (find state *state-table* :key #'first))
          (evlist (rest (rest state-entry)))
          (ev-entry (find event evlist :test #'member :key #'first)))
      (rest ev-entry)))
```

The first call to find gets the entry for state, a simple linear search. The first two elements of the entry can be discarded, and the remainder searched for event, then the remainder of that is the two element list desired.

Action Funtions

Altogether the DICOM standard defines 28 different action functions, including 8 related to the dialog for establishing an association, two related to data transfer, 10 related to association release, and 8 to handle various abort related conditions. The actions may just involve signaling an event, constructing and sending a PDU, starting or stopping a timer to handle failure or non-response from the remote program, opening and closing the transport (TCP) connection, and some other actions. We will just provide some examples here. Since we have not yet explained how to write the code that parses or generates PDUs or PDVs, we will encapsulate the problem by assuming we can simply call functions to perform those operations.

Both server and client code will start out in state-01. A server will at this point be waiting for an incoming TCP connection on port 104 (or another port if the configuration specifies an alternate). When a connection is made, that is, the call to accept-connection returns, this is event-05, action function ae-05 is called and the next state will be state-02. In state-02 the server should post a read, awaiting a PDU, and then depending on what happens with that process, a new event will be posted, a new action function will be called, and a next state transition will take place. In most cases these action functions will be very simple. As we will see, some merely signal an event and others do nothing at all. Their presence only serves to provide a uniform way in which the state machine will work, even when specific actions are not necessary beyond updating the state variables.

You will notice that the lists in DICOM Part 8 (Tables 9.6 through 9.9) also specify the next state. This is not modular, though it is accurate. Instead we represent the action functions as making events happen, which is how the next state is determined.

The first action function pertaining to server operation is ae-05, which is invoked to indicate that the server has received an incoming connection. This function simply returns nil, and that means the environment will be reset to an empty one. Nothing else need be done at this point except possibly to write a message to the log file. After a connection is established, the next thing to happen is to read a PDU (in the main state machine loop). That will in turn signal an event and the state machine will proceed to the next iteration.

```
(defun ae-05 (env tcp-stream)
   "Issue connection open message"
   nil) ;; return a fresh, empty environment
```

The next action function, ae-06, does a lot more work. It needs to examine the environment to see if an A-ASSOCIATE-RQ PDU came in with a request acceptable to the server. If it is, the action is to signal event-07, an "accept" response indication, and if not, then signal event-08.

Now, if you look carefully in DICOM Part 8 at the state table and the association establishment action function table, you will see that the designers of DICOM tried to short circuit what should happen, and made the state table appear non-deterministic. What should be in the state table is that in `state-02` with `event-06` the next state is always `state-03`. The event that is signaled will be handled in `state-03` and the next transition, to either `state-06` or `state-13`, will be determined by which event (7 or 8) was signaled by `ae-06`. And this is how we will implement it.

What determines if the proposed association is acceptable? Here there are many design choices. At a minimum, if the request includes a proposed SOP Class UID, identifying the operation and data that are requested (it is actually not required at this point), it is a good idea for the server to check if it is capable of providing that service. Most DICOM implementations do nothing more than this. It would be better to determine if the requestor was known to the server (this could be done with a configuration file with a list of allowed remote clients), and if the remote client is using a DICOM protocol version the server supports. One could check against the IP address of the remote, and the AE Title. This kind of check is the job of the functions `remote-id-ok` and `SOP-Class-UID-OK`, which we won't describe here.

Another job of this function is to examine the proposed Presentation Contexts, to see if any are supported by the server. Since the DICOM default transfer syntax is required to be supported and always available, this should always succeed. The option to offer (and for the server to accept) other transfer syntaxes is available, and a well-designed server should examine what was sent, to check this. The Presentation Context items will be in a list under the `:A-ASSOCIATE-RQ` tag in the environment, as shown in the sample environment list earlier in this section.

```
(defun ae-06 (env tcp-stream)
   "Determine acceptability of A-ASSOCIATE-RQ"
   (cond ((and (remote-id-ok env tcp-stream)
               (SOP-Class-UID-OK env))
          (setq *event* 'event-07)
          (setq *presentation-contexts*
               (check-presentation-contexts env)))
         (t (setq *event* 'event-08)
            (setq *reject-reasons* (reject-reasons env)))))
   env) ;; return env unchanged
```

The values stored in the global variables will be used by the actions in the next iteration of the state machine. The two functions `check-presentation-contexts` and `reject-reasons` will not be developed here, but for this level of detail, you can refer to the full implementation at the author's web site.

We need action functions that will send association accept or reject messages. These are ae-07 and ae-08, respectively. They each will send the appropriate PDU. There is no need to signal an event, since the next state will require a PDU read operation. When accepting an association, the server accepts the client's proposed maximum PDU size and caches the smaller of it and the server's internal parameter specifying maximum PDU buffer size. The server then is assured of not exceeding what the client can handle.

```
(defun ae-07 (env tcp-strm)
   "Issue A-Associate-AC message and send associated PDU"
   (let ((limit (path-find '(:A-Associate-RQ
                             :User-Information-Item
                             :Max-DataField-Len-Item
                             Max-DataField-Len)
                           env)))
      (setq *max-datafield-len*
        (if (typep limit 'fixnum) (min limit #.PDU-Bufsize)
           #.PDU-Bufsize))
      (apply #'send-pdu :A-Associate-AC env tcp-strm *args*)
      (setq *args* nil)))

(defun ae-08 (env tcp-strm)
   "Issue A-Associate-RJ message and send associated PDU"
   (apply #'send-pdu :A-Associate-RJ env tcp-strm *args*)
   (setq *args* nil))
```

We defer developing the actions for sending and receiving data for now.

The Association Release actions are all straightforward, and require either sending an appropriate PDU (constructed from the current environment), or signaling an event, or closing the TCP connection. The code should match closely with the description in DICOM Part 8, Table 9–8. Just one of the Association Release action functions is shown here. The rest should follow the same general design.

```
(defun ar-01 (env tcp-strm)
   "Send A-Release-RQ PDU"
   (send-pdu :A-Release-RQ env tcp-strm)
   nil)
```

Conditions occur which require termination of communication, and these are handled by the Abort actions. These conditions include:

- receipt of an unexpected PDU, one that is inappropriate for the current state,
- expiration of a timer, indicating that the remote system has become inaccessible or some other remote or network problem,
- the TCP connection is closed unexpectedly by the remote system.

The Abort actions are also simple. They either attempt to send an A-Abort PDU or simply close the TCP connection on the local end of the communication, or simply ignore the unexpected PDU. Here again, the reader may try implementing these or find the full original code at the Prism DICOM web site mentioned earlier.

So far, we have shown mostly server side actions, so now we look at some more examples of client side actions. Function ae-01 is defined as "Issue transport connect request to local transport service" (this refers to TCP, although in the future other protocols such as OSI could be supported). So, this function opens an active connection to a specified remote server. This is the client action in state-01. The variable tcp-strm is not used because a new connection is being opened, but all action functions are called with the same argument list, so it needs to be in the parameter list.

```
(defun ae-01 (env tcp-strm)
(open-connection (path-find '(remote-hostname) env)
                 (path-find '(remote-port) env))
(setq *event* 'event-02)
env)
```

The remote host name and port should already be in the environment. This is one of the differences between client and server. The open-connection function will perform the call to make-socket that we originally put in the outermost layer.

Function ae-02 is defined as "Send A-Associate-RQ PDU." It is also a client-only function. It is simply:

```
(defun ae-02 (env tcp-strm)
  (send-pdu :A-Associate-RQ env *tcp-buffer* tcp-strm)
  env)
```

The global variable *tcp-buffer* should be an array of bytes, created at startup, big enough to hold the largest expected PDU.

Note that no event is signaled here, so env is returned unchanged. You might be wondering why these two functions are client-only, while ae-05, ae-06, etc. are server only. It reflects the convention that only clients request associations, while servers only accept or reject association requests.

5.4.4.2 Parsing and Generation of PDUs

Now we turn to reading, parsing, generating, and writing PDUs. We treat each PDU as a sentence in a DICOM language. The DICOM specification may be viewed as defining the grammar and vocabulary for these sentences. A particular DICOM message then can be treated as a sequence of tokens whose content and ordering conforms to this grammar. The vocabulary is defined by the data dictionary. The structure or sequencing of the tokens is expressed in rules, expressions that specify what sequences of tokens constitute valid messages.

Since a PDU can contain a variable number of items and there is no way to determine how many by simply parsing tokens sequentially, every PDU has a length field, that tells the reader how many bytes to read (and subsequently process). The length field is a 32-bit binary quantity encoded in bytes 2–5 (counting the first byte as byte 0). So first, we read the first 6 bytes, then using the length information, read the remaining bytes, all into a buffer that will be passed to a parser. A function which we will call decode-fixnum is needed to convert the four byte sequence into an integer data type.

Several things might happen at this point, including error conditions like "connection closed or reset" at the remote end, or an end-of-file condition, or the number of bytes read does not match with the number of bytes expected. In those cases, the data are ignored and the appropriate event signaled. If all is well and the read completes as requested, the rule-based-parser function (to be defined shortly) is called to translate the contents of the PDU into entries in the environment list. Then depending on what type of PDU it was, the appropriate event is signaled. Here is all that in the read-pdu function.

```
(defun read-pdu (tcp-stream env)
  (let ((tail 6) ;; read only 6 bytes to start
        (pdu-end 0) ;; will be set when PDU length is known
        (connection-reset nil)
        (timeout nil)
        (eof? nil))
    (unless
      (ignore-errors
        (cond ((= tail (read-sequence *tcp-buffer* tcp-strm
                                      :start 0 :end tail))
               (setq pdu-end ;; encoded PDU length + 6 bytes
                 (+ (decode-fixnum *tcp-buffer* 2 4) tail))
               (setq tail (read-sequence *tcp-buffer* tcp-strm
                                         :start tail :end pdu-end))
               t)
              (t (setq eof? t))))
      (setq connection-reset t))
    (setq *event*
      (cond
        (connection-reset 'event-17)
        (timeout 'event-18)
        (eof? 'event-17)
        ((< tail pdu-end) 'event-17)
```

```
                     (t (let ((rule (get-rule (aref *tcp-buffer* head):parser)))
                         (if rule ;; try to parse the message
                             (multiple-value-bind (next-byte new-env)
                                 (rule-based-parser rule
                                            env *tcp-buffer* (+ head 6) pdu-end)
                               (cond
                                 ((eql new-env: Fail) 'event-19)
                                 ((= next-byte pdu-end)
                                  (setq env new-env)
                                  (case (first rule)
                                    (:A-Associate-AC 'event-03)
                                    (:A-Associate-RJ 'event-04)
                                    (:A-Associate-RQ 'event-06)
                                    (:P-Data-TF 'event-10)
                                    (:A-Release-RQ (if (eq *mode*: Client)
                                                         'event-12A
                                                         'event-12B))
                                    (:A-Release-RSP 'event-13)
                                    (:A-Abort 'event-16)))
                                   (t 'event-15) ;; inconsistent length PDU
                             ))
                             ;; unrecognized or invalid PDU, or no matching rule
                             'event-19)))))
                 env))
```

After the read is done we test for errors, connection closed, or end of file (EOF). If an error happens in either of the two `read-sequence` calls, it most likely is a *socket-reset* error. This indicates "Stream-Closed by Remote Host," which the DICOM protocol handles the same as EOF. One matter we do not address here is that the read should be done with provision for timeout, which is referred to in the DICOM specification as the ARTIM timer.

Parsing of PDUs

We implement parsing of incoming PDU data and generation of outgoing PDU data via rules which define the elements that must be matched in an incoming stream or placed in an outgoing stream. The rules contain variable references, allowing the substitution of data values for named variables during the parsing and generation process.

The same rule format will also be used for commands contained in a :P-Data-TF PDU. There is one unfortunate difference. For a PDU, the first byte already tells us the type of message, so we can fetch the rule for it and apply it. For a command, the particular type of command is embedded in the byte stream, so we have to try each command rule in turn, until one works or the parse fails (in the case that the message is erroneous).

```
(:A-Associate-RQ
  #x01                    ; A-Associate-RQ PDU Type tag
  =ignored-byte       ; Reserved field [1 byte]
  (=ignored-bytes 4)    ; PDU Length [4 bytes]
  (>decode-var Protocol-Version fixnum 2 :Big-Endian)
  (=ignored-bytes 2) ; Reserved field -- not used
  (>decode-var Called-AE-Title string 16 :Space-Pad)
  (>decode-var Calling-AE-Title string 16 :Space-Pad)
  (=ignored-bytes 32)   ; Reserved field -- not used
  :Application-Context-Item
  (:Repeat (1 :No-Limit) :Presentation-Context-Item-RQ)
  :User-Information-Item)
```

Figure 5.6 Parse rule for A-Associate-RQ PDU.

In our rule language each rule is a list whose first term is a keyword symbol naming that rule and whose second term is the numeric code for it in the DICOM standard. The rest will be the details of that PDU's structure. Figure 5.6 shows the rule for an :A-Associate-RQ PDU. The rule-based-parser function simply goes through the descriptions in the rule, one by one and calls parse-term to handle each. If the parse completes successfully, the only thing remaining is to see if anything new was found. If so, it is tagged and added to the environment.

```
(defun rule-based-parser (rule env buffer head tail)
  (let ((init-head head)
        (init-env env))
    (dolist (term (rest rule))
      (multiple-value-bind (next-byte new-env)
            (parse-term term env buffer head tail)
        (if (eq new-env:Fail)
            (return (values init-head :Fail))
            (setq head next-byte env new-env))))
    ;; If nothing added to environment, return it unchanged.
    ;; If anything added, package items added during parse
    ;; into a tagged structure and add it at front.
    (unless (eq env init-env)
      (do ((item env (rest item))
           (next (rest env) (rest next)))
          ((eq next init-env)
           (setf (rest item) nil)
           (setq env
                 (cons (first rule) (nreverse env)))))
```

```
(setq env
  ;; If environment additions duplicate
  ;; items already there, ignore them.
  (if (equal env (first init-env)) init-env
    ;; Otherwise prepend new material.
    (cons env init-env))))))))
(values head env)))
```

The `parse-term` function will use a representation in the parse rules that codifies the PDU structure. The elements of these grammar rules are *tokens* (symbols naming constants or other rules) or *expressions* (an expression is a parenthesized list of tokens). Expressions can represent *variables* (data values that can be referred to by name in other rules), *function calls* (data values computed from simpler expressions), or *operators* (instructions to generate particular values, such as computing the length of the PDU currently being constructed).

For parsing, a rule describes *conditions* for matching the incoming byte stream and the *variable bindings* to be added to the environment as a result of such matches. The rule shown in Figure 5.6 would be used to parse an `A-Associate-RQ` message, which has a structure similar to that shown in Figure 4.7. It illustrates several element types, including tags, constants, ignored fields, variables, recursive rule invocations, and optional items.

The symbol `:A-Associate-RQ` is a *tag* and is returned as the result of decoding this PDU. It also serves to name the rule, allowing rules to be looked up by name in the list of rules.

The hexadecimal value `#x01` is a constant that must be matched by a single byte in the input stream. This value identifies the PDU as an `A-Associate-RQ` PDU.

The entries `=ignore-byte` for a single byte and `(=ignored-bytes 32)` for a field of 32 bytes (and similar entries) indicate that the parser should account for byte-field lengths but should otherwise ignore the content of those fields.

`(>decode-var Called-AE-Title string 16 :Space-Pad)` indicates that the data in the next 16 bytes should be interpreted as a character string, with space-padding stripped, and stored into a *variable* named `Called-AE-Title`.

`:Application-Context-Item` means that a rule by that name should be used *recursively* to parse the next sequence of bytes (that is, rules can use other rules recursively to define their own components).

`(:Repeat (1 :No-Limit) :Presentation-Context-Item)` refers to an item that may appear one or more times in the input data stream. *Which* item is indicated by its tag, and the minimum and maximum number of allowable repeats (including a minimum of zero for *optional* items) is indicated by the limits `(1 :No-Limit)` (one to infinity in this case).

Since the rule interpreter for PDU parsing works recursively, components can contain other components (nested arbitrarily deeply). The rule for the outer component need simply refer to the name of the subrule (the keyword symbol used to look up the subrule in the ruleset stored

```
(:Abstract-Syntax-Item-RQ
  #x30                  ; Abstract Syntax Item type tag
  =ignored-byte         ; Reserved field [1 byte]
  ;; Abstract Syntax Name field length [2 bytes]
  (>decode-var ASN-Len fixnum 2 :Big-Endian)
  ;; Abstract Syntax Name string - variable-length
  (>decode-var ASN-Str
        string
        (<lookup-var ASN-Len)
      :No-Pad))
```

Figure 5.7 Parse rule for abstract syntax items.

as an association list). Variables in the environment can pass information between subrules as well as storing information obtained during parsing of an incoming transmission for use in formatting an outgoing reply.

As an example, the rule for an Abstract Syntax Item is shown in Figure 5.7. This item is contained within a Presentation Context Item which itself is contained within an A-Associate-RQ PDU.

Parsing of the data using this subrule works exactly as described above. The additional concept illustrated here is that function calls can be embedded within rules. In the last expression, the (<lookup-var ASN-Len) term tells the parser that the length of the string to be decoded is obtained by calling the function <lookup-var with the variable ASN-Len as an input. The function <lookup-var is simply an environment accessor. The reason for this mechanism is that the rule cannot specify a constant length (as does the Called-AE-Title slot in the A-Associate-RQ PDU rule) because that length is unknown at the time the rule was written—it is determined at run-time from the message. Essentially, data parsed from one part of the message is used in decoding another part of the same message.

The parse-term function is quite large, several pages of code when all is included. It will have a big conditional expression, with a clause for each type of term, such as > decode-var, =ignored-bytes, a fixnum (small integer), a keyword symbol, etc. However, once all the types of terms are implemented, the parser is complete, and independent of the actual rules. So, another protocol, whose rules use the same language, that is, the same types of terms, but in different order, would work fine with this parser.

The last byte of the PDU in the buffer is known, but the parser may be called recursively so the first byte to process may vary. We call one head and the other tail. If a parse succeeds, the next byte to read is returned along with the new environment, but if it fails, the parse returns the unchanged start point and the keyword :fail. Remember that the pointer *increases* as you proceed through the buffer, so the head always has to be less than the tail, until you are at the end.

```
(defun parse-term (term env tcp-buffer head tail)
  (let ((init-head head)))
```

```
(cond
  ((typep term 'fixnum) ;; required fixed byte value
   (cond ((>= head tail) (setq env :Fail))
         ((= term (aref tcp-buffer head))
          (setq head (1+ head)))
         (t (setq env :Fail))))
  ((eq term '=ignored-byte)
   (when (> (setq head (1+ head)) tail)
     (setq head init-head env :Fail)))
  ((keywordp term) ;; a sub-item with own rule, call parser
   (if (>= head tail) (setq env :Fail)
       (multiple-value-setq (head env)
           (rule-based-parser
             (get-rule item :parser)
             env tcp-buffer head tail))))
  ((eq (first term) '>decode-var)
   ...)
  ...<more clauses>)))
```

New kinds of messages can be supported by just adding rules and state table entries rather than adding to the code. Similarly, modifications to the DUL protocol can be accommodated by making the corresponding alterations in the rules and state table.

Generation of PDUs

The PDU generation code can use rules in almost the same format as the parse rules to describe the generation and structure of outgoing messages. Generation rules can contain references to the same variables as do the parse rules, providing a simple mechanism to echo received values on transmission (or to output values computed from previously received data). The main differences are that the generation rules encode variable values instead of decoding them, and the lengths of encoded items are not known in advance and must be computed.

Once we have a rule-based PDU parser and a table-driven data-object parser, it is easy to add generation of PDUs (including the transmission of embedded data objects). We simply run the parsers "backwards" as instantiators. We create generation rules rather than using the same rules for both kinds of tasks because there are a few semantic differences in generating messages vs. parsing, which were most easily accommodated with slight variations in the rules.

Figure 5.8 shows the generation rule for the A-Associate-RQ PDU. It is a reflection of the corresponding rule for the parser, but instead of creating variables and values to add to the environment, the PDU is constructed from variables and values in the environment. This rule tells the generator to output a byte stream consisting of:

```
(:A-Associate-RQ
  #x01                  ; A-Associate-RQ PDU Type tag
  #x00                  ; Reserved field [1 byte]
  :Place-Holder         ; PDU Length [4 bytes]
  ;; Protocol Version - fixed value
  (=fixnum-bytes #x0001 2 :Big-Endian)
  (=constant-bytes #x00 2)  ; Reserved field
  (<encode-var Called-AE-Title string 16 :Space-Pad)
  (<encode-var Calling-AE-Title string 16 :Space-Pad)
  (=constant-bytes #x00 32) ; Reserved field
  :Application-Context-Item
  :Presentation-Context-Item-RQ ; we only provide one
  :User-Information-Item-RQ)
```

Figure 5.8 Generate rule for `A-Associate-RQ` **PDU.**

- one byte—the constant value 01 (hex), indicating the PDU type,
- one byte—the constant value 00 (hex),
- four bytes reserved as a place-holder where the length of the entire PDU will be inserted (we don't know that length until we finish generating the entire PDU, but space must be reserved at the beginning for it),
- a fixnum constant of length 2 bytes encoding the value 0001 (hex),
- a string of length 2 of repeated bytes, each with value 00 (hex),
- the value of the variable "Called-AE-Title" (retrieved from the environment as described above), encoded as a string with padding by Space characters out to 16 bytes,
- the value of the variable "Calling-AE-Title," likewise encoded as a 16-byte Space-padded string,
- a string of length 32 of repeated bytes, each with value 00 (hex),
- an Application Context Item (here the generator calls itself recursively using the rule indexed by `:Application-Context-Item` and uses the environment to supply any needed variable bindings),
- a Presentation Context Item, similarly generated by a recursive call,
- finally, a User Information Item, similarly recursively generated.

The "-RQ" in the keyword symbol names in Figure 5.8 refers to rules for items specialized slightly for use inside the `A-Associate-RQ` PDU. The rules might vary slightly for the same item used in other contexts, as in the `A-Associate-AC` PDU.

Since the rule interpreter for PDU generation works recursively, just as for parsing, generated components can contain other components (nested arbitrarily deeply). The rule for the outer component simply refers to the name of the subrule (the keyword symbol used to look up the subrule in the ruleset stored as an association list). As for parsing, variables in the environment can pass information between subrules, and lookup functions can be used to obtain information for use in formatting an outgoing message.

The generator code may be found in the full DICOM package.

5.4.4.3 Parsing and Generation of Commands and Data

The parser for PDUs containing commands uses the mechanism described above, with the same language components as in the parser rules just described and implemented by the same recursive-descent parser using these rules. One component of the P-Data-TF PDU is the "message," that is, the combination of commands and/or data objects that are being communicated between DICOM entities. The PDU parser treats the message as a single entity which is stored in the environment as the value of the variable PDV-Message. To the PDU parser, this message is simply an uninterpreted stream of bytes. The DICOM Upper Layer (DUL) protocol specifies a header for each message encoding the length of the following data field and some flags indicating whether the data field is a command message or a data object (and additionally whether the data field contains the complete object or is a portion of a long data field that has been fragmented into multiple messages). So, for the message *data* we need a somewhat different parser and generator, which will still be table driven.

Parsing and Generation of Commands

The rules for parsing commands will use the same conventions and language as for parsing any other type of PDU. Figure 5.9 shows a command rule.

The handling of command parsing is different in one way from PDU parsing. In read-pdu, where the parser (rule-based-parser) is called, we know which rule to use, because the first byte identifies the PDU type. The commands are not encoded this way. Where the command identifier appears depends on the command, so we will have to try each of the supported commands until one matches or we exhaust the supported commands list. This is simple iteration, calling rule-based-parser with the same data and each command rule.

```
(defun parse-command (env *tcp-buffer* head tail)
  (dolist (msgtype *Message-Type-List*)
    (multiple-value-bind (next-byte new-env)
      (rule-based-parser (get-rule msgtype :parser)
                         env *tcp-buffer* head tail)
      (unless (eq new-env :Fail)
        (return (values msgtype new-env))))))
```

If a command matches the message data, the function returns the command type and the new environment. If none match, the function returns the keyword :fail and the second value is nil.

```
(:C-Store-RQ
   (=fixnum-bytes #x0000 2 :Little-Endian)   ;; Tag (0000,0000)
   (=fixnum-bytes #x0000 2 :Little-Endian)
   (=fixnum-bytes 4 4 :Little-Endian)        ;; Length
   (>decode-var Group-Len fixnum 4 :Little-Endian) ;; Value
   (=fixnum-bytes #x0000 2 :Little-Endian)   ;; Tag (0000,0002)
   (=fixnum-bytes #x0002 2 :Little-Endian)
   (>decode-var Store-SOP-Class-UID-Len fixnum 4 :Little-Endian)
   (>decode-var Store-SOP-Class-UID-Str     ;; Value
               string
               (<lookup-var Store-SOP-Class-UID-Len)
               :Null-Pad)
   (=fixnum-bytes #x0000 2 :Little-Endian) ;; Tag (0000,0100)
   (=fixnum-bytes #x0100 2 :Little-Endian)
   (=fixnum-bytes 2 4 :Little-Endian) ;; Length
   (=fixnum-bytes #x0001 2 :Little-Endian) ;; Value
   ... ;; other items in here elided
   (=fixnum-bytes #x0000 2 :Little-Endian) ;; Tag (0000,1000)
   (=fixnum-bytes #x1000 2 :Little-Endian)
   (>decode-var Store-SOP-Instance-UID-Len fixnum 4 :Little-Endian)
   (>decode-var Store-SOP-Instance-UID-Str
               string
               (<lookup-var Store-SOP-Instance-UID-Len)
               :Null-Pad)
   (:Repeat (0 1) :Move-Originator-AE)
   (:Repeat (0 1) :Move-Originator-ID))
```

Figure 5.9 A rule describing the structure of the C-STORE-RQ command.

Parsing and Generation of Data

Here we describe generally how to parse the actual data elements, images, and so on. If the message contained in a P-DATA-TF PDU is a data object, the buffer (and begin/end pointers) are passed to a different parser. This parser is also data-driven, using the data dictionary to decode arbitrary data objects. It is different from the PDU parser because the structure of the data messages is specified by the structure of the data types, rather than an explicit message structure.

The data dictionary is a text file, containing Lisp expressions. It can be read by a human reader and can be written and extended by using a text editor. But it is also readable by the program, and functions as executable code, though in a new declarative language. Each entry in our version of the data dictionary is a list containing the tag, value representation, and name. This list maps data-object tags (both fields, group and element numbers) to symbols naming the object's datatype and a string describing the object. An excerpt from this table is shown in Figure 5.10. For example,

```
((#x0020 . #x0032) DS "Image Position Patient")
```

is the entry for the item mentioned in Section 4.3.3, Image Position (Patient).

```
(defparameter *group/elemname-alist*
  ;;-------------------------------------------------
  ;; Group 0000: CMD "Command"
  '(((#x0000 . #x0000) UL "Group Length")
    ((#x0000 . #x0002) UI "Affected SOP Class UID")
    ((#x0000 . #x0003) UI "Requested SOP Class UID")
    ((#x0000 . #x0100) US "Command Field")
    ((#x0000 . #x0110) US "Message ID")
    ...
    ;;-------------------------------------------------
    ;; Group 0010: PAT "Patient Information"
    ((#x0010 . #x0000) UL "Group Length")
    ((#x0010 . #x0010) PN "Patient Name")
    ((#x0010 . #x0020) LO "Patient ID")
    ((#x0010 . #x0021) LO "Issuer of Patient ID")
    ((#x0010 . #x0030) DA "Patient's Birthdate")
    ...
    ;;-------------------------------------------------
    ;; Group 0028: IMG "Image"
    ((#x0028 . #x0000) UL "Group Length")
    ((#x0028 . #x0002) US "Samples Per Pixel")
    ((#x0028 . #x0005) RET "Image Dimensions (RET)")
    ((#x0028 . #x0006) US "Planar Configuration")
    ((#x0028 . #x0010) US "Rows")
    ((#x0028 . #x0011) US "Columns")
    ...
    ))
```

Figure 5.10   An excerpt from the DICOM data dictionary table.

Value representations, such as DS, are defined in another table, also an association list. An excerpt from this table is shown in Figure 5.11. This association list maps data types (identified by the first symbol in each sublist) to descriptive information (the string naming it, the second element) and information needed to parse objects of that type (the optional third and fourth elements). If these optional elements are present, they specify the Lisp data type of the object. The types and included information are:

- STRING—minimum and maximum length, whether and how the string is padded,
- FIXNUM—integer, including number of bytes,
- DOUBLE-FLOAT—8-byte floating-point reals,
- SINGLE-FLOAT— 4-byte floating-point reals,
- (UNSIGNED-BYTE 8)—Uninterpreted string of bytes used to encode 8-bit data values,
- (UNSIGNED-BYTE 16)—Uninterpreted string of bytes used to encode 16-bit data values.

Certain data types are used to delineate nested data objects. These include: IT "Item in Sequence", ITDL "Item Delimiter", SQ "Sequence of Items", and SQDL "Sequence Delimiter". In such cases, the containing object is a "Sequence of Items," whose end is marked

```
(defparameter *datatype-alist*
  '((AE "Application Entity" (string 0 16) :Space-Pad)
    (AS "Age String" (string 4) :No-Pad)
    (AT "Attribute Tag" (fixnum 4))
    (CS "Code String" (string 0 16) :Space-Pad)
    (DA "Date" (string 8) :No-Pad)
    (DS "Decimal String" (string 0 16) :Space-Pad)
    (FD "Floating-Point Double" (double-float 8))
    (FL "Floating-Point Single" (single-float 4))
    (IT "Item in Sequence")
    (ITDL "Item Delimiter")
    (OB "Other Byte" ((unsigned-byte 8) 0 *) :Null-Pad)
    (OW "Other Word" ((unsigned-byte 16) 0 *) :No-Pad)
    (SL "Signed Long" (fixnum 4))
    (SQ "Sequence of Items")
    ...
    ))
```

Figure 5.11   An excerpt from the value representation table, part of the DICOM data dictionary.

by a "Sequence Delimiter." Each element composing the sequence is an "Item in Sequence" with its end marked by an "Item Delimiter." Thus an item in a sequence can itself consist of one or more sequences. Data objects can be nested to arbitrary depth in this manner, and our parser uses a simple recursive-descent algorithm to maintain the state of the parse as it accumulates the subelements and packages them into the containers.

The basic object-parsing algorithm is simple. All data objects are represented as a sequence of elements. Each element (whether it is the outermost container for a large data structure or the innermost scalar element of a larger structure) is represented by a three-component byte-field in the data stream. The first field is the Data Element Tag, an ordered pair of 16-bit unsigned integers, the Group and Element numbers. Indexing by this pair of numbers into the Group/Element-name association list gives the symbol naming the data type of the following data value. (It also yields the descriptive string useful for printing debugging messages and for dumping arbitrary dataobjects.) The data type symbol then serves as an index into the data type association list, giving information needed to parse the data object (number byte-field lengths, string lengths, and padding).

The second field for each element is "Value Length," a fixed-length integer giving the byte-string length of the encoding of the value of the particular element. The third field is the "Value Field," the string of bytes (of length given by "Value Length") holding the actual data. Our parser simply reads the appropriate number of bytes and interprets them according to the specifications for the given data type (numerical, string with or without padding, etc.).

For data objects whose value field contains a sequence, an alternative specification of "Value Length" is a special code for "Undefined Length." In that case, the Value Field for that element contains a sequence of items, each represented by its own Tag-Length-Value structure. The parser decodes each element in turn, until detecting an element of type "Sequence Delimiter." That is the indication that the "Undefined Length" field is now coming to an end.

The first element of any group of data objects (i.e., whenever the Group Number portion of the Element Tag increments) is an element giving the length of the entire group, encoded as a fixed-length integer. The value decoded from this element can be used to decode the rest of the elements in the group (it specifies the length in bytes of the entire group). This element is not strictly necessary, but having it simplifies the reconstruction of data objects whose representations have been fragmented into multiple P-DATA-TF PDUs (images are the prime example of this).

5.4.4.4 *Deployment Experience and Observations*

In [200] we reported on the experience we had with our initial testing and deployment of the Prism DICOM System. The original code is still in operation as of this writing, accepting images and radiation planning structure sets from various sources, as well as sending radiation treatment plan data to our radiation treatment machines. We have had no problems traceable to coding errors for many years.

The one part we have not addressed, as with HL7, is what is done with the data once it is parsed into the environment association list, and conversely for the client software, how is the environment association list generated from which the client code can encode messages and send them out. The C-STORE operation on the server side has a function that is proprietary to the Prism system. It sifts through the environment, using path-find, to collect data elements corresponding to data structures (defined as CLOS classes) in the Prism Radiation Treatment Planning System. These are then written out to files in the Prism system, using the put-object function described in Chapter 1.

Similarly, in the Prism program there is a user interface component that is used to select a treatment plan to be sent to a specified treatment machine control computer. This user interface component then calls a function that goes through the Prism data structure for a treatment plan, and constructs the required environment association list. Then that component calls a dicom-client function, which works exactly like the basic client code in Section 5.4.4.

In addition to verification by inspection, we tested our implementation with others, sending and receiving data both locally and long distance (some test nodes were located as far away from Seattle as England and Israel). We discovered errors in our own implementation and errors in some of the other implementations.

Moreover, we found that many implementations are tolerant of violations of the DICOM specification. The CTN code was never intended to be a validation test suite. Even implementations intended to function as validation tests, such as the AGFA test suite used by many vendors, cannot be relied on for this purpose. We found errors in our code that were tolerated by some commercial implementations. We also found at least one commercial implementation with an encoding error that has apparently been tolerated by many other implementations for many years, including the AGFA software.

To my knowledge, *contrary to HL7*, there are no inconsistencies in the DICOM specification, and although it is often obscure and arcane, the prose description and tables did in every case provide enough information to decide what we needed and what to do with it. The specification is encyclopedic. From a software developer's point of view, this is adequate to ensure that systems will work interoperably. Indeed, our system *did* work correctly with all other systems, except in a few cases, noted above, where either our system or theirs clearly violated the DICOM specification.

The DICOM specification does not provide any guidance about which of the many related data elements is the best to use for any given application. It is very important to be familiar with the context, specifically, what do radiologists, radiation oncologists, and other related medical staff do with the data. Seemingly equivalent data elements may not at all behave the same or represent the same semantics.

It was unfortunate that the ACR-NEMA committee did not separate the data buffering problem from the definition of messages (PDUs are sometimes complete messages and sometimes message fragments). This is the most significant design flaw in DICOM. It would clean up the design considerably for the standard to remove this. However, it is most likely that existing implementations will undergo a major rewrite to accommodate such a change. There is no obvious way to make this backward compatible. The design we described here goes to some length to minimize the impact of the buffering scheme, and it would not be difficult to strip it out. The original 50 pin wiring design was motivated by a desire for higher bandwidth than 10 MB ethernet could provide, but in practice this judgment call turned out to be wrong. Presumably if the ACR-NEMA committee had skipped version 1 entirely and relied on already well-established world-wide network protocols and hardware, the buffering scheme would not have been included.

5.5 Stochastic Processes

The computational models described in previous chapters provide systematic descriptions of biological structure, molecular processes, pathology, pharmacology, and other areas of biology, medicine, and health. Although these representations provide potentially powerful ways to infer relationships, predict and analyze outcomes, they do not provide an explicit description of how things change over time. Modeling how systems change with the passage of time can be done in many ways. A model can consist of variables that are continuous functions of time, such as drug concentrations in the blood stream, or discrete variables, such as the on and off states of genes and regulatory elements, which typically would be modeled by time steps rather than continuously varying time.

An important theme in Chapter 3 has been the idea of computing or predicting the *probability* of occurrence of a particular value for an observable variable, rather than the value itself. When the value of a variable (whether discrete or continuous, numerical or other) may change with time, one can imagine predicting the sequence of observed values, or alternatively, the sequence of probabilities. Thus, the probability of observing a particular event or variable value may be a function of time. So, the probability density function, $p(x)$, for a random variable X at the point $X = x$, becomes $p(x, t)$, referring to the probability density at a particular time, t. Typically, time is considered in discrete steps, $t_0, t_1, \ldots, t_n$, though you should keep in mind that time is not itself a random variable, but a parameter.

To model a biological process that involves variables changing with time, one can hypothesize that the probability distribution function changes with time according to some formula or formulas, which can be considered the axioms of a theory. A simple example is the population size of cells in a cell culture, such as a Petri dish containing bacteria. The reproduction of bacteria is a random process, but that process has some probability of occurring within a given time interval, so the number of cells present will be a random variable with a probability distribution depending on time. Another example is the cell cycle itself. With four states, G1, S, G2, and M, there is a (discrete) probability distribution with a probability for being in each state. Each of these numbers changes as time progresses.

To be precise, each observable entity at a particular time is modeled as a random variable, so that taking the whole sequence over time, we have a set of random variables that depend on each other. Such a set or model is called a *stochastic process*. The dependency of each successive random variable on the previous ones is an expression of what we believe about the underlying biological processes, so a stochastic model is a kind of theory of dynamic biological systems. The dependency of a state at a particular time on the previous states or history may be very complex or relatively simple. We will consider only discrete time steps, so that we have a discrete set of random variables indexed by (numbered) time steps.

For the case where the random variable itself is discrete, one can define a set of conditional probabilities called the "transition matrix." These numbers are the probabilities of being in a particular state at time t_{n+1} given the state at time t_n. For m states, the transition probability matrix is a m by m matrix, each of whose entries may in turn depend on time as well as on the current and previous states. These kinds of dependencies mean that in general the probability for a particular next state is *not* a simple linear combination of the probabilities for the current states, even though the matrix representation makes it appear so. This is because the matrix values themselves may depend on the current state (and previous states) in some complex way.

5.5.1 Markov Chains

When the probability distribution for a random variable at the next time step depends only on the probability distribution at the current time and not on any further history, it is called a Markov process and the sequence of random variables for the various time steps is called a Markov chain. Although this may seem like a drastic simplification, such sequences have wide applicability and useful properties. This section presents just a few ideas about Markov chains, and some biological/medical application examples. For more details of basic concepts and proofs of theorems, the reader is referred yet again to Wasserman [430, Chapter 23].

The definition of a Markov chain can be stated mathematically as a restriction on the conditional probability of a random variable X_n having value x given the values of all the previous (in time or in the sequence) random variables.

$$P(X_n = x | X_0, \ldots, X_{n-1}) = P(X_n = x | X_{n-1}). \tag{5.1}$$

In modeling biological processes it is often possible to make a further simplification, that these conditional probabilities do not change with time, so, $P(X_n = x | X_{n-1})$ is in such a case independent of n. The conditional probabilities then constitute a transition matrix, p_{ij}, the probability of transition to state j given that the current state is i. Since they are probabilities, they all must be non-negative. Also, since the observable must have *some* value after each transition, the sum of all the values in each row must add up to 1 (i is the row index and j is the column index).

$$\sum_j p_{ij} = 1.0. \tag{5.2}$$

These two assertions are from Axioms 3.1 and 3.3.

The probability mass distribution for the observable at a particular time can be represented as a vector, the state at time n, $s_i(n)$. Using Equation (3.8), the transition process can be represented as a matrix-vector product,

$$s_j(n+1) = \sum_i p_{ij} s_i(n) \tag{5.3}$$

and thus by repeated application of the transition matrix, the probability distribution at any future time can be computed from an initial probability distribution.

A simple example will serve to illustrate.[12] Consider a system with two states A and B, so that we can define the following transition probabilities

$$p_{AB} = 0.8,$$
$$p_{AA} = 0.2,$$
$$p_{BA} = 0.1,$$
$$p_{BB} = 0.9,$$

where p_{AB} is the probability of state A changing to state B after 1 time step. Alternatively, we can express these transitions in a transition matrix

$$\begin{bmatrix} p_{AA} & p_{AB} \\ p_{BA} & p_{BB} \end{bmatrix} = \begin{bmatrix} 0.2 & 0.8 \\ 0.1 & 0.9 \end{bmatrix}.$$

---

[12]This example and the discussion of simulation of prostate cancer were contributed by Edward Samson, and were originally part of a project report by Mr. Samson for my class, *Life and Death Computing*.

The initial state can be represented as a column vector, with an initial probability of being in each of the two states. For example,

$$\begin{bmatrix} S_{0_A} \\ S_{0_B} \end{bmatrix} = \begin{bmatrix} 0.3 \\ 0.5 \end{bmatrix},$$

where S_{0_A} and S_{0_B} are the initial distributions between the A and B states. To find the distribution after one time step, we apply the transformation matrix to the distribution vector, to get

$$\begin{bmatrix} S_{1_A} \\ S_{1_B} \end{bmatrix} = \begin{bmatrix} 0.2 & 0.8 \\ 0.1 & 0.9 \end{bmatrix} \begin{bmatrix} 0.3 \\ 0.5 \end{bmatrix} = \begin{bmatrix} 0.46 \\ 0.48 \end{bmatrix}.$$

An interesting phenomenon occurs if we try the initial state

$$\begin{bmatrix} S_{0_A} \\ S_{0_B} \end{bmatrix} = \begin{bmatrix} 0.5 \\ 0.5 \end{bmatrix}.$$

Applying the transition matrix to this state vector gives

$$\begin{bmatrix} S_{1_A} \\ S_{1_B} \end{bmatrix} = \begin{bmatrix} 0.2 & 0.8 \\ 0.1 & 0.9 \end{bmatrix} \begin{bmatrix} 0.5 \\ 0.5 \end{bmatrix} = \begin{bmatrix} 0.5 \\ 0.5 \end{bmatrix}.$$

The operation returns the initial state. This state is referred to as a steady-state solution, or fixed point. In general we define a steady-state solution, S to be a state that satisfies the linear equations $AS = S$.

For any transformation matrix that has only positive entries, there must exist a unique steady-state solution that all trajectories approach over time. A Markov chain satisfying these properties is called a Regular Markov chain.

In fact, we can use an even weaker requirement. Given a transformation matrix, A, if there exists a power $n > 0$, such that all elements of A^n are positive, then the matrix A is a regular matrix and A^n approaches the steady-state matrix, T, where every column of T is the steady-state solution that satisfies the steady-state property defined above.

5.5.2 Example: Prostate Cancer

Samson proposed that the progression of prostate cancer could be modeled by a Markov chain. He offered the following considerations as a starting point for such a model.

Prostate cancer is a common disease which affects almost all men in their later ages [261]. The diagnosis and treatment of prostatic carcinoma depends on a number of factors. Here we will focus primarily on the classification and grading of prostatic carcinoma. The goal is to find at what points the prostatic carcinoma is likely to metastasize and spread to distant other parts of the body. A model based on Markov chains could be used to predict the progression of the cancer, possibly identify the best time to treat the carcinoma, and minimize the probability of distant metastasis. There are several interesting questions to ask about this process, such as what are the possible trajectories, and do any steady-states exist in this process? Even more, how do the answers to these questions change as the defined transitions change?

5.5.2.1 Histologic Grading

The Gleason grading method is a commonly accepted method to determine the prognosis of a patient with suspected prostate cancer. The Gleason grading system determines a numerical score which corresponds to the severity of the cancer. These scores can provide a basis for the Markov chain to be constructed.

A Gleason grade is determined by analyzing a stained sample of the prostate tissue, retrieved from a biopsy. The structure of glands inside the prostate tissue along with the shape of the stained cells is used to determine a score in the range of 1–5 for the two major glandular patterns [106]. A score close to 1 is representative of normal tissue, with easily identifiable, healthy glands. A score of 5 indicates the tissue has no recognizable glands. Since the first two major patterns in the tissue are scored, a score is given in the range from 2 to 10. The higher the score, the more aggressive the cancer.

5.5.2.2 Surgical Staging

In addition to the Gleason grading scheme, there is an additional scheme to describe the progression of the carcinoma. This system of describing the location, size, and distant metastasis of the carcinoma is called the Whitmore stage [261].

In stage A, the carcinoma is only clinically detectable, with no physical indications upon physical examination. Stage A is further separated into stage A_1 and stage A_2, where A_2 indicates a more decentralized tumor involving several regions of the prostate.

In stage B, the carcinoma is detectable, usually through rectal examination. Once again stage B_1 indicates a carcinoma that is relatively localized. Stage B_2 indicates a much more spreadout carcinoma that involves a large quantity of glands.

In stage C, the cancer is no longer confined to the prostate and starts affecting and growing into the surrounding organs. However, the tumor has not metastasized, and only exists relatively close to the exterior of the prostate. In stage D, metastasis of the tumor is evident. Stage D can be separated into further stages: stage D_1 indicates the pelvic region is affected, while D_2 indicates metastasis of other regions.

The issue with the current surgical staging is that it is difficult to estimate how a metastasis has occurred during the time of the examination [261]. Therefore, it is not possible to solely use the surgical staging to determine the severity of the cancer. A combination of the surgical and histologic staging is required. To describe the different combinations of both of these metrics, a 4 stage system was developed. Stage 1 corresponds to clinically detectable carcinoma with no metastasis, a Gleason grade of 2 is required. Stage 4 represents a highly metastasized tumor with any Gleason grade and any surgical stating. It seems necessary to consider both transitions in the surgical staging and overall grade as defined above.

The difficulty is to determine the appropriate probabilities for the transition matrix between the four general severity levels of prostatic carcinoma and the probabilities of distant metastasis during each level.

Consider a generic male patient, age 55, with an elevated PSA. An elevated PSA is one clinical sign that a prostate cancer is present. If the prostate biopsy confirms that a cancer exists in level 1, in 10 years we expect 1 in 10 cases to metastasize in 10 years. In 10 years the cancer will also progress to level 2 severity. Then, from a level 2 severity, the chance of metastasis in the next 10 years is almost 0.90. From there, it requires only 5–10 years to develop into a level 3 prostatic carcinoma. Finally, the transition to a level 4 tumor can occur in as little time as 1 year. Although it may be possible to estimate transition probabilities for the elements of the transition matrix, the probabilities are time-dependent, which is beyond the scope of the simple introduction to Markov chains given here. The next section presents a somewhat different treatment of the progression of cancer and metastatic spread in which the assumption of time independence seems to hold, but the model requires linking a series of Markov chains, in order to represent the connectivity of the lymphatic system, an important path of spread for some kinds of tumors.

5.5.3 Example: Radiotherapy Clinical Target Volume

Section 2.2.8 described the problem of treating cancer with radiation. It is an elegant treatment modality because modern radiation treatment machines can aim radiation beams of any size or shape anywhere in a person's body, so potentially a radiation treatment can be more precise than surgery or chemotherapy, with less debilitating side effects. However, a key to the process is determining just where the target is, what is the extent of the tumor, both visible (in images) and not visible, as well as allowing for factors that necessitate allowing for motion and variability in patient positioning. That section described how a logic model can be used to determine the motion and variability allowance. This section will show how Markov chains can be used together with the FMA and path search to help determine the Clinical Target Volume (CTV) for radiation therapy.

The Clinical Target Volume is a region that does not correspond to a visible organ or actual structure, but instead is a region defined by where the radiation oncologist believes the tumor cells have spread. Therefore, it cannot be generated automatically from image data alone. However, many tumors are known to proliferate (metastasize) through the lymphatic system, so a model of the lymphatic system can potentially provide a means for computationally predicting the regional spread of these cancers.

The lymphatic ducts and the lymph nodes that drain tissues of the head and neck are initial sites of metastatic spread. Therefore, these sites serve as important targets for prognosis. Upon initial diagnosis, the number of affected nodes provides insight into how likely it is that a patient will develop metastatic disease. With a small number of affected nodes, it is less likely for disease status of a patient to go beyond localized spread of cancer. If several nodes are involved, it is much more likely that distant sites in the body harbor deposits of metastatic cells.

5.5.3.1 Anatomic Model of Lymphatic Drainage

Using the Foundational Model of Anatomy (FMA) developed by the University of Washington Structural Informatics Group [353], we can construct a small program that will map the lymphatic drainage pathways from any anatomical site. A query to the FMA can tell us what lymphatic chains drain a given anatomical entity (and its parts). Further queries to the FMA can then tell us what lymphatic chains are downstream, or *efferent to*, the initial drainages, as shown in Figure 5.12. A recursive query can trace these paths all the way back to the final lymph

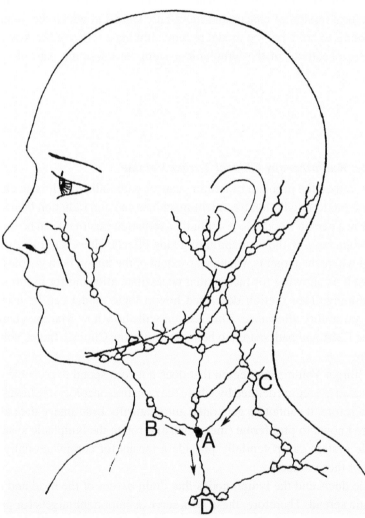

Figure 5.12 A diagram showing some of the lymphatic chains and nodes in the head and neck region, illustrating the "efferent to" and "afferent to" relations. In this diagram, node A is efferent to node D, and is afferent to nodes B and C. The arrows just show the direction of flow of lymphatic fluid. (Drawing kindly provided by Prof. Mark Whipple, University of Washington)

vessels, the thoracic duct and the right lymphatic duct, which then drain into the blood stream. Once tumor cells reach the blood stream, they can be dispersed very great distances from their origins, but that is not going to be modeled here.

Getting the information from the FMA in order to construct the paths is an information access and search problem. The information access part is described in Chapter 4 and general methods for searching graphs such as the connectivity relations in the FMA are developed in this chapter. In the development here, to query the FMA for the pathways of lymphatic drainage of the entity referred to by a particular term, we will use the `lymphatic-paths` code described in Section 4.3.5 of Chapter 4, without concern at this point for the details. This function returns a list of lists. Each of the lists is a sequence of lymphatic chains (in reverse order) from the starting structure downstream eventually to either the thoracic duct or the right lymphatic duct. As an example of the results of the computation, one of the lymphatic paths for the soft palate is the following:

```
("Right lymphatic duct" "Right jugular lymphatic trunk"
 "Right superior deep lateral cervical lymphatic chain"
 "Right retropharyngeal lymphatic chain")
```

The right retropharyngeal lymphatic chain directly drains the soft palate. From there, one path for lymph is to the right superior deep lateral cervical lymphatic chain, then into the right jugular lymphatic trunk, and finally into the right lymphatic duct, and then into the veins. The path is generated in reverse, because, as the search progresses, new nodes are put on the *front* of the list. Now our challenge is how to use this information to create a CTV.

5.5.3.2 Lymphatics and Tumor Spread

The next step in computing the CTV is to determine how far a given tumor is likely to have spread along the pathways computed in the previous section. One way to do this is to assign probabilities based on surgical data from head and neck resections. Unfortunately there is no direct way to do this. Instead, we can create a simple computational model with only a few parameters, assume uniform behavior, and see how well this can predict the rates of actual observed positive nodes in patients.

Given the lymphatic drainage paths, and the pretty well-established idea that tumors in the head and neck (and some other areas) spread through the lymphatics, it should be possible to estimate the probability of downstream spread of tumor cells, depending on how big the primary is. In a recent project [26], Noah Benson created a Markov model structure to predict these probabilities computationally. The model hypothesizes that each lymphatic chain can be in a series of states, no tumor, a small amount, and progressively more. The transition of a chain between the possible states is a Markov process, involving establishment of tumor in a chain if there is tumor upstream, and local growth once some tumor is established in a chain. Much to our surprise, with reasonable guesses at transition probabilities and assumption of uniform behavior, the model gave predictions in surprising agreement with the published surgical data mentioned above.

The modeling strategy uses the idea of a Markov chain, described earlier. Each lymphatic chain[13] can be described in terms of the probability for it to be in any of five possible states, numbered 0 through 4. State 0 means no tumor present. State 1 means a little tumor, state 2 somewhat more, and so on, to state 4 which is a big palpable node. The only way a lymphatic chain can go from state 0 to state 1 is by some tumor cells migrating downstream from the upstream chain, but once some tumor cells are there, they can transition to the states of larger growth on their own. The simplest way to think about the spread of tumor cells is to imagine that the process happens in discrete time steps. The tumor is in some initial state, and each lymphatic chain is as well. The states change with each time step.

The probabilities of being in any of the states at a particular time will add up to 1.0, because the states are mutually exclusive and exhaustive (Equation (3.3) from Chapter 3). Each lymphatic chain has to be in *some* state. However, each probability will change with time, as the tumor grows and cells migrate and spread through the lymphatic paths. We will number the chains in a path sequentially, 1, 2, 3, etc. Then we designate the probability of chain j being in state k at time t as $p(j, k, t)$. First we can express the requirement of all the probabilities summing to 1.0 as Equation (5.4).

$$\sum_{k=0}^{4} p(j, k, t) = 1.0. \tag{5.4}$$

This is an invariant, independent of time t and true for all chains, that is, for all j. When we start with a set of initial probabilities we need to choose numbers that satisfy this requirement and we need to insure that any updates done by the model preserve this relationship.

The idea of a Markov chain is that a system can undergo state transitions, so the lymphatic chain can go from state 1 to state 2, and so on, with some *transition probability*, and the probability that the system is in any state at time $t + 1$ depends only on the *present* state, the state of everything at time t, and *not* the history of how it got there.

If each lymphatic chain were independent of all the others, the probabilities would form a state *vector* with five components, one probability for each state the lymphatic chain could be in. For each possible transition we would have some probability, p_{ij}, the conditional probability that the system is in state j at time $t + 1$ given that it is in state i at time t. Strictly speaking, this transition probability can depend on time. Processes in which the transition probabilities do *not* depend on time are called *homogeneous*. In such cases, the transition probabilities form a matrix, and the relation between the state vector at time $t + 1$ and the state vector at time t is just multiplication of the state vector by the transition matrix, to get the new state vector.

For tumor growth at a lymphatic chain, we could postulate for example, that at any state where there is tumor, that is, the state is in the range 1–4, the probability of growing bigger, to the next state, is 0.5 (and of course the probability of staying the same will be 0.5, so the

---

[13]We are using the term "chain" in two ways here. A lymphatic chain is so called because it typically consists of a series of vessels connecting several lymph nodes in a linear string. That unit as a whole is modeled as a Markov chain, meaning its state is represented as a stochastic process, or sequence over time.

probability of something happening will add up to 1.0). This of course excludes the possibility that the tumor grows so fast that it skips a state, and assumes that it never shrinks (we are not considering the effects of treatment or immune system response). In this case, the transition matrix, P, would be as shown in Equation (5.5).

$$P = \begin{pmatrix} 1.0 & 0.0 & 0.0 & 0.0 & 0.0 \\ 0.0 & 0.5 & 0.5 & 0.0 & 0.0 \\ 0.0 & 0.0 & 0.5 & 0.5 & 0.0 \\ 0.0 & 0.0 & 0.0 & 0.5 & 0.5 \\ 0.0 & 0.0 & 0.0 & 0.0 & 1.0 \end{pmatrix}. \tag{5.5}$$

The state at time $t + 1$ is computed from the state at time t by the formula in Equation (5.6).

$$p(j, k, t + 1) = \sum_{i=0}^{4} p(j, i, t) P(i, k). \tag{5.6}$$

In this formula, the index i in $P(i, k)$ is a row label and the index k is a column label. So, for example, starting in state $(0, 1, 0, 0, 0)$, the next state will be $(0, 0.5, 0.5, 0, 0)$, the state after that will be $(0, 0.25, 0.5, 0.25, 0)$, and so on.

For the primary tumor, this works just fine (except that the probabilities we used in the real model were 0.9 to stay the same and 0.1 to grow). Note that if there is no tumor initially the system will always be in state 0, since we have not modeled spontaneous appearance of tumor cells. That is the 1.0 in the first row and column, with 0 entries in the rest of the first column and the first row. However, for the lymphatic chains, this is not quite the whole story. In addition to transitioning through growth, the way that tumor cells get there in the first place is from the upstream chain or the primary tumor. So, state 0 and state 1 do not follow Equation (5.6), though it should be reasonable to use Equation (5.6) for states 2, 3, and 4. If lymphatic chain j is in state 0, but chain $j - 1$ is in state 1, 2, 3, or 4, there is some finite probability of a metastatic process. Let's call those numbers $q(k)$, the probability that if chain $j - 1$ is in state k and chain j is in state 0, that chain j will transition to state 1. Then the probability, $p'(j, t + 1)$ of chain j getting metastatic cells at time $t + 1$, is given by multiplying the probability of chain $j - 1$ being in each state by the probability that metastasis will occur from that state, as shown in Equation (5.7).

$$p'(j, t + 1) = \sum_{i=0}^{4} p(j - 1, i, t) q(i). \tag{5.7}$$

The numbers $q(k)$ will be bigger for bigger values of k. The bigger the lesion the more likely it is that cells will break off and go downstream. To get the probability for chain j to be in state 1 at time $t + 1$, then, we have to multiply $p'(j, t + 1)$ by the probability of chain j being in state 0, and add that to the simple probability from Equation (5.6). So, the proper update formula for state 1 should be Equation (5.8).

$$p(j, 1, t+1) = \sum_{i=0}^{4} p(j, i, t) P(i, 1) + p'(j, t+1) p(j, 0, t). \tag{5.8}$$

Finally, we need to make the complementary modification of Equation (5.6) for state 0. If there is some probability of transitioning from state 0 to state 1, that means the probability of being in state 0 at time $t + 1$ will be less than it was at time t. Here then is the update formula for state 0.

$$p(j, 0, t+1) = \sum_{i=0}^{4} p(j, i, t) P(i, 0) - p'(j, t+1) p(j, 0, t). \tag{5.9}$$

The metastasis from chain $j - 1$ could also contribute to transitions of chain j from state 1 to state 2, and so on. This effect should be very small compared to the local growth once tumor is already present, and would require guessing more numbers, so we will just leave it out.

Now, it is an algebraic exercise to prove that these update formulas preserve probability, namely, if $p(j, k, t)$ satisfies Equation 5.4 then $p(j, k, t + 1)$ does also. This of course also requires that the rows of P each have to sum to 1.0.

So, this is a little more complicated than the simple example of a two state Markov chain presented earlier. It is really a series of coupled Markov chains, in which the series represents a particular path through the lymphatic system from a given starting point. This representation of a lymphatic chain by a state vector that behaves as a Markov process is called a hidden Markov model (HMM). So, each position in the lymphatic drainage pathway is modeled individually as an HMM, and the update formulas for state 0 and state 1 connect the individual HMMs to each other. Figure 5.13 illustrates how the HMMs, each representing a position along the lymphatic drainage pathway of the primary site, are connected.

5.5.3.3 Applying the Model with Actual Parameter Values

When we created this model, we chose some arbitrary but reasonable sounding numbers for the initial states of the tumor, the lymphatic chains, and the transition probabilities. For the primary tumor, we started it off with probability 1.0 in one of the four "tumor present" states, depending on how big it was. In cancer staging terminology this is called the T stage. It is usually a number between 1 and 4. All the lymphatic chains started out with no tumor, that is, their state was $(1, 0, 0, 0, 0)$. Then we computed $p(j, k, t)$ for each chain for a series of time steps (the number is arbitrary; we decided to use four times the T stage).

Explicitly, for the primary tumor, the transition matrix, P_{tumor}, was

$$P_{tumor} = \begin{pmatrix} 1.0 & 0.0 & 0.0 & 0.0 & 0.0 \\ 0.0 & 0.9 & 0.1 & 0.0 & 0.0 \\ 0.0 & 0.0 & 0.9 & 0.1 & 0.0 \\ 0.0 & 0.0 & 0.0 & 0.9 & 0.1 \\ 0.0 & 0.0 & 0.0 & 0.0 & 1.0 \end{pmatrix}. \tag{5.10}$$

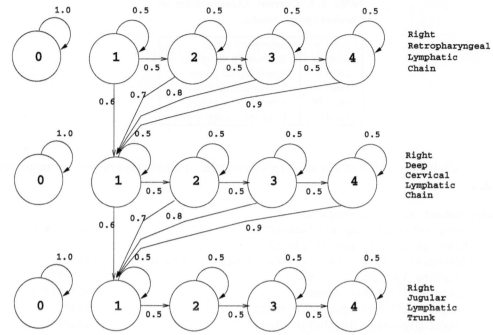

Figure 5.13 Diagram of connected hidden Markov models for lymphatic chains.

For the vector q, for metastasis from the tumor, we used

$$q_{tumor} = (0.0\ 0.2\ 0.4\ 0.6\ 0.8) \tag{5.11}$$

and for metastasis from one lymphatic chain to the next, we used

$$q_{chain} = (0.0\ 0.6\ 0.7\ 0.8\ 0.9). \tag{5.12}$$

Implementing a small program to turn a given path or set of paths into a set of models according to these formulas is not a hard problem. It takes a few pages of code, which we will not reproduce here. However, a few more comments are needed in order to make sense of the output of such a program in terms of reported surgical data. The surgical data are described in relation to nodal *regions*, which somewhat overlap the lymphatic drainage system. So we need to have a way to map from lymphatic chains to these regions.

The development of standardized surgical approaches to the neck (such as radical neck dissection, and selective neck dissection) has paralleled development of standardized nomenclature for the various nodal regions ("levels") in the neck [344,390]. Table 5.1 shows how the levels correspond to groups of lymphatic chains.

A mapping from the lymphatic drainage entities in the FMA to the clinical region system used by surgeons and radiation oncologists is not readily available. This had to be done by hand. Here is an example table mapping a portion of the lymphatic drainage of the head and neck.

Table 5.1 Current Classification of Cervical Lymph Nodes

Level Ia	Submental group
Level Ib	Submandibular group
Level II	Upper jugular group
Level III	Middle jugular group
Level IV	Lower jugular group
Level V	Posterior triangle group

This was a collaborative effort of a head and neck surgeon, a radiation oncologist, an anatomist, and Noah.

```
(defconstant *clinical-regions*
  '(("Deep cervical lymphatic chain"                        "Va" "Vb")
    ("Deep parotid lymphatic chain"                         "P")
    ("Inferior deep lateral cervical lymphatic chain"       "IV")
    ("Jugular lymphatic chain"                              "VI" "IV")
    ("Jugular lymphatic trunk"                              "VI" "IV")
    ("Jugulo-omohyoid lymphatic chain"                      "III")
    ("Jugulodigastric lymphatic chain"                      "IIa")
    ("Left deep cervical lymphatic chain"                   "Va" "Vb")
    ("Left inferior deep lateral cervical lymphatic chain"  "IV")
    ("Left jugular lymphatic tree"                          "IV");VI?
    ("Left jugular lymphatic trunk"                         "VI" "IV")
    ("Left retropharyngeal lymphatic chain"                 "RP")
    ("Left submandibular lymphatic chain"                   "Ib")
    ("Left superficial cervical lymphatic chain"            )
    ("Left superior deep lateral cervical lymphatic chain"  "Va")
    ("Right deep cervical lymphatic chain"                  "Va" "Vb")
    ("Right inferior deep lateral cervical lymphatic chain" "IV")
    ("Right jugular lymphatic tree"                         "IV");VI?
    ("Right jugular lymphatic trunk"                        "VI" "IV")
    ("Right retropharyngeal lymphatic chain"                "RP")
    ("Right submandibular lymphatic chain"                  "Ib")
    ("Right superficial cervical lymphatic chain"           )
    ("Right superior deep lateral cervical lymphatic chain" "Va")
    ("Submandibular lymph node"                             "Ib")
    ("Submandibular lymphatic chain"                        "Ib")
    ("Submental lymphatic chain"                            "Ia")
    ("Superficial cervical lymphatic chain"                 )
    ("Superior deep lateral cervical lymphatic chain"
                                            "IIa" "III" "IV")))
```

Several published series of surgically treated patients have provided valuable pathologic data allowing statistical estimation of the risk for subclinical nodal involvement according to these neck subdivisions [56,58,59,372]. The example data shown in Table 5.2 are adapted from Shah

Table 5.2 Prevalence of Subclinical Lymph Node Involvement According to Tumor Site

Tumor Location	Nodal Level				
	I	II	III	IV	V
Oral tongue	14%	19%	16%	3%	0%
Floor of mouth	16%	12%	7%	2%	0%
Alveolar ridge	27%	21%	6%	4%	2%
Retromolar trigone	19%	12%	6%	6%	0%
Buccal mucosa	44%	11%	0%	0%	0%

[372]. Noah Benson's work used a slight variation on the way we computed the probability of migration of cells downstream to the next lymphatic chain. His article [26] provides the actual comparison. The predictions of the model, with the arbitrary guesses for transition probabilities, came surprisingly close to the data reported.

This project is interesting from the point of view that it combines symbolic reasoning and probabilistic reasoning in a straightforward way, to lay the foundation for what may in the future become a powerful clinical tool for radiation therapy planning.

5.6 Summary

- Dynamic systems (changes over time) can be modeled with symbolic representations in addition to quantitative modeling.
- From the difficulties we uncovered in trying to understand HL7 and DICOM, it should be apparent that the development of protocols for handling medical data communication requires:
  - a deep understanding of the structure of medical data, medical knowledge representation, and medical terminology,
  - a commitment to develop consistent, complete, and unambiguous specifications,
  - an understanding of the technical aspects of computer networking and communications, and
  - an understanding of modular design.

  Even a seemingly technical and low level activity like the development of a data transmission protocol involves most of the foundational principles in biomedical informatics. As with all software to be used in actual medical practice environments, considerable attention should be given to design reuse, software correctness (precise conformance to specifications), and consistency.
- The general idea of modeling biologic systems as state machines has many other examples, and there are other formalisms for modeling state machines. One example is the management of patients with Type II Diabetes, described in a book chapter [391], which also includes an introduction to the use of the Unified Modeling Language (UML) and statecharts, UML's

formal representation of state machine models. The Object Management Group maintains a web site for UML [308], which is a more extensive introduction to this widely used formalism.

5.7 Exercises

1. Sequence search

   The legendary bioinformaticist, Gene Ome, has created a data repository of proteins in which each protein's primary sequence is a list of amino acid symbols. Write a function that will examine a protein sequence and determine the closest proximity of an amino acid to another amino acid, that is, (`proximity prot 'tyr 'leu`) will return the smallest distance between `tyr` and `leu` in the sequence. In case the two amino acids are the same, the function should return the smallest distance between two distinct occurrences of the amino acid, and not the value 0.

2. Touching the Bases

   Molecular biologist Lotta Jeans wants to use the pathway search code in this chapter to determine what happens in a particular organism, comparing normal metabolism with various cases where there are genetic variants and their corresponding gene products are missing.

   (a) What would the annotations have to include?
   (b) Would any of the code have to be changed?
   (c) What would you add to the code or change in the code to make a connection with a list of annotated genes?

3. Aligning protein sequences

   In Section 5.2.5 you saw how to use generalized search methods as a way to do sequence alignment. The example in this section uses DNA and a weight or penalty matrix as part of a scoring function. How would you modify this method to apply it to protein sequence alignments? One possibility is to replace the DNA penalty matrix with something like the BLOSUM62 matrix (which can be found in [24, page 197] or other books on bioinformatics). What other modifications would be necessary? What problems are associated with this strategy?

4. Search takes time

   Why does protein sequence alignment take so much more time and stack space than DNA sequence alignment using the generic search methods? How does the search computation scale with the length of the sequences being aligned and the size of the penalty matrix?

5. Searching the FMA

   In Chapter 4 a network server, the Foundational Model Server (FMS), that provides query access to the FMA is briefly described. Using the FMS we can send Lisp-like expressions and get back lists of results for queries about lymphatic chains, organs, and other anatomical objects. The full API (the valid expressions you can send to the server) is described at.

```
http://sig.biostr.washington.edu/share/docs/ts-browse-api.html
```

The Foundational Model of Anatomy provides a symbolic representation of the connectivity of the lymphatic system. It also includes the blood vessels, the veins, and arteries. Chapter 4, Section 4.3.5 describes some functions for querying the FMS and tracing the lymphatics from any organ or organ part, all the way back to the thoracic duct or the right lymphatic duct.

(a) Building on the functions in Section 4.3.5, add a function, `downstream`, that takes names of two lymphatic chains as inputs, and returns `t` if its first input is directly or indirectly downstream from its second input.

(b) For blood vessels, the FMA uses the two relations `tributary of` and `branch of` for veins and arteries, respectively. Write a function like the one for lymphatics to do the same for veins and arteries.

(c) The FMA class structure is accessible by using the link types that return class information, namely, `:DIRECT-TYPE`, `:DIRECT-SUPERCLASSES`, and `:DIRECT-SUBCLASSES`. Write a function that will handle any of the three types of vessels, veins, arteries, or lymphatics, by inquiring about the types of its arguments, and calling one of your functions above. This is a kind of "poor man's" object oriented programming; the interface does not support any higher level operations, so you need to be able to do it explicitly.

6. State transition tables

Rewrite the `box-transition` function in Section 5.4.1 using a state table instead of explicit logic in the code. Include in your function some text output showing information about the transitions, as shown in the sample run, using `format`.

7. Progress reporting

From the sample transcript of the hand box code, you can see that some output operations were added in the running code to print progress or state transition messages. Where in the code should these output operations (presumably using `format`) be put? Can they be encapsulated in a progress reporting function that would be an input to the state machine? If so, modify the state machine code to provide for this. If not, explain why it would not be practical.

8. More than one way to start

The code that implements the yeast cell cycle state machine in this chapter has a fixed initialization, setting the `size` signal initially on, then off. Modify the code so that it is possible to specify a set of genes to be initially on, and then turned off after the first cycle.

9. More on state machines

The state machine code is in this chapter not restricted to a finite number of states, only a finite number of variables, although the two systems we considered, the hand box and the yeast cell cycle are indeed finite state machines. Of the various methods for implementing state transitions, which would not be possible if one or more of the variables could take on any of an infinite number of values? From what you know, do cells appear to have such properties? Certainly the number of genes, proteins, and other molecules is large, but not infinite.

CHAPTER 6

Biomedical Data, Knowledge, and Systems in Context

Chapter Outline

So far we have addressed how to make biomedical data and knowledge computable, and how to design software and systems that do interesting things. This chapter will now turn to the challenges of safety, security, and the context in which computers, software, and data are deployed and managed.

Two significant advances in computing technology have raised concerns about safety and security in building, buying, and using computers in medical practice. One is the invention of the microprocessor, or more generally, the microminiaturization of electronics, including digital electronics. The other is the invention of the Internet in the decades between 1965 and 1985, and its rapid expansion in the decades following. Failure to build in adequate safety engineering in the design and use of a radiation treatment machine led to several deaths and injuries [250]. Many other device-related accidents with computer controlled medical devices have since been

Principles of Biomedical Informatics. http://dx.doi.org/10.1016/B978-0-12-416019-4.00006-8

reported. The possibility of such failures is exacerbated by the fact that these medical devices are (and need to be) connected to hospital networks and the Internet. More recently, embedded medical devices such as pacemakers and insulin infusion pumps present challenges involving wireless access [85,268,318]. These devices present problems that show that safety, privacy, and function can often be in conflict.

Beyond the issues around computerized medical devices, the adoption of electronic medical record systems has opened up problems with identity theft, inability to access critical systems because of intrusions or disruptions, theft of data, and a host of other problems. In this chapter we will examine a few examples of these problems, as well as the general problem of implementing a modern computerized health care institution with a high degree of access and connectivity, while maintaining adequate security and other controls.

6.1 A Bit of Personal History and Perspective

When I was a teenager, a computer was a room full of large cabinets connected together by wiring that ran under a raised floor. The computers of that time (late 1950s, through the 1960s) were slow, large, had limited storage capacity, and were accessible only to a few people. I felt very privileged to have access to the Burroughs 220 computer at Cornell University as an undergraduate in 1962, where I was paid $1.00 per hour as a student computer operator during nights and weekends. The picture at the beginning of the Preface to the first edition shows me at that time sitting at the Cornell Burroughs 220 console. The Burroughs was a great advance in its day, with an all-vacuum tube CPU and magnetic core memory. It used binary coded decimal (BCD) digits, where each of the digits 0–9 was stored in direct binary form, 4 bits per decimal digit. To encode alphabetic characters, the code was extended to 6 bits (the encoding was actually called "sixbit"), to accommodate upper case characters and a few special characters. A version of this was used for encoding information on Hollerith punched cards.[1] Each machine word in the Burroughs 220 was 10 such BCD digits. It was capable of doing arithmetic in milliseconds, much faster than I could do on paper. The program counter was a four digit register, and so were the other address registers, so the total memory of the machine was 10,000 of these 10 digit words. There were no "hard disks" or interactive displays in those days; programs and data were punched into Hollerith punched cards, and read by a card reader. Program source code and data, as well as output produced by the program was printed on a very large, very slow, and very noisy line printer. For those who were willing to learn sophisticated operations, it was also possible to copy a stack of punched cards to a magnetic tape, and subsequently load the program from the tape. I was one of the lucky few to be issued my very own personal reel of tape. This installation cost six hundred thousand dollars (1960

---

[1] The early Control Data Corporation computers used 60 bit binary words, which conveniently encoded 10 sixbit characters in each machine word. Later (1963 or 1964) IBM extended it further to 8 bits to accommodate lower case and a variety of other standard characters. This encoding was known as EBCDIC, or Extended Binary Coded Decimal Interchange Code. IBM used EBCDIC in its internal machine byte representation, and pretty much everyone else adopted ASCII (see page 21).

dollars!). For more history, refer to "The History of Computing at Cornell University," at the web site, `http://www.cit.cornell.edu/computer/history/`.

Today, of course, things are different. We have devices that include over a million times more memory than the Burroughs 220, are a million times faster, cost only a few hundred dollars (2012 dollars!), and fit in a shirt pocket. The significance of this for medicine is vast. Although the most visible manifestation is that physicians and nurses are now very involved in exploring how small handheld computers can be used to help them in their practice, a far more significant change has happened with very little visibility. Practically every machine, or device, that the physicians and nurses are using in medical practice has some kind of embedded computer or computers in it.[2] The most obvious are the digital radiologic imaging systems, Computed Tomography (CT), Magnetic Resonance Imaging (MRI), Digital Radiography (DR), to mention a few. However, blood pressure measurements are now done with digital devices, and vital signs monitors also have embedded computers. All radiation machinery that is used to plan and treat cancer is now computer controlled. Even the intravenous (IV) infusion pumps that are connected directly to patients' veins are computer controlled, and have wireless network interfaces as well. These are the devices on poles that pump powerful medications into patients at carefully controlled rates. This is not even to mention the billions of dollars being spent to install electronic medical record (EMR) systems. When an EMR system is unavailable, serious problems can occur. Critical information may not be available that is needed for medical decision making on the hospital wards, and recording of critical information has to be done with alternative methods (paper records). Many articles and books have been written about the significance of EMR systems; I'll just refer to one very recent one, which has many citations to others [427].

So, the wonderful advances in technology that have made powerful devices possible and inexpensive have also introduced new dangers and points of failure. The engineers who designed medical machinery before the digital age were well versed in safety engineering for devices that relied on electromechanical relays and other such devices. They were not at all prepared to deal with the requirements of safety engineering for software controlled devices. Many people have claimed that it is hopeless, that software is too complex to use safety engineering methods, and that we just have to rely on the brilliance and talent of programmers. This claim, if true, would imply that we should not attempt to build such inherently unsafe devices. Fortunately the claim is false. It is possible to do a reasonable job of safety engineering with computer-controlled devices and there is a growing literature describing methods, tools, and processes for accomplishing the task. This chapter includes some pointers to this literature. Unfortunately,

---

[2] As this trend accelerated, and in the UW Cancer Center we became involved in our own device engineering projects, my mentor, the late Peter Wootton, remarked that soon everything, even the toilets, would be computer controlled. He intended to be humorous, but I notice that these days in most places the toilets have infrared sensors that automatically flush the toilet, and are likely small digital devices, hotel rooms have doorknobs that read magnetic strip cards and automobile tires have RFID chips, as do pet dogs and cats. Even cell phone and other device batteries (notably cordless power tools) are not really just batteries but have their own microchips, requiring that the device that uses them have the proper "driver" software.

medical device manufacturers still have a long way to go in achieving the state of the art in this area.

So much for the impact of transitioning to the world of digital devices. What of the rapid growth of the Internet in the most recent decades? In 1978, when I started a fellowship in medical physics at the University of Washington, the Internet only involved a few departments in the mathematical sciences (including Computer Science and Engineering, of course). However, within the Cancer Center, we began to acquire computer-controlled machinery, and our own minicomputer. By the early 1980s it was clear that we would need to connect our growing computer collection together locally, and the Clinical Neutron Therapy System (CNTS) needed to be connected to the departmental computer cluster. Initially this arrangement was a proprietary one, DECnet, because all the computers were products of Digital Equipment Corporation, including the PDP-11 that ran the CNTS. Later, as Internet connections became more widely available through the campus, we connected all the Radiation Oncology Department computers to the Internet, using TCP/IP. This was necessary to support research projects [61,182,198] that involved joint work on comparative radiation treatment planning, as well as exchange of software and medical image data.

During this time, many clinical applications of computing had become established, as mentioned earlier. There was a clear need to provide access to systems and data throughout the medical center. Early efforts at EMR systems were based on the idea that these "ancillary systems" would feed data to a central repository, and the central repository would be accessible by terminal devices connected in some proprietary way to the central system. This failed. Instead, around 1990, the hospital began a project to extend the campus network into the medical center, with uniform TCP/IP networking and Ethernet for physical connections of devices. The project was a great success. The wards had access to a variety of data resources through X terminals, and later through desktop computers that could act as peers on a TCP/IP network. Now, the Internet is ubiquitous and necessary. Essential to hospital operation is access from off site by providers and other staff, exchange of data between organizational components throughout the city, and access between the UW Medical Center and other completely independent organizations involved in our patients' health care, throughout a five state region served by the UW School of Medicine and its health care components, as well as nationally and even internationally. Health care today cannot operate with isolated hospital networks. Connectivity is now a fact of medical practice.

The most recent version of the burgeoning use of network technology is the incorporation of wireless communications in medical devices. This provides yet another level of challenge, as wireless networking makes controls more difficult, while also providing truly useful new capabilities.

The use of computers in medical devices, and for managing critical data in medical practice, has developed rapidly. The demand for worldwide connectivity and interoperability of diverse systems and networks has outpaced the ability of vendors and IT leadership in medical practice to respond with appropriate safety and security measures. The focus of this chapter is on the nature of these challenges and how we might respond.

6.2 Medical Device Safety

Examples of actual problems in medical software system design should serve to illustrate the principles and the consequences of failure to apply them. Except for the Therac problems, all of these examples come from my personal experience and work with colleagues in the Radiation Oncology Department at the University of Washington Medical Center, and in my role as IT Security Director for UW Medicine.[3] Other excellent sources for related information include [176,249,250]. A broader text addressing medical device software safety [92] is one of many that are now available.

6.2.1 Radiation Therapy Planning Systems

When I began a postdoctoral fellowship in medical physics at the University of Washington in 1978, the Radiation Oncology Department was using some simple dose calculation programs for planning radiation treatments. These programs ran on a "mainframe" computer at the UW Academic Computer Center. The charges for use of the computer service, which at best could give a 1 day turnaround of results for a submitted batch job, amounted to around $1500 per month. At that time, some of the department staff became involved in a software development project to produce a commercial RTP system. At that point we decided to abandon the mainframe computer programs, and invest in a minicomputer and RTP software. However, instead of buying a commercial product, we bought a VAX11/780 minicomputer, and I was committed to writing RTP software for use in our clinic. Jonathan Jacky joined the project at that time, and the two of us set out to put something together as fast as possible. Inspired by our colleagues in the Computer Science and Engineering (CSE) Department, we experimented with radically different approaches to software design than was the practice in our field [178,179,181,202]. We were able to get substantial functionality implemented in much less time than was typical for projects at other institutions, and with substantially fewer lines of code.

These projects had their flaws. Here I will relate just a few. One involved the commmercial project mentioned earlier, in which I was involved. Another was a very remarkable problem that could be classified as software or hardware, depending on your viewpoint.

The commercial software that I wrote was accepted, and became part of a popular commercial product. After a few years I heard from the company sales staff that a customer had discovered a problem. My program modeled the radiation from sealed capsules of radium, cesium, or other isotopes used in treatment of cervical cancer. It was designed to be able to model treatments combining capsules of different types. However, if different capsule types were combined in a certain way, the dose calculation would use only one type, and clinically significant errors could result. I examined the source code, found the error, and sent a report back to the company, with detailed instructions for how to fix it. Some time later I noticed that the user manual for the program noted this feature, stated that it does not work correctly, and directed the customer not

---

[3]The term UW Medicine encompasses the University of Washington School of Medicine, the two major medical centers it operates (University of Washington Medical Center and Harborview Medical Center), a network of outpatient clinics, some specialty centers, and several additional community hospitals and medical centers.

to use it. Apparently it was determined that putting this notice in the manual was preferred over actually fixing the program, even though the fix only involved modifying two lines of code.

The second incident happened early in the life of our VAX11/780 minicomputer. The dosimetrist, in the course of using one of our programs, noticed that the dose calculation results for a more or less standard cervical cancer treatment plan appeared to be different than usual. A calculation of a few points by hand suggested that the results were about 20% different from what we would expect. We could not find any errors in the input data. I recompiled the program, since we had done some updates recently to the operating system software. This did not change the output. I examined the source code. There had been no recent changes. Finally, I thought to check that the VAX itself was running correctly. Now, the VAX11/780, like many computers of its time, had a core processor built from smaller components called bit slices. These 4 bit microprocessors (a set of 8 altogether) were programmable, and the program is called *microcode*. The microcode implements the basic instructions of the VAX computer, the instructions we usually think of as the hardware or machine instruction set. So, the VAX hardware is really software. The microcode is stored on a diskette in a drive connected to a separate computer called the *console* computer, in this case a PDP-11. When the VAX starts up, the PDP-11 copies the microcode into the VAX CPU. The PDP-11 also has test programs that run to test the operation of the VAX internal "hardware" and various components. So I ran the arithmetic instruction set exerciser, and discovered that the microcode had somehow changed, so that the floating point multiply (FMUL) instruction was not working properly. Reloading a backup copy of the microcode fixed the problem.

The most disturbing thing about this second incident was that the failure of the floating multiply instruction to work correctly did *not* cause a catastrophic failure (which would have been much better) but actually caused only a subtle problem. It was enough to get clinically erroneous results, but the rest of the computer seemed to be operating just fine. In terms of safety, it is often better to have a system fail "hard" rather than be off just a little, enough to cause safety problems, but not enough to be sure to be noticed. There were no clinical consequences of this error, because the dosimetrist routinely reviewed the output of the program for reasonability. We continue this practice today. We rely on computer software and hardware to do much of the routine work for us, but we perform checks on the machine operations and software behavior to catch possible malfunctions.

It is extremely important to note that we are *not* checking the accuracy with which the formulas implemented by the program can predict the radiation dose that a source will produce. That is a matter of physics, *not* a matter of whether the software is correct. Despite this, there is still today a widespread notion that when checking the operation of an RTP system, you do measurements and verify that the program is calculating the dose accurately. This is *also* necessary, but does *not* provide any assurance of correct functioning of the program. The accuracy of the knowledge implemented by a computer program must be validated separately from the verification that the computer program is correctly implementing the intended procedure.

At the UW, we had developed a completely new program suite, as noted. Jon and I believed very strongly that it was important to check our programs for correctness, independently of the question of whether the programs implemented an accurate dose computation model. This is a point of confusion for many people. A program is a device. It has functions, inputs and outputs, and operating instructions. A well-designed device starts its life cycle with a clear set of specifications for what the designer intends it to do. The first thing that needs to be done in terms of quality is to insure that the program does what was intended. We created a large set of test data, specifically to verify that the program operated as intended, and performed the tests on our own programs when we made changes or additions [180,187].

Despite our best efforts, we still found and fixed many errors in our programs. One was particularly striking. A part of the radiation dose calculation involved simple linear interpolation based on the intersection of a line with a segment of a polygon representing an internal organ whose density differed from the surrounding tissue. This was used to compensate the basic formula, which assumed a uniform tissue density. Our program appeared to give satisfactory agreement with a simpler program we obtained from another medical physics group, but there was a very tiny discrepancy. Because we implemented the same algorithm, we thought the two programs should produce identical, not approximately equal, results. We created an extreme test case, which indeed magnified the discrepancy. After some study, we found an implementation error in our code. Once corrected, the two programs indeed gave identical results.

The idea that comparing the outputs of two programs that compute the same thing can detect errors has its limits. In many cases, one would indeed expect small differences. Ultimately, the only way to know for sure that a program is correct is to have a formal specification of what the program can do, and to do a proof either manually or with a computer program designed to do consistency checking or other kinds of analysis. Such programs and methods exist and have been used in practical projects. See [177] for more about this and formal methods for program development more generally.

6.2.2 Who is Master, the Human or the Computer?

In 1980, as we began the process of acquiring new radiation treatment machinery at the UW, vendors were starting to provide attachments to the machines, to handle digital data about the treatment and the patient. Such systems were called "Record and Verify" systems, or sometimes "Check and Confirm" systems. The purpose of these systems was to have a computer screen and keyboard entry facility by which the operator can select the patient's treatment plan data, set up the machine and then the computer would compare the manual setup with the digital data to insure that there are no setup errors. This is not very effective [248] but was popular for some time.

The installation of record and verify systems at the UW Medical Center at that time created an interesting human-computer interface problem. Soon after installation, the little digital data tape drives failed. The tapes were used for transfer of data from the treatment planning systems or were created separately by an offline process. The service engineers fixed the tape units, and restored the Record and Verify system to normal operation. This was not critical since the machines could be run in the more usual way completely independently of the Record and Verify systems. However, after a short time, the tape drives failed again. After a few cycles of this the Record and Verify system was taken out of operation. Though there was never a formal announcement about why the tape systems appeared to be failing so frequently, it was pretty well known that the therapists did not want to use the system.

For many years following, I argued in discussions at medical physics meetings both locally and nationally, that this arrangement was exactly backwards. It should be set up so that the RTP system and treatment machine use data transfer to reduce the tedium of manual data entry and setup, so that some useful function is performed by the system. Then the human should check the machine, not the other way around. In fact, we had exactly this configuration with the neutron therapy facility (CNTS), and it has worked well for nearly 30 years [204]. Eventually as the treatment operations themselves became too complex for manual operations, vendors and users of radiation treatment machines adopted the idea of using automated data transfer and as much as possible, automated setup, with the operator checking the machine setup. The gain in efficiency by network transfer of data using DICOM is very large. Now there are computer systems for data management that integrate the RTP system, the machines, treatment record keeping, management of image data, appointment scheduling, and other things.

6.2.3 The Therac-25: A Real-Time Disaster

Experiments with complex electromechanical control of radiation therapy machinery have been going on for some time. An example of pioneering work is the project started by Takahashi in 1965 to do conformal radiation therapy [406]. Radiation treatment machines built with computer-based control systems have been available since at least the mid-1970s. For more of this history, I refer the reader to the review article of Fraass [114].

Modern machines are designed to provide both high intensity electron beams and X-ray beams. Switching between these modes is complex, requiring coordination between multiple subsystems. One such machine, the Therac-25, was designed with software that erroneously allowed a sequence of control panel keystrokes that put the machine in an unsafe state. In this state, the electron beam was at the extremely high intensity required for producing X-rays, but it was allowed to exit the machine as if the machine were in electron mode. There were several such malfunctions in different cancer treatment centers, resulting in several injuries and deaths. Some time later the machines were shut down, and the model withdrawn from the market. The definitive review of this history is an article by Nancy Leveson [250], who also published a book on software safety [249].

Many factors combined to cause the deaths and injuries in the Therac-25 incidents. Software design flaws were only a part of the problem. The machine often produced error or malfunction alerts, some of which did not appear to require any corrective action. In any case, the error messages were cryptic and not explained in the user manuals. It appears that operators were used to routinely overriding or proceeding despite the messages.

The Therac software was large and complex. It may seem daunting to require a safety analysis, but it seems foolish not to do it. At the time, the FDA had (and still has) a role in certifying and approving medical devices for marketing in the US. There are a growing number of products that come into direct contact with patients, like the infusion pumps mentioned here, as well as instruments such as heart monitors, that are computer controlled.

A conservative response to these problems might be to not design safety critical systems with software. That begs further analysis. A properly designed system that is implemented with software can provide important capabilities that improve safety and effectiveness, as well as making devices more serviceable. The legal system recognizes this. A significant liability judgment was awarded to a patient who had been paralyzed from the neck down by an unsatisfactory radiation treatment [97]. The court found that the practitioner was liable because he did not use an RTP system to plan the treatment but based his plan only on X-ray films. This happened well before RTP systems were in most radiation therapy clinics. The expectation of the court was that when technology becomes available and is demonstrated to be effective, it should be used, even if it is not yet uniformly or generally in use.[4]

The challenge is to achieve the level of quality in software that we expect of hardware. It is possible, though rare. Even today, injuries and deaths attributable to radiation mistreatments are occurring and are the subject of successful malpractice lawsuits. It is important also to recognize that the design flaws in software are not the sole cause in these cases. Inadequate training of hospital staff, understaffing so that there is not sufficient time to perform periodic reviews to insure that software configurations are correct, and protective reactions by both medical practitioners and vendor management all continue to be significant factors in treatment failures. Much remains to be done.

6.2.4 Prism

In my own experience I have seen the same user behavior of ignoring malfunctions described in the Therac reports. In the initial months of operation in 1994 of the Prism RTP system (described in Section 2.2.8 of Chapter 2), operators (dosimetrists) experienced many continuable error conditions. They occurred so frequently that the dosimetrists did not routinely report them to me. The errors occurred during communications between the Prism program and the X server program that handled the local display on the user's workstation. They were asynchronous, so the usual debugging tools and strategies were useless. Systematic source code analysis was the only way to track down and eliminate these problems.

---

[4]A legal precedent cited in this finding is a famous case where a tugboat collision occurred in fog in New York Harbor, in 1932. Justice Learned Hand, in the US Second Circuit Court of Appeals, ruled that the tugboat should have been equipped with a radio, even though it was not common practice. For more on this and other legal issues around medical devices, see [338].

Once I found out what was happening, I began a systematic analysis of the program source code, to identify and correct every instance I could find of certain classes of errors. These involved places in the program where some software resource in the display part of the program was allocated and referenced but never properly deallocated or dereferenced. It took several months to conduct this review. After completion of the review and correction of every discovered instance of these errors, the seemingly random errors never occurred again. The problem of memory management, not only at the level of allocating and deallocating dynamic memory, but at the application level, of having pointers to things that have been removed, is a widespread problem. The discipline I used was that every object or reference creation must have a corresponding deallocation, in the same software module, and this pairing must be provably correct, independent of the rest of the program.

6.2.5 The UW Neutron Therapy Facility

In 1979 the University of Washington contracted with the National Cancer Institute to build and operate a neutron radiation therapy facility to perform clinical trials and evaluate the efficacy of neutron therapy in the treatment of cancer [101]. Early studies had indicated some real potential for neutron therapy of tumors that were resistant to normal radiation [316,441]. This resistance was attributed to the oxygen effect, where poorly oxygenated tumors were more resistant to radiation than normal tissue which was well perfused. It appeared that with neutrons there was no oxygen effect, and thus neutrons would have a therapeutic advantage. We contracted with Scanditronix AB in Sweden to build the facility. The division of Scanditronix that worked on this project had experience building cyclotrons but not much with medical facilities.

The control console they provided had software for monitoring the thousands of signals, control lines, voltages, and currents in the various subsystems. There are many subsystems, each with its own status display. The initial design had a separate program implementing each display, with an interface to an array of buttons on a control panel, from which the operator could select a status display. The programs used unreliable synchronization methods to detect a button press and transfer control from one program to another. It was possible to get the status display to "freeze" by pressing a few buttons within a short time. At first the engineers were reluctant to redesign the programs as a single program without any need for interprocess synchronization (my original recommendation). It took a demonstration in front of the project manager to get this change implemented.

Overall, the interaction between the UW staff and the Scanditronix engineers was sufficient to produce a reliable system. It has operated since 1984 without any safety incidents [204], even with several changes to the RTP system from which the treatment machine control data are sent, and through a complete replacement of the therapy room equipment control with a locally developed system [185,186]. The control system project incorporated formal methods and specifications as a way to insure safe designs and to ease the checking that the resulting system would be safe for use with patients [184]. A by-product of this was a book by Jonathan Jacky [177] on the Z formal specification language.

6.2.6 Safety by Design: Radiotherapy Machines

In this section I'll present a simple approach to software design with declarative safety conditions. It should serve to illustrate the idea of a safety analysis.[5]

As we described in Chapter 2, Section 2.2.8, the goal of radiation therapy is to:

- create a plan for aiming radiation beams to irradiate all the visible and invisible tumor cells in the region of the primary tumor, which will *not* irradiate any normal tissues unnecessarily, and
- deliver the planned treatment in a safe and reliable manner.

The focus here is on the design characteristics of the machine that accomplishes the delivery of treatment.

Typical radiation treatment machines can produce electron or X-ray (photon) beams, that can be made narrow or wide, aimed in most any direction, and turned on for varying amounts of time at varying rates of dose delivery. All the parameters of such treatments are selectable at the console, and are typically specified in a treatment plan using an RTP system. Figure 2.4 in Chapter 2, Section 2.2.8 shows a typical radiotherapy machine setup. The process for planning radiation treatments is also described there. Here we will describe the design of a simplified hypothetical radiation treatment machine, to illustrate the principles of safety in treatment machine design.

Radiation therapy machine safety issues include the following:

- Underdose of tumor $\Rightarrow$ treatment failure
- Overdose of tumor or critical structures $\Rightarrow$ injury or death
- Machine collision with patient $\Rightarrow$ injury or death
- Patient falls off couch $\Rightarrow$ injury or death

We will focus only on the overdose issue in our simplified machine design. Figure 6.1 shows a simplified schematic of how a radiation therapy machine works. The name *linac* is usually used, as an abbreviation for "linear accelerator," which describes the basic design principle. An electron gun produces a large quantity of free electrons, inside a waveguide, a linear metal duct. Very intense microwave energy is pumped into the waveguide and the traveling radio (electromagnetic) waves accelerate the electrons to the far end, where they pass through a window to the outside. The electron beam can be allowed to exit, through a collimating system (not shown), or it can be made to strike a heavy metal target, thus producing a beam of X-rays.

When the electron beam is used with the target in place to produce X-rays, the intensity of the electron beam must be very much higher than it would be if it were used directly. This is because almost all of the energy of the electron beam is dissipated in heat in the target. Thus when the target is in, the electron beam intensity could be as much as 100 times what is needed for direct electron treatment. If the beam intensity is set this high, and the target is for some

---

[5]This material was first developed and presented by Jonathan Jacky as a guest lecture for a class I taught, MEBI 531, "Life and Death Computing." The examples originally used C notation.

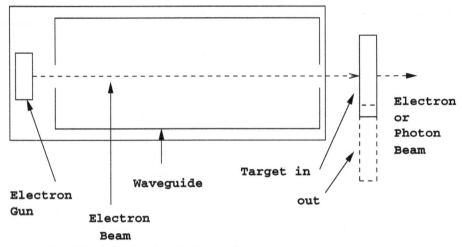

Figure 6.1 **A simplified diagram showing how a linear accelerator works.**

reason not in place, the intensity of the electron beam will be sufficient to cause very serious injury or death. This is the basis for the Therac-25 accidents mentioned earlier. In the next section we will see how a formal description of the control logic of a linac can provide a means for analyzing whether such accidents can happen or whether the design is intrinsically safe.

6.2.6.1 A Simple Linac Control System

We will model the linac as a finite state machine with discrete state variables. The idea is similar to what we did for the cell cycle control genes in Chapter 5, Section 5.4.3. However, we will not implement the state machine itself. Instead we will write formal descriptions of the states, the transitions, and their properties. Rather than run simulations, we will analyze the transition descriptions to see what can be learned from them.

The state variables for our abstract linac are (with initial values):

```
(defvar beam 'off "Allowed values: off on")

(defvar current 'high "Allowed values: high low")

(defvar target 'in "Allowed values: in out")
```

Altogether there are eight possible states. However, some of them have special significance. We will define predicate functions that return t when the machine is in the named states or sets of states.

```
(defun electron-mode ()
    (and (eql current 'low) (eql target 'out)))

(defun X-ray-mode ()
    (and (eql current 'high) (eql target 'in)))
```

Note that each of these really describes a set of states, since there are two values of the beam variable, and in either case, the mode predicates return the same result.

There are other states besides these, corresponding to all the other combinations of values for the state variables. Some are safe and some are unsafe. So, next, we need a way to describe the safety properties of the machine. The way we specify the safe states is to define a safety predicate function, `safe-mode`. This function will have the value t when the state is safe, and nil otherwise.

```
(defun safe-mode ()
   (or (eql beam 'off) ;; safe for sure
       (eql target 'in) ;; safe even with current high
       (eql current 'low) ;; safe even with target out
       ))
```

The machine may start up in some well-defined starting state, but to operate, it will need to be able to make state transitions. So, we have to define functions that implement state transitions, which correspond to actually operating the machine. A state transition is an operation that changes the value of one or more state variables. Each state transition will have preconditions and state variable updates. If the preconditions are not met, the state transition does not take place, so the values of the variables do not change.

The operator will need to press buttons to turn the beam on and turn it off. If the preconditions are met, the transition is *enabled*, but it does not happen unless an appropriate event, like a button press, takes place. Here are transition functions for the beam to go on and off.

```
(defun beam-on ()
   (if (eql beam 'off) (setf beam 'on)))

(defun beam-off ()
   (if (eql beam 'on) (setf beam 'off)))
```

The operator will also need to be able to change from X-ray mode to electron mode and back again. These mode changes should only be done while the beam is off, so that is a precondition.

```
(defun select-electrons ()
   (if (eql beam 'off)
       (setf current 'low target 'out)))

(defun select-X-rays ()
   (if (eql beam 'off)
       (setf target 'in current 'high)))
```

The central idea of the analysis is that certain conditions change but other conditions are required to remain the same. Conditions that do not change when an operation is performed are *invariants* with respect to that operation. Something that does not change under any conditions is an invariant of the system. The safety requirements of our simple system are that it should initially be in a state in which the `safe-mode` predicate evaluates to `t` and it should be an invariant of the system, so that regardless of what transitions are enabled or actually take place, `safe-mode` remains `t`. You should be able to now analyze whether the above design satisfies the invariant.

You might think it does, but as described, it is not a safe system. The state with `beam off`, and `current high` and `target out` is a safe state, because the beam is `off`. However, the `beam-on` transition is enabled, and thus the system can make a transition to an unsafe state. You might think, how can it start out in such a state? We did not specify anything except that the machine starts in a *safe* state. So, one way to fix this problem is to define `safe-mode` more narrowly.

```
(defun safe-mode ()
   (or (electron-mode)
       (X-ray-mode)
       ))
```

Then if we require that the system starts in a safe state, the system will remain in a safe state. Note that there is no "mode" state variable, because there are states that are not included in electron mode or X-ray mode. It could even be misleading to design the system with such a variable, because it would suggest that there are no other states. What would the value of a mode variable be in a state which was neither `electron` or `X-ray`? Keeping the value consistent with the other state variables adds complexity to the system and thus additional opportunities for errors.

Now, it may be that we need additional state transitions to make the system actually work, even though they do not correspond directly to user initiated operations. for example, we may need to be able to move the target in or out, or change the current setting, as part of a larger process. Here are transitions for the `target` variable,

```
(defun target-in ()
   (if (eql target 'out) (setf target 'in)))

(defun target-out ()
   (if (eql target 'in) (setf target 'out)))
```

and similarly for the `current` variable. Now we can repeat the safety analysis. Can the system get into an unsafe state?

1. Start with beam off, target in, current high. This is safe because the beam is off. It also corresponds to X-ray mode.
2. Move the target out. This is allowed and the new state is still safe.

3. The `beam-on` state transition is enabled, so do it.
4. Now the system is in an unsafe state, since none of the safety conditions are met.

So, with these lower level transitions, the system no longer satisfies the safety invariant.

How can we prevent unsafe transitions? The principal safety concern is that one must not allow the beam to be turned on if that leads to an overdose. We can make this system safe by putting a guard condition on the `beam-on` operation:

```
(defun beam-on ()
    (if (and (eql beam 'off)
             (or (electron-mode) (X-ray-mode)))
        (setf beam 'on)))
```

If the machine is in any other state, including ones that may be safe, but could become unsafe if the beam were turned on, the beam cannot be turned on.

This example is so simple that it may not be clear what the value of this analysis actually is. The point here is that the declarative description of the states and state transitions is the starting point for an effective analysis. This style of description enables us to examine each part, each state transition, to determine its effect, regardless of what other states might be enabled, and regardless of how the system arrived in any particular state. An alternative, to create flowcharts and procedural code, makes the analysis much more difficult, because of the large numbers of paths through the flowchart that need to be considered.

6.2.6.2 Adding a Dosimetry System

Let's make the system a little more complicated by adding a dosimetry system. Typically, the machine has ionization chambers that monitor the amount of radiation going out of the machine, and as each machine pulse is generated (the machine works in pulses rather than continuous streams of electrons), the dose counter counts up by an amount measured by the ionization chamber. When the accumulated dose equals or exceeds the prescribed dose, the machine should shut off automatically. So, we need a few more state variables.

```
(defvar accum 0.0)
```

```
(defvar prescribed 0.0)
```

```
(defvar tolerance 0.1)
```

The `accum` variable has a value that is the dose delivered so far. This variable's value will be updated as the beam is turned on to deliver some radiation. The `prescribed` variable has a value that is the total to be delivered. It is set by the machine operator. The `tolerance` variable is a small amount that `accum` is allowed to exceed `prescribed` without harm. We can express this with the predicate `safe-dose`, which is true as long as `accum` does not exceed the sum of `prescribed` and `tolerance`. Taking this as another safety invariant, we can also write the combined safety condition, `safe-linac`.

```
(defun safe-dose ()
  (< accum (+ prescribed tolerance))))

(defun safe-linac ()
  (and (safe-mode) (safe-dose)))
```

What remains in the design with dosimetry is to write the transition that describes an increment of dose delivered when a machine pulse takes place. The dose is only delivered when the beam is on, so that is a precondition.

```
(defun deliver-dose ()
  (if (eql beam 'on)
      (incf accum delta)))
```

The parameter `delta` is the amount of dose delivered by the machine with each pulse. It represents the dose *rate* which is adjustable within some range on most machines. It is also easy to write the transition that turns the beam off when the prescribed dose is reached. The preconditions are that the beam is on and the accumulated dose has equaled or exceeded the prescribed dose.

```
(defun beam-off-dosim ()
  (if (and (eql beam 'on)
           (>= accum prescribed))
      (setf beam 'off)))
```

Note that we don't call the `beam-off` function. In a real design, instead of `defun` we would use an operator that created *transition* objects, and we would create a real state machine that would perform the enabled transitions, by examining their structure, rather than just calling functions. This could also be something like the rule interpreter code where a rule is not a callable object but an expression to examine and interpret.

Now, is the `safe-dose` invariant always satisfied? Although it seems that the beam would go off when the required dose is reached, two things can go wrong. First, after the machine turns off, it is still possible to turn the beam on again. That means another `delta` of dose will be delivered, before the beam then shuts off. This can go on indefinitely. We can easily fix this by providing an additional guard in `beam-on`.

```
(defun beam-on ()
  (if (and (eql beam 'off)
           (or (electron-mode) (X-ray-mode))
           (and (< accum prescribed)))
      (setf beam 'on)))
```

That takes care of one problem. However, there is a second problem, that `accum` may be less than `prescribed` by a very small amount, and `delta` may be big enough that the result of one pulse may produce an `accum` that is more than `prescribed` plus `tolerance`. So, if the dose rate is too high or the sampling interval is too long, problems can occur. It is left as an exercise to fix this last problem.

6.2.6.3 Formal and Automated Analysis

In this problem there are few enough states and variables that we were able to do the analysis manually. When systems get bigger, one might reasonably ask if automated procedures are possible. There are indeed many tools and methods for analyzing formal descriptions of designs like this one. One way to think of it is to use graph search as described in Section 5.2. Iterate through the states satisfying the safety condition, apply all the transitions to determine if you reach any unsafe states. If not, the system satisfies the safety invariants. If so, the design is unsafe. This is a bit too simplistic to scale well, but more sophisticated methods are available, and are mentioned in [177].

6.3 System and Data Security

Now we turn to a very different topic, that of system security. Until now, we have considered the internal design of systems, and how things might go wrong. Since we said it is a fact of life that hospitals are connected to the Internet, another very big problem is the proliferation of large-scale malicious attempts to break into computer systems. This was in fact a problem at least 20 years ago. The most famous such incident is reported in an article in Communications of the ACM [398], and a book by the same author [399].

6.3.1 Real Life Examples

As with device safety, my experience dealing with system and network security spans about 30 years. My last 5 years before retiring in 2011 I served as the Director of Networking and Security for UW Medicine, as explained at the beginning of this chapter. During that time I became active again in adult amateur ice hockey. Since several team members of the newly formed team (2005) were involved, as I was, in various aspects of computers and software, we named the team "The Hackers." I played (and still play) defense, and served as the team captain. As with computers, the equipment has improved significantly and the skill level of the attackers has increased, but many of the challenges of the game remain the same as ever. The picture on page xi at the beginning of the Preface to this second edition shows me in 2005, ready for whatever may come up. A few recent examples should serve as an introduction to some of the issues around security and medical computing.

6.3.1.1 DICOM and Internet Security

While DICOM does not raise directly any safety issues, erroneous transmission of image data can have implications for diagnosis and treatment. In addition, there is the question of system security, which we will take up shortly. The DICOM specification did not, until recently, address security or encryption, although it does include details about methods of image compression.

Even with the recent adoption of Part 15, many questions are left up to the implementation. We had to address security and authorization questions early on with our DICOM project at the UW. What is the appropriate response to clients that request connections from unknown IP addresses? How much tolerance should an application provide for matching of Application Entity titles? We found variations on both of these in other implementations. Some commercial DICOM implementations provided no security or authentication whatsoever. Others had requirements that seemed ineffective.

Because of these wide variations, we decided in our implementation to make IP address checking configurable, so that our server could be set to accept TCP/IP connections only from known source IP addresses and AE titles. Initially we required AE titles to match exactly. The server could also be operated in a "promiscuous" mode where it accepted connections from anywhere. The server logs all connection requests, including details for those that are rejected, so that we can document possible break-in attempts or other problems. We logged no incidents of break-in attempts, but we did get some transmissions of images for unknown patients from known sources. They were stored in the temporary repository as described above, and later discarded by hand, after verifying that the data were simply sent to the wrong destination. We also logged some port scans done by University of Washington network security staff.

One vendor of radiation treatment machinery has implemented their DICOM radiation treatment plan storage server so that it is even more restrictive. It requires that different IP addresses must have distinct AE titles. This makes it difficult to install a DICOM client in a clustered computer environment and just use any computer in the cluster to send data. We therefore had to invent a scheme by which our DICOM client can generate a unique AE title based on the IP address on which it is currently running, in order to solve this problem. We don't believe this restriction adds security, but the vendor has committed to it.

6.3.1.2 Intravenous Infusion Pumps on the Internet

These days, IV pumps come with a wireless network interface. They are used with a portable computer and wireless access point to load range values into the pumps. Ultimately it is a wonderful innovation, as it could allow for remote monitoring and faster response than the audible alarm alone could provide. It did not occur to the hospital administration (or to the vendor apparently) that other wireless computer equipment in the vicinity (carried by visitors on the wards, or the patient's personal laptop computer) could possibly cause disruption of the operation of the pumps.

Before the hospital placed a major order of IV infusion pumps, our security engineers got a test unit from the vendor and performed penetration tests. We were able to put the pump into alarm mode without any authorization or wireless encryption information. We discovered other vulnerabilities in the device. The vendor's response to our report was to address all of the concerns to our satisfaction, and the pumps that were delivered met our requirements. We also were able to include security terms and conditions governing the devices in our purchase contract.

One of the biggest surprises was the vendor's comment that no other customers had expressed the level of concern that we had. It is interesting to find out from my counterparts in the Seattle area, that vendors have made similar comments to other security officials at other health care organizations here. All I can conclude from this is that in Seattle, "all the security people are above average."

6.3.2 Security Management – the Context

High level management in hospitals, as well as in research organizations, needs to make difficult decisions about how network technology is deployed, what the rules are regarding connection of systems to the network, and what should be done to protect systems. Earlier I mentioned the decision to make the UW Medical Center network an extension of the campus network. That decision was a radical one at the time, and still is today. Many people believe that hospitals should establish closed networks with single points of entry, and establish strict perimeter control. Moreover, medical system manufacturers also expect that this is the model on which hospital IT is based. There are many problems and implications of such a closed design. We will just mention a few here.

In my opinion, access controls and protective measures are better placed close to the devices being protected. One of the biggest reasons for this is that it causes less disruption to the departments and owners of systems that need broad connectivity to get their work done. Another very important one is that it puts limits on the ability of intruders to broaden their base of compromised systems. In an environment where owners depend on perimeter control, when a single system gets compromised, the compromise easily and quickly spreads to hundreds of systems. If strong controls are in place at the host level, and "trusts" are very limited, such proliferation will not happen. This is consistent with my actual experience at the University of Washington Medical Center.

Until recently, most medical software and system manufacturers gave no attention whatsoever to system security. They assumed that there is no need for these systems to communicate with other systems at all or at least not with other systems outside the hospital "perimeter." This assumption is wrong in just about every case I have dealt with in the last 20 years. Eventually awareness of this issue became widespread. Subsequently, the FDA issued an advisory [420, January, 2005], clarifying their position on Internet security and medical products based on commercial off the shelf software and systems (products that use standard PC or other generic computers, and commercial operating systems from Microsoft, Sun, or open source systems such as Linux).

The issue was that customers (like me) were demanding that vendors apply operating system security patches to systems that are classified as medical devices (like RTP systems), or at least that vendors should allow customers to install the patches if the vendor did not want to invest the time. Vendors typically claimed that they could not allow this because the device was an "approved" device and installing the patch would invalidate their FDA approval, requiring the vendor to repeat the approval process. The FDA guidance advises that, contrary to such claims, the vendor is responsible to install security patches, and in most cases this should not require any recertification.

The problem really seems to be that the patches break the vendor's software. There is no nice way to put this, but in my opinion this raises serious concerns about software design. It should be possible to design software in such a way that it is not so fragile. However, this is a new requirement on design, and it will take some time before most vendors can adapt.

Vendors often are amenable to (or even insist on) providing support services via the Internet. Many vendors have implemented a closed environment like the one they imagine is in use at their customer site. They expect the customer to provide a virtual private network (VPN) connection, presumably to get past the customer's firewall. In an open environment in which the controls are not primarily network based but are host based, this means we have to install and maintain extra devices, not for our security, but to match the configuration of the vendor's site.

The myth of the perimeter as the key to security is being challenged at the international corporate level. The Open Group (http://www.opengroup.org/), a voluntary standards consortium, sponsors an activity known as the Jericho Forum. It brings together major interests from banking, aviation, the IT industry, and many others, to investigate alternatives to the idea of a hard perimeter and large internal webs of "trust." The URL is http://www.opengroup.org/jericho/.

One cannot talk about security without at least a few words about the Health Insurance Portability and Accountability Act of 1996 (HIPAA). The extent of fervent misunderstanding and ignorance about this law and its associated rules is astounding to me. Here is the actual text of the prolog as passed by Congress [87]:

> *To amend the Internal Revenue Code of 1986 to improve portability and continuity of health insurance coverage in the group and individual markets, to combat waste, fraud, and abuse in health insurance and health care delivery, to promote the use of medical savings accounts, to improve access to long-term care services and coverage, to simplify the administration of health insurance, and for other purposes.*

It took many years and lots of discussion before the HIPAA Final Security Rule was adopted in 2003. The law calls for security standards to protect health information. The Final Rule, published in the Federal Register [86], has exactly one column of text (one third of a page) for technical security requirements. In general the HIPAA security requirements are the same kinds of reasonable measures that anyone would have taken in 1988 if they understood the threats facing any Internet connected systems. The challenge of HIPAA is that it asks organizations to actually do risk assessments, and develop controls to address the results of the risk assessments. With few exceptions it does not dictate the specific controls that must be in place.

One of the few specifics of the HIPAA Security Rule is that access to protected health information must be by individuals using their own individual user IDs. Generic logins or the use of other individuals' logins is one of the few clear-cut violations, but it is a widespread problem.

A reasonable approach to HIPAA compliance from a security perspective is to build a sound security program based on principles and basic business requirements. Some guidance is provided for this in the next section, based on the author's 50 years of experience with computers, including a recent 5-year term as the Security Director for University of Washington Medicine.

There is some reason for optimism. My most recent engagement with vendors indicates that many are starting to face the challenge of the Internet, and medical device safety. They will solve these problems because there is no other choice. We will need to work cooperatively to succeed. At the American Medical Informatics Association Fall Symposium in 2007 for the first time, an entire session was devoted to security and privacy.

For a wealth of information on IT security practices and standards an excellent source is the National Institute of Standards and Technology (NIST), URL `http://www.nist.gov/`. Another important resource is Carnegie Mellon University's Computer Emergency Response Team (CERT), at the CMU Software Engineering Institute. The CERT home page is URL `http://www.cert.org/`.

Security has become a recognized research subfield of computer science. The research literature is rapidly expanding, and spans both the technical issues around increasing sophistication in exploits of network software and the social issues around management of access controls and management of systems, some of which we have touched on here. The University of Washington Security and Privacy Research Group is representative of such leading edge research, addressing operating systems, mobile devices, Web services, and embedded medical devices.[6]

An often overlooked point on security is that back in 1988, people like Cliff Stoll, who managed computer systems, had an integrated view of ownership. Owning a system meant you were responsible for its upkeep, its security, backup, software maintenance, etc. Some time after that, it seems that security got split out as a separate activity, and became a property of the environment, and assigned to a Security Team. For 5 years I headed such a team. This is not a recipe for success. In the future, security will have to be reintegrated into system ownership. Users, too, have to follow sound practices, with the biggest security threats nowadays being social engineering attacks, to convince unwitting users to reveal information to intruders. There will still be a role for a security team, to do monitoring, incident response, consulting to system owners, etc., but the bottom line is that security is everyone's responsibility.

6.3.3 Ideas for a Security Program

In brief, a sound approach to IT security in a medical center or academic research environment would start with a statement of goals that the security program should achieve. Goals can be inadequate to meet business needs of the health care system, but they can also be overly ambitious, resulting in unnecessary expense and effort. The following is a list of goals that should be considered.

---

[6]See Web site `http://www.cs.washington.edu/research/security/`

1. Reliable, disruption-free 24/7 operation,
2. Safety, particularly with respect to computer-controlled devices that connect to or operate on patients,
3. Assurance that authorized individuals do not make inappropriate use of systems and data,
4. Protection from access by unauthorized individuals or entities,
5. Assurance of integrity of all important digital information,
6. Assurance of privacy of protected health information (as defined by HIPAA, for example) and other confidential information,
7. Record keeping as required by commonly accepted practices and by law (such as HIPAA), and
8. Compliance with all other aspects of applicable law, both State and Federal.

The biggest challenge in gaining support for a health care institutional security program is that if successful it is totally invisible. Systems that provide new diagnostic or treatment capabilities have visible and palpable return on the investments they require. Improving the efficiency of a billing system promises to substantially increase revenues. In contrast, investment in security is about prevention of problems, not about adding new desired functionality.

A successful Security Program will be invisible to patients, but failure can be dramatic. Patients expect:

- Confidentiality—their personal data will be protected from inappropriate disclosure (identity theft, etc.),
- Integrity—medical devices and systems needed will be functioning correctly and data will not be missing or corrupted,
- Availability—devices and systems will not be unavailable due to security breaches.

If a patient asks (and they do!) if your systems are appropriately protected from intrusion, your staff should be able to answer confidently that you have a sound, well-designed and comprehensive security program covering all your IT investment.

Once a set of goals are agreed upon, institutional leadership would develop policies, strategies, and an implementation plan that are designed to accomplish the goals. A critical part of the implementation plan is risk assessment, to identify the threats and obstacles to achieving the goals, so that a mitigation plan can be developed. It is also critical to create an evaluation plan through which you can determine if the goals have been met, and if not, what corrective action is needed. NIST, the US National Institute for Standards and Technology, among others, has published documents addressing the development of metrics for security.

Key principles to keep in mind are:

- If you don't know where your problems are, you can't fix them.
- Security costs money, staff, and time.
- Security is an *ongoing structured process*, not a one time or occasional quick fix. It is also not sufficient to simply deal with crises as they occur. As in human health, prevention generally turns out to be much less expensive than treatment. As with health, most administrators do

not invest in security as a preventive measure because they believe the risk is small, as in "It won't happen to me."

Balancing the risks and costs is an executive responsibility. Don't expect perfection, but don't shortchange the plan.

6.4 Summary

Safety and security of computer systems and software are an integral and necessary part of the research and health care enterprise. Along with the technical aspects, it is essential to understand the organizational and management aspects of these topics. As with everything else we have examined, much work remains to be done.

6.5 Exercises

1. Radiotherapy machines

   In this chapter we sketched a formal model for a radiation therapy machine, with electron and X-ray capabilities. To make the model a little more like a real machine, we added a dosimetry system, something that represents the preset dose to be delivered, and monitors the cumulative dose delivered. We also include a specification of the tolerance by which the dose may safely exceed the preset.

   a. The system is initialized in electron mode with the beam off and the `preset-dose` and `accum-dose` initialized to any values that satisfy `safe-dose`. Does the system satisfy the overall system safety requirement? If not, describe what can go wrong. We are considering *only* matters described in the model, and not extraneous things like machine hardware failures, and other such stuff. We want to know if this software design is safe.

   b. As noted, it is possible that the result of one pulse may produce an `accum` that is more than `prescribed` plus `tolerance`, for example if the dose rate is too high or the sampling interval is too long. How would you fix these problems?

2. Proof of Safety

   Radiation therapy linac engineer Randy Rad has written a design for a new radiation machine, but it is too complex for a manual safety analysis. He would like to have an automatic safety checker to test his design. Medical informaticist Patty Prolog claims that it is easy; you just rewrite all the design functions as rules and assertions, and make the query `safe-machine` with a specified initial state.

   a. Sketch what this would look like for the machine model described in this chapter. Don't rewrite everything—just do enough to illustrate the idea, no more than five or six expressions. You can use Prolog syntax, or Graham's inference syntax.

   b. Will this actually work? Explain why or why not. If not, is there an alternative that would work?

3. Remote Pump Control

   An IV pump is a device that mounts on a pole, with a tube running through it. On one end of the tube is some fluid in a bag. Typically it could be a nutrient solution, or could contain antibiotics, or chemotherapy drugs, or other substances. On the other end of the tube a small catheter is inserted into a vein in a patient's arm. The purpose of the pump is to infuse the fluid into the patient's blood stream at a controlled rate for a preset time (or until the bag runs out). These devices are used all over the hospital. The poles have wheels. Patients who are able to walk (ambulatory) sometimes are free to leave their rooms and go to other areas of the hospital for a period of time, perhaps to go to a reading room, or look at art work in the lobby, or go to the cafeteria. The devices have batteries so they can continue to work in this case. The newest pumps have wireless network interfaces and can transmit and receive data (such as the patient-specific settings, alarms, battery status, etc.) to/from computers on the wards.

   a. Is this a good idea? What are the merits and liabilities of using wireless networks for this purpose? Be specific about safety issues.

   b. Should the communication between the pumps and the base station use TCP or UDP? Explain your choice.

   c. Sketch a protocol that would support a few of the functions mentioned above, such as what messages would be sent/received and in what order.

4. Remote controlled coffee pots

   Internet standards are called RFCs. They are managed by the Internet Engineering Task Force (IETF). The main page of the IETF is at `http://www.ietf.org/`.

   A few RFCs were issued on April 1 some years to poke fun at the process. One you might enjoy is RFC 2324, Hyper Text Coffee Pot Control Protocol (HTCPCP/1.0). It is directly accessible at `http://www.ietf.org/rfc/rfc2324.txt`.

   The HTCPCP protocol is a spoof, but if it were real, what elements of this would have to be made more precise in order to really implement remote coffee brewing services? Now, consider the problem of intravenous infusion pumps, which are manufactured with network interfaces (wireless, actually). Describe some of the safety and security issues in this case. Sketch a design that might deal with one or more of these safety issues.

CHAPTER 7

Epilogue

Chapter Outline

> *No epilogue, I pray you, for your play needs no excuse.*
>
> *—Theseus, Duke of Athens*
> *A Midsummer Night's Dream, Act V, William Shakespeare*

Now we have reached the end of this lengthy exploration of the nature of biomedical informatics. Yet, it seems that it has only just begun. Several chapters end with a sentence like "Much remains to be done." I started this book with lofty goals, to create a unified exposition of principles that make up the field, and to describe and demonstrate exemplars of how those principles play out in various application areas with real working code. Has this exposition succeeded? It falls far short of the vision I had when I began, but that is a good thing. Not only is there more for me to do, but lots more for anyone else who would like to take up the task.

The code in this book may seem small and simple. Most programming projects these days seem to grow inexplicably large; one hundred thousand lines of code is not unusual. But I would implore the reader, keep things small, avoid building horrifically complicated user interfaces, or alternately, keep the user interface code completely isolated from the really interesting code. Write code that you would be proud to have others read, that is clear enough that others actually *could* read it. Pay as much attention to the quality of the code you write as to the prose.

Some of the insights I gained in this writing project and in my own research were the result of a hard wrestling match, trying to understand, and not getting it. Trying again, simplifying, asking others, and struggling again to work out the details, all these are vital to success. Here is an expanded version of the quote from Alexander Pope, the beginning of which starts out Chapter 2.

> A little learning is a dangerous thing;
> Drink deep, or taste not the Pierian spring;
> There shallow draughts intoxicate the brain,
> And drinking largely sobers us again.

Principles of Biomedical Informatics. http://dx.doi.org/10.1016/B978-0-12-416019-4.00007-X

7.1 The Halting Problem—A Metaphoric Version

Section 2.2.10 of Chapter 2 pointed out that first order logic (FOL) has limits in terms of what can be computed. In particular, FOL is semi-decidable, meaning that no algorithm that searches for a proof of a formula can be guaranteed to terminate. There are a number of other computational questions for which no algorithmic solution exists. The prototype of this is the Halting Problem, where we seek an algorithm to determine if a program terminates or loops forever on its various inputs. It can be proven that no such algorithm exists.

The technical part of the science of biomedical informatics could be described as the search for a complete formal (computational) description of all of biology, medicine, and health, a set of computer programs or other formalisms that account for all biomedical structures, functions, etc. For example, the FMA claims to be (or will be) a complete description of human anatomy. Does this process terminate? Put more generally, will we eventually be able to create the complete catalog or theory of biology and medicine? (At this point, of course, research funding will no longer be necessary or available.).

The fact that there might be an infinite number of possible instances of a biological system, or even an infinite number of systems is not an argument against termination. Physicists already deal with infinite-dimensional spaces and states. So if infinity is a concern here, you would have to explain how the infinity in biology and medicine differs from that of physics.

In Chapter 2 we noted that Einstein viewed the process of theory creation and testing as a way to discern how the universe really works. The idea here is that as the predictions of our theories match more and more closely what is actually observed, and the theories are the simplest ones (in some vague sense) that do so, the theories can be said to represent fundamental processes and structures, even if we cannot observe them directly. Another view, held by Mach, that the theories only represent the most efficient possible summary of the observable data, was also mentioned. Both of these views raise an interesting question. If two separate, isolated civilizations conduct biological research along these lines, as each gets better theories, will they eventually arrive at a common theory, or might there be independent but non-equivalent descriptions or summaries? A fanciful discussion of this question for physics was published some time ago [354].

7.2 Where Do New Ideas Come From?

One topic I have not addressed much is that of data mining (also called "machine learning"). I confess to a certain lack of enthusiasm for this subject. Certainly much can be gleaned from data, and important insights come from careful analysis. However, the great breakthroughs in biology and medicine, as in physics, will not come from automated procedures, but will come from a strong dose of imagination. Sometimes it is the imagination to put old ideas together in a new way. For a lucky few, it will be a truly great new idea. It is essential, however, to be able to take such ideas and make them precise, put them in computational form, to squeeze out of them their entailments, to then measure their worth. If that seems poetic, then you should write computer programs that are poetic. Again, the words of Theseus, from the beginning of Act V of A Midsummer Night's Dream:

> The poet's eye in a fine frenzy rolling,
> Doth glance from heaven to earth, from earth to heaven,
> And as imagination bodies forth.
> The forms of things unknown, the poet's pen.
> Turns them to shapes, and gives to airy nothing.
> A local habitation and a name.

More than anything else I know, mathematical formulas and computer programs provide a medium in which to realize this vision. And no where as much as in biology, medicine, and health are we so preoccupied with naming things in a comprehensive and consistent way.

7.3 Making A Difference

More than many other areas of scientific research, biomedical informatics is motivated, influenced, and even shaped by the context in which problems are formulated, ideas are developed, and innovations are put into practice. It is necessary, as I suggested in the Preface to the first edition, that for biomedical informatics to stand as a legitimate and viable academic discipline, there must be sound and powerful foundational principles, ideas, theories, and methods. Being useful does not make a science. But it must also be said that even when "making a difference" is furthest from our thoughts and feelings, even when we pursue new ideas and seek answers to deep questions simply out of intellectual curiosity or the desire to be creative and imaginative, the impact of this work will eventually be felt. There is no way to predict what that impact might be.

My favorite example comes from personal experience. I spent my first year as a theoretical physics graduate student at Princeton University (1965–1966) with Professor John Wheeler, fellow student Kip Thorne, Charles Misner, and others working on very difficult questions related to Einstein's General Theory of Relativity, merging Quantum Field Theory with it to arrive at a Quantum Theory of Gravity, to push the limits of fundamental physics. In 1916, when Einstein published his work on Gravity and General Relativity, the observable effects that could be measured to test the theory were so small that only a few experiments had been attempted. Unlike nuclear and elementary particle physics, it appeared that still in 1965 General Relativity had little or no practical use, and made no difference in ordinary life. Now, nearly a century after the publication of Einstein's paper on General Relativity, millions of people carry pocket-sized devices whose accuracy depends on our understanding of General Relativity, namely the Global Positioning System. Even though knowing where on Earth you are has been recognized for millenia as vitally important for commerce, exploration, warfare, and many other human activities, no one in 1916 could have anticipated the impact of the idea that the presence of matter ever so slightly bends and stretches the fabric of space and time. Einstein wasn't doing this research to "make a difference." But if we have learned anything, it is that this wanting to know how things work just for the sake of curiosity is a tremendously important part of human existence and meaning. We abandon it at our peril. With billions of dollars going into a mad rush to implement electronic medical record systems and massive data repositories, and the emphasis on practical immediate payoff, it is more important than ever to defend, protect and maintain our nation's (and the world's) investment in basic research on the nature of biological and medical information and knowledge, even as we devote (as we also must) huge resources to meet immediate needs in health care.

Research in biomedical informatics, as much as in any other field, is a long-term commitment, as is the solution of important practical problems in the application of biomedical informatics principles to build tools for biology, to build systems to help in health care practice, public health, and consumer health. You'll only survive if you enjoy what you are doing, along the way. I hope that Pete Szolovits is right and that you will have some fun with this material. I certainly have.

Notes on Lisp, Other Languages, and Resources

Chapter Outline

The purpose of this Appendix is to prorvide some references for help with programming, Lisp in particular, and for obtaining software and data mentioned in various chapters. In addition I have written a few notes on Lisp to expand a little on topics only briefly introduced in the main text. This Appendix does not constitute anything even close to an exhaustive literature survey. I have provided my own opinions on books and software in a few cases, but these are not authoritative and listing a reference to a software system or product here does *not* constitute an endorsement. The most important thing is to try things out and come to your own conclusions.

Lisp is the second oldest programming language in use today, only a year younger than FORTRAN. It will likely outlast most of the more recently invented languages, though it has never been "mainstream." It has been said (Greenspun's tenth rule[1]), "Any sufficiently complicated C or Fortran program contains an ad hoc, informally-specified, bug-ridden, slow implementation of half of Common Lisp." It seems that Lisp will continue to be rediscovered over the generations, a point of view celebrated by the cartoon in Figure A.1.

In addition I have provided short descriptions of some interesting projects, mostly to confirm that indeed serious research and development projects are using Lisp and Prolog. In regard to information in biology, medicine, and public health, these are each vast subject areas, and it is beyond the scope of this book to do much more than mention a few standard texts, which I have done in the main text.

---

[1]For more information, see the Wikipedia page, `http://en.wikipedia.org/wiki/Greenspun's_tenth_rule`.

Principles of Biomedical Informatics. http://dx.doi.org/10.1016/B978-0-12-416019-4.00016-0

Figure A.1 Cartoon, "Lisp Cycles," from the on-line comic strip, XKCD, authored by Randall Munroe, and used here with permission from the author's web site, `http://xkcd.com/`. **The permanent URL for this particular comic is** `http://xkcd.com/297/`.

A.1 Lisp Notes

> *Lisp has jokingly been called "the most intelligent way to misuse a computer." I think that description is a great compliment because it transmits the full flavor of liberation: it has assisted a number of our most gifted fellow humans in thinking previously impossible thoughts.*
>
> *– Edsger Dijkstra [93]*

In the Preface to the first edition (page xxii) there is a brief explanation of the choice of Lisp over the more familiar programming languages. This section contains some further explanation of Lisp programming ideas, as well as some examples of the use of declarations to optimize numerical processing. It is not a substitute for the excellent resources described in Section A.1.3, but should help the beginner by providing some additional insights. For the expert, it may be interesting to see another programmer's view of this venerable and robust programming language.

In Chapter 1 I explain functions or other operators built into Common Lisp, as I introduce their use. In later chapters, some operators are used without much explanation. If a function or macro (or other symbol usage) appears in code and has not been defined previously, it is likely to be a built-in part of the Common Lisp language specification. To check this and find out more about such functions, macros, or other entities, the reader should look them up, for example, in Appendix D of Graham [137], which has a complete and concise summary of all functions, macros, special forms, variables, and parameters defined in the ANSI Common Lisp specification. More detailed explanation can be found in Steele [394]. Both books have excellent and thorough indexing, with respect to these definitions; I strongly recommend using the index in each of these books, to find the definitions and explanations of functions and other operators.

For other symbols, which may not be part of the ANSI standard but are defined elsewhere, a reference is usually included where they first appear. Examples of such additional sources for predefined code are: The Common Lisp Object System Meta-Object Protocol (MOP) [229], which is not yet part of the ANSI standard, and the CLX library [363], which implements a Common Lisp binding to the X window system protocol.

Most of the time, the way in which Lisp is used and programs are run is in the interactive Lisp environment, the `read-eval-print` loop explained in Chapter 1. You can type in expressions to be evaluated, including ones that define functions, and have other side effects. You can also cause the interpreter to read a file as if you typed it, using the `load` function. Evaluating the expression (`load "filename"`), where `filename` is replaced by an actual file name like `frames.cl`, or just `frames`, will cause the interpreter to read everything in that file as if you typed it. By saving your code in files using a text editor, you can build up a large program made of many functions, and get it into the interactive environment with one `load` call.

It is also possible to write the entire Lisp environment out to a binary file in such a way that it can be executed by typing a command. This is analogous to the way in which C or FORTRAN programs can be compiled to produce runnable binaries. The directions for creating such binaries are different for different Lisp systems, so if you want to do this you will need to read the documentation for your Lisp system.

Sometimes people say that Lisp is slow because it is interpreted rather than compiled. This is not an accurate description. The interactive Lisp environment can accommodate either interpreted or compiled code or any mix, that is, some functions can be interpreted and some can be compiled. They all will work together correctly. The compiled version of a function generally runs much faster than the interpreted version. From source files, the compiler can generate compiled files, which can then be loaded just as the source code can. This means you have a lot of flexibility for debugging and testing, that is not available with other programming languages. It does *not* mean you have to give up speed of execution. When you are satisfied your code is correct, compile everything and it will run fast. This works in both the interactive environment and in creating the stand-alone binaries mentioned in the previous paragraph.

A.1.1 CLOS and MOP

The Common Lisp Object System (CLOS) can be thought of in several ways, but most importantly it is an extension of the Common Lisp *type* system. In Lisp, data have type associated with them. Variables can have any kind of data as their values, so type is *not* usually associated with variables. This contrasts sharply with so-called strongly typed languages like Pascal, FORTRAN, C, C++, and others. In those languages a variable is a name that the compiler associates with some storage preallocated for some type of datum. So, an integer, short or long, will occupy a certain number of bytes, as will a floating point number. User-defined structures too will have predefined sizes and storage allocations. In Lisp, on the other hand, symbols used as variables each just have a slot for a pointer to an actual piece of data. The *datum itself* has preallocated storage, and a tag to identify its type.

Although much of what happens with Lisp variables, lists, and other structures that contain pointers can be emulated in a language like C, it has to be done explicitly. This leads to many coding errors, for example, when storage was allocated but never properly deallocated, or storage was deallocated but somewhere in the code there is still a pointer to it that is now invalid. In

Lisp, such storage management is automatic. The programmer can think and write on a higher level of abstraction.

Since classes extend the type system, Lisp is object oriented at its core. Integers are objects, as are strings, lists, and so on. The types (and classes) of which they are instances are *built in*. Nevertheless they can be used in the same way as objects that are instances of standard classes. In particular, generic functions can be made to dispatch different methods depending on the type of their inputs, and the type that the methods apply to can be any type. Graham [137] illustrates this beautifully with the made up `combine` function. The idea is that `combine` should take two inputs, and depending on what types they are, do something appropriate that we might think of as combination. For numbers, we should add them.

```
(defmethod combine ((x number) (y number))
  (+ x y))
```

Note that this method will apply when both arguments are any type of numbers, and it will work OK because the + operator can add different subtypes of numbers.

If the arguments are strings, it would make sense to just concatenate them, so here is a method for that.

```
(defmethod combine ((x string) (y string))
  (concatenate 'string x y))
```

Of course it would be a good idea to have a default method in case there is no more specific one. This method will always be applicable and will just put the two arguments together in a list.

```
(defmethod combine (x y)
  (list x y))
```

We have not said anything about classes. Object-oriented programming is really about types and their relationships, not particularly about *user defined* types. Of course, without user-defined types, there is far less expressivity.

Thus, the second way in which the Common Lisp Object System extends basic Lisp is that it generalizes the idea of a function or procedure. Functions, too, are objects, instances of the various *function* types. Generic functions provide a way to build up composite functions from parts (methods) as illustrated above. An important way in which Lisp differs from other programming languages is that expressions that make up programs (code) uses exactly the same forms as other types of data. Therefore functions can be written that construct other functions, analyze function expressions, and transform code from one form to another (macros). It makes Lisp an *extensible* programming language, to which new features can be added without having to redefine the language at its base or rewrite the compiler. Indeed, CLOS was first implemented as a collection of macros, that worked in any ANSI standard Common Lisp environment.

The idea that methods are associated with generic functions rather than classes is very powerful. First of all, it makes dispatch on multiple arguments very easy and uniform. We can have methods that apply to arguments of different types. For example, here is a `combine` method that applies when the first argument is of type ice-cream and the second is of type topping, again taken from Graham. We assume that both types have a name attribute, but these types do not need to be subtypes (or subclasses) of some "base" type.

```
(defmethod combine ((x ice-cream) (y topping))
   (format nil "~A ice cream with ~A topping." (name x) (name y)))
```

Does this method belong to the ice-cream class or to the topping class? The answer is, "neither". It belongs to the generic function, combine.

Slot accessors that are provided by defclass using the :accessor slot option are *also* generic functions. The methods that are provided dispatch only on the one argument, so you might think of them as associated with the class containing that slot. However, if a generic function of that name is already defined, it is not a problem, and if there are methods for doing some other kind of computation on a single argument, there will be no conflict. Depending on the input, the right method will be used.

In several places, we use functions in CLOS that look up the class of an object. Every class definition also adds a type to the type hierarchy. Similarly, every type has a corresponding class, even the built in types. There are two distinct functions that return different things. The type-of function returns a type specifier, which is a symbol naming a type, or a list specifying a type. On the other hand, the class-of function returns a *class metaobject*, the class of the input datum. In Common Lisp, a class is a real object in the system, not just a name referring to information used by the compiler to generate appropriate code. The class object contains information about the class, analogously to the class frame idea in the mini-frame system described in Chapter 2. This is just as true for built in classes like numbers, strings, conses, and so on, as it is for classes defined by defclass.

```
> (type-of 7)
FIXNUM
> (class-of 7)
#<BUILT-IN-CLASS FIXNUM>
> (describe (class-of 7))
#<BUILT-IN-CLASS FIXNUM > is an instance of
    #<STANDARD-CLASS BUILT-IN-CLASS>:
 The following slots have :INSTANCE allocation:
   PLIST                   NIL
   FLAGS                   0
   DIRECT-METHODS          (NIL)
   NAME                    FIXNUM
   DIRECT-SUPERCLASSES     (#<BUILT-IN-CLASS INTEGER>)
   DIRECT-SUBCLASSES       NIL
   CLASS-PRECEDENCE-LIST   (#<BUILT-IN-CLASS FIXNUM>
                            #<BUILT-IN-CLASS INTEGER>
```

```
                                   #<BUILT-IN-CLASS RATIONAL>
                                   #<BUILT-IN-CLASS REAL>
                                   #<BUILT-IN-CLASS NUMBER>
                                   #<BUILT-IN-CLASS T>)
         WRAPPER                   #(23164783 T NIL NIL
                                     #<BUILT-IN-CLASS FIXNUM>
                                     0 NIL)

    >
```

This output of the describe function shows that even built-in types like integer, real, and number are classes, and that these classes are actual entities in the environment, with slots and values in those slots. The class-precedence-list slot, for example, shows the inheritance hierarchy for the class fixnum, that it is a subclass of integer, which is a subclass of rational, and so on.

Just as understanding Lisp itself can give you deep insight into what is a function, and what a computation really is, understanding the Meta-Object Protocol will give deep insight into what object-oriented programming really means, and a deep insight especially into the power of abstraction.

A.1.2 Optimization

Lisp is often wrongly criticized as being very slow in execution, especially for numerical computations. While it is true that many people write Lisp programs that are horribly inefficient, it is not hard to write programs that are just the opposite, that run just as fast as the equivalent computation written in C, and usually much faster than the equivalent computation written in java. The key to writing Lisp programs that run fast is to understand what compilers do and what exactly are you trying to compute.

When the types of data and the operations being performed on them are fixed in advance, the compiler can determine what they are, and the compiler can often generate binary code that is very efficient. It is pretty clear that most of the time generic functions will run slower than ordinary functions because the run-time code must do some extra work to determine what the types are of the inputs, then find the most specific applicable method, and then finally call the method. If the type of the input to a function is known by the compiler, the compiler can generate just the right code, with no dispatch overhead. Similarly, if the type of a value that will be attached to a variable is known to the compiler, the compiler does not need to use pointers, but can store the data efficiently and use direct addressing instead of pointer lookup. The way you, the programmer, can help the compiler generate such fast code, is to provide *declarations* in your source code.

Often in code you will see *declarations*, for example, (declare fixnum start). Except for special declarations, no declaration should have any effect on the correctness or meaning of a program. It can be *useful*, even though unnecessary, to declare the type of object that will always be the value of a symbol, if you have made sure your code will work this way. Such declarations can be used by a good compiler to generate very fast compiled code. This is important when writing code for numerically intensive computations, when using arrays and vectors (where all the elements are of the same type), and in other circumstances as well.

For example, if you (the programmer) can guarantee that a certain array's elements are always exactly of type (unsigned-byte 16) and that the array is a simple array, the compiler can generate very fast array accesses, and store the array very compactly. If a body of arithmetic expressions have elements that are of a guaranteed type and don't require type conversions, similarly the compiler can generate code that uses registers effectively and does not waste time or space on managing generic types.

In Chapter 1, Section 1.1.5, we defined a function, map-image, to convert a gray scale image to an adjusted gray scale image according to a mapping array. Here is that same function with declarations added. We can add these declarations because we can guarantee from the rest of the program that the inputs will always be the specified types.

```
(defun map-image (raw-image window level range)
  (declare (fixnum window level range)
           (type (simple-array (unsigned-byte 16) 2) raw-image))
  (let* ((x-dim (array-dimension raw-image 1))
         (y-dim (array-dimension raw-image 0))
         (new-image (make-array (list y-dim x-dim)
                                :element-type '(unsigned-byte 8)))
         (map (make-graymap window level range)))
    (declare (type fixnum x-dim y-dim))
    (declare (type (simple-array (unsigned-byte 8) 2) new-image))
    (declare (type (simple-array (unsigned-byte 8)) map))
    (dotimes (j y-dim)
      (declare (fixnum j))
      (dotimes (i x-dim)
        (declare (fixnum i))
        (setf (aref new-image j i)
          (aref map (aref raw-image j i)))))
    new-image))
```

In this version, we use the fact that we know that the input image will always be an array of 16-bit integers, that window, level, and range will always be small integers (fixnums), and that the new mapped image will always be an array of 8-bit bytes. Providing the declarations does not change the meaning of the code, but it allows the compiler to generate binary compiled code that runs 100 times faster than the version without declarations.

The code to read binary image data from files, implemented in the function read-bin-array in Section 1.2.3, can also benefit from declarations. Here too, it is important to specify the type of data stored uniformly in each element of the array as well as the dimensions of the array, so that the compiler can generate a really efficient call to the read-sequence function. Although the code for reading multi-dimensional arrays will be the same for both 2 and 3 (and higher) dimensional arrays, the declarations will be slightly different since they need to be a bit more explicit than the code. Here is what the code for `read-bin-array` looks like with these declarations:

```
(defun read-bin-array (filename dimensions)
   "read-bin-array filename dimensions
reads an array of dimensions specified by dimensions from a binary file
named 'filename' into an array of (unsigned-byte 16). Arrays of
1 through 3 dimensions are currently supported."
  (let* ((bin-dim (if (numberp dimensions) (list dimensions)
                    dimensions))
         (num-dim (length bin-dim)))
    (with-open-file (infile filename :direction :input
                                     :element-type '(unsigned-byte 16))
      (case num-dim
        (1 (let ((bin-array (make-array bin-dim
                              :element-type
                              '(unsigned-byte 16))))
             (declare (type (simple-array (unsigned-byte 16) (*))
                            bin-array))
             (read-sequence bin-array infile)
             bin-array))
        (2 (let* ((bin-array (make-array bin-dim
                               :element-type
                               '(unsigned-byte 16)))
                  (disp-array (make-array (array-total-size bin-array)
                                :element-type
                                '(unsigned-byte 16)
                                :displaced-to bin-array)))
             (declare (type (simple-array (unsigned-byte 16) (* *))
                            bin-array)
                      (type (simple-array (unsigned-byte 16) (*))
                            disp-array))
             (read-sequence disp-array infile)
             bin-array))
        (3 (let* ((bin-array (make-array bin-dim
                               :element-type
                               '(unsigned-byte 16)
                               #+allegro :allocation
                               #+allegro :old))
                  (disp-array (make-array (array-total-size bin-array)
                                :element-type
                                '(unsigned-byte 16)
                                :displaced-to bin-array)))
             (declare (type (simple-array (unsigned-byte 16) (* * *))
                            bin-array)
```

```
                   (type (simple-array (unsigned-byte 16) (*))
                         disp-array))
            (read-sequence disp-array infile)
            bin-array))))))
```

There are tradeoffs. Providing declarations and compiling with a high optimization setting can eliminate important information that would otherwise be available when debugging. Error messages become much more obscure, and often entire sections of code may be restructured by the compiler so it is harder to determine where the error occurred. The best strategy is first and foremost to design your code to be correct, and only optimize it *after* it is determined to be correct.

A.1.3 Books on Lisp

We have referred throughout this book to the two books on Common Lisp that I consider to be essential.

- Graham, *ANSI Common Lisp* [137] is the book I find the most clear and compact. Graham gets right to the point. However he only covers the essentials.
- Guy Steele's *Common Lisp: the Language, 2nd ed.* [394] is the canonical reference volume for the serious Common Lisp programmer. When you want to know the full range of options for a function or other element of Common Lisp, this is the place to look. There are some differences with the final ANSI standard, which was adopted later.

Other books that provide a thorough introduction to the elements of Common Lisp include the following:

- Winston, *Lisp* [440], is an older book that provides more examples, and goes more slowly than Graham, but does not get to the depth that Graham does.
- Some excellent AI books include an introduction to Lisp:
  - Tanimoto, *The Elements of Artificial Intelligence Using Common Lisp* [408] and
  - Norvig, *Paradigms of Artificial Intelligence Programming* [305], which I already mentioned many times. This book has a particularly good discussion of how to write efficient programs.
- A more recent book by Seibel [371] includes some practical applications with database access, web programming, and other extensions.
- A book by Lamkins [246] provides yet another very thorough introduction, and covers also how to call functions in other languages, how to design event-driven interfaces and other interesting topics.
- An old but still very useful book by Touretzky [416] is due to be reissued as a Dover book in 2013.

For some of the more obscure functions it may also pay to consult the ANSI standard for Common Lisp (available from ANSI). An HTML document derived from the ANSI standard is available online at

```
http://www.lispworks.com/reference/HyperSpec/
```

(the Common Lisp HyperSpec). Some commercial Lisp systems include it as well in their software distributions.

My recommendation for a place to start, to get a basic introduction to Common Lisp is:

- the first three chapters of Norvig [305], and
- the first six chapters of Graham [137].

For a concise introduction to the Common Lisp Object System (CLOS), see Graham [137], Chapter 11. Norvig [305], Chapter 13 also has a useful treatment of CLOS and object-oriented programming ideas. Another nice introduction to CLOS is an old but still relevant book by Keene [222]. For really ambitious students who are interested in learning more about the Common Lisp Object System Meta-Object Protocol (MOP), the book by Kiczales et. al. [229] is still the best.

A.1.4 *Common Lisp Systems, Libraries, and Resources*

For lists of Common Lisp programming systems that you can download or purchase for installation on your own computer, the best list is at the web site of the Association of Lisp Users (ALU), at URL `http://www.alu.org/alu/home`. Of the various systems listed there, I personally have used only GNU CLISP, CMUCL, and Franz, Inc.'s Allegro Common Lisp. These are all fine systems. The first two are free. Allegro CL is available in a free trial edition for evaluation. The main advantage of a commercial product like Allegro CL is the support and stability. Allegro is very well documented. The other two are also very well documented, and can be used very effectively, too. More work is involved in getting them installed and using them.

In addition to both free and commercial Common Lisp systems, a very large body of Common Lisp code, including general application support libraries and many interesting projects, is available from the community under one or another Open Source style license. Here we provide some web links for these resources:

- URL `http://clocc.sourceforge.net/` is the web site of the Common Lisp Open Code Collection (CLOCC). It includes lots of small, medium and large functions to do tasks that you would otherwise have to write your own code for. It also includes a link to the ANSI Common Lisp Hyperspec, and more importantly a link to the CMU Common Lisp repository. The CMU CL repository includes a very powerful mathematics library, as well as extensive tools for parsing and pattern matching.
- URL `http://common-lisp.net/` is another one among many gateways to Common Lisp code collections.
- URL `http://www.cl-user.net/`, the Common Lisp Directory, another gateway to Common Lisp resources, contains 562 links to libraries, tools, and software, and 177 links to documents and informational web sites.

- URL `http://www.cliki.net/index`, known as "CLiki," is a Common Lisp wiki. It contains yet more links to resources and free software implemented in Common Lisp.

A.2 Other Programming Languages

A 2009 article in PLOS Computational Biology by Dudley and Butte [98] provides an excellent presentation of why it is important to develop programming skills for bioinformatics. This article also provides pointers to resources for general programming skills and other programming languages such as Perl, Python, and Ruby.

A.2.1 The `Factorial` Function: A Snapshot Language Comparision

The factorial function computes n!, the product of the first *n* integers. Here is how it looks in Common Lisp:

```
(defun factorial (n)
  "for any integer n, returns n!"
  (if (= n 0) 1
    (* n (factorial (1- n))))))
```

The two lines of code correspond pretty directly to the two lines of the mathematical definition.

This function definition is linear, $O(n)$ in both time and space, since it needs a stack frame for each recursive call. However, writing it as a tail-recursive function, a compiler can turn it into an interative process, which is also linear in time, but $O(1)$ in space. A tail-recursive function is one in which there is no computation remaining after the recursive call returns its result. In that case, the stack frames are not needed. The compiled code can just return the value immediately to the top level with a single "jump" instruction. Here is a tail-recursive version:

```
(defun factorial (n)
  "tail recursive, returns n!"
  (labels
    (fact-iter (n accum)
      (if (= n 0) accum
        (fact-iter (1- n) (* n accum)))))
    (fact-iter n 1)))
```

The `labels` form allows the definition of `fact-iter` to be local, so there is no concern about name conflicts elsewhere, and the product is accumulated on the way in rather than on the way out as in the original.

The Factorial Function in C can be written simply as an iterative function, a using so-called *for* loop, but C does not provide an interactive environment. To use a C function, you need to write a "main" function, compile the whole thing, and then run the resulting binary.

```
#include <stdio.h>
main()
```

```
{
    int n;
    scanf ("%d", &n)
    printf ("n! =%d\n" factorial(n))
}
/* for any small integer n, returns n! */
long factorial (int n)
{
    int i, p;
    p = 1;
    for (i = 1; i <= n; ++i)
        p = p * i;
    return p;
}
```

In the java programming environment, everything is written in the form of class definitions, which compile to runnable code in an interactive environment. Here is how the factorial function would look in java (taken from "Java in a Nutshell," by David Flanagan, with end of line comments omitted):

```
/**
 * This program computes factorial of a number
 */
public class Factorial {
    public static void main (String[] args) {
        int input = Integer.parseInt (args[0]);
        double result = factorial (input);
        System.out.println (result);
    }
    public static double factorial (int x) {
        if (x < 0)
            return 0.0;
        double fact = 1.0;
        while (x > 1) {
            fact = fact * x;
            x = x - 1;
        }
        return fact;
    }
}
```

The while construct takes the place of the for loop in the C version, and there is a lot of extra syntax in the java class construct, much of which is to handle name space management.

Significant work in bioinformatics and medical informatics has also been done using Prolog. Examples include a system for support of computational analysis using information about expressed genes in erythropoesis [397], a molecular biology qualitative simulation system [152], and more recently a system for modeling genetic regulation networks [450].

Full native Prolog implementations are available from two sources that I have looked at and successfully installed on Linux.

- URL `http://www.swi-prolog.org/` is the web site for SWI Prolog. SWI Prolog is free software and provides a full function Prolog, including coverage of Part 1 of the ISO standard and the de facto Edinburgh Prolog standard.
- URL `http://www.gprolog.org/` is the web site for GNU Prolog, also free software, covering the ISO standard and many extensions.

The canonical introduction to Prolog is the book of Clocksin and Mellish [68], which originally set a de facto standard for the language. Another very readable introduction to Prolog is the book of Bratko [48].

A.3 Other Resources

Most Linux distributions include some version of the Emacs text editor. For those that don't, as well as for users of MS Windows and the Macintosh, the official GNU Emacs web site is at URL `http://www.gnuemacs.org/`.

The following is a list of the web sites of various biomedical informatics projects that use Common Lisp or Prolog.

- CHIP (Children's Hospital Informatics Program) Bioinformatics Tools includes a variety of software tools built on LispWeb, a web server written in Common Lisp by Alberto Riva [341]. They include tools for searching important public databases such as the dbSNP database of single nucleotide polymorphisms, the Swiss-Prot database of proteins, a browser for the Gene Ontology database of gene products, and a tool for analysing data output by the `phred` program. The `phred` program translates input in "Standard Chromatogram Format" (experimental sequencing data) into a variety of output files listing bases [108]. The URL is `http://www.chip.org/`
- Marco Antoniotti and colleagues at NYU Courant Institute have developed a simulation system that incorporates numerical and symbolic modeling for biological systems [16]. This system provides a flexible approach to modeling systems with large amounts of data, and for which there are large regulatory networks. The underlying system, `xssys`, is implemented in Common Lisp. For this and the parent project, Simpathica, see Antoniotti's home page, URL `http://www.bioinformatics.nyu.edu/marcoxa/`.

Finally, an extremely important resource, already mentioned many times, is the National Library of Medicine (NLM), which includes the National Center for Biotechnology Information

(NCBI). The other NIH institutes also have significant biomedical informatics components, notably the National Human Genome Research Institute (NHGRI).

A.4 Exercises

I hear and I forget. I see and I remember. I do and I understand.
– Confucius (551 BC-479 BC)
attributed

One who studies in order to teach, is given the means to study and to teach; and one who studies in order to practice, is given the means to study and to teach, to observe and to practice.
– Rabbi Yishmael bar Rabbi Yose
Pirkei Avot 4:6

The exercises are provided so that you don't just remember but also understand and can practice. Even more than mathematics, learning computer programming is an interactive process. You don't learn it by memorizing or organizing a body of factual material. You learn by writing programs and running them.

1. Write a function that takes a list of lists as input, and returns a new list of lists in which each list is the reverse of the one in the same position in the original list, so for example, if the input is

    ((g c c a t a t) (a a t t) (g g c c t))

    the result will be

    ((t a t a c c g) (t t a a) (t c c g g))

    Write a version that uses the built-in function `mapcar`, and one that does not use `mapcar`.

2. In the previous exercise you will likely use the built-in function `reverse`. Write a definition for this function as if it did not exist already (call it `alt-reverse` to avoid name conflicts).

3. Write a function of two inputs called `add` that does a kind of generalized addition, depending on the types of its inputs:

    *   if both are numbers, it does arithmetic addition;
    *   if both are symbols, it returns a list of the symbols;
    *   if both are strings, it returns a concatenated string;
    *   if both are lists, it returns a merged list containing the contents of both lists, and
    *   if the inputs are of incompatible types or other types, it returns `nil`.

    You can do this in two ways: one is to make it a generic function and write methods for the different types of inputs; the other is to use *type testing* functions, such as `type-of`, `typecase`, `typep`.

(Note: this is the core of what in other languages is called run-time dispatch, a key to being able to write so-called "methods." In Lisp it is just a consequence of every datum having a type, information that can be used directly by a program, not indirectly by a compiler.)

4. The `mapcar` function takes a function and uses it to transform a list of data elements into a new list. Write a kind of upside down version of this (call it `mapfun`), that takes a piece of data, and a list of functions, and produces a new list in which each element is the result of using each function in the list on the piece of data. For example, suppose the first argument is a list, and the second argument is a list of functions that operate on lists, we get a list of results:

```
> (mapfun '(a b c)
          (list #'length #'reverse #'first #'rest #'listp))
(3 (C B A) A (B C) T)
```

5. The `bank-account` function described in Norvig [304, page 92], and reproduced here

```
(defun bank-account (balance)
  #'(lambda (action &optional (amount 0.0))
      (case action
        (deposit (setf balance (+ balance amount)))
        (withdraw (setf balance (- balance amount)))))))
```

only recognizes the `deposit` request and the `withdraw` request. Modify the `bank-account` function so that it returns a function that recognizes the `balance` request, which just returns the current balance in the account. This is simple but may be a bit confusing. The purpose of this exercise is to help clarify when symbols are evaluated and when they are used without evaluation.

6. In Chapter 11 of ANSI Common Lisp, Graham [137] provides classes for two types of geometric shapes, `rectangle` and `circle`, along with methods for a generic function, `area`, more or less as follows:

```
(defclass circle ()
  (radius:accessor radius))

(defclass rectangle ()
  (height:accessor height)
  (width:accessor width))

(defmethod area ((x circle))
  (* pi (expt (radius x) 2)))

(defmethod area ((x rectangle))
  (* (height x) (width x)))
```

Define a class representing generalized cylindrical objects, in which there are two slots, the cross section (an *instance* of one of the geometrical shape classes above), and the length of the cylinder. Then define a function that takes a cylinder instance and computes its volume. This function can be an ordinary function; it does *not* need to be a generic function. Explain why.

7. What is needed in order to add a triangle shape to the above code, and still have the volume function work with a cylinder whose cross section is a triangle instance?

8. Write a function that takes two inputs; the first is a list of functions, and the second is a list of items. Each function in the list is a function of one input. All the items in the second list are in the domains of all the functions. This function should return a flat list of all the results of applying each function to every item (so if you have a list of four functions and five items, the resulting list will have 20 elements). You can use iteration, recursion, or any other approach you can think of. Write a brief reason for your choice of algorithmic method.

9. Write a function that takes one input, a two-dimensional array (not necessarily square), and returns a one-dimensional array in which each element is the average of all the elements in each column of the input array. The number of elements in this new array will be equal to the number of columns in the original array. The rows are indexed by the first dimension of the two-dimensional array, and the columns are indexed by the second dimension.

10. For each function defined in the following, state whether using the function could cause side effects or not. Give a short statement of why or why not.

```
(defun add-entry (name project)
   (setf project-table
      (cons (list name project) project-table)))

(defun bank-account (balance)
   #'(lambda (action &optional (amount 0.0))
       (case action
          (deposit (setf balance (+ balance amount)))
          (withdraw (setf balance (- balance amount)))
          (balance balance))))

(defun static-value (board)
   (let ((result 0)
         (dim (array-dimension board 0)))
    (dotimes (m dim result)
     (dotimes (n dim)
       (cond ((eq 'white (aref board m n)) (setf result (+ 1 result)))
             ((eq 'black (aref board m n)) (setf result (- Result 1)))
             ((eq 'white-king (aref board m n)) (setf result (+ Result 2)))
             ((eq 'black-king (aref board m n))
                                  (setf result (- Result 2))))))))
```

11. Minnie Com Pewter is still a beginning programmer. She wrote the following function, which is not syntactically correct. The interpreter gives an error message when she tries to use it. Identify the parts that are not OK and state why each is not correct. The idea

of the function is to return a piece of useful information depending on the value of the `weather` variable. There may be more than one way to fix this function. Any reasonable and syntactically correct version will be accepted.

```
(defun seattle (weather)
    (cond (equal weather 'rain) 'jacket
            (if weather 'sunny) 'sunglasses
            else 'stay-tuned))
```

12. Write a function to do a generalized version of multiplication. If its inputs are numbers, it returns the arithmetic product. If they are symbols, it returns a list containing the two symbols, and if they are lists, it returns a list of pairs, with each pair containing one element from each of the original lists, that is, `(multiply '(a b c) '(d e f))` would give `((a d) (b e) (c f))`. If the lists are of unequal length, the remaining elements in the longer list result in one element lists in the result. If the arguments are not all of the same type, the function should return `nil`.

Here are examples (you can call it `multiply`, since there is no function built in named `multiply`).

```
> (multiply 2 3 4)
24
> (multiply 'ira 'raven 'peter 'jim)
(IRA RAVEN PETER JIM)
> (multiply '(x y z) '(foo bar))
((X FOO) (Y BAR) (Z))
> (multiply 'apples '(oranges grapefruit))
NIL
```

13. The multiply function in the previous exercise just handles its inputs directly. Write a different version that does the same thing as before, for numbers and symbols, but for lists, it recursively applies to the content of the lists, that is, `(multiply '(2 3 4) '(5 6 7))` gives `(10 18 28)`, not `((2 5) (3 6) (4 7))`.

14. What is wrong with the following function definition? This function is supposed to shorten lists, in other words, given a list and a number n, it returns a list without the first n elements and without the last n elements.

```
(defun clip-both-ends (some-list n)
    (let ((list-size (length some-list)))
        (if list-size > (2 * n)
                (nthcdr n (butlast some-list (- list-size n)))
            list-size)))
```

Write a corrected definition for this function.

15. Write a function that takes a symbol and a list as input, and returns how many times the symbol occurs in the list. Your function should be able to search nested lists, so that

    ```
    (count-symbol 'a '(b g (a c a) (c a) (b (d a)) a))
    ```

    should return 5, not 1.

16. Define a structure called person, that includes the person's name, age, gender, spouse, and a list of names of children. Assume all the people are in a list. Write a function that for a given person's name, will find his/her parents in the list (if they are there).

17. Write a function, cousin-p, to determine if two instances of the person structure (above) represent people who are cousins. This is a bit complicated. Two people are cousins if a parent of one is the brother or sister of a parent of the other.

    To create some test data for this, you will need to do a little family tree.

18. Write a family tree program, with which the user can create entries and answer queries from it. You will need a structure definition for a family tree entry, an interactive procedure for making a new entry by prompting the user, a function for deleting an entry specified by name, and some functions to find various kinds of relatives, for example, grandchildren, uncles, and aunts. You may use the solution to the previous exercise as a starting point. There are many possible variants, including having the actual children structures in the children slot instead of the names. Include a function to print the contents of any subtree of the family tree, in a nice way, not just as S-expressions.

19. Write a version of your solution to the above using classes and generic functions instead of structures.

20. A bibliographic database search can return lists of results that match keywords. However, sometimes you don't know what you are looking for, but only the type of thing you want, for example, a list of journals cited, together with the number of citations of each, ordered by that number. Suppose a simple bibliographic database were in the form of a list of entries, where each entry is a multi-element list, starting with a symbol for the journal name, then the article's authors as a string, and the title as another string. Write a function that will search the list and count all the entries, returning a list of all the journal names, each with its frequency in the list, ordered by frequency. Don't count exact duplicates, in case the database might contain duplicate entries.

    The database will look something like this:

    ```
    ((jamia "Shortliffe and Musen"
            "A new medical informatics course")
     (jamia "Kalet"
            "Use of medical images in radiotherapy planning")
     (radiology "Cohen and Levi"
              "Nondestructive identification of citrons")
    ```

```
(meth-info-med "Fagan" "ICU monitoring expert system")
(jamia "Jones and Smith"
        "Standardisation of names of medical technologists")
(jamia "Kalet"
        "Use of medical images in radiotherapy planning")
(meth-info-med "Brinkley and Rosse"
                "Location, location, location")
...)
```

Make up a sizeable database (at least 20 entries) for testing. Put it in a file with a `setq` form so it can be read in using the `load` function.

21. On page 269, Section 9.1 of Norvig [305] a method called memoization is described, by which a function can be redefined so that it saves computed results for the next time the function is called with the same input. The idea is to replace the function definition with a version that has a local hidden variable that caches results in a hash table. Here you will do something simpler.

    Define a function that "countifies" a function that is input to it, something like `memoize`. The input will be a symbol naming a function, and the `countify` function will arrange for it to be redefined as a new function that returns the same result as the original function, but keeps a running count of how many times it was called, and returns the current count as a second return value. This will give you some intuition about how the built in `trace` facility works, and how to build a profiler that tells you where your program is spending its time.

22. Define a function that searches for an integer solution of the Pythagorean formula, $c^2 = a^2 + b^2$ up to a specified limit, n for a and b, that is, check all possibilities for a, b, each up to but not greater than n. Your function should print the results to a file in some nice way, as well as accumulating and returning a list of the triples of successful numbers.

23. Define a macro named `with` that takes a list of variables and a body of code that references them, and arranges for their original values to be restored when the call to `with` returns. For example,

```
> (setq x 5 y 12 z 5)
5
> (with (z)
        (setq z (sqrt (+ (* x x) (* y y))))
        (format t "For now, Z = A" z))
For now, Z = 13.0
nil
> z
5
```

Write the result of applying `macroexpand-1` with your macro to the example above.

24. Write a simple version of `trace` that takes a symbol as input and replaces the function definition for that symbol with another function that prints out the inputs (using `format`), calls the original function, and prints out the return value (assuming a single value is returned). Don't worry about being able to *untrace* the function. Hint: it's the same general idea as `memoize` and `countify`.

Bibliography

[1] Abelson, H., and G. J. Sussman. *Structure and Interpretation of Computer Programs*. MIT Press, Cambridge, Massachusetts, second edition, 1996.

[2] Adami, C. Information theory in molecular biology. *Physics of Life Reviews*, 1, 2004. URL: <http://arxiv.org/abs/q-bio/0405004>.

[3] Adami, C., C. Ofria, and T. C. Collier. Evolution of biological complexity. *Proceedings of the National Academy of Sciences*, 97(9):4463–4468, 2000.

[4] Adamusiak, T., and O. Bodenreider. Quality assurance in LOINC using description logic. In *Proceedings of the American Medical Informatics Association Annual Symposium*, pages 1099–1108, Bethesda, MD, 2012. American Medical Informatics Association.

[5] Adlassnig, K.-P, C. Combi, A. K. Das, E. T. Keravnou, and G. Pozzi. Temporal representation and reasoning in medicine: Research directions and challenges. *Artificial Intelligence in Medicine*, 38:101–113, 2006.

[6] Aikins, J. S. Prototypical knowledge for expert systems. *Artificial Intelligence*, 20:163–210, 1983.

[7] Aikins, J. S., J. C. Kunz, E. H. Shortliffe, and R. J. Fallat. PUFF: An expert system for interpretation of pulmonary function data. *Computers in Biomedical Research*, 16:199–208, 1983.

[8] Aitken, J. S., B. L. Webber, and J. B. L. Bard. Part-of relations in anatomy ontologies: A proposal for RDFS and OWL formalisations. In Russ B. Altman, A. Keith Dunker, Lawrence Hunter, Tiffany A. Jung, and Teri E. Klein, editors, *Proceedings of the Pacific Symposium on Biocomputing 2004*, pages 166–177, Hackensack, New Jersey, Singapore, London, 2004. World Scientific.

[9] Akgül, C. B., D. L. Rubin, S. Napel, C. F. Beaulieu, H. Greenspan, and B. Acar. Content-based image retrieval in radiology: Current status and future directions. *Journal of Digital Imaging*, 24(2):208–222, 2011.

[10] Alberts, B., A. Johnson, J. Lewis, M. Raff, K. Roberts, and P. Walter. *Molecular Biology of the Cell*. Garland Publishing, Incorporated, fourth edition, 2002.

[11] Allen, J. F. Time and time again: The many ways to represent time. *International Journal of Intelligent Systems*, 6(4):341–355, 1991.

[12] American College of Radiology and National Electrical Manufacturers Association. Digital imaging and communications in medicine (DICOM), 2003. the latest version is available from the NEMA DICOM web site, at <http://medical.nema.org/dicom/>.

[13] American Medical Association (AMA). CPT (current procedural terminology) web page, 2008. URL: <http://www.ama-assn.org/ama/pub/category/3113.html>.

[14] Ang, C., and K. Tan. The interval B-tree. *Information Processing Letters*, 53:85–89, 1995.

[15] ANSI. Information retrieval (Z39.50): Application service definition and protocol specification. Technical report, American National Standards Institute, 2003.

[16] Antoniotti, M., F. Park, A. Policriti, N. Ugel, and B. Mishra. Foundations of a query and simulation system for the modeling of biochemical and biological processes. In Russ B. Altman, A. Keith Dunker, Lawrence Hunter, Tiffany A. Jung, and Teri E. Klein, editors, *Pacific Symposium on Biocomputing 2003*, pages 116–127, Hackensack, New Jersey, Singapore, London, 2003. World Scientific.

[17] Antoniou, G., and F. van Harmelen. Web ontology language: OWL. In Steffen Staab and Rudi Studer, editors, *Handbook on Ontologies*, pages 67–92, Springer-Verlag, Berlin, Heidelberg, New York, 2004.

Principles of Biomedical Informatics. http://dx.doi.org/10.1016/B978-0-12-416019-4.00017-2

[18] Ashburner, M., C. A. Ball, J. A. Blake, D. Botstein, H. Butler, J. M. Cherry, A. P. Davis, K. Dolinski, S. S. Dwight, J. T. Eppig, M. A. Harris, D. P. Hill, L. Issel-Tarver, A. Kasarskis, S. Lewis, J. C. Matese, J. E. Richardson, M. Ringwald, G. M. Rubin, and G. Sherlock. Gene Ontology: Tool for the unification of biology. the Gene Ontology Consortium. *Nature Genetics*, 25:25–29, 2000.

[19] Ashburner, M., C. A. Ball, J. A. Blake, et al. Creating the Gene Ontology resource: Design and implementation. the Gene Ontology Consortium. *Genome Research*, 11:1425–1433, 2001.

[20] Au, A., X. Li, and J. Gennari. Differences among cell-structure ontologies: FMA, GO, & CCO. In *Proceedings of the American Medical Informatics Association Fall Symposium*, pages 16–20, 2006.

[21] Austin-Seymour, M., I. Kalet, J. McDonald, S. Kromhout-Schiro, J. Jacky, S. Hummel, and J. Unger. Three-dimensional planning target volumes: A model and a software tool. *International Journal of Radiation Oncology Biology Physics*, 33(5):1073–1080, 1995.

[22] Baeza-Yates, R., and B. Ribeiro-Neto. *Modern Information Retrieval*. Addison-Wesley Longman, Reading, Mass, 1999.

[23] Barke, A. Use of Petri nets to model and test a control system. Master's thesis, University of Washington, Seattle, Washington, 1990.

[24] Baxevanis, A. D., and B. F. Francis Ouellette, editors. *Bioinformatics: A Practical Guide to the Analysis of Genes and Proteins*. Wiley-Interscience, New York, third edition, 2004.

[25] Beck, R., and S. Schulz. Logic-based remodeling of the digital anatomist foundational model. In *Proceedings of the American Medical Informatics Association Fall Symposium*, pages 71–75, 2003.

[26] Benson, N., M. Whipple, and I. Kalet. A Markov model approach to predicting regional tumor spread in the lymphatic system of the head and neck. In *Proceedings of the American Medical Informatics Association (AMIA) Fall Symposium*, pages 31–35, 2006.

[27] Bentley, R., and J. Milan. An interactive digital computer system for radiotherapy treatment planning. *British Journal of Radiology*, 44(527):826–833, 1971.

[28] Berman, J. J. *Biomedical Informatics*. Jones and Bartlett Publishers, Sudbury, MA, 2007.

[29] Berra, Y. *The Yogi Book*. Workman Publishing Company, Inc., New York, NY, 1998.

[30] Bidgood, W. D. Jr., S. C. Horii, F. W. Prior, and D. E. Van Syckle. Understanding and using DICOM, the data interchange standard for biomedical imaging. *Journal of the American Medical Informatics Association*, 4(3):199–212, 1997.

[31] BioCyc Research Group. The BioCyc knowledge library. URL: <http://www.biocyc.org/>.

[32] Blaha, M. R., and J. R. Rumbaugh. *Object Oriented Modeling and Design With UML*. Prentice Hall, Engelwood Cliffs, NJ, second edition, 2004.

[33] Blois, M. S. *Information and Medicine: The Nature of Medical Descriptions*. University of California Press, Berkeley, CA, 1984.

[34] Blois, M. S. Medical information science as 'science'. *Medical Informatics (London)*, 9:181–183, 1984.

[35] Blois, M. S. What is it that computers compute? *MD Computing*, 4(3):30–33, 1987.

[36] Blois, M. S. Medicine and the nature of vertical reasoning. *New England Journal of Medicine*, 318:847–851, 1988.

[37] Bodenreider, O. The unified medical language system (UMLS): Integrating biomedical terminology. *Nucleic Acids Research*, 32:Database issue D267–D270, 2004.

[38] Bogan-Marta, A. A., Hategan, and I. Pitas. Language engineering and information theoretic methods in protein sequence similarity studies. In Arpad Kelemen, Ajith Abraham, and Yulan Liang, editors, *Computational Intelligence in Medical Informatics*, volume 85 of *Studies in Computational Intelligence*, pages 151–183, Springer-Verlag, Berlin, 2008.

[39] Bower, J. M., and H. Bolouri. *Computational Modeling of Genetic and Biochemical Networks*. MIT Press, Cambridge, Mass, 2001.

[40] Boyce, R. D., C. Collins, J. Horn, and I. J. Kalet. Qualitative pharmacokinetic modeling of drugs. In *Proceedings of the American Medical Informatics Association Fall Symposium*, pages 71–75, 2005.

[41] Boyce, R. D., C. Collins, J. Horn, and I. J. Kalet. Modeling drug mechanism knowledge using evidence and truth maintenance. *IEEE Transactions on Information Technology in Biomedicine*, 11(4):386–397, 2007.

[42] Bracewell, R. N. Image reconstruction in radio astronomy. In Gabor T. Herman, editor, *Image Reconstruction from Projections: Implementation and Applications.* Springer, 1979. *Topics in Applied Physics,* 32(87).

[43] Bracewell, R. N. Strip integration in radio astronomy. *Australian Journal of Physics,* 9, 1956.

[44] Bracewell, R. N. *The Fourier Transform and its Applications.* McGraw-Hill, New York, 1978.

[45] Bracewell, R. N., and A. C. Riddle. Inversion of fan beam scans in radio astronomy. *Astrophysical Journal,* 150, 1967.

[46] Brachman, R. J., and H. J. Levesque. *Knowledge Representation and Reasoning.* Elsevier (Morgan-Kaufmann), San Francisco, California, 2004.

[47] Bradburn, C., J. Zeleznikow, and A. Adams. FLORENCE: Synthesis of case-based and model-based reasoning in a nursing care planning system. *Computers in Nursing,* 11(1):20–24, 1993.

[48] Bratko, I. *Prolog Programming for Artificial Intelligence.* Pearson Education Limited, Edinburgh, third edition, 2001.

[49] Breasted, J. H. *The Edwin Smith Papyrus.* New York Historical Society, 1922.

[50] Breiman, L. *Statistics: With a View Toward Applications.* Houghton-Mifflin Co., Boston, Massachusetts, 1969.

[51] Brin, S., and L. Page. The anatomy of a large-scale hypertextual Web search engine. *Computer Networks and ISDN Systems,* 30(1–7):107–117, 1998.

[52] Brinkley, J. F. A flexible, generic model for anatomic shape: Application to interactive two-dimensional medical image segmentation and matching. *Computers and Biomedical Research,* 26(2):121–142, 1993.

[53] Brinkley, J. F., S. W. Bradley, J. W. Sundsten, and C. Rosse. The digital anatomist information system and its use in the generation and delivery of web-based anatomy atlases. *Computers and Biomedical Research,* 30:472–503, 1997.

[54] Brinkley, J. F., and C. Rosse. The digital anatomist distributed framework and its applications to knowledge-based medical imaging. *Journal of the American Medical Informatics Association,* 4(3):165–183, 1997.

[55] Buchanan, B., and E. H. Shortliffe, editors. *Rule-Based Expert Systems: The MYCIN Experiments of the Stanford Heuristic Programming Project.* Addison-Wesley, Reading, Massachusetts, 1984.

[56] Byers, R. M., P. F. Wolf, and A. J. Ballantyne. Rationale for elective modified neck dissection. *Head and Neck Surgery,* 10:160–167, 1988.

[57] Cadag, E. M. High-throughput inference supported protein characterization utilizing the biomediator data integration platform. Master's thesis, University of Washington, 2006.

[58] Candela, F. C., K. Kothari, and J. P. Shah. Patterns of cervical node metastases from squamous carcinoma of the oropharynx and hypopharynx. *Head and Neck,* 12:197–203, 1990.

[59] Candela, F. C., J. P. Shah, and D. P. Jaques. Patterns of cervical node metastases from squamous carcinoma of the larynx. *Archives of Otolaryngology—Head and Neck Surgery,* 116:432–435, 1990.

[60] Capurro, D. *Secondary use of clinical data: barriers, facilitators and a proposed solution.* PhD thesis, University of Washington, 2012.

[61] Carrick, N. M. Radiotherapy treatment planning tools. National Cancer Institute RFP NCI-CM-97564-23, March 1988.

[62] Carroll, L. What the tortoise said to achilles and other riddles. In *The World of Mathematics* [302], pages 2402–2405, reprinted by Dover Publications, 2000.

[63] Ceusters, W., B. Smith, A. Kumar, and C. Dhaen. Mistakes in medical ontologies: Where do they come from and how can they be detected? In Domenico M. Pisanelli, editor, *Ontologies in Medicine,* pages 145–163, IOS Press, Amsterdam, 2004.

[64] Chang, B. L., K. Roth, E. Gonzales, D. Caswell, and J. DiStefano. CANDI. A knowledge-based system for nursing diagnosis. *Computers in Nursing,* 6(1):13–21, 1988.

[65] Charniak, E., C. Riesbeck, D. McDermott, and J. Meehan. *Artificial Intelligence Programming.* Psychology Press (Taylor and Francis Group), New York and London, second edition, 1987.

[66] Chaudhri, V. K., A. Farquhar, R. Fikes, P. D. Karp, and J. P. Rice. OKBC: A programmatic foundation for knowledge base interoperability. In *Proceedings of the 15th National Conference on Artificial Intelligence (AAAI-98),* pages 600–607, 1998.

[67] Cho, P. S., S. Lee, S. Oh, S. G. Sutlief, and M. H. Phillips. Optimization of intensity modulated beams with volume constraints using two methods: Cost function minimization and projection onto convex sets. *Medical Physics*, 25:435–443, 1998.

[68] Clocksin, W. F., and C. S. Mellish. *Programming in Prolog: Using the ISO Standard*. Springer-Verlag, Berlin, Heidelberg, New York, fifth edition, 2003.

[69] Codd, E. F. A relational model of data for large shared data banks. *Communications of the ACM*, 13(6):377–387, 1970.

[70] Codd, E. F., S. B. Codd, and C. T. Salley. Providing OLAP to user-analysts: An IT mandate, 1993, Technical report, originally published by Hyperion Solutions, Sunnyvale, California, `<http://www.hyperion.com/>`, now owned by Oracle, Inc.

[71] Collins, C., and R. Levy. Drug-drug interaction in the elderly with epilepsy: Focus on antiepileptic, psychiatric, and cardiovascular drugs. *Profiles in Seizure Management*, 3(6), 2004.

[72] Combi, C., E. Keravnou-Papailiou, and Y. Shahar. *Temporal Information Systems in Medicine*. Springer, Berlin, Heidelberg, New York, 2010.

[73] Combi, C., L. Missora, and F. Pinciroli. Supporting temporal queries on clinical relational databases: The S-WATCH-QL language. In James J. Cimino, editor, *Proceedings of the 1996 American Medical Informatics Association (AMIA) Fall Symposium*, pages 527–531, Philadelphia, 1996. Hanley and Belfus.

[74] Combi, C., and Y. Shahar. Temporal reasoning and temporal data maintenance in medicine: Issues and challenges. *Computers in Biology and Medicine*, 27(5):353–368, 1997.

[75] Comer, D. E., and D. L. Stevens. *Internetworking with TCP/IP: Client-Server Programming and Applications, BSD Socket Version*, volume III. Prentice-Hall, Upper Saddle River, NJ, second edition, 1996.

[76] Cormack, A. M. Representation of a function by its line integrals, with some radiological applications. *Journal of Applied Physics*, 34:2722–2727, 1963.

[77] Cormack, A. M. Early two dimensional reconstruction and recent topics stemming from it. *Medical Physics*, 7(4):277–282, 1980.

[78] Cormack, A. M. 75 years of Radon transform. *Journal of Computer Assisted Tomography*, 16(5):673, 1992.

[79] Cuddigan, J., J. Norris, S. Ryan, and S. Evans. Validating of the knowledge in a computer-based consultant for nursing care. In *Proceedings of the 11th Annual Symposium on Computer Applications in Medical Care*, Washington, DC, 1987.

[80] Cunningham, J. R., P. N. Shrivastava, and J. M. Wilkinson. Program IRREG—calculation of dose from irregularly shaped radiation beams. *Computer Programs in Biomedicine*, 2:192–199, 1972.

[81] Dahl, O.-J., and K. Nygaard. Simula—an algol-based simulation language. *Communications of the ACM*, 9:671–678, 1966.

[82] D'Ambrosio, C. P. *Computational Representation of Bedside Nursing Decision-Making Process*. Ph.D., University of Washington, Seattle, Washington, 2003.

[83] Date, C.J. *An Introduction to Database Systems*. Addison-Wesley, Reading, Massachusetts, eighth edition, 2003.

[84] Dean, A. G. Computerizing public health surveillance systems. In Steven M. Teutsch and R. Elliott Churchill, editors, *Principles and Practice of Public Health Surveillance*, pages 229–252, Oxford University Press, New York, NY, second edition, 2000.

[85] Denning, T., A. Borning, B. Friedman, B. T. Gill, T. Kohno, and W. H. Maisel. Patients, pacemakers, and implantable defibrillators: Human values and security for wireless implantable medical devices. In *Proceedings of the ACM Conference on Human Factors in Computing Systems (CHI 2010)*, pages 917–926, New York, NY, April, 2010. Association for Computing Machinery.

[86] Department of Health and Human Services. Health insurance reform: Security standards; final rule. Federal Register, 45 CFR Parts 160, 162, and 164, 2003. `<http://www.cms.hhs.gov/SecurityStandard/Downloads/securityfinalrule.pdf>`.

[87] Department of Health and Human Services. DHHS office of civil rights, HIPAA web site, 2008. `<http://www.hhs.gov/ocr/hipaa/>`.

[88] Department of Health and Human Services, Office of the Secretary. Health insurance reform: Security standards; final rule. Federal Register, 45 CFR Parts 160, 162, and 164, 2003.

[89] DePinho, R. A., R. A. Weinberg, V. T. DeVita, T. S. Lawrence, and S. A. Rosenberg, editors. *DeVita, Hellman, and Rosenberg's Cancer: Principles & Practice of Oncology*. Lippincott Williams & Wilkins, Philadelphia, Pennsylvania, eighth edition, 2008.

[90] Detwiler, L. T., D. Suciu, and J. F. Brinkley. Regular paths in SparQL: Querying the NCI thesaurus. In *Proceedings of the American Medical Informatics Association Fall Symposium*, pages 161–165, 2008.

[91] Dewey, M. A classification and subject index for cataloguing and arranging the books and pamphlets of a library. Available from Project Gutenberg at <http://www.gutenberg.org/ebooks/12513>.

[92] Dhillon, B. S. *Medical Device Reliability and Associated Areas*. CRC Press, Boca Raton, FL, 2000.

[93] Dijkstra, E. W. The humble programmer. *Communications of the ACM*, 15(10):859–866, 1972. Turing Award Lecture.

[94] Doctor, J. N., H. Bleichrodt, J. M. Miyamoto, N. R. Temkin, and S. A. Dikmen. New and more robust test of QALYs. *Journal of Health Economics*, 23:353–367, 2004.

[95] Domingos, P., S. Kok, H. Poon, M. Richardson, and P. Singla. Unifying logical and statistical ai. In *Proceeedings of the 21st National Conference on Artificial Intelligence (AAAI-06). 18th Innovative Applications of Artificial Intelligence Conference (IAAI-06)*, pages 2–7, Menlo Park, CA, 2007. AAAI Press.

[96] Donnelly, M. On parts and holes: The spatial structure of the human body. In *Proceedings of MEDINFO 2004*, pages 351–355, Amsterdam, 2004. IOS Press.

[97] Dow, H. Large malpractice award to california teenager reported. Vanderbilt University Television News Archive, 1978. CBS Evening News report by Harold Dow, hosted by Walter Cronkite, Friday, October 20, 1978.

[98] Dudley, J. T., and A. J. Butte. A quick guide for developing effective bioinformatics programming skills. *PLoS Computational Biology*, 5(12), 2009. http://dx.doi.org/10.1371/journal.pcbi.1000589

[99] Duncan, J. S., and N. Ayache. Medical image analysis: Progress over two decades and the challenges ahead. *IEEE Transactions on Pattern Analysis and Machine Intelligence*, 22(1):85–106, 2000. <http://dx.doi.org/10.1109/34.824822>.

[100] Dybvig, K. R. *The Scheme Programming Language*. The MIT Press, Cambridge, Massachusetts, third edition, 2003.

[101] Eenmaa, J., I. Kalet, R. Risler, and P. Wootton. The University of Washington clinical neutron therapy facility. *Medical Physics*, 9:620, 1982.

[102] Einstein, A. Physics and reality. *The Journal of the Franklin Institute*, 221(3):349–382, 1936.

[103] Einstein, A. *Ideas and Opinions*. Bonanza Books, New York, 1954.

[104] Einstein, A. *Out of My Later Years*. Citadel Press (Carol Publishing Group), 1995. Revised reprint of the original 1950 edition from Philosophical Library, Inc.

[105] Elmasri, R., and S. B. Navathe. *Fundamentals of Database Systems*. Addison-Wesley, Reading, Massachusetts, fifth edition, 2006.

[106] Epstein, J. I. An update of the gleason grading system. *The Journal of Urology*, 183(2):433–440, 2010.

[107] Evans, S. A computer-based nursing diagnosis consultant. In *Proceedings of the Eighth Annual Symposium on Computer Applications in Medical Care*, Washington, DC, 1984.

[108] Ewing, B., L. Hillier, M. C. Wendl, and P. Green. Base-calling of automated sequencer traces using phred. I. Accuracy assessment. *Genome Research*, 8(3):175–185, 1998.

[109] Farquhar, A., R. Fikes, and J. Rice. The ontolingua server: A tool for collaborative ontology construction. *International Journal of Human-Computer Studies*, 46:707–727, 1997.

[110] Floridi, L. *The Philosophy of Information*. Oxford University Press, Oxford, UK, 2011.

[111] Foley, J. D., A. Van Dam, S. K. Feiner, and J. F. Hughes. *Fundamentals of Interactive Computer Graphics*. Addison Wesley, Reading, Massachusetts, second edition, 1990.

[112] Forbus, K. D., and J. de Kleer. *Building Problem Solvers*. MIT Press, Cambridge, Massachusetts, 1993.

[113] Forrey, A. W., C. J. McDonald, G. DeMoor, S. M. Huff, D. Leavelle, D. Leland, T. Fiers, L. Charles, B. Griffin, F. Stalling, A. Tullis, K. Hutchins, and J. Baenziger. Logical observation identifier names and codes (LOINC) database: A public use set of codes and names for electronic reporting of clinical laboratory test results. *Clinical Chemistry*, 42(1):81–90, 1996.

[114] Fraas, B. A., D. L. McShan, M. L. Kessler, G. M. Matrone, J. D. Lewis, and T. A. Weaver. A computer-controlled conformal radiotherapy system. I: Overview. *International Journal of Radiation Oncology Biology Physics*, 33(5):1139–1157, 1995.

[115] Fraass, B. A., and D. L. McShan. 3-d treatment planning I. Overview of a clinical planning system. In I. A. D. Bruinvis, P. H. van der Giessen, H. J. van Kleffens, and F. W. Wittkämper, editors, *Proceedings of the Ninth International Conference on the Use of Computers in Radiation Therapy*, pages 273–277, Amsterdam, North-Holland, 1987.

[116] Fraass, B. A. The development of conformal radiation therapy. *Medical Physics*, 22(11):1911–1921, 1995.

[117] Franz, Inc. Allegro common lisp documentation web site, 2004. URL: <http://www.franz.com/support/documentation/>.

[118] Franz, Inc. Allegro prolog documentation, 2012. URL: <http://www.franz.com/support/documentation/current/doc/prolog.html>.

[119] Franz, Inc. Allegrocache—object persistence in lisp, 2012. URL: <http://www.franz.com/products/allegrocache/>.

[120] Fujita, S. $X^Y M T_E X$—*Typesetting Chemical Structural Formulas*. Addison-Wesley, Tokyo, Japan, 1997. URL: <http://imt.chem.kit.ac.jp/fujita/fujitas3/xymtex/indexe.html>.

[121] Furrie, B. *Understanding MARC Bibliographic: Machine-Readable Cataloging*. Cataloging Distribution Service, Library of Congress and Follett Software Company, 2003. Available online at URL: <http://www.loc.gov/marc/umb/>.

[122] Galper, A. R., D. L. Brutlag, and D. H. Millis. Knowledge-based simulation of DNA metabolism: Prediction of action and envisionment of pathways. In Lawrence Hunter, editor, *Artificial Intelligence and Molecular Biology*, chapter 10, pages 365–395, American Association for Artificial Intelligence Press, Menlo Park, California, 1993.

[123] Gamow, G. *One, Two, Three...Infinity*. Dover Publications, Inc., Mineola, NY, 1988. Reprint. Originally published by Viking Press, New York, 1961.

[124] Gasteiger, E., C. Hoogland, A. Gattiker, S. Duvaud, M. R. Wilkins, R. D. Appel, and A. Bairoch. Protein identification and analysis tools on the ExPASy server. In John M. Walker, editor, *The Proteomics Protocols Handbook*, pages 571–607, Humana Press, Totowa, New Jersey, 2005.

[125] Gat, E. Lisp as an alternative to java. *Intelligence*, Winter 2000. ACM SIGART quarterly newsletter.

[126] The Gene Ontology Consortium. Gene Ontology project web site, 2008. URL: <http://www.geneontology.org/>.

[127] Genesereth, M. R. Knowledge interchange format: Draft proposed American National Standard (dpANS) NCITS.T2/98-004. Technical report, Knowledge Systems Laboratory, Stanford University, 1998.

[128] Gerstein, M. B., C. Bruce, J. S. Rozowsky, D. Zheng, J. Du, J. O. Korbel, O. Emanuelsson, Z. D. Zhang, S. Weissman, and M. Snyder. What is a gene, post-ENCODE? history and updated definition. *Genome Research*, 17:669–681, 2007.

[129] Gödel, K. *On Formally Undecidable Propositions of Principia Mathematica and Related Systems*. Dover Publications, Inc., New York, NY, 1992. An unabridged and unaltered republication of the work first published by Basic Books, Inc., New York, 1962.

[130] Goitein, M., and M. Abrams. Multi-dimensional treatment planning: I. Delineation of anatomy. *International Journal of Radiation Oncology Biology and Physics*, 9:777–787, 1983.

[131] Goitein, M., M. Abrams, D. Rowell, H. Pollari, and J. Wiles. Multi-dimensional treatment planning: II. Beam's eye-view, back projection, and projection through CT sections. *International Journal of Radiation Oncology Biology and Physics*, 9:789–797, 1983.

[132] Goitein, M., M. Urie, J. E. Munzenrider, R. Gentry, J. T. Lyman, G. T. Y. Chen, J. R. Castro, M. H. Maor, P. M. Stafford, M. R. Sontag, M. D. Altschuler, P. Bloch, J. C. H. Chu, and M. P. Richter. *Evaluation of Treatment Planning for Particle Beam Radiotherapy*. National Cancer Institute, 1987. A report prepared for the National Cancer Institute.

[133] Goldfarb, C. F. *The SGML Handbook*. Oxford University Press, London, 1990.

[134] Gómez-Pérez, A., M. Fernández-López, and O. Corcho. *Ontological Engineering*. Springer-Verlag, London, 2004.

[135] Gorbach, I., A. Berger, and E. Melomed. *Microsoft SQL Server 2008 Analysis Services Unleashed*. Sams Publishing, Indianapolis, Indiana, 2008.

[136] Graham, P. *On Lisp: Advanced Techniques for Common Lisp*. Prentice-Hall, Englewood Cliffs, New Jersey 07632, 1994.

[137] Graham, P. *ANSI Common Lisp*. Prentice-Hall, Englewood Cliffs, New Jersey 07632, 1996.

[138] Grenon, P., B. Smith, and L. Goldberg. Biodynamic ontology: Applying BFO in the biomedical domain. In Domenico M. Pisanelli, editor, *Ontologies in Medicine*, pages 20–38, IOS Press, Amsterdam, 2004.

[139] Grosse, I., S. V. Buldyrev, and H. E. Stanley. Average mutual information of coding and non-coding DNA. In *Proceedings of the Pacific Symposium on Biocomputing, 2000*, pages 611–620, 2000. URL: <http://psb.stanford.edu/>.

[140] Grossman, D. A., and O. Frieder. *Information Retrieval Algorithms and Heuristics*. Springer-Verlag, Berlin, Heidelberg, New York, second edition, 2004.

[141] Hall, E. J. *Radiobiology for the Radiologist*. Lippincott Williams & Wilkins, Philadelphia, PA, fifth edition, 2000.

[142] Halperin, E. C., C. A. Perez, L. W. Brady, D. E. Wazer, and C. Freeman, editors. *Perez and Brady's Principles and Practice of Radiation Oncology*. Lippincott Williams & Wilkins, Philadelphia, Pennsylvania, fifth edition, 2007.

[143] Hanlon, J. T., K. E. Schmader, C. M. Ruby, and M. Weinberger. Suboptimal prescribing in older inpatients and outpatients. *Journal of the American Geriatrics Society*, 49:200–209, 2001.

[144] Hansten, P. D. Drug-drug interactions, 2003. Section in Human Biology 543 Syllabus Autumn 2003.

[145] Hansten, P. D., and J. R. Horn. *The Top 100 Drug Iteractions, A Guide to Patient Management*. H&H Publications, Edmonds, Washington, 2004.

[146] Hansten, P. D., and J. R. Horn. *Managing Clinically Important Drug Interactions*. Lippincott, Williams & Wilkins, Philadelphia, Pennsylvania, fifth, revised edition, 2006.

[147] Hansten, P. D., J. R. Horn, and T. K. Hazlet. ORCA: OpeRational ClassificAtion of drug interactions. *Journal of the American Pharmacy Association*, 41(2):161–165, 2001.

[148] Hardman, J. G., and L. E. Limbird, editors. *Goodman and Gilman's the Pharmacological Basis of Therapeutics*. McGraw Hill, New York, 10th edition, 2001. Consulting editor, Alfred Goodman Gilman.

[149] Harold, E. R., and W. S. Means. *XML in a Nutshell*. O'Reilly Media, Sebastopol, California, third edition, 2004.

[150] Hawking, S. *A Brief History of Time*. Bantam Books, 10th anniversary edition, 1998.

[151] Hazlet, T. K., T. A. Lee, P. D. Hansten, and J. R. Horn. Performance of community pharmacy drug interaction software. *Journal of the Americal Pharmacy Association*, 41(2):200–204, 2001.

[152] Heidtke, K. R., and S. Schulze-Kremer. BioSim—a new qualitative simulation environment for molecular biology. In *Proceedings of the International Conference on Intelligent Systems in Molecular Biology*, volume 6, pages 85–94, 1998.

[153] Hendee, W. R., G. S. Ibbott, and E. G. Hendee. *Radiation Therapy Physics*. Wiley-Liss, New York, third edition, 2004.

[154] Hendee, W. R., and R. E. Ritenour. *Medical Imaging Physics*. Wiley-Liss, New York, fourth edition, 2002.

[155] Hersh, W. R. *Information Retrieval: A Health and Biomedical Perspective*. Springer-Verlag, New York, second edition, 2003.

[156] Hersh, W. R., and R. A. Greenes. SAPHIRE—an information retrieval system featuring concept matching, automatic indexing, probabilistic retrieval, and hierarchical relationships. *Computers and Biomedical Research*, 23:410–425, 1990.

[157] Hiller, L., and L. Isaacson. *Experimental Music: Composition with an Electronic Computer*. McGraw-Hill, Boston, Massachusetts, 1959.

[158] Hirsch, M., B. L. Chang, and S. Gilbert. A computer program to support patient assessment and clinical decision-making in nursing education. *Computers in Nursing*, 7(4):157–160, 1989.

[159] Hofstadter, D. R. *Gödel, Escher, Bach: An Eternal Golden Braid*. Basic Books, Inc., New York, 1979.

[160] Holzner, S. *Differential Equations for Dummies*. For Dummies, a branded imprint of Wiley, 2008.

[161] Horrocks, I. R. Using an expressive description logic: FaCT or fiction? In *Proceedings of the Sixth International Conference on Principles of Knowledge Representation and Reasoning (KR'98)*, pages 636–647, 1998.

[162] Horvitz, E., and G. Rutledge. Time-dependent utility and action under uncertainty. In *Proceedings of Seventh Conference on Uncertainty in Artificial Intelligence, Los Angeles, CA*, pages 151–158, San Mateo, CA, July, 1991. Morgan Kaufman.

[163] Hounsfield, G. N. Computed medical imaging. *Science*, 210(4465):22–28, 1980.

[164] Hripcsak, G. Writing arden syntax medical logic modules. *Computers in Biology and Medicine*, 24(5):331–363, 1994.

[165] Hripcsak, G., P. Ludemann, T. A. Pryor, O. B. Wigertz, and P. D. Clayton. Rationale for the arden syntax. *Computers and Biomedical Research*, 27(4):291–324, 1994.

[166] Hunink, M., P. Glasziou, J. E. Siegel, J. C. Weeks, J. S. Pliskin, A. S. Elstein, and M. C. Weinstein. *Decision Making in health and medicine: Integrating evidence and values*. Cambridge University Press, Cambridge, the United Kingdom, 2001.

[167] Hunter, L. Molecular biology for computer scientists. In Lawrence Hunter, editor, *Artificial Intelligence and Molecular Biology*, chapter 1, pages 1–46. American Association for Artificial Intelligence Press, Menlo Park, California, 1993. available online at <http://compbio.uchsc.edu/hunter/01-Hunter.pdf>.

[168] Hunter, L. Life and its molecules: A brief introduction. *AI Magazine*, 25(1):9–22, 2004.

[169] Hunter, L. *The Processes of Life: An Introduction to Molecular Biology*. MIT Press, Cambridge, Massachusetts, 2009.

[170] IEEE. IEEE 802.3 LAN/MAN CSMA/CD access method, 2005. URL: <http://standards.ieee.org/getieee802/802.3.html>.

[171] IEEE. IEEE 802.11 LAN/MAN wireless LANS, 2007. URL: <http://standards.ieee.org/getieee802/802.11.html>.

[172] International Commission on Radiation Units and Measurements. *Prescribing, Recording and Reporting Photon Beam Therapy*. International Commission on Radiation Units and Measurements, Bethesda, MD, 1993. Report 50.

[173] International Commission on Radiation Units and Measurements. *Prescribing, Recording and Reporting Photon Beam Therapy (Supplement to ICRU Report 50)*. International Commission on Radiation Units and Measurements, Bethesda, MD, 1999. Report 62.

[174] Isern, D., and A. Moreno. Computer-based execution of clinical guidelines: A review. *International Journal of Medical Informatics*, 77(12):787–808, 2008.

[175] Ito, K., H. S. Brown, and J. B. Houston. Database analyses for the prediction of in vivo drug drug interactions from in vitro data. *British Journal of Clinical Pharmacology*, 57(4):473–486, 2003.

[176] Jacky, J. Programmed for disaster: Software errors that imperil lives. *The Sciences*, 29(5):22–27, 1989.

[177] Jacky, J. *The Way of Z: Practical Programming with Formal Methods*. Cambridge University Press, Cambridge, the United Kingdom, 1997.

[178] Jacky, J., and I. Kalet. A general purpose data entry program. *Communications of the ACM*, 26(6):409–417, 1983.

[179] Jacky, J., and I. Kalet. An object-oriented approach to a large scientific application. In Norman Meyrowitz, editor, *OOPSLA '86 Object Oriented Programming Systems, Languages and Applications Conference Proceedings*, pages 368–376, Association for Computing Machinery, 1986 (also *SIGPLAN Notices*, 21(11), November 1986).

[180] Jacky, J., and I. Kalet. Assuring the quality of critical software. In I. A. D. Bruinvis, P. H. van der Giessen, H. J. van Kleffens, and F. W. Wittkämper, editors, *Proceedings of the Ninth International Conference on the Use of Computers in Radiation Therapy*, pages 49–52, Amsterdam, North-Holland, 1987.

[181] Jacky, J., and I. Kalet. An object-oriented programming discipline for standard Pascal. *Communications of the ACM*, 30(9):772–776, 1987.

[182] Jacky, J., I. Kalet, J. Chen, J. Coggins, S. Cousins, R. Drzymala, W. Harms, M. Kahn, S. Kromhout-Schiro, G. Sherouse, G. Tracton, J. Unger, M. Weinhous, and D. Yan. Portable software tools for 3-D radiation therapy planning. *International Journal of Radiation Oncology, Biology and Physics*, 30(4):921–928, 1994.

[183] Jacky, J., and M. Patrick. Modelling, checking, and implementing a control program for a radiation therapy machine. In Rance Cleaveland and Daniel Jackson, editors, *AAS '97: Proceedings of the First ACM SIGPLAN Workshop on Automated Analysis of Software*, pages 25–32, 1997.

[184] Jacky, J., and J. Unger. Designing software for a radiation therapy machine in a formal notation. In M. Wirsing, editor, *ICSE-17 Workshop on Formal Methods Application in Software Engineering Practice*, pages 14–21, 1995.

[185] Jacky, J., and J. Unger. From Z to code: A graphical user interface for a radiation therapy machine. In J. P. Bowen and M. G. Hinchey, editors, *ZUM '95: The Z Formal Specification Notation*, Ninth International Conference of Z Users, Springer-Verlag, pages 315–333, Berlin, Heidelberg, New York, 1995. Lecture Notes in Computer Science 967.

[186] Jacky, J., J. Unger, M. Patrick, D. Reid, and R. Risler. Experience with Z developing a control program for a radiation therapy machine. In Jonathan P. Bowen, Michael G. Hinchey, and David Till, editors, *ZUM '97: The Z Formal Specification Notation*, pages 317–328, Berlin, Heidelberg, New York, 1997. Tenth International Conference of Z Users, Springer-Verlag. Lecture Notes in Computer Science 1212.

[187] Jacky, J., and C. P. White. Testing a 3-D radiation therapy planning program. *International Journal of Radiation Oncology, Biology and Physics*, 18:253–261, 1990.

[188] Jain, N. L., and M. G. Kahn. Ranking radiotherapy treatment plans using decision-analytic and heuristic techniques. *Computers in Biomedical Research*, 25:374–383, 1992.

[189] Jain, N. L., M. G. Kahn, R. E. Drzymala, B. Emami, and J. A. Purdy. Objective evaluation of 3-D radiation treatment plans: A decision-analytic tool incorporating treatment preferences of radiation oncologists. *International Journal of Radiation Oncology Biology and Physics*, 26(2):321–333, 1993.

[190] Jain, N., and M. Kahn. Ranking radiotherapy plans using decision-analytic and heuristic techniques. In *Proceedings of the 15th Annual Symposium on Computer Applications in Medical Care*, pages 1000–1004, Silver Spring, Maryland, 1991. IEEE Computer Society Press.

[191] James, J., N. DeClaris, W. Kohn, and A. Nerode. Intelligent integration of medical models: A rigorous approach to resolving the dichotomy in medical models. In *Proceedings of the IEEE International Conference on Systems, Man and Cybernetics*, volume 2, pages 1291–1296, New York, NY, 1994. IEEE.

[192] Jensen, F. V., and T. D. Nielsen. *Bayesian Networks and Decision Graphs*. Springer-Verlag, Berlin, Heidelberg, New York, second edition, 2007.

[193] Jensen, K., and N. Wirth. *PASCAL User Manual and Report*. Springer-Verlag, Berlin, Heidelberg, New York, third edition, 1985. Revised for the ISO Pascal standard.

[194] Jirapaet, V. A computer expert system prototype for mechanically ventilated neonates. Development and impact on clinical judgement and information access capability of nurses. *Computers in Nursing*, 19(5): 194–203, 2001.

[195] Johns, H. E., and J. R. Cunningham. *The Physics of Radiology*. Charles C. Thomas, Springfield, Illinois, fourth edition, 1983.

[196] Kahn, M. G., L. M. Fagan, and S. Tu. Extensions to the time-oriented database model to support temporal reasoning in medical expert systems. *Methods of Information in Medicine*, 30(1):4–14, 1991.

[197] Kalet, I., and J. Unger. Behavioral abstraction in design of radiation treatment planning software. In A. R. Hounsell, J. M. Wilkinson, and P. C. Williams, editors, *Proceedings of the 11th International Conference on Computers in Radiotherapy*, pages 164–165, Manchester, UK, 1994. Christie Hospital.

[198] Kalet, I. J. Designing radiotherapy software components and systems that will work together. *Seminars in Radiation Oncology*, 7(1):11–20, 1997.

[199] Kalet, I. J., D. Avitan, R. S. Giansiracusa, and J. P. Jacky. DICOM conformance statement, Prism radiation therapy planning system, version 1.3. Technical Report 2002-02-01, Radiation Oncology Department, University of Washington, Seattle, Washington, 2002.

[200] Kalet, I. J., R. S. Giansiracusa, J. P. Jacky, and D. Avitan. A declarative implementation of the DICOM-3 network protocol. *Journal of Biomedical Informatics*, 36(3):159–176, 2003.

[201] Kalet, I. J., R. S. Giansiracusa, C. Wilcox, and M. Lease. Radiation therapy planning: An uncommon application of Lisp. In Richard Gabriel, editor, *Lisp in the Mainstream: Proceedings of the Conference on the 40th Anniversary of Lisp*, Berkeley, CA, 1998. Franz, Inc.

[202] Kalet, I. J., and J. P. Jacky. A research-oriented treatment planning program system. *Computer Programs in Biomedicine*, 14:85–98, 1982.

[203] Kalet, I. J., J. P. Jacky, M. M. Austin-Seymour, S. M. Hummel, K. J. Sullivan, and J. M. Unger. Prism: A new approach to radiotherapy planning software. *International Journal of Radiation Oncology, Biology and Physics*, 36(2):451–461, 1996.

[204] Kalet, I. J., J. P. Jacky, R. Risler, S. Rohlin, and P. Wootton. Integration of radiotherapy planning systems and radiotherapy treatment equipment: 11 years experience. *International Journal of Radiation Oncology, Biology and Physics*, 38(1):213–221, 1997.

[205] Kalet, I. J., and D. R. Kennedy. A comparision of two radiological path length algorithms. *International Journal of Radiation Oncology, Biology and Physics*, 13:1957–1959, 1987.

[206] Kalet, I. J., and W. Paluszyński. A production expert system for radiation therapy planning. In Allan H. Levy and Ben T. Williams, editors, *AAMSI Congress 1985*, pages 315–319, Washington, DC, 1985. American Association for Medical Systems and Informatics.

[207] Kalet, I. J., and W. Paluszyński. Knowledge-based computer systems for radiotherapy planning. *American Journal of Clinical Oncology*, 13(4):344–351, 1990.

[208] Kalet, I. J., J. M. Unger, J. P. Jacky, and M. H. Phillips. Experience programming radiotherapy applications in Lisp. In Dennis Leavitt and George Starkschall, editors, *Proceedings of the 12th International Conference on Computers in Radiotherapy(ICCR)*, pages 445–448, Madison, WI, 1997. Medical Physics Publishing.

[209] Kalet, I. J., M. Whipple, S. Pessah, J. Barker, M. M. Austin-Seymour, and L. Shapiro. A rule-based model for local and regional tumor spread. In Isaac S. Kohane, editor, *Proceedings of the American Medical Informatics Association Fall Symposium*, pages 360–364, Philadelphia, Pennsylvania, 2002. Hanley & Belfus, Inc.

[210] Kalet, I. J., J. Wu, M. Lease, M. M. Austin-Seymour, J. F. Brinkley, and C. Rosse. Anatomical information in radiation treatment planning. In Nancy M. Lorenzi, editor, *Proceedings of the American Medical Informatics Association Fall Symposium*, pages 291–295, Philadelphia, Pennsylvania, 1999. Hanley & Belfus, Inc.

[211] Kalet, I. J., G. Young, R. Giansiracusa, J. Jacky, and P. Cho. Prism dose computation methods, version 1.2. Technical Report 97-12-01, Radiation Oncology Department, University of Washington, Seattle, Washington, 1997.

[212] Kapetanovic, I. M., and H. J. Kupferberg. Stable isotope methodology and gas chromatography mass spectrometry in a pharmacokinetic study of phenobarbital. *Biomedical Mass Spectrometry*, 7(2):47–52, 1980.

[213] Kapetanovic, I. M., H. J. Kupferberg, R. J. Porter, W. Theodore, E. Schulman, and J. K. Penry. Mechanism of valproate-phenobarbital interaction in epileptic patients. *Clinical Pharmacology and Therapeutics*, 29(4):480–486, 1981.

[214] Karp, P. D., M. Riley, S. M. Paley, and A. Pellegrini-Toole. The metacyc database. *Nucleic Acids Research*, 30(1):59–61, 2002.

[215] Karp, P. D., M. Riley, M. Saier, I. T. Paulsen, J. Collado-Vides, S. M. Paley, A. Pellegrini-Toole, C. Bonavides, and S. Gama-Castro. The ecocyc database. *Nucleic Acids Research*, 30(1):56–58, 2002.

[216] Karp, P. D. A qualitative biochemistry and its application to the regulation of the tryptophan operon. In Lawrence Hunter, editor, *Artificial Intelligence and Molecular Biology*, chapter 8, pages 289–324, American Association for Artificial Intelligence Press, Menlo Park, California, 1993.

[217] Karp, P. D. An ontology for biological function based on molecular interactions. *Bioinformatics*, 16(3):269–285, 2000.

[218] Karp, P. D. Pathway databases: A case study in computational symbolic theories. *Science*, 293:2040–2044, 2001.

[219] Karp, P. D., K. Meyers, and T. Gruber. The generic frame protocol. In *Proceedings of the International Joint Conference on Artificial Intelligence*, pages 768–774, San Mateo, California, 1995. Morgan Kauffman.

[220] Kasner, E. and J. R. Newman. Paradox lost and paradox regained. In *The World of Mathematics* [302], pages 1937–1955, reprinted by Dover Publications, 2000.

[221] Kasner, E. and J. R. Newman. *Mathematics and the Imagination*. Dover Publications, 2001. originally published by Simon and Schuster, 1940.

[222] Keene, S. E. *Object-Oriented Programming in Common LISP: A Programmer's Guide to CLOS*. Addison-Wesley, Reading, Massachusetts, 1989.

[223] Keeney, R. L., H. Raiffa, and R. Meyer. *Decisions with Multiple Objectives: Preferences and Value Tradeoffs.* Cambridge University Press, Cambridge, the United Kingdom, 1993.

[224] Kernighan, B. W., and D. M. Ritchie. *The C Programming Language.* Prentice-Hall, Englewood Cliffs, New Jersey, second edition, 1988.

[225] Ketting, C. H., M. M. Austin-Seymour, I. J. Kalet, J. P. Jacky, S. E. Kromhout-Schiro, S. M. Hummel, J. M. Unger, and L. M. Fagan. Evaluation of an expert system producing geometric solids as output. In Reed M. Gardner, editor, *Proceedings of the 19th Annual Symposium on Computer Applications in Medical Care*, pages 683–687, Philadelphia, PA, 1995. Hanley and Belfus, Inc.

[226] Ketting, C. H., M. M. Austin-Seymour, I. J. Kalet, J. P. Jacky, S. E. Kromhout-Schiro, S. M. Hummel, J. M. Unger, L. M. Fagan, and T. W. Griffin. Automated planning target volume generation: An evaluation pitting a computer-based tool against human experts. *International Journal of Radiation Oncology, Biology and Physics*, 37(3):697–704, 1997.

[227] Ketting, C. H., M. M. Austin-Seymour, I. J. Kalet, J. M. Unger, S. M. Hummel, and J. P. Jacky. Consistency of three-dimensional planning target volumes across physicians and institutions. *International Journal of Radiation Oncology, Biology and Physics*, 37(2):445–453, 1997.

[228] Khan, F. M. *The Physics of Radiation Therapy.* Lippincott Williams & Wilkins, Philadelphia, PA, third edition, 2003.

[229] Kiczales, G., J. des Rivières, and D. G. Bobrow. *The Art of the Metaobject Protocol.* The MIT Press, Cambridge, Massachusetts, 1991.

[230] Kim, H. S. Structuring the nursing knowledge system: A typology of four domains. *Scholarly Inquiry for Nursing Practice: An International Journal*, 1(2):99–110, 1987.

[231] Kirschke, H. *Persistenz in objekt-orientierten Programmiersprachen.* Logos Verlag, Berlin, 1997.

[232] Kirschke, H. Persistent lisp objects, 2005. URL: <http://http://plob.sourceforge.net/>.

[233] Klasko, R. K. DRUGDEX system, 2003, Thomson, Colorado.

[234] Klimov, D., Y. Shahar, and M. Taieb-Maimon. Intelligent visualization and exploration of time-oriented data of multiple patients. *Artificial Intelligence in Medicine*, 49:11–31, 2010.

[235] Knowledge Systems Laboratory, Stanford University. Ontolingua home page, 2008. URL: <http://www.ksl.stanford.edu/software/ontolingua/>.

[236] Koch, B., and J. McGovern. EXTEND: A prototype expert system for teaching nursing diagnosis. *Computers in Nursing*, 11(1):35–41, 1993.

[237] Kohn, W., J. James, A. Nerode, and N. DeClaris. A hybrid systems approach to integration of medical models. In *Proceedings of the 33rd IEEE Conference on Decision and Control*, volume 4, pages 4247–4252, New York, NY, 1994. IEEE.

[238] Koller, D., and A. Pfeffer. Probabilistic frame-based systems. In *Proceedings of the 15th National Conference on Artificial Intelligence (AAAI)*, pages 580–587, 1998.

[239] Kromhout-Schiro, S. *Development and evaluation of a model of radiotherapy planning target volumes.* PhD thesis, University of Washington, 1993.

[240] Kromhout-Schiro, S., M. Austin-Seymour, and I. Kalet. A computer model for derivation of the planning target volume from the gross or clinical tumor volume. In A. R. Hounsell, J. M. Wilkinson, and P. C. Williams, editors, *Proceedings of the 11th International Conference on Computers in Radiotherapy*, Manchester, UK, 1994. Christie Hospital.

[241] Kuipers, B. J. Qualitative simulation. *Artificial Intelligence*, 29:289–338, 1986.

[242] Kuo, T-H., E. A. Mendonça, Jianrong Li, and Yves A. Lussier. Guideline interaction: A study of interactions among drug-disease contraindication rules. In *Proceedings of the AMIA 2003 Annual Symposium*, page 901, 2003.

[243] Kuperman, G. J., D. W. Bates, J. M. Teich, J. R. Schneider, and D. Cheiman. A new knowlege management structure for drug-drug interactions. In Judy G. Ozbolt, editor, *Proceedings of the 18th Annual Symposium on Computer Application in Medical Care*, pages 836–840, Philadelphia, Pennsylvania, 1994. Hanley & Belfus, Inc.

[244] Kutcher, G. J., C. Burman, L. Brewster, M. Goitein, and R. Mohan. Histogram reduction method for calculating complication probabilities for three-dimensional treatment planning evaluations. *International Journal of Radiation Oncology Biology and Physics*, 21(1):137–146, 1991.

[245] Kutcher, G. J., and C. Burman. Calculation of complication probability factors for non-uniform normal tissue irradiation: The effective volume method. *International Journal of Radiation Oncology, Biology and Physics*, 16:1623–1630, 1989.

[246] Lamkins, D. B. *Successful Lisp: How to Understand and Use Common Lisp*. bookfix.com, 2004. This book is also available online.

[247] Lamport, L. *LATEX: A Document Preparation System: User's Guide and Reference Manual*. Addison-Wesley, Reading, Massachusetts, 1994. Revised editon for LATEX 2_ε.

[248] Leunens, G., J. Verstraete, W. Van den Bogaert, J. Van Dam, A. Dutreix, and E. van der Schueren. Human errors in data transfer during the preparation and delivery of radiation treatment affecting the final result: Garbage in, garbage out. *Radiotherapy and Oncology*, 23:217–222, 1992.

[249] Leveson, N. G. *Safeware: System Safety and Computers*. Addison-Wesley, Reading, Massachusetts, 1995.

[250] Leveson, N. G., and C. S. Turner. An investigation of the Therac-25 accidents. *IEEE Computer*, 26(7):18–41, 1993.

[251] Li, F., T. Long, Y. Lu, Q. Ouyang, and C. Tang. The yeast cell-cycle network is robustly designed. *Proceedings of the National Academy of Sciences of the USA*, 101(14):4781–4786, 2004.

[252] Li, H. *A model driven laboratory information management system*. PhD thesis, University of Washington, 2011.

[253] Li, H., J. H. Gennari, and J. F. Brinkley. Model driven laboratory information management systems. In *Proceedings of the American Medical Informatics Association (AMIA) Fall Symposium*, pages 484–488, 2006.

[254] Liddy, E. How a search engine works. *Searcher*, 9(5), 2001. online publication, <http://www.infotoday.com/searcher/may01/liddy.htm>.

[255] Lin, H. J., S. Ruiz-Correa, L. G. Shapiro, M. L. Cunningham, and R. W. Sze. A symbolic shaped-based retrieval of skull images. In *Proceedings of the American Medical Informatics Association Fall Symposium*, page 1030, 2005.

[256] Lin, H. J., S. Ruiz-Correa, L. G. Shapiro, M. L. Speltz, M. L. Cunningham, and R. W. Sze. Predicting neuropsychological development from skull imaging. In *Conference Proceedings of the IEEE Engineering in Medicine and Biology Society*, pages 3450–3455, 2006.

[257] Lin, J. H. Sense and nonsense in the prediction of drug-drug interactions. *Current Drug Metabolism*, 1(4):305–332, 2000.

[258] Lindberg, D. A., Betsy L. Humphreys, and Alexa T. McCray. The Unified Medical Language System. *Methods of Information in Medicine*, 32(4):281–291, 1993.

[259] Lipman, D. J., and W. R. Pearson. Rapid and sensitive protein similarity searches. *Science*, 227:1435–1441, 1985.

[260] Long, W. Temporal reasoning for diagnosis in a causal probabilistic knowledge base. *Artificial Intelligence in Medicine*, 8:193–215, 1996.

[261] Longo, D. L., D. L. Kasper, J. L. Jameson, A. S. Fauci, S. L. Hauser, and J. Loscalzo, editors. *Harrison's Principles of Internal Medicine*. McGraw Hill, New York, 18 edition, 2011.

[262] Lyman, J. T. Complication probability as assessed from dose-volume histograms. *Radiation Research*, 104(Supplement 8):S-13–S-19, 1985.

[263] Lyman, J. T., and A. B. Wolbarst. Optimization of radiation therapy III: A method of assessing complication probabilities from dose-volume histograms. *International Journal of Radiation Oncology, Biology and Physics*, 13:103–109, 1987.

[264] Lyman, J. T., and A. B. Wolbarst. Optimization of radiation therapy, IV: A dose-volume histogram reduction algorithm. *International Journal of Radiation Oncology, Biology and Physics*, 17:433–436, 1989.

[265] MacGregor, R. A description classifier for the predicate calculus. In *Proceedings of the 12th National Conference on Artificial Intelligence, (AAAI94)*, pages 213–220, 1994.

[266] MacGregor, R., and R. Bates. The LOOM knowledge representation language. Technical Report RS-87-188, Information Sciences Institute, University of Southern California, 1987.

[267] Maes, F., A. Collignon, D. Vandermeulen, G. Marchal, and P. Suetens. Multimodality image registration by maximization of mutual information. *IEEE Transactions on Medical Imaging*, 16(2):187–198, 1997.

[268] Maisel, W. H., and T. Kohno. Improving the security and privacy of implantable medical devices. *New England Journal of Medicine*, 362(13):1164–1166, 2010.

[269] Mattes, D., D. R. Haynor, H. Vesselle, Thomas K. Lewellen, and W. Eubank. Nonrigid multimodality image registration. In Milan Sonka and Kenneth M. Hanson, editors, *Proceedings of the SPIE Conference on Medical Imaging 2001: Image Processing*, volume 4322, 2001.

[270] Mavrovouniotis, M. L. Identification of qualitatively feasible metabolic pathways. In Lawrence Hunter, editor, *Artificial Intelligence and Molecular Biology*, chapter 9, pages 325–364, American Association for Artificial Intelligence Press, Menlo Park, California, 1993.

[271] McBride, B. The resource description framework (RDF) and its vocabulary description language RDFS. In Steffen Staab and Rudi Studer, editors, *Handbook on Ontologies*, pages 51–66, Springer-Verlag, Berlin, Heidelberg, New York, 2004.

[272] McCarthy, J. Recursive functions of symbolic expressions and their computation by machine, part I. *Communications of the ACM*, 3(4):184–195, 1960.

[273] McDonald, C. J., S. M. Huff, J. G. Suico, G. Hill, D. Leavelle, R. Aller, A. Forrey, K. Mercer, G. DeMoor, J. Hook, W. Williams, J. Case, and P. Maloney. LOINC, a universal standard for identifying laboratory observations: A 5-year update. *Clinical Chemistry*, 49(4):624–633, 2003.

[274] McKinnell, R. G., R. E. Parchment, A. O. Perantoni, and G. B. Pierce. *The Biological Basis of Cancer*. Cambridge University Press, Cambridge, the United Kingdom, 1998.

[275] McShan, D. C., S. Rao, and I. Shah. PathMiner: Predicting metabolic pathways by heuristic search. *Bioinformatics*, 19(13):1692–1698, 2003.

[276] McShan, D. C., M. Updadhayaya, and I. Shah. Symbolic inference of xenobiotic metabolism. In *Proceedings of the Pacific Symposium on Biocomputing*, pages 545–556, 2004.

[277] McShan, D. L., B. A. Fraass, and A. S. Lichter. Full integration of the beam's eye view concept into computerized treatment planning. *International Journal of Radiation Oncology, Biology and Physics*, 18:1485–1494, 1990.

[278] Meadow, C. T., B. R. Boyce, and D. H. Kraft. *Text Information Retrieval Systems*. Academic Press (Elsevier), Amsterdam, 2007.

[279] Merge Technologies, Inc. *MergeCOM-3 Advanced Integrator's Tool Kit Reference Manual*, 2001.

[280] Metcalfe, R. M., and D. R. Boggs. Ethernet: Distributed packet switching for local computer networks. *Communications of the ACM*, 19(7):395–404, 1976.

[281] Meyer, J., M. H. Phillips, P. S. Cho, I. J. Kalet, and J. N. Doctor. Application of influence diagrams to prostate IMRT plan selection. *Physics in Medicine and Biology*, 49(9):1637–1653, 2004.

[282] Microsoft Corporation. OLE DB for online analytical processing (OLAP), 2012. URL: <http://msdn.microsoft.com/en-us/library/windows/desktop/ms714949(v=vs.85).aspx>.

[283] Miller, P. L., S. J. Frawley, F. G. Sayward, W. A. Yasnoff, L. Duncan, and D. W. Fleming. Combining tabular, rule-based, and procedural knowledge in computer-based guidelines for childhood immunization. *Computers and Biomedical Research*, 30:211–231, 1997.

[284] Miller, R. A., H. E. Pople, Jr., and J. D. Myers. INTERNIST-1, an experimental computer-based diagnostic consultant for general internal medicine. *New England Journal of Medicine*, 307:468–476, 1982.

[285] Min, F. D., B. Smyth, N. Berry, H. Lee, and B. C. Knollmann. Critical evaluation of handheld electronic prescribing guides for physicians. *Clinical Pharmacology & Therapeutic*, 75:P92, 2004.

[286] Minsky, M. L. A framework for representing knowledge. In Patrick H. Winston, editor, *The Psychology of Computer Vision*, pages 211–277, McGraw-Hill, New York, 1975.

[287] Mittelbach, F., and M. Goossens. *The LATEX Companion*. Addison-Wesley, Reading, Massachusetts, second edition, 2004. with Johannes Braams, David Carlisle and Chris Rowley.

[288] Moore, S. M., D. E. Beecher, and S. A. Hoffman. DICOM shareware: A public implementation of the DICOM standard. In R. Gilbert Jost, editor, *Proceedings of the SPIE Conference on Medical Imaging 1994—PACS: Design and Evaluation*, volume 2165, pages 772–781, Bellingham, Washington, 1994. Society of Photo-optical Instrumentation Engineers (SPIE).

[289] Musciano, C., and B. Kennedy. *HTML & XHTML: The Definitive Guide*. O'Reilly Media, Inc., Sebastopol, California, sixth edition, 2006.

[290] Musen, M. A., J. H. Gennari, H. Eriksson, S. W. Tu, and A. R. Puerta. PROTÉGÉ-II: Computer support for development of intelligent systems from libraries of components. In *Proceedings of the World Congress on Medical Informatics, MEDINFO'95*, pages 766–770, 1995.

[291] Musen, M. A. Medical informatics: Searching for underlying components. *Methods of Information in Medicine*, 41:12–19, 2002.

[292] Nagel, E. Symbolic notation, haddocks' eyes and the dog-walking ordinance. In *The World of Mathematics* [302], pages 1878–1900, reprinted by Dover Publications, 2000.

[293] National Library of Medicine. Visible human project. RFP NLM-90-114/SLC, June 1990.

[294] National Library of Medicine. Medical subject headings (MeSH) official web site, 2008. URL: <http://www.nlm.nih.gov/mesh/>.

[295] National Library of Medicine. Unified medical language system official web site, 2008. URL: <http://www.nlm.nih.gov/research/umls/>.

[296] National Library of Medicine. Visible human project official web site, 2008. URL: <http://www.nlm.nih.gov/research/visible/visible_human.html>.

[297] Naur, P. Revised report on the algorithmic language Algol-60. *Communications of the ACM*, 6:1–20, 1963.

[298] Neapolitan, R. E. *Learning Bayesian Networks*. Pearson Prentice Hall, Upper Saddle River, NJ, 2004.

[299] Nease, R. F., and D. K. Owens. Use of influence diagrams to structure medical decisions. *Medical Decision Making*, 17:263–276, 1997.

[300] Nelson, D. L., and M. M. Cox. *Lehninger Principles of Biochemistry*. W. H. Freeman, New York, fourth edition, 2004.

[301] Nester, E. W., C. E. Roberts, B. J. McCarthy, and N. N. Pearsall. *Microbiology: Molecules, Microbes and Man*. Holt, Rinehart and Winston, Inc., New York, 1973.

[302] Newman, J. R. *The World of Mathematics*. Simon and Schuster, New York, 1956. reprinted by Dover Publications, 2000.

[303] Nilsson, N. J. *Artificial Intelligence: A New Synthesis*. Morgan-Kaufmann, San Mateo, California, 1998.

[304] Norvig, P. A retrospective on *Paradigms of AI Programming*. URL: <http://www.norvig.com/Lisp-retro.html>.

[305] Norvig, P. *Paradigms of Artificial Intelligence Programming: Case Studies in Common Lisp*. Morgan Kaufmann, San Mateo, California, 1992.

[306] Noy, N. F., M. Crubézy, R. W. Fergerson, H. Knublauch, S. W. Tu, J. Vendetti, and M. A. Musen. Protégé-2000: An open-source ontology-development and knowledge-acquisition environment. In *Proceedings of the American Medical Informatics Association (AMIA) Fall Symposium*, page 953, 2003. URL: <http://protege.stanford.edu/>.

[307] Noy, N. F., M. A. Musen, J. L. V. Mejino, and C. Rosse. Pushing the envelope: Challenges in a frame-based representation of human anatomy. *Data and Knowledge Engineering*, 48(3):335–359, 2004.

[308] Object Management Group. Unified modeling language (UML) resource page, 2012. URL: <http://www.uml.org>.

[309] O'Carroll, P. W., W. A. Yasnoff, M. E. Ward, L. H. Ripp, and E. L. Martin, editors. *Public Health Informatics and Information Systems*. Health Informatics. Springer-Verlag, New York, NY, 2003.

[310] Oliver, D. E., D. L. Rubin, J. M. Stuart, M. Hewett, T. E. Klein, and R. B. Altman. Ontology development for a pharmacogenetics knowledge base. In *Proceedings of the 2002 Pacific Symposium on Biocomputing*, pages 65–76, 2002.

[311] OpenGL Architecture Review Board, Mason Woo, Jackie Neider, Tom Davis, and Dave Shreiner. *OpenGL Programming Guide*. Addison-Wesley, Reading, Massachusetts, third edition, 1999.

[312] Overby, C. L., P. Tarczy-Hornoch, J. I. Hoath, I. J. Kalet, and D. L. Veenstra. Feasibility of incorporating genomic knowledge into electronic medical records for pharmacogenomic clinical decision support. *BMC Bioinformatics*, 11(Supplement 9), 2010.

[313] Owens, D., R. Schachter, and R. Nease. Representation and analysis of medical decision problems with influence diagrams. *Medical Decision Making*, 17:241–262, 1997.

[314] Paluszyński, W. *Designing radiation therapy for cancer, an approach to knowledge-based optimization.* PhD thesis, University of Washington, 1990.

[315] Paluszyński, W., and I. Kalet. Design optimization using dynamic evaluation. In N. S. Sridharan, editor, *Proceedings of the 11th International Joint Conference on Artificial Intelligence*, pages 1408–1412, San Mateo, California, 1989. Morgan-Kaufmann.

[316] Parker, R. G., H. C. Berry, J. B. Caderao, A. J. Gerdes, D. H. Hussey, R. Ornitz, and C. C. Rogers. Preliminary clinical results from US fast neutron teletherapy studies. *Cancer*, 40(4):1434–1438, 1977.

[317] Patashnik, O. BIBTEX*ing*, 1988. Documentation for general BIBTEX users.

[318] Paul, N., T. Kohno, and D. C. Klonoff. A review of the security of insulin pump infusion systems. *Journal of Diabetes Science and Technology*, 5(6):1557–1562, 2011.

[319] Pawsey, J. L., and R. N. Bracewell. *Radio Astronomy.* Clarendon Press, Oxford, United Kingdom, 1955.

[320] Pearson, W. R., and D. J. Lipman. Improved tools for biological sequence comparison. *Proceedings of the National Academy of Sciences*, 85:2444–2448, 1988.

[321] Peleg, M., A. A. Boxwala, O. Ogunyemi, Q. Zeng, S. Tu, R. Lacson, E. Bernstam, N. Ash, P. Mork, L. OhnoMachado, E. H. Shortliffe, and R. A. Greenes. GLIF3: The evolution of a guideline representation format. In *Proceedings of the American Medical Informatics Association Annual Symposium*, pages 645–649, 2000.

[322] Peleg, M., S. Tu, A. Manindroo, and R. B. Altman. Modeling and analyzing biomedical processes using work-flow/petri net models and tools. In *Proceedings of MEDINFO 2004*, pages 74–78, Amsterdam, 2004. IOS Press.

[323] Peleg, M., I. Yeh, and R. B. Altman. Modeling biological processes using workflow and petri net models. *Bioinformatics*, 18(6):825–837, 2002.

[324] Perlis, A. J. Epigrams on programming. *SIGPLAN Notices*, 17(9):7–13, 1982.

[325] Petrucci, K. E., A. Jacox, K. McCormick, P. Parks, K. Kjerulff, B. Baldwin, et al. Evaluating the appropriateness of a nurse expert system's patient assessment. *Computers in Nursing*, 10(6):243–249, 1992.

[326] Pham, D. L., C. Xu, and J. L. Prince. Current methods in medical image segmentation. *Annual Review of Biomedical Engineering*, 2:315–337, 2000.

[327] Phillips, M. H., J. Meyer, P. S. Cho, I. J. Kalet, and J. N. Doctor. Bayesian decision making applied to IMRT. In B. Yi, S. Ahn, E. Choi, and S. Ha, editors, *Proceedings of the XIV International Conference on the Use of Computers in Radiation Therapy (ICCR)*, pages 108–111, Seoul, Korea, 2004. Jeong Publishing.

[328] Photon Treatment Planning Collaborative Working Group. Three-dimensional display in planning radiation therapy: A clinical perspective. *International Journal of Radiation Oncology, Biology and Physics*, 21:79–89, 1991.

[329] Pierce, J. R. *An Introduction to Information Theory: Symbols, Signals and Noise.* Dover Publications, Inc., New York, NY, second, revised edition, 1980.

[330] Plaisant, C., S. Lam, B. Schneiderman, M. S. Smith, D. Roseman, G. Marchand, M. Gillam, C. Feied, J. Handler, and H. Rappaport. Searching electronic health records for temporal patterns in patient histories: A case study with Microsoft Amalga. In *Proceedings of the American Medical Informatics Association (AMIA) Fall Symposium*, pages 601–605, 2008.

[331] Post, A. R., and J. H. Harrison, Jr. PROTEMPA: A method for specifying and identifying temporal sequences in retrospective data for patient selection. *Journal of the American Medical Informatics Association*, 14(5):674–678, 2007.

[332] Quinlan, J. R. Induction of decision trees. *Machine Learning*, 1(1):81–106, 1986.

[333] Quinlan, J. R. *C4.5: Programs for Machine Learning.* Morgan-Kaufmann, San Mateo, California, 1993.

[334] Ramakrishnan, R. *Database Management Systems.* McGraw-Hill, Boston, Massachusetts, 1998.

[335] Rector, A. L., S. Bechhofer, C. A. Goble, I. Horrocks, W. A. Nowlan, and W. D. Solomon. The GRAIL concept modelling language for medical terminology. *Artificial Intelligence in Medicine*, 9(2):139–171, 1997.

[336] Rector, A. L., and W. A. Nowlan. The GALEN project. *Computer Methods and Programs in Biomedicine*, 45:75–78, 1994.

[337] Reilly, C. A., R. D. Zielstorff, R. L. Fox, E. M. O'Connell, D. L. Carroll, K. A. Conley, et al. A knowledge-based patient assessment system: Conceptual and technical design. In *Proceedings of the 24th Annual Symposium on Computer Applications in Medical Care*, Los Angeles, CA, 2000.

[338] Reiser, S. J., and M. Anbar. *The Machine at the Bedside*. Cambridge University Press, Cambridge, the United Kingdom, 1984.

[339] Repenning, A., and A. Ioannidou. X-expressions in XMLisp: S-expressions and extensible markup language unite. In *Proceedings of the ACM SIGPLAN International Lisp Conference (ILC 2007)*, Cambridge, England, 2007. ACM Press.

[340] Richardson, M., and P. Domingos. Markov logic networks. *Machine Learning*, 62:107–136, 2006.

[341] Riva, A., R. Bellazzi, G. Lanzola, and M. Stefanelli. A development environment for knowledge-based medical applications on the world-wide web. *Artificial Intelligence in Medicine*, 14(3):279–293, 1998.

[342] Riva, A. Implementation and usage of the CL-DNA library, 2002. Presented at the 2002 International Lisp Conference, `<http://www.international-lisp-conference.org/2002/>`. The CL-DNA code and documentation are available at Alberto Riva's web site, `<http://www.chip.org/alb/lisp.html>`.

[343] Robb, R. A. *Three-Dimensional Biomedical Imaging*. CRC Press, Boca Raton, FL, 1985.

[344] Robbins, K. T., J. E. Medina, G. T. Wolfe, et al. Standardizing neck dissection terminology. *Archives of Otolaryngology—Head and Neck Surgery*, 117:601–605, 1991. Official report of the Academy's committee for head and neck surgery and oncology.

[345] Rogers, J., and A. Rector. GALEN's model of parts and wholes: Experience and comparisions. In *Proceedings of the American Medical Informatics Association (AMIA) Fall Symposium*, pages 714–718, 2000.

[346] Röntgen, W. C. On a new kind of rays (preliminary communication). *British Journal of Radiology*, 4:32, 1931. Translation of a paper read before the Physikalische-medicinischen Gesellschaft of Würzburg on December 28, 1895.

[347] Rosenman, J., S. L. Sailer, G. W. Sherouse, E. L. Chaney, and J. E. Tepper. Virtual simulation: Initial clinical results. *International Journal of Radiation Oncology, Biology and Physics*, 20:843–851, 1991.

[348] Rosenman, J., G. W. Sherouse, H. Fuchs, S. Pizer, A. Skinner, C. Mosher, K. Novins, and J. Tepper. Three-dimensional display techniques in radiation therapy treatment planning. *International Journal of Radiation Oncology, Biology and Physics*, 16:263–269, 1989.

[349] Rosse, C., J. L. Mejino, B. R. Modayur, R. Jakobovits, K. P. Hinshaw, and J. F. Brinkley. Motivation and organizational principles for anatomical knowledge representation: The Digital Anatomist symbolic knowledge base. *Journal of the American Medical Informatics Association*, 5:17–40, 1998.

[350] Rosse, C., A. Kumar, J. L. Mejino, D. L. Cook, L. T. Detweiler, and B. Smith. A strategy for improving and integrating biomedical ontologies. In *Proceedings of the American Medical Informatics Association (AMIA) Fall Symposium*, pages 639–643, October 2005.

[351] Rosse, C., and J. L. Mejino. A reference ontology for biomedical informatics: The foundational model of anatomy. *Journal of Biomedical Informatics*, 36(3):478–500, 2003.

[352] Rosse, C., and J. L. Mejino. The foundational model of anatomy ontology. In A. Burger, D. Davidson, and R. Baldock, editors, *Anatomy Ontologies for Bioinformatics: Principles and Practice*, chapter 4, pages 59–117, Springer-Verlag, New York, 2008.

[353] Rosse, C., L. G. Shapiro, and J. F. Brinkley. The digital anatomist foundational model: Principles for defining and structuring its concept domain. In *Proceedings of the American Medical Informatics Association (AMIA) Fall Symposium*, pages 820–824, November 1998.

[354] Rothstein, J. Wiggleworm physics. *Physics Today*, 15(9):28–38, 1962.

[355] Rothwell, D. J., R. A. Cote, J. P. Cordeau, and M. A. Boisvert. Developing a standard data structure for medical language—the SNOMED proposal. In *Proceedings of the 17th Annual Symposium on Computer Applications in Medical Care SCAMC*, pages 695–699, Boston, Massachusetts, McGraw-Hill, November 1993.

[356] Rubin, D. L., F. Shafa, D. E. Oliver, M. Hewett, and R. B. Altman. Representing genetic sequence data for pharmacogenomics: An evolutionary approach using ontological and relational models. In *Proceedings of the 10th International Conference on Intelligent Systems for Molecular Biology*, pages 207–215, London, 2002. Oxford University Press.

[357] Rubin, P., editor. *Clinical Oncology for Medical Students and Physicians, a Multidisciplinary Approach*. American Cancer Society, Atlanta, Georgia, sixth edition, 1983.

[358] Rumbaugh, J., M. Blaha, W. Premerlani, F. Eddy, and W. Lorensen. *Object-Oriented Modeling and Design*. Prentice-Hall, Englewood Cliffs, New Jersey, 1991.

[359] Russell, S. J., and P. Norvig. *Artificial Intelligence: A Modern Approach*. Prentice Hall, Engelwood Cliffs, New Jersey, 1995.

[360] Russell, S. J., and P. Norvig. *Artificial Intelligence: A Modern Approach*. Prentice Hall, Engelwood Cliffs, New Jersey, second edition, 2003.

[361] Sagreiya, H., and R. B. Altman. The utility of general purpose versus specialty clinical databases for research: Warfarin dose estimation from extracted clinical variables. *Journal of Biomedical Informatics*, 43(5):747–751, 2010.

[362] Samwald, M., K. Fehrec, J. de Bruina, and K.-P. Adlassnig. The arden syntax standard for clinical decision support: Experiences and directions. *Journal of Biomedical Informatics*, 45(4):711–718, 2012.

[363] Scheifler, R. W. et al. CLX: Common LISP X interface. Technical report, Texas Instruments, Inc., 1989.

[364] Schneider, T. Molecular information theory and the theory of molecular machines, 2008. URL: <http://www.lecb.ncifcrf.gov/~toms/>.

[365] Schultheiss, T. E. Treatment planning by computer decision. In J. Van Dyk, editor, *Proceedings of the Eighth International Conference on the Use of Computers in Radiation Therapy*, pages 225–229, Silver Spring, MD, 1984. IEEE Computer Society Press.

[366] Schultheiss, T. E., and C. G. Orton. Models in radiotherapy: Definition of decision criteria. *Medical Physics*, 12(2):183–187, 1985.

[367] Schulz, S., and U. Hahn. Parts, locations and holes—formal reasoning about anatomical structures. In S. Quaglini, P. Barahona, and S. Andreassen, editors, *Proceedings of the Eighth Conference on Artificial Intelligence in Medicine in Europe*, pages 293–303, Berlin and Heidelberg, Germany, 2001. Springer-Verlag.

[368] Schwid, H. A. Anesthesia simulators—technology and applications. *Israel Medical Association Journal*, 2:949–953, 2000.

[369] Schwid, H. A. Components of an effective medical simulation software solution. *Simulation and Gaming*, 32(2):240–249, 2001.

[370] Searls, D. B. The computational linguistics of biological sequences. In Lawrence Hunter, editor, *Artificial Intelligence and Molecular Biology*, chapter 2, pages 47–120, American Association for Artificial Intelligence Press, Menlo Park, California, 1993.

[371] Seibel, P. *Practical Common Lisp*. Apress, Berkeley, California, 2005.

[372] Shah, J. P., F. C. Candela, and A. K. Poddar. The patterns of cervical node metastases from squamous carcinoma of the oral cavity. *Cancer*, 66:109–113, 1990.

[373] Shahar, Y., S. Miksch, and P. Johnson. The Asgaard project: A task-specific framework for the application and critiquing of time-oriented clinical guidelines. *Artificial Intelligence in Medicine*, 14:29–51, 1998.

[374] Shannon, C. E., and W. Weaver. *The Mathematical Theory of Communication*. University of Illinois Press, Urbana, Illinois, paperback edition, 1963.

[375] Shaw, M., L. T. Detwiler, N. F. Noy, J. F. Brinkley, and D. Suciu. vSPARQL: A view definition language for the semantic web. *Journal of Biomedical Informatics*, 44(1):102–117, 2011.

[376] Sherouse, G. W. 3D computerized treatment design using modern workstations. *Physics in Medicine and Biology*, 33(Supplement 1):140, 1988. (abstract).

[377] Shortliffe, E. H. *Computer-Based Medical Consultations: MYCIN*. American-Elsevier, New York, 1976.

[378] Shortliffe, E. H., and J. Cimino, editors. *Biomedical Informatics: Computer Applications in Health Care and Biomedicine*. Springer-Verlag, Berlin, Heidelberg, New York, third edition, 2006.

[379] Shortliffe, E. H. et al. ONCOCIN: An expert system for oncology protocol management. In *Proceedings of the Seventh International Joint Conference on Artificial Intelligence*, pages 876–881, Los Altos, California, 1981. William Kaufmann, Inc.

[380] Shrager, J. BioBike web site. URL: <http://biobike.csbc.vcu.edu/>.

[381] Shrager, J. Hacking the Gene Ontology. <http://nostoc.stanford.edu/jeff/jeff/mbcs/go.html>.

[382] Shrager, J. Simple intelligent microarray interpretation. <http://nostoc.stanford.edu/jeff/jeff/mbcs/array1.html>.

[383] Simms, A. M. *Mining mountains of data: Organizing all atom molecular dynamics protein simulation data into SQL*. PhD thesis, University of Washington, Biomedical and Health Informatics, 2011.

[384] Simms, A. M., and V. Daggett. Protein simulation data in the relational model. *The Journal of Supercomputing*, 62:150–173, 2012. <http://dx.doi.org/10.1007/s11227-011-0692-3>.

[385] Smith, B. Mereotopology: A theory of parts and boundaries. *Data and Knowledge Engineering*, 20:287–303, 1998.

[386] Smith, B. Beyond concepts: Ontology as reality representation. In *Proceedings of the International Conference on Formal Ontology and Information Systems FOIS*, 2005. Turin, November 2004.

[387] Smith, B. HL7 watch, 2008. URL: <http://hl7-watch.blogspot.com/>.

[388] Smith, B., and W. Ceusters. HL7 RIM: An incoherent standard. *Studies in Health Technology and Informatics*, 124:133–138, 2006.

[389] Smith, W. P., J. Doctor, J. Meyer, I. J. Kalet, and M. H. Phillips. A decision aid for IMRT plan selection in prostate cancer based on a prognostic bayesian network and markov model. *Artificial Intelligence in Medicine*, 46(2):119–130, 2009.

[390] Som, P. M., H. D. Curtin, and A. Mancuso. An imaging-based classification for the cervical nodes designed as an adjunt to recent clinical based nodal classification. *Archives of Otolaryngology—Head and Neck Surgery*, 125:388–396, 1999.

[391] Sordo, M., T. Hongsermeier, V. Kashyap, and R. A. Greenes. A state-based model for management of type II diabetes. In [444] pages 27–61.

[392] Spellman, P. T., G. Sherlock, M. Q. Zhang, V. R. Iyer, K. Anders, M. B. Eisen, P. O. Brown, D. Botstein, and B. Futcher. Comprehensive identification of cell cycle-regulated genes of the yeast *Saccharomyces cerevisiae* by microarray hybridization. *Molecular Biology of the Cell*, 9:3273–3297, 1998.

[393] Spokoiny, A., and Y. Shahar. A knowledge-based time-oriented active database approach for intelligent abstraction, querying and continuous monitoring of clinical data. In *Proceedings of the 2004 World Congress on Medical Informatics (MEDINFO)*, pages 84–88, 2004.

[394] Steele, G., Jr. *COMMON LISP, the Language*. Digital Press, Burlington, Massachusetts, second edition, 1990.

[395] Stein, J., editor. *The Random House Dictionary of the English Language*. Random House, Inc., New York, NY, unabridged edition, 1971.

[396] Sterling, T. D., H. Perry, and G. K. Bahr. A practical procedure for automating radiation treatment planning. *British Journal of Radiology*, 34:726–733, 1961.

[397] Stoeckert, C. J. Jr., Fidel Salas, Brian Brunk, and G. Christian Overton. EpoDB: A prototype database for the analysis of genes expressed during vertebrate erythropoesis. *Nucleic Acids Research*, 27(1):200–203, 1999.

[398] Stoll, C. Stalking the wily hacker. *Communications of the ACM*, 31(5):484–497, 1988.

[399] Stoll, C. *The Cuckoo's Egg: Tracking a Spy Through the Maze of Computer Espionage*. Doubleday, New York, 1989.

[400] Stonebraker, M., L. A. Rowe, B. G. Lindsay, J. Gray, M. J. Carey, M. L. Brodie, P. A. Bernstein, and D. Beech. Third-generation database system manifesto. *SIGMOD Record*, 19(3):31–44, 1990.

[401] Studholme, C., D. L. G. Hill, and D. J. Hawkes. An overlap invariant entropy measure of 3D medical image alignment. *Pattern Recognition*, 32:71–86, 1999.

[402] Sullivan, K. J., I. J. Kalet, and D. Notkin. Evaluating the mediator method: Prism as a case study. *IEEE Transactions on Software Engineering*, 22(8):563–579, 1996.

[403] Suppes, P. *Introduction to Logic*. Van Nostrand Reinhold Company, Princeton, New Jersey, 1957. Republished as a Dover book, 1999.

[404] Sussman, G. J., and J. Wisdom, M. E. Mayer. *Structure and Interpretation of Classical Mechanics*. MIT Press, Cambridge, Massachusetts, 2001.

[405] Sutton, D. R., and J. Fox. The syntax and semantics of the PRO*forma* guideline modeling language. *Journal of the American Medical Informatics Association*, 10(5):433–443, 2003.

[406] Takahashi, S. Conformation radiotherapy rotation techniques as applied to radiography and radiotherapy of cancer. *Acta Radiological Supplementum*, 242, 1965.

[407] Tang, P. C., J. S. Ash, D. W. Bates, J. M. Overhage, and D. Z. Sands. Personal health records: Definitions, benefits, and strategies for overcoming barriers to adoption. *Journal of the American Medical Informatics Association*, 13(2):121–126, 2006.

[408] Tanimoto, S. *The Elements of Artificial Intelligence Using Common Lisp.* W. H. Freeman, New York, second edition, 1995.

[409] Tarczy-Hornoch, P., T. S. Kwan-Gett, L. Fouche, J. Hoath, S. Fuller, K. N. Ibrahim, D. S. Ketchell, J. P. LoGerfo, and H. Goldberg. Meeting clinician information needs by integrating access to the medical record and knowledge resources via the web. In *Proceedings of the American Medical Informatics Association (AMIA) Fall Symposium*, pages 809–813, 1997.

[410] Teng, C.-C., M. M. Austin-Seymour, J. Barker, I. J. Kalet, L. Shapiro, and M. Whipple. Head and neck lymph node region delineation with 3D CT image registration. In Isaac S. Kohane, editor, *Proceedings of the American Medical Informatics Association Fall Symposium*, pages 767–771, Philadelphia, Pennsylvania, 2002. Hanley & Belfus, Inc.

[411] Teng, C.-C., L. G. Shapiro, and I. J. Kalet. Automatic segmentation of neck ct images. In D.J. Lee, Brian Nutter, Sameer Antani, Sunanda Mitra, and James Archibald, editors, *Proceedings of the 19th IEEE International Symposium on Computer-based Medical Systems*, pages 442–445, Silver Spring, Maryland, 2006. IEEE Computer Society.

[412] Teng, C.-C., L. G. Shapiro, and I. J. Kalet. Head and neck lymph node region delineation using a hybrid image registration method. In *Proceedings of the IEEE International Symposium on Biomedical Imaging*, pages 462–465, Silver Spring, Maryland, 2006. IEEE Computer Society.

[413] Teng, C.-C., L. G. Shapiro, I. J. Kalet, C. Rutter, and R. Nurani. Head and neck cancer patient similarity base on anatomical structural geometry. In *Proceedings of the 2007 IEEE International Symposium on Biomedical Imaging (ISBI): From Nano to Macro*, pages 1140–1143, Piscataway, New Jersey, 2007. IEEE Engineering in Medicine and Biology (EMB) Society.

[414] Tisdall, J. D. *Beginning Perl for Bioinformatics.* O'Reilly Media, Inc., Sebastopol, California, 2001.

[415] Tisdall, J. D. *Mastering Perl for Bioinformatics.* O'Reilly Media, Inc., Sebastopol, California, 2003.

[416] Touretzky, D. S. *LISP—A Gentle Introduction to Symbolic Computing.* Harper and Row, 1984.

[417] Tracton, G., E. Chaney, J. Rosenman, and S. Pizer. MASK: Combining 2D and 3D segmentation methods to enhance functionality. In *Mathematical Methods in Medical Imaging III: Proceedings of 1994 International Symposium on Optics, Imaging, and Instrumentation*, volume 2299, Bellingham, Washington, July 1994. Society of Photo-optical Instrumentation Engineers (SPIE).

[418] Tsien, C. L., I. S. Kohane, and N. McIntosh. Multiple signal integration by decision tree induction to detect artifacts in the neonatal intensive care unit. *Artificial Intelligence in Medicine*, 19:189–202, 2000.

[419] Tsien, K. C. The application of automatic computing machines to radiation treatment planning. *British Journal of Radiology*, 28:432–439, 1955.

[420] US Food and Drug Administration. Guidance for industry—cybersecurity for networked medical devices containing off-the-shelf (OTS) software, January 2005. <http://www.fda.gov/cdrh/comp/guidance/1553.html>.

[421] US Food and Drug Administration. FDA guideline: Drug interaction studies—study design, data analysis, and implications for dosing and labeling, September 2006. <http://www.fda.gov/Cber/gdlns/interactstud.htm>.

[422] USC Information Sciences Institute. LOOM project home page, 2006. URL: <http://www.isi.edu/isd/LOOM/LOOM-HOME.html>.

[423] van Bemmel, J. H., and M. A. Musen. *Handbook of medical informatics.* Springer-Verlag, Berlin, Heidelberg, New York, 1997.

[424] van de Geijn, J. Computational methods in beam therapy planning. *Computer Programs in Biomedicine*, 2(3):153–168, 1972.

[425] Viola, P., and W. M. Wells III. Alignment by maximization of mutual information. *International Journal of Computer Vision*, 24(2):137–154, 1997.

[426] W3C. SPARQL query language for RDF, 2008. URL: <http://www.w3.org/TR/rdf-sparql-query/>.

[427] Walker, J. M., P. Carayon, N. Leveson, R. A. Paulus, J. Tooker, H. Chin, A. Bothe, and W. F. Stewart. EHR safety: The way forward to safe and effective systems. *Journal of the American Medical Informatics Association*, 15(3):272–277, 2008.

[428] Walker, J. M., editor. *The Proteomics Protocols Handbook*. Humana Press, Totowa, New Jersey, 2005.

[429] Wang, D., M. Peleg, S. W. Tu, A. A. Boxwala, O. Ogunyemi, Q. Zeng, R. A. Greenes, V. L. Patel, and E. H. Shortliffe. Design and implementation of the GLIF3 guideline execution engine. *Journal of Biomedical Informatics*, 37:305–318, 2004.

[430] Wasserman, L. *All of Statistics: A Concise Course in Statistical Inference*. Springer-Verlag, Berlin, Heidelberg, New York, 2004.

[431] Watt, D. A. *Programming Language Syntax and Semantics*. Prentice-Hall, Engelwood Cliffs, New Jersey, 1991.

[432] Webb, C., M. Russo, and A. Ferrari. *Expert Cube Development with Microsoft SQL Server 2008 Analysis Services*. Packt Publishing, Birmingham, United Kingdom, 2009.

[433] Webb, S., and A. Nahum. A model for calculating tumour control probability in radiotherapy including the effects of inhomogeneous distribution of dose and clonogenic cell density. *Physics in Medicine and Biology*, 38:653–666, 1993.

[434] Webb, S. *The Physics of Three Dimensional Radiation Therapy: Conformal Radiotherapy, Radiosurgery and Treatment Planning*. Institute of Physics (IOP) Publishing, Bristol, United Kingdom, 1993.

[435] Webb, S. *Intensity-Modulated Radiation Therapy*. Institute of Physics (IOP) Publishing, Bristol, United Kingdom, 2001.

[436] Weinkam, J., and T. Sterling. A versatile system for three-dimensional radiation dose computation and display, RTP. *Computer Programs in Biomedicine*, 2(3):178–191, 1972.

[437] Weiss, M. A. *Data Structures and Algorithm Analysis in Java*. Addison-Wesley, Reading, Massachusetts, third edition, 2011.

[438] Wen, X., J. S. Wang, K. T. Kivisto, P. J. Neuvonen, and J. T. Backman. In vitro evaluation of valproic acid as an inhibitor of human cytochrome P450 isoforms: Preferential inhibition of cytochrome P450 2C9 (CYP2C9). *British Journal of Clinical Pharmacology*, 52(5):547–553, 2001.

[439] Whitehorn, M., R. Zare, and M. Pasumansky. *Fast Track to MDX*. Springer, New York and London, second edition, 2006.

[440] Winston, P. H., and B. P. K. Horn. *LISP*. Addison-Wesley, Reading, Massachusetts, third edition, 1989.

[441] Wootton, P., K. Alvar, H. Bichsel, J. Eenmaa, J. S. R. Nelson, R. G. Parker, K. A. Weaver, D. L. Williams, and W. G. Wyckoff. Fast neutron beam radiotherapy at the University of Washington. *Journal of the Canadian Association of Radiologists*, 26:44, 1975.

[442] World Health Organization (WHO). International classification of diseases (ICD), 2008. URL: <http://www.who.int/classifications/icd/en/>.

[443] Xu, D., J. M. Keller, M. Popescu, and R. Bondugula. *Applications of Fuzzy Logic in Bioinformatics*. Imperial College Press, London, UK, 2008.

[444] Yoshida, H., A. Jain, A. Ichalkaranje, L. C. Jain, and N. Ichalkaranje, editors, *Advanced Computational Intelligence Paradigms in Healthcare—1*, volume 48 of *Studies in Computational Intelligence*, pages 27–61, Springer-Verlag, Berlin, 2007.

[445] Yu, Y. Multiobjective decision theory for computational optimization in radiation therapy. *Medical Physics*, 24(9):1445–1454, 1997.

[446] Yu, Y. Multiobjective optimization in radiation therapy: Applications to stereotactic radiosurgery and prostate brachytherapy. *Artificial Intelligence in Medicine*, 19:39–51, 2000.

[447] Zadeh, L. A. Fuzzy sets. *Information and Control*, 8:338–353, 1965.

[448] Zhou, L., and G. Hripcsak. Tempporal reasoning with medical data—a review with emphasis on medical natural language processing. *Journal of Biomedical Informatics*, 40:183–202, 2007.

[449] Zumdahl, S. S. *Chemical Principles*. Houghton-Mifflin Company, Boston, Massachusetts, fifth edition, 2006.

[450] Zupan, B., I. Bratko, J. Demsar, P. Juvan, T. Curk, U. Borstnik, J. R. Beck, J. Halter, A. Kuspa, and G. Shaulsky. GenePath: A system for inference of genetic networks and proposal of genetic experiments. *Artificial Intelligence in Medicine*, 29:107–130, 2003.

[451] Zupan, B., J. Demsar, M. W. Kattan, J. R. Beck, and I. Bratko. Machine learning for survival analysis: A case study on recurrence of prostate cancer. *Artificial Intelligence in Medicine*, 20:59–75, 2000.

Printed and bound by CPI Group (UK) Ltd, Croydon, CR0 4YY

03/10/2024

01040323-0009